HANDBOOK OF GENDER AND TECHNOLOGY

INTERNATIONAL HANDBOOKS ON GENDER

Founding Editor: the late Sylvia Chant *FRSA, FAcSS, formerly Professor of Development Geography, London School of Economics and Political Science, UK*

International Handbooks on Gender is an exciting *Handbook* series under the general editorship and direction of Sylvia Chant. The series comprises high quality, original reference works offering comprehensive overviews of the latest research within key areas of contemporary gender studies. International and comparative in scope, the *Handbooks* are edited by leading scholars in their respective fields, and comprise specially commissioned contributions from a select cast of authors, bringing together established experts with up-and-coming scholars and researchers. Each volume offers a wide-ranging examination of current issues to produce prestigious and high-quality works of lasting significance.

Individual volumes will serve as invaluable sources of reference for students and faculty in gender studies and associated fields, as well as for other actors such as NGOs and policy makers keen to engage with academic discussion on gender. Whether used as an information resource on key topics, a companion text or as a platform for further study, Elgar International Handbooks on Gender will provide a source of definitive scholarly reference.

Titles in the series include:

Handbook on Gender and Health
Edited by Jasmine Gideon

Handbook on Gender in World Politics
Edited by Jill Steans and Daniela Tepe-Belfrage

Handbook on Gender and War
Edited by Simona Sharoni, Julia Welland, Linda Steiner and Jennifer Pedersen

Handbook on Gender and Social Policy
Edited by Sheila Shaver

Handbook on Gender and Violence
Edited by Laura J. Shepherd

Handbook on Gender, Diversity and Federalism
Edited by Jill Vickers, Joan Grace and Cheryl N. Collier

Handbook on Gender in Asia
Edited by Shirlena Huang and Kanchana N. Ruwanpura

Handbook of Feminist Governance
Edited by Marian Sawer, Lee Ann Banaszak, Jacqui True and Johanna Kantola

Handbook of Gender and Technology
Environment, Identity, Individual
Edited by Eileen M. Trauth and Jeria L. Quesenberry

Handbook of Gender and Technology

Environment, Identity, Individual

Edited by

Eileen M. Trauth

Professor Emeritus, The Pennsylvania State University, University Park, USA

Jeria L. Quesenberry

Teaching Professor of Information Systems, Carnegie Mellon University, USA

INTERNATIONAL HANDBOOKS ON GENDER

Edward Elgar
PUBLISHING

Cheltenham, UK • Northampton, MA, USA

Published by
Edward Elgar Publishing Limited
The Lypiatts
15 Lansdown Road
Cheltenham
Glos GL50 2JA
UK

Edward Elgar Publishing, Inc.
William Pratt House
9 Dewey Court
Northampton
Massachusetts 01060
USA

A catalogue record for this book
is available from the British Library

Library of Congress Control Number: 2022950533

This book is available electronically in the **Elgar**online
Business subject collection
http://dx.doi.org/10.4337/9781800377929

ISBN 978 1 80037 791 2 (cased)
ISBN 978 1 80037 792 9 (eBook)

Printed and bound by CPI Group (UK) Ltd, Croydon, CR0 4YY

We dedicate this book to our families. The women in Eileen's family – her mother, Martha, and sisters Charlene, Denise, Suzanne, Jeanette, Patricia and Kathleen – have been role models, mentors, and a source of inspiration throughout her career. Jeria's parents (Ray and Glenna) always encouraged their children to "reach for the stars." Most of all, we dedicate this book to our spouses, Kathy and Steve, and to Jeria's daughter, Ella. They supported us, daily, as we gave life to this book.

Contents

Contributors

Monica Adya (she/her) is a Professor of Management at Rutgers School of Business – Camden, USA, conducts research in women's participation in IT and design and use of decision aids. She has published in *Information Systems Research*, *Decision Support Systems*, *Information Technology & People*, and *Human Resource Management*, among others. Decision Science Institute and American Council of Education have recognized her pedagogical innovations. She has received funding from 3M Foundation and Naval Surface Warfare Center. Monica received her PhD in MIS from the Weatherhead School of Management.

Manju K. Ahuja (she/her) is Frasier Family Chair of Information Systems at the University of Louisville, USA. Manju is currently a Senior Editor at Information Systems Research and was previously a Senior Editor at the *MIS Quarterly* and the *Journal of AIS*. She was named an AIS Fellow in 2021 and is the recipient of a Lifelong Service Award from Academy of Management's OCIS division for 2020. She has received four NSF grants for over two million dollars for her research on IT workforce issues. She is actively involved in research on issues related to impacts and use of IT, innovation, virtual communities and teams, effects of mobile technologies, IT workforce, and future of work.

Patience Akpan-Obong (she/her) is an Associate Professor at Arizona State University, USA. She holds degrees in political science, journalism and mass communication and previously practiced journalism in Nigeria and Canada. Her research centers on gender, technology and development, science, technology and society, and e-governance adoption in emerging democracies. She has authored several books including *Information and Communication Technologies in Nigeria: Prospects and Challenges for Development* (2009), and *Letters to Nigeria: Journal of an African Woman in America* (2013).

Hala Annabi (she/her) is an Associate Professor in the Information School at the University of Washington, USA. Her research focuses on creating and maintaining inclusive learning organizations. Dr. Annabi investigates diversity and inclusion interventions in the technology industry aimed at retaining and advancing women, as well as recruiting, retaining, and advancing individuals with autism. Dr. Annabi earned a BS in Business Administration and Management Information Systems and an MBA from Le Moyne College. She earned a PhD in Information Science and Technology from The Information School at Syracuse University in 2005.

Deborah J. Armstrong (she/her) is a Professor of Management Information Systems at Florida State University, USA. Her research interests cover issues at the intersection of IS personnel and cognition, involving the human aspects of technology, change, and learning. Many of the research problems that she finds interesting involve gender-related IS workforce issues. Deb has published articles in *Management Information Systems Quarterly*, the *Journal of Management Information Systems*, the *Journal of the Association for Information Systems*, and the *European Journal of Information Systems*.

Jenine Beekhuyzen (she/her) OAM advocates for equity for all. She is Australia's leading evidence-based expert on STEM, entrepreneurship, digital inclusion and educational design.

As the founder and CEO of the Tech Girls Movement Foundation and Adroit Research, she is a highly sought-after professional speaker and mentor, and a prolific author with her Tech Girls Are Superheroes series of books in every Australian school library. Jenine works with the biggest brands including Google, Amazon Web Services, Xero, Accenture, TPG Telecom, Deloitte, Blackboard and Salesforce. She received an Order of Australia Medal in 2020 for her contributions to Information Technology, and to Women.

Bettina Berendt (she/her) is a Professor in the Artificial Intelligence/Machine Learning and Data Mining group at the Department of Computer Science at the University of Leuven, Belgium. Her research interests are web and text mining and in particular the interactions with how people make decisions faced with the artificial and human intelligence they find online and via other digital media. This means investigating how data mining affects and interacts with privacy and data protection, how it can contribute to liberating people and increasing diversity – or to discriminating against individuals and groups. It also investigates what ethical choices stakeholders face when dealing with data and data science. These questions require both a multidisciplinary and a multi-stakeholder approach. Within this range of topics and methods, Bettina Berendt has concentrated on combining methods from data science, HCI, and behavioural economics and investigated questions arising for data subjects, researchers, institutional decision makers, and regulators. Through a range of projects and other initiatives, she collaborates with teachers, news media, and EU institutions.

Mari W. Buche (she/her) is the Associate Dean in the College of Business at Michigan Technological University, USA and Professor of Management Information Systems. In addition, she is the Director for both the TechMBA® and Master's in Engineering Management graduate programs, and was the inaugural Director for the interdisciplinary Data Science graduate program from 2014–2017. She earned her doctorate at the University of Kansas and joined the College of Business in 2003. Dr. Buche's research investigates the disruptive impact of changes in technology and information systems on IT professionals intimately involved in developing, implementing, securing, and supporting those systems. Her research is primarily of an applied nature, enabling organizations to effectively implement improvements in workforce management and system optimization. Another stream of her research focuses on providing recommendations to improve the gender balance and diversity within STEM fields (Science, Technology, Engineering and Mathematics). She has developed expertise in both qualitative and quantitative methodologies, analyzing primary data from surveys and interviews. She has served on editorial review boards for academic journals, assisting in the development and maintenance of high-quality publications.

Curtis C. Cain (he/his) is an Associate Professor in the Department of Information Systems and Supply Chain Management at Howard University, USA and an Affiliate Professor of Computer Science and Electrical Engineering in the College of Architecture and Engineering. Broadly, his research interests over the last decade are computer science education and broadening participation in computing. Specifically, he studies pathways, barriers, and success stories of underrepresented groups in informatics, computing, and engineering. He is the recipient of the prestigious National Science Foundation (NSF) CAREER award, has received NSA funding to increase capacity and knowledge generation in cybersecurity, and a $9,000,000 grant from the Department of Defense and Office of Naval Research to develop workforce skills in Human-Centered AI. Additionally, he has the distinction of receiving an Excellence in

Teaching award from The Institute for Citizens and Scholars (Woodrow Wilson Foundation). During his PhD studies, he received the National Science Foundation (NSF) Graduate Research Fellowship (GRFP) to study Black men's underrepresentation in their pursuit of a computing degree.

Shiya Cao (she/her) is a MassMutual Assistant Professor in Statistical and Data Sciences at Smith College, USA. Her research centers around disability inclusion and broader social inclusion topics using quantitative, qualitative, and design science methods. She has published multiple papers in high-ranking journals and conferences including *Behavior & Information Technology* and the Hawaii International Conference on System Sciences.

Florence M. Chee (she/her) is an Associate Professor in the School of Communication and Program Director of the Center for Digital Ethics and Policy (CDEP) at Loyola University Chicago, USA. She is also founding director of the Social & Interactive Media Lab Chicago (SIMLab), devoted to the in-depth study of social phenomena at the intersection of society and technology. Her research examines the social, cultural, and ethical dimensions of emergent digital lifestyles with a particular focus on the examination of artificial intelligence, games, social media, mobile platforms, and translating insights about their lived contexts across industrial, governmental, and academic sectors.

Namjae Cho (he/his) is a Professor of Management Information Systems at the School of Business of Hanyang University, Seoul, Korea. He received his bachelor's degree in Industrial Engineering from Seoul National University, master's degree in Management Science from Korea Advanced Institute of Science and Technology, and doctoral degree in MIS from Boston University, USA. He has published research papers in journals including *Industrial Management and Data Systems*, *Computers and Industry*, *International Journal of Information Systems and Supply Chain*, *Contemporary Management Review*, and the *Journal of Data and Knowledge Engineering*. He also published several books including *Supply Network Coordination in the Dynamic and Intelligent Environment* (2010, IGI Global), *Innovations in Organizational Coordination Using Smart Mobile Technology* (2013, Springer Verlag), *Technology Planning and Road Mapping* (2014, Sigma Press) and *Business Design Thinking* (2021, Book Star). He consulted government organizations and several multinational companies. His research interest includes technology planning and innovation, analysis of IT impacts, knowledge management, industrial ICT policy, design thinking, and the management of family business.

Regina Connolly (she/her) is a Professor of Information Systems at Dublin City University, Ireland. Her research focuses on use of digital technology for organizational and societal transformation. Her work has guided national policy regarding digital inclusion of underrepresented communities and she has conducted longitudinal research examining Irish women's participation in the IT workforce. She leads the Athena SWAN initiative in DCU Business School and is a member of Women in Technology and Science, Ireland.

Elena Gorbacheva (she/her) is a Data Scientist and a Visiting Lecturer at universities in Germany and Vietnam. She completed her PhD thesis at the Department of Information Systems, University of Münster, where she focused on the phenomenon of underrepresentation of women in the IT profession. Afterwards Elena worked as a postdoctoral researcher and a project lead. She also served as the equal opportunities officer at the School of Business and Economics. Elena has published over 30 scholarly papers.

K.D. Joshi (she/her) is a Professor of Information Systems in the College of Business at the University of Nevada, Reno, USA. Previously, she was the Philip L. Kays Distinguished Professor of Information Systems and Department Chair at Washington State University. Her published research is cited over 7,300 times (h-index of 33; i10Index of 63). She has received grants totaling over $5 million from the National Science Foundation (NSF). Her research has appeared in journals such as *MIS Quarterly*, *Information Systems Research*, *Journal of the AIS*, *Information Systems Journal*, and *Decision Support Systems*, among others. She is the editor-in-chief of the Data Base for Advances in Information Systems.

Gyeung-Min Kim (she/her) is a Professor of Management Information Systems in the College of Business Administration at Ewha Womans University, Korea. She received her MS and PhD degrees in Management Information Systems from Texas Tech University. She earned her BS degree in Computer Science from Ewha Womans University in Korea. Her research interests include platform ecosystem, global IS sourcing, and gender issues in the IS area. Her publications have appeared in the *Journal of Organizational Computing and Electronic Commerce* and *Journal of Systems and Software* among others. She is a member of Association for Information Systems (AIS) and Korea Society of Management Information Systems (KMIS).

Hee-Sun Kim (she/her) is a member of the Hyundai Autoever IT Service Innovation Team. Prior to that she worked as General Manager of Digital Transformation Center, in DB Inc., Korea for 18 years. She received her PhD in Management Information System from Sogang University, Korea. Her research fields are IT outsourcing, IT Governance, ITSM and Data analytics. She is serving as ISO/IEC SC40 Korean national body HoD (Head of Delegates).

Eleanor T. Loiacono (she/her) is Professor of Business Analytics at William and Mary, USA. Over the past 20 years, she has focused on how people feel about the technology they use and how technologies, such as mobile apps and social media, can improve people's experiences. She is considered an expert in how those with differing abilities interact with technologies. Her research has expanded into diversity of the IT workforce and women in IT. She is the Principal Investigator on a three-year, $1 million National Science Foundation ADVANCE grant, "ImPACT: Increasing the Participation and Advancement of Women in Information Technology" (https://impactit.pages.wm.edu).

Tricia Massey (she/her) is a Consultant with Adroit Research and a Drama Specialist at Ormiston College, Australia. Her research interests include theatre for social change, post-colonial and multicultural theatre, intercultural communication, leadership, and identity construction.

Adanna Nedd (she/her) is a Millennium Scholar pursuing a major in Information Sciences and Technology and a minor in English at The Pennsylvania State University, USA. She is also an i3 Research Fellow at the University of Pittsburgh and has published research at the iConference that explores diversity and identity representation in video gaming communities.

Sue Nielsen (she/her) has degrees in linguistics, library science and information systems and more than 40 years' experience teaching in research methods, information systems and business analysis. She has published extensively in the areas of gender and IT and the design of applications for deaf children. She was co-founder of the WinIT (Women in IT) project, a foundation member of the Tech Girls Movement Foundation, and contributed to the Tech Girls are Superheros books. Since retirement in 2014 she has focused on editing and reviewing academic papers and dissertations and becoming fluent in Auslan (Australian sign language).

Sangeeta Parashar (she/her), an Associate Professor of Sociology and Program Director of the International Studies minor at Montclair State University, USA, has research interests in occupational segregation, gender disparities in health, and aging. She has published in *Monthly Labor Review*, *International Journal of Sociology and Social Policy*, and *Social Science & Medicine*, among others, and has also co-authored a book, *Divisions and Integrations: The Expansion of Global Capitalism* (Kendall Hunt, 2014). Sangeeta received her PhD from the University of Maryland, College Park.

Eric Patridge (he/him) leads the Computational Systems Biology team at Viome, USA. As an interdisciplinary researcher, Eric is passionate about infrastructure and advancing healthcare, especially in the clinical and life sciences. His experience includes artificial intelligence, data engineering, clinical studies, and both medical and nutritional ontologies. Before his technology pivot, Eric coordinated research at Yale, focusing on natural products, pharmacology, enzymology, and microbiology. He holds a PhD in Life Sciences from The Pennsylvania State University and a BA from Skidmore College. Eric also founded oSTEM Incorporated (Out in Science, Technology, Engineering, & Mathematics), a non-profit serving LGBTQA+ people in STEM.

Jeria L. Quesenberry (she/her) is a Teaching Professor of Information Systems and an Associate Dean of Faculty in the Dietrich College of Humanities and Social Sciences at Carnegie Mellon University, USA. Her research interests are directed at the study of cultural influences on information technology students and professionals, including topics of social inclusion, broadening participation, career values, organizational interventions, and work–life balance. Along with Carol Frieze, Quesenberry is co-author of two books on women in computing. *Cracking the Digital Ceiling: Global Views of Women in Computing* is an edited collection of global perspectives that challenge commonly held western views and stereotypes about women in computing. *Kicking Butt in Computer Science: Women in Computing at Carnegie Mellon University* explores how computer science curricula can enable women's successful participation without becoming "pink" or "female friendly."

Isabel Ramos (she/her) is an Associate Professor (Habilitation) in the Department of Information Systems (DSI) of the School of Engineering of the University of Minho, Portugal and a researcher at the Algoritmi Research Center. Isabel is President of the Portuguese Association of Information Systems and the Chair of the Technical Committee 8 of the International Federation of Information Processing (IFIP). She was granted the IFIP's Outstanding Service Award (2009) and Silver Core Award (2013), and the IIAKM – Lifetime Academic Achievement Award (2021).

Ita Richardson (she/her) is a Professor of Software Quality at the University of Limerick, Ireland. Her research focuses on improving software quality through process improvement. An advocate for gender equality in academia, she held the Equality, Diversity and Inclusion portfolio in UL for three years to 2020. She leads the Athena SWAN initiative in UL's Department of Computer Science and Information Systems and Lero, sits on the UL Faculty of Science and Engineering EDI board and on the UL Athena SWAN committee. She is a member of Women in Technology and Science, Ireland.

Cynthia K. Riemenschneider (she/her) is a Professor of Information Systems in the Hankamer School of Business at Baylor University, USA. She holds the Helen Ligon Professorship in Information Systems. Her publications have appeared in *MIS Quarterly*, *Information Systems Research*, *Journal of Management Information Systems*, *Journal of the Association for Information Systems*, *Information Systems Journal*, *European Journal of Information Systems*, and others. She currently researches IT workforce issues, women and minorities in IT, IT professional turnover, and turnaway intention.

Geoffrey Martin Rockwell (he/his) is a Professor of Philosophy and Digital Humanities and Director of the Kule Institute for Advanced Study at the University of Alberta, Canada. He received his MA and PhD in Philosophy from the University of Toronto where he worked in computing as a Senior Instructional Technology Specialist. From 1994 to 2008 he was at McMaster University where he directed the Humanities Media and Computing Centre and helped found the Department of Communication Studies and Multimedia. Rockwell publishes on video games, textual visualization, text analysis, ethics of technology and on digital humanities including a co-edited book on *Right Research: Modelling Sustainable Research Practices in the Anthropocene* (Open Book Publishers, 2021) and a co-authored book *Hermeneutica: Computer-Assisted Interpretation in the Humanities* (MIT Press, 2016). He is co-developer of Voyant Tools (voyant-tools.org), an award-winning suite of text analysis tools.

Minna Salminen-Karlsson (she/her) is an Associate Professor at the Centre of Gender Research and a gender equality specialist in the HR division at Uppsala University, Sweden. Her main research areas are gender in STEM education and gender in organizations, in particular in high-tech workplaces, including the academe. In addition to her research, she has been involved in several European action-oriented projects for enhancing gender balance in STEM education and improving gender equality in the academe.

Courtney Smith (she/her) is a Millennium Scholar pursuing a major in Security and Risk Analysis at The Pennsylvania State University, USA. She is also an i3 Research Fellow at the University of Pittsburgh and has published research at the iConference that explores how queer people of color use technology to cope with stress. In 2020, she was awarded an Erickson Discovery Grant to support an independent research project on parasocial relationships on Tumblr.

Todd Suomela (he/his) is the Associate Director for Digital Pedagogy and Scholarship at Bucknell University, USA. He works with faculty and students to incorporate technology into their research, teaching, and learning. Before moving to Bucknell he worked as a CLIR/DLF Postdoctoral Fellow in Data Curation for the Social Sciences and Humanities at the University of Alberta. He received a PhD, in communication and information, from University of Tennessee in 2014 for research into science communication and citizen science. He completed an MIS from University of Michigan in 2007. He is currently conducting research on web archives, research ethics, library management, and history of science and technology.

Eileen M. Trauth (she/her) is Professor Emeritus of Information Sciences & Technology, and Gender, Women's & Sexuality Studies at The Pennsylvania State University, University

Park, USA. Her research, writing and outreach focus on barriers to inclusion in the information technology profession experienced by women and other marginalized groups. Her work in Europe, Africa, the Asia Pacific, and the USA has been supported by the Fulbright Foundation, the National Science Foundation, and Science Foundation Ireland. Her findings have been published in journals such as *MIS Quarterly*, *European Journal of Information Systems*, *Information Systems Journal*, *Information Technology & People*, *Information and Organization*, *The Data Base for Advances in Information Systems*, and *Communications of the ACM*, as well as in several books, and a play, *iDream*. She has advised EU gender projects, served as co-editor-in-chief of *Information Systems Journal*, and was founding president of the AIS special interest group on social inclusion.

Roli Varma (she/her) is Carl Hatch Endowed Professor in the School of Public Administration and an Associate in the School of Engineering at the University of New Mexico, Albuquerque, USA. She has published on women and minorities in information technology education, Asian immigrants in the science and engineering workforce, and return migration of Asians from the United States. Her research has been supported by the National Science Foundation.

Lynette Yarger (she/her) holds a joint appointment as Associate Professor at the College of Information Sciences and Technology and Associate Dean for Equity and Inclusion at the Schreyer Honors College at The Pennsylvania State University, USA. She has published extensively on critical research approaches to computing that explore the intersection of race and gender, with a particular emphasis on the reproduction of inequality.

Acknowledgments

This book is edited by two individuals; the finished product represents the contribution of many more. First and foremost, we want to thank the authors of the chapters in this book. We invited them in Fall 2020 when the world was in chaos; little had changed by Spring 2022 when this book project ended. Throughout the entire time they worked on their chapters these authors were coping with switching between remote and in-person teaching, significant challenges in their research agendas, and major disruptions in their personal lives. Some of them lost loved ones or were seriously ill themselves.

We would also like to acknowledge the members of our "research family" who helped to further the gender and technology agenda at The Pennsylvania State University, College of Information Sciences and Technology. Along with Jeria, others who as graduate students worked with Eileen include Dr. Kayla Booth, Dr. Allison Morgan Bryant, Dr. Curtis Cain, Dr. Lisa Erickson, Dr. Haiyan Huang, and Dr. Benjamin Yeo.

We would like to thank the colleagues in the global gender and technology community with whom we have interacted and collaborated over the years. Feedback received at the ACM SIGMIS Computers and People community, and from the AIS SIG Social Inclusion community helped us to refine our perspectives on gender and technology. Several of them have contributed chapters to this book. In particular we would like to thank our colleagues Dr. Lynette Kvasny Yarger, Dr. K.D. Joshi, and Dr. Carol Frieze with whom we worked on several research projects.

The chapters in this book were reviewed by fellow authors and by external reviewers. We thank them for making the individual chapters and the overall book much stronger. Finally, we would like to thank Matthew Pittman and Alex O'Connell of Edward Elgar Publishing who invited us to edit this book.

1. Introduction to the *Handbook of Gender and Technology*

Eileen M. Trauth and Jeria L. Quesenberry

INTRODUCTION

We are so pleased that Edward Elgar Publishers invited us to edit a book on gender and technology entitled *Handbook of Gender and Technology: Environment, Identity, Individual.* The book takes a comprehensive look at societal, organizational, and individual factors that affect gender in relationship to technology use and technology careers. The structure of the book reflects the source of influences on people: the external societal and organizational environments; one's intersectional identity; and individual factors. Between the two of us we have edited six books on the topics of gender, social inclusion, and research methods, along with several conference proceedings. We have drawn upon those experiences in this book project.

After Eileen completed her PhD in information science, she spent half of her academic career working in business schools and the other half working in information technology (IT) departments. But her abiding interest that remained across various academic appointments has been the human impact of the information society that emerged in the second half of the twentieth century and came to fruition in the early part of the twenty-first. Her particular focus has been the IT workforce. She began by studying the changing skills and knowledge requirements associated with technological change (Niederman et al., 2016). She then moved on to examining the role of IT work in economic development, something that provided opportunities for international research and teaching. Against this backdrop she embarked upon her explorations of the haves and have nots in the information society. Her earliest work consisted of field studies of women in the IT sectors of several countries. This led to development of a theory to explain the IT gender gap, *the individual differences theory of gender and IT* (IDTGIT). Following qualitative studies of women in the IT field in several countries, she broadened her scope and research methods to examine social inclusion issues related to gender minorities based on identity characteristics such as race, ethnicity, disability, and sexuality (Trauth, 2017). In addition to her gender research and publishing, she has also developed and taught gender courses in several countries. Throughout her career, she has looked at gender issues from a variety of vantage points as: an American in other countries, a researcher, a professor, an administrator.[1]

Jeria's pathway into academia began with a career in IT consulting. Having worked in industry for several years, she had become accustomed to being one of the only women in the office, leading to feelings of isolation and non-belonging. She frequently wondered "where are all the women ...?" In 2002, Jeria decided to pursue a PhD in information sciences and technology while working as a research assistant for Eileen. During this time, she began to study career values and motivations of women in the IT workforce. In 2007, Jeria began her faculty career in information systems at Carnegie Mellon University. For the past two decades, her research interests have remained directed at the study of cultural influences on women in IT, including topics of social inclusion, broadening participation, career values, and organizational interventions. Her efforts have an action-oriented focus including a 15-year longitudinal intervention study of students in computer science at Carnegie Mellon (Frieze and Quesenberry, 2015) and challenging commonly held western views and stereotypes about women in computing (Frieze and Quesenberry, 2020). Throughout her career, she has also looked at gender issues from a variety of vantage points as: an IT professional, a researcher, a professor, and an administrator.[2]

The purpose of this chapter is to provide some background and context for the chapters that follow. In this first section of the chapter, we discuss our motivation for editing this book. We speak to why we believe it is important to have a book that looks at the issues confronting women and other gender minorities in today's information-intensive and technological society. We also present a high-level consideration of the issues. In the second section, we discuss the intended audience for this book, its distinctive features, the scope, and the book's conceptual framing. Then in the third section of the chapter we provide brief overviews of each chapter in this book.

Why Should We Care About the "Gender Problem" in STEM?

We begin with the question of why we should care about the haves and the have nots in the information society. In particular why should we care about a gender gap among those who are *constructing* this society? Despite the progress made by women and other gender minorities, in breaking down barriers that have held them back, they are still significantly underrepresented in the technology fields.

More than ever before, girls are studying and excelling in post-secondary education. Take for instance, in the United States, women earn approximately 57 percent of all undergraduate bachelor's degrees. Yet these gains have not been matched in technical areas of study. Only 22 percent of computer and information science undergraduate degree recipients were women. This is a figure down from a high of 37 percent in 1985 (DuBow and Gonzalez, 2020). The gender divide is also evident at the graduate level where in 2019–20 in the United States and Canada (combined), 21.7 percent of all doctoral computing degree recipients were women (Zweben and Bizot, 2020).

This discrepancy persists in the workplace and in some ways is more problematic when considering retention and professional development trends. In the United States, women comprise 57 percent of the professional workforce, but only 26 percent of the IT workforce (DuBow and Gonzalez, 2020). Moreover, this figure is substantially lower than where we were 30 years ago. In 1991, women comprised 36 percent of the IT workforce (Ashcraft et al., 2016), by 2004 the number dropped to only 32 percent (ITAA, 2005), to where we are today at 26 percent. In terms of retention, 56 percent of women left the United States IT workforce within the first five years of employment, which is twice the turnover rate of men in IT and women in other professional fields (Ashcraft et al., 2016; Glass et al., 2013). The numbers show women still hold a small share of IT leadership roles. Women account for only 16 percent of senior level technology jobs, 10 percent of executive positions (Sigacheva, 2018), and 13 percent of chief technology officer (CTO) positions (DuBow and Gonzalez, 2020). Annual salaries for men in science, technology, engineering and mathematics (STEM) are nearly $15,000 higher per year than women ($85,000 compared to $60,828) (Pew Research Center, 2018).

The data is even more alarming when considering the participation of women minorities in the United States IT workforce. Black women comprise around 13 percent of the American population of women but only 3 percent of the IT workforce. Hispanic or Latina women comprise 18 percent of the American population of women but only 7 percent of the IT workforce (DuBow and Gonzalez, 2020; United States Census Bureau, 2020). And Black and Latina women in STEM earn around $33,000 less than men (at an average of around $52,000 a year) (Pew Research Center, 2018). Clearly, women, and in particular minority women, are seriously underrepresented in the United States IT workforce.

When we examine the data from different countries, we find that women are largely underrepresented in IT post-secondary education in many parts of the world. The Organisation for Economic Co-operation and Development (OECD) collects gender-disaggregated data on distribution of tertiary degrees awarded in information communication technology (ICT). In 2015, the average percentage of women for all OECD countries was slightly below 25 percent. In only *six* countries for which OECD collects data – India (over 40 percent), Mexico, Indonesia, Turkey, Estonia, and Canada – women made up more than 30 percent of graduates in ICT disciplines. India is one of the few notable examples where women comprise 40 percent, 65 percent, and 50 percent of students in computer science and engineering at the undergraduate, master's and doctorate levels, respectively (Huyer, 2019).

Throughout the world, women are also largely underrepresented in the IT workforce. Chow and Charles (2019) analyzed data from the International Labour Organization and report that (ICT) professional work is the most man-dominated profession in the economically developed regions of Western Europe and North America (16.9 percent women) and Eastern Europe (19.8 percent women). Many countries in these regions have less than 15 percent women's representation including the Netherlands, Austria, Poland, Germany, Belgium, Switzerland, Hungary, Greece, and the Czech Republic. The figures for Latin American countries and the Caribbean are somewhat better (21.1 percent women). The highest representation of women in ICT work can be found in the Asia-Pacific region (30.4 percent women) and Africa (31.3 percent women).

As the data shows, women are underrepresented across the IT career pathways, from enrollment in secondary and postsecondary programs, to employment in the IT workforce throughout most parts of the world. Yet, data can only tell us part of the story. To get a better understanding of the underrepresentation of women in IT we need to pay immediate and close attention to the factors that might be enabling or deterring women's participation.

This chapter considers the question of why it is important to address the underrepresentation of women and gender minorities in STEM fields, and why we were motivated to edit this book. A good starting point for the discussion of gender and technology is to address the question of why it is important to change this situation. This question continues to be raised. In 2002, while Eileen was giving a talk at an American university about the IT gender imbalance, a man on the faculty wanted to know why this was a problem. In 2007, during a lecture she was giving in Spain, a man in the audience asked the same question. In 2019, an anonymous reviewer on one of her gender papers asked it again. In 2019, a senior faculty man expressed surprise in a public forum to learn that there were gender issues in IT academia. Jeria has experienced these questions as well. In 2006, while giving a poster presentation at a graduate student symposium, a woman faculty evaluator asked why we should care if women are underrepresented in the technology field. In 2022 (as this book was being developed), during a conference workshop on women in technology the presenters encouraged scholars to be prepared "to take on this persistent question" should they choose to study gender and technology issues. As Wajcman (2006) observed, it is "imperative that we examine the extent to which existing societal patterns of gender inequality are transformed or reproduced in a new technological guise" (p. xxii).

To answer this question about why we should care, we begin with a consideration of the global information economy. During the second half of the twentieth century a shift occurred in the nature of work and the basis for economic development. Globally, there began a transition to societies in which information has become a key economic resource. Daniel Bell (1973), who was one of the first individuals to characterize this shift termed it the "post-industrial society." It has also been characterized as the "information society" or "knowledge society." Despite the label, the important point is that information and associated technology are significant factors in the economic viability of a society.

In this society the production, manipulation and use of information engages a significant portion of work. Thus, underlying this information society is its "information economy" (Trauth, 2000). The information economy consists of the workers and work engaged in the processing of information and the production of information technology. Porat (1977) characterized the information economy as composed of two parts. The primary information sector is comprised of individuals engaged in work associated with: information and communication hardware, software, processing, and services; and information content. The secondary information sector consists of individuals engaged in some other work (such as health care, transportation or finance) that requires them to process information in order to carry out their work. The combined primary and secondary information sectors have produced a twenty-first century global society in which the production and consumption of information is a significant part.

In the twenty-first century the task of ensuring a supply of qualified IT personnel is increasingly bound up with issues of diversity (Sorensen et al., 2011; Zorn et al., 2007). As the appetite for IT continues to grow, the IT profession is challenged with meeting the demand to enlarge the IT workforce by recruiting and retaining personnel from historically underrepresented groups such as women and other gender minorities. Given the size and importance of the information economy, it is incumbent upon societies to open their doors, completely, to all potential workers. It is a matter of both economic necessity and social justice. So, with respect to the question of why it is important to address the gender imbalance in the IT field, there are several answers (Glover, 2000; Trauth, 2011; Trauth et al., 2006).

The first is that an information economy is characterized as an innovation economy. As technology becomes commodified its production moves to lower wage, typically developing, economies. Consequently, developed nations increasingly focus on continuous innovation related to new information products and services. At the same time, less developed economies look to technological innovation as a key way to move more quickly up the ranks.[3] In such an economy human capital is prized because it is brainpower and creativity that fuels innovation. And the "best brains" are not particular to a certain gender, race, ethnicity or any other identity. Hence, the ability of regions and workplaces to attract and retain diverse talent is growing in importance (Florida, 2002; Trauth et al., 2008a).

The second answer is that in an information society in which virtually everyone is engaged in the consumption of information products, it is crucial that the varying needs of this diverse consumer base be represented. Part of this understanding of diversity needs to be the inclusion of a gender dimension in the development of technology (Trauth, 2019). The air bag story is now given as a classic example of the failure to include a gender dimension in design consider-ations. The problem with the automobile air bag is that it was designed with a western man as the generic "person" without sufficient consideration given to the effect of the deployed airbag on someone of a lighter and slighter build than the average western man – with the tragic consequences that have resulted (Smith, 2009). Hence, a diversity of people with different identities and experiences are needed to bring different perspectives to better understand the needs of a diverse consumer base.

The third answer about the need for gender balance comes from demography. In many countries demographic trends add urgency to the need to create a more diverse IT labor force. The retirement of the baby boom generation, coupled with employment shifts following the COVID-19 pandemic, have added stress to a growing IT sector. This IT labor force demand cannot be satisfied by white men alone. Yet, as previously described, the statistics show that women are woefully underrepresented. They make up 26 percent of the United States IT labor force (DuBow and Gonzalez, 2020) and less than 25 percent of the IT labor force of many countries (Chow and Charles, 2019; Panko, 2008). Even more troubling is that this gender gap is not just a function of recruitment into STEM fields, as studies of women's attrition have revealed (Hewlett et al., 2008; National Academy of Sciences, 2006; Tattersall et al., 2006).

The economic answer to the gender gap question has several dimensions. Recent recessions and the pandemic-induced economic upheavals have driven home the lesson that it is incum-bent upon all individuals to be prepared for economic uncertainties. Thus, with the possibility of spouses and partners losing jobs, everyone needs to be prepared to work. As whole industries reshape and sometimes disappear, two-income families are becoming increasingly important. Hence, there is economic vulnerability in a heterosexual household unit that allocates earning capacity based on gender. Given the potential for a husband to become displaced from work, a wife would become the major bread winner. And the economic vulnerability of a household unit with no men in it is even greater. Since women represent half of the labor force of most countries, any society that disenfranchises half of its labor force puts itself at a distinct com-petitive disadvantage in the competition for brainpower in a knowledge society. Further, there is a productivity dimension. The constant effort to recognize and overcome biases drains one's time and energy and leads to frequent turnover costs to an organization.

There is also a social equity response to this question. A democratic society espouses fairness to all, providing equal opportunity to achieve wealth. Hence, we need to address societal barriers to women for the same reasons that we do so with regard to race, disability, etc. (Adam, 2005). That is, women need to have the same access as men to jobs that pay well, such as those in the IT field. Also, in today's information society power derives from control of information and IT. So, women deserve to have equal access to this source of power. Indeed, the recent UNESCO report on artificial intelligence (AI) and gender equality (UNESCO, 2020) adopts a gender equality lens with respect to artificial intelligence. It recognizes that gender biases found in AI training data sets, algorithms and devices have the potential of spreading and reinforcing harmful gender stereotypes. Further, the growing ubiquity of AI puts women at risk of being left behind in all realms of economic, political, and social life, since the majority of workers holding jobs that face a high risk of automation, such as clerical, administrative, bookkeeping, and cashier positions are women.

Finally, there is a policy answer. Addressing the gender imbalance in IT is part of a global gender equity movement at both the national and organizational levels. For example, the European Union (EU) requires the inclusion of "gender mainstreaming" – integrating a gender equality perspective – at all stages and levels of policies, programs, and projects. Some countries have instituted gender quotas, such as Germany's 2016 law requiring that 30 percent of executive and supervisory positions be held by women. Other countries, while not bound by law, have established aspirational policies represented by membership in the "30 Percent Club." Awareness about gender and technology is steadily moving from the periphery to the core of concerns in the field.

How Should We Think About the "Gender Problem" in STEM?

Notwithstanding those who continue to question the technology gender gap, growing numbers of organizations and individuals have expressed a desire to address this imbalance. They seem to care but don't always know what to do, demonstrating a need to both understand and frame the "gender problem." The earliest research on the gender gap in the STEM disciplines was conceptualized as "gender differences" research, employing quantitative data to focus on *whether* and *where* gender differences existed with respect to the use of technology – sometimes characterized as a "digital divide" (Cooper and Weaver, 2003) – and participation in its careers. The focus of this research was on providing evidence of a gender gap but not on understanding the reasons for it. But what is needed now (whether using quantitative or qualitative methods), is to get at the story behind the statistics in order to explore *why* these differences exist and *how* they came about. A concerted effort is needed to both understand and address gender barriers affecting engagement with STEM fields. This is achieved by taking a critical look at what it is that needs to be fixed, which will shed light on *who or what is broken*.

A common explanation that has been given for the gender imbalance is that women are not cut out for technical work. This explanation assumes a universal gender binary and suggests that the world of work can be dichotomized into fixed masculine and feminine domains. Hence, there are types of work that are natural to each of these two genders, with STEM being "natural" to men only. But critiques of the gendered nature of STEM disciplines from a feminist perspective have, for decades, taken on prevailing societal assumptions about STEM being an exclusively masculine domain (e.g., Harding, 1986; Schiebinger, 1999).

This stereotype about gender-technology relations is that women are soft, nurturing, and ill-suited to technological pursuits. They are too emotional, irrational, and illogical to excel in STEM fields. Men, on the other hand, have an inherent fascination with machinery enabling them to excel in STEM fields. Societal stereotypes about hegemonic masculinity, the culturally dominant form of masculinity that include traits such as virility, exploitation, dominance, and aggressiveness, further reinforce this stereotype (Connell, 2005; Wajcman, 1998).

The association of this hegemonic masculinity with technology derives from a similarly stereotypical and narrow view of STEM, leading to damaging stereotypes about technology careers. Stereotypical views of STEM suggest that to work in these fields is to disdain interacting with other people. Fields such as IT are stereotyped as solitary, its work done by individuals with little or no interest in group interactions. Those who enter such professions are assumed to be antisocial workaholics lacking in social skills. This reflects a tendency to think of "technology" as simply an artifact. A broader view is that technology is the embodiment of those who invent, develop, and use it.

> As we have seen repeatedly, technology is more than a set of physical objects or artifacts. It also fundamentally embodies a culture or set of social relations made up of certain sorts of knowledge,

beliefs, desires and practices … Treating technology as a culture has enabled us to see the way in which technology is expressive of masculinity and how, in turn, men characteristically view themselves in relation to these machines. (Wajcman 1991, 149)

The inclusion of gender biases, assumptions, and stereotypes in the conceptualization of technology can be a negative influence on women's participation in STEM. For example, the perception that technology work must necessarily be all consuming in one's life can be a deterrent to those women who expect to assume primary or full responsibility for children and the domestic realm.

People often use observations about the STEM gender gap as the explanation for it. But this explanation doesn't stand up to the facts. If women are not suited to technical work, then how could there be *any* women in technical fields? Further, if we look at different points in history, we will see that there is no consistency in women's participation in technical work. Programming was originally a feminine pursuit, as evidenced in recent visibility of the "ENIAC[4] girls" – women who programmed the first general purpose electronic digital computer in 1946. Further, during World War II women took on the majority of manufacturing jobs while men were in military service.

Other explanations suggest that the gender gap is due to the fact that women – and other gender minorities – are somehow "broken." Hence, while they might have the potential to work in STEM, the reasoning goes, they must first undergo technical remediation in order to be capable of technological careers. Such an explanation reproduces stereotypes of those who are underrepresented in STEM as technologically ignorant. Accompanying this explanation is the view that things are changing, the gender gap will disappear; it just takes time. However, this "fix the women and wait" explanation also does not stand up to the data. The gender gap has not been significantly reduced during the time that such explanations have been put forth. Hence, in the absence of some fundamental interventions, the gender gap will continue.

This book is motivated by an interest in alternative explanations that are not grounded in misogyny and gender stereotypes. We wanted to look beneath the surface of superficial explanations in order to examine deeper causes for the gender imbalance, rejecting the simple assumption that lack of representation means lack of interest. We wanted to question the unstated link between hegemonic masculinity and technology. Accounting for the gender–technology gap in this way directs us to situate gender-technology issues within the larger landscape of diversity and inclusion. We wanted to examine the social shaping of women and gender minorities with respect to technology: cultural messages about who "fits" into a particular role, which often result in people imposing limits on themselves regarding what they can do in life (Lerman et al., 2003; Wyer et al., 2001).

This social shaping takes various forms. Women and gender minorities can be the *target* of biases – beliefs, attitudes, norms, and stereotypes – visited upon them by *agents* – individuals and organizations in the larger society. These biases can be consciously held beliefs about the role of gender in society, in STEM, or about gender-based intellectual capacity to work in STEM fields. Or they might be unconsciously held beliefs such as that women should be nurturers, that leadership is not feminine, or that women are less committed in the workplace: they have jobs, men have careers. These attitudes present underrepresented people with *subtle barriers* that can hold them back. Valian (1999) refers to these as "gender schemas." Sometimes *explicit barriers* are imposed through discriminatory behaviors. This might be gender-differentiated advising or treatment in a classroom, being left out of after-work socializing, or having one's contributions overlooked. While these behaviors can be either intentional or unintentional, they nevertheless serve as a barrier to women and gender minorities. Sometimes these beliefs and stereotypes are unconsciously *internalized*, so that what might appear as a girl not being interested in STEM, is really her succumbing to a negative gender-technology discourse.

To understand and rectify the issues underlying the gender gap in STEM we need to develop a better understanding of socio-cultural factors that serve as barriers to and facilitators of the recruitment and retention of women and gender minorities. This, in turn, will aid comprehension of why some individuals are able to overcome barriers while others cannot. Hence, this book includes chapters on theoretical developments to enhance our understanding of the gender imbalance as well as consideration of interventions that flow from research. The research presented in this book speaks to gender equality as it applies to recruitment and retention into technology fields. Chapters also highlight intervention projects to raise awareness about gender diversity, promote understanding about gender dynamics, and question assumptions about gender, masculinity, femininity, and STEM.

The Intervention Problem

If we adopt the perspective that what is "broken" is society and its organizations, not those who are underrepresented, then we must look to both educational institutions and the workplace to address the situation (e.g., AAUW, 2000). In furtherance of these needs, this book presents research with the ultimate goal of using it to inform interventions that will make a positive contribution to addressing the gender imbalance. This perspective also suggests that interventions need to affect both recruitment into and retention within STEM careers.[5]

Several approaches to interventions have been taken. One is to use quotas to increase the representation of women. This is the case in EU countries that are striving to have a critical mass of women in managerial positions. However, having quotas alone can be a problematic intervention. It could be misinterpreted as hiring unqualified women and gender minorities just to satisfy a quota, which could lead to tokenism. Further, having a few high-profile women in technology roles does not mean that it is "normal" for any woman to work with technology. Adding more women, alone, does not address the issue of why there are so few women to begin with. Interventions also need to go deeper, to be tailored to specific issues encountered by specific individuals in specific contexts. Another approach has been to base interventions on an intuitive sense of both what "the gender problem" is that needs remediation, and the best way to address it. But the risk is that the interventions are not addressing the actual problems. For example, women who are facing sexual harassment in the workplace will probably not be helped by maternity leave policies.

What we need, instead, is a clear understanding of intervention need that is based on data and theoretical insights in order to frame "the problem," and to use this insight in the development, ongoing implementation, and assessment of interventions (e.g., Adya and Kaiser, 2005; Annabi and Lebovitz, 2018; Blickenstaff, 2005; Clayton et al., 2021; Craig, 2016; Gorbacheva, 2019; Klawe et al., 2009; Panteli, 2012; Quesenberry and Trauth, 2012; Ridley and Young, 2012; Trauth, 2010). A significant amount of time, money and effort is being spent on activities to address the gender imbalance in the technical workplace. Yet, the gender imbalance persists. This suggests that in addition to a "gender problem" there is also an "intervention problem." But why does it persist? We believe there is a disconnect between gender scholarship and the efforts of well-intended people who develop interventions. We argue for more "evidenced-based interventions."[6]

For example, consider the different explanations for the underrepresentation of women and gender minorities that have been considered thus far in this chapter. One is that they do not have the ability and skills to be successful in a technical career. A second is that they are uninterested in technical careers. A third is that members of these gender groups perceive technical careers as outside the scope of someone with their particular identity. But a fourth reason is that they have career opportunities elsewhere that they perceive to be more rewarding in terms of status, salary, job security, etc. Finally, their families may not want them to pursue technical careers for some reason. So, before undertaking interventions for a particular group of underrepresented individuals, research is needed to determine which causes to address.

One size does not fit all. For example, motherhood accommodations are a good thing for those who have children but wouldn't address other issues encountered by women. And the diversity climate in an educational institution for straight, white women might be perceived as very good, while LGBTQ[7] students might experience the climate very differently, and negatively. What this means is that in the absence of data, a generic approach to diversity, equity and inclusion interventions might not address the real problems at hand.

Interventions need to be implemented at all levels: societal, institutional, organizational. They need to be implemented by policy makers, parents, educators, and employers. UNESCO (2020) offers an example of societal-level interventions in the form of proposed elements of a gender equality framework related to artificial intelligence. So that women are not left behind, it argues for a whole societal-level view of issues, positioning gender equality as part of AI ethics principles. It also offers possible approaches for operationalizing AI and gender equality principles. The EU project, Female Empowerment in Science and Technology Academia (FESTA) is a good example of institutional interventions, in this case in research universities and institutes (Salminen-Karlsson, 2016). In the United States the AAUW has produced two important documents that put forth interventions to address the gender gap in STEM (Hill and Corbett, 2010; Corbett and Hill, 2015). An example of organizational interventions can be seen in the efforts of Carnegie Mellon University to attract and retain women students in STEM (e.g., Frieze and Quesenberry, 2020; 2019; 2015; Margolis and Fisher, 2002). The percentage of women enrolling and graduating in undergraduate computer science and information systems at Carnegie Mellon has consistently been near or above 50 percent. In a nutshell this was achieved by changing the culture in four critical areas: (1) building institutional support (deans, faculty, staff, administrators, funding, and values); (2) revising admissions criteria while maintaining an academically challenging curriculum; (3) offering student leadership opportunities through creative women's organizations (providing leadership, mentoring, encouragement, and peer-to-peer programs); and (4) leveling the playing field (to ensure women, and others, do not miss out on valuable social, academic, and professional opportunities and experiences).

One purpose of this book is to examine the tension between the general and the specific. The chapters in this book demonstrate the way in which gender barriers (and the dismantling of them) result from a combination of societal influences, structural barriers, and varied individual responses to them. They also address the interaction between research and intervention by considering one or more of the following goals: raising awareness about stereotypes, biases, and barriers; changing attitudes about who can participate in STEM careers; and motivating behavioral change to dismantle stereotypes and other barriers.

BACKGROUND

Audience

The goal of this book is to offer a single source to those interested in knowing more about the topic of gender and technology, and about state-of-the-art research and interventions. The intended audience for this book consists of people who want to delve more deeply into the nuances of gender diversity in the technology fields. Collectively, this book represents many years of research, lecturing and teaching about gender and STEM by the contributing authors. The chapters in this book provide the reader with access to theory and results of research on the topic of gender and technology written by some of the best minds globally. While the contributing authors have written a considerable number of academic research papers, a wider audience is intended for this book. We wanted to edit a book that could be understood by general readers without specialized knowledge of science and technology or gender studies, but who are interested in knowing more about the gender imbalance in the STEM fields, and what can be done about it. Great care was taken with the language and writing of the chapters to ensure that the information contained in them would be *accessible* to both academic and non-scholarly readers.

Hence, this book can support general awareness as well as research, teaching, and interventions. It can help people who want to better understand the diversity of colleagues with whom they are working. This might be managers who want to understand how to go about addressing gender issues in their companies, or who are charged with implementing gender mainstreaming policies. It can also support gender studies academics looking at the STEM context, or technology professors and graduate students interested in conducting gender research.

This book can also support teaching. Both of us have taught courses that this book could have supported were it available. Eileen has developed and taught a number of courses (both in the United States and in several other countries) in which gender was the sole focus or was part of a larger focus on diversity.[8] Similarly, Jeria has included gender as topics in her social issues teaching and in several undergraduate research projects. This experience with course development and teaching has influenced our approach to editing this book and the structure into which the chapters are arranged: environmental, identity, and individual influences on gender.

Finally, this book can be helpful to organizations that are interested in developing interventions to redress the gender imbalance. This is because the interventions considered in these chapters are informed by research and not gender stereotypes. Often, we hear about failed interventions – perhaps well intentioned – that were not found to be effective or well executed. The ongoing gender gap in IT coupled with these issues have prompted a call for action from the academic community for theoretically grounded research that can inform more effective interventions through data-driven practices (e.g., Annabi and Lebovitz, 2018; Trauth, 2013; von Hellens et al., 2012).

Distinctive Features

We were motivated to take on this book project out of our commitment to advancing gender equity in the technology fields. We see this book as a foundational source for anyone interested in gender and social inclusion in STEM and IT, whether a new or an established scholar, who wants to undertake gender research. The number of requests we have received from scholars asking for advice about identifying relevant literature, and the number of journal special issues on gender and social inclusion[9] are indicative of this need. This book is also an essential source for those interested in teaching a course on gender, social inclusion, or adding these topics to an existing course. This is the readable book we wish had existed when we undertook our gender research and teaching!

This book has several distinctive features. First, there are many authors but unlike papers in special issues of journals this book has a conceptual unity. This is because chapters are organized according to three themes representing constructs of a theory that both editors have used in research: environment, identity, individual. Each chapter is connected to one of these three themes. Second, while there is thematic unity, the approaches taken by the authors vary. They include a variety of qualitative and quantitative methods. Some are positivist, testing models and hypotheses. Others are interpretive, seeking to understand the subjective reality of those who are experiencing gender and STEM barriers. Still other authors take a critical approach, looking at wider systems of oppression to challenge existing assumptions about gender and technology, asking whose interests are being privileged by the existing power structure (Howcroft and Trauth, 2008). Third, the book has representational unity. Despite the wide variation in topics and approaches, each chapter has the same format so as to make it easier for readers to understand: *Introduction* (which includes the purpose and chapter roadmap); *Background* (what is necessary about theory, prior literature, etc. for the reader to understand the chapter); *Main part of chapter*; *Conclusion* (stating chapter contribution). Fourth, the writing style is welcoming to a wide variety of readers. The chapters employ storytelling and reflexive modes, not typically found in scientific research reporting (Trauth, 2011). Some authors also employ a memoir format in which they reflect on their body of gender research. The tone of chapters is honest, straightforward and transparent, emphasizing accessibility of content to non-experts, and a global audience.

As a whole, this book takes a comprehensive look at societal, organizational, and individual factors that are affecting the relationship between gender and technology. The structure of the book reflects the source of influences on people: the external societal and organizational environments; one's intersectional identity; and individual influences. Each of the three parts is focused on one of these sources of influence. Chapters focus on specific factors that can explain/predict the enhancement or inhibition of underrepresented groups' engagement with STEM and information technology. Collectively, the chapters raise issues, discuss findings, and introduce interventions.

Scope

We are both scholars in the information technology field, hence our perspective on "technology" and STEM is very much oriented in that direction, and most of the chapters have that perspective on technology. For this reason, readers will notice that we have used the terms "information technology" and "STEM" almost interchangeably in this chapter. However, we believe findings about gender and IT would arguably apply more broadly to STEM. Nevertheless, there is a need for more work that looks at gender and technology from other STEM disciplinary perspectives.

Therefore, the focus of chapters in this book ranges from STEM, broadly understood, to information technology in particular. They look at both developers and users of technology. The book takes a broad view of gender that includes not just women but also gender minorities based on intersection with identities such as race, ethnicity, nationality, sexual orientation, and disability. It is also broad in geographical scope. Authors from around the world were recruited to ensure that the resulting book would offer a global perspective on these topics.

The issues addressed in these chapters speak to both recruitment and retention of women and gender minorities in STEM. Recruitment focuses primarily on educational and environmental factors in one's educational development that enhance or inhibit subsequent entrance into the STEM professions. Retention, on the other hand, focuses primarily on workplace factors that enhance or inhibit progression up the career ladder as well as job turnover behavior and intent. While some chapters report on research so that the reader can better understand the issues, others focus on interventions intended to address them.

Conceptual Framing

We chose the factors represented in the individual differences theory of gender and information technology as the conceptual framework for organizing the chapters in this book. We did so because this theory takes into account factors along the continuum from societal to individual, consistent with calls for multiple approaches to and perspectives on women, gender and technology from different levels of analysis: culture and societies, institutions, organizations, and individuals (Fox et al., 2006). Hence, the chapters seek the answer to the gender gap by examining societal factors, structural barriers and varied individual responses to them. In doing so, they address factors at both individual and group levels of analysis. This is not to say that all of the chapters in this book use IDTGIT as the theory guiding the research. Rather, the themes explored in each of the chapters touch on aspects of IDTGIT.

Environmental influences are contextual factors in society, institutions, and organizations, which affect the interaction of women/gender minorities and technology. These include cultural, economic, infrastructural, and policy influences that can facilitate or discourage interest in a technical career. These influences include both societal-level (e.g., Trauth et al., 2008b) and organization-level (e.g., Quesenberry and Trauth, 2012) influences. Individual variation in the effect of these environmental influences occurs in two ways. *Individual identity* refers to demographic characteristics such as ethnicity, age, socio-economic class, and motherhood status, which result in different gender discourses and biases affecting an individual. It also includes IT identity, which considers the effects of working in different aspects of the IT field. *Individual influences* incorporate the effect of significant people and experiences along with personal characteristics, on women, girls, and other gender minorities. That is, significant people in one's life (role models, mentors, sponsors) are conduits for the transmission of both societal barriers and ways to resist them. Hence, individual influences serve to either inhibit or enhance barriers to participation in STEM fields. Within gender-group variation occurs because the group-level influence of environmental factors (e.g., in a particular culture) can be either reinforced or mitigated by one's identity and individual influences.[10]

CHAPTER SUMMARIES

One of the most interesting and rewarding aspects of preparing this book was the collaboration with our authors who brought their diverse perspectives to the work. We are delighted to include 21 chapters from 34 experts, practitioners, researchers, educators, and activists who are leaders in understanding gender and technology. The authors raise interesting questions and use a variety of approaches and methods to present their findings. The authors also represent a variety of different disciplines – such as computer science, data science, information sciences, management information systems, communications, gender studies, political science, and public administration – which lends richness to this work and its conclusions. Each chapter was reviewed by at least one other author in addition to the two editors.

The book begins with a chapter by Eileen Trauth, which provides an overview of the IDTGIT and how it was born out of a desire to balance the role of societal and individual factors in explaining the underrepresentation of women in the IT field. As described earlier, the theory's premise is that women are not a monolithic group, all experiencing the same biases and barriers in the same ways, and therefore requiring uniform interventions. Rather, the gender gap results from environmental influences that are moderated or enhanced by one's intersectional identity, and individual factors. Her chapter demonstrates how the reasons for this disparity are to be found in the interaction of societal influences and individual variation in response to them. She also describes how the theory can be applied to social inclusion more generally and extended by the addition of a dynamic dimension, which explains the ways in which these factors exert their influence over time and context.

The following chapters are then organized into three parts according to their focus on one of the three constructs of the IDTGIT: environmental influences, identity influences, or individual influences.

Part I: Environmental Influences

Environmental influences refer to societal, institutional, and organizational factors that are external to the individual and generally beyond one's control. Culture can be considered a pattern of shared group assumptions that is broadly shaped by the environment. Cultural definitions of femininity that place IT outside the boundary of "feminine" may depict the technological profession as non-feminine. This part of the book explores the cultural, economic, infrastructural, and policy influences that can facilitate or discourage interest in a STEM career.

The first two chapters in the environmental influences part focus on societal-level perspectives. Patience Akpan-Obong shows how policies in many African countries often frame the role of women in the development process as an afterthought leading to problematic consequences. Drawing from secondary research and personal interviews with women in Nigeria and Cameroon, her chapter traces the evolution of women's interactions with technologies from the nascent stage as users of basic information communication technologies to the forefront as entrepreneurs and developers. Her work showcases innovative practices by women in gendered systems of exclusion and frames African women as agents of development. Gyeung-min Kim, Namjae Cho, and Hee-Sun Kim employ a contextual hierarchy perspective to investigate how the IS service industry's project-based work practice influences the work lives and career development of women professionals in Korea. Their work uses a thematic interview analysis of women employees in large IS service firms in Korea conducted in 2010 and again in 2021. The findings show that very little change occurred and that masculine discourses originally found in the IS service industry are continuously repeated and reproduced a decade later. Although these two regional contexts are quite different, their work demonstrates how environmental influences shape gender norms and expectations for women in technology careers.

There is a need to consider environmental factors such as national culture and policy when seeking to progress women's workforce participation and advancement in STEM. The environmental influences part of the book explores several projects funded by the European Union that aimed to improve gender equality policies at the organizational, country, and multinational levels. Minna Salminen-Karlsson provides an analysis of a five-year Female Empowerment in Science and Technology Academia (FESTA) project which aimed to advance gender equality in STEM departments in seven European countries. The focus of the project was primarily on improving the working environment of junior researchers. The FESTA actions centered on collecting and presenting relevant metrics, improving PhD supervision, enhancing informal decision making and communication, and managing resistance. Regina Connolly and Ita Richardson describe why environmental scaffolding must be accompanied by micro-level organizational initiatives that focus on systematically embedding gender equality. Their chapter focuses on one such initiative, Athena SWAN, which supports the systematic embedding of gender equality best practices in third-level educational institutions and supersedes previous episodic initiatives undertaken in pockets around universities and led by individuals. Drawing on the perspective of the University of Limerick, their chapter details the benefits and contribution of Athena SWAN to supporting women's career progression, presenting relevant data from STEM, and outlining associated challenges. Elena Gorbacheva and Isabel Ramos provide an analysis of the accomplishments and the challenges encountered by the Departments of Information Systems at the University of Muenster in Germany and the University of Minho in Portugal during the EQUAL-IST project entitled "Gender Equality Plans for Information Sciences and Technology Research Institutions." The project worked to implement structural interventions that changed organizational rules, regulations, processes, and cultures to enhance gender equality, diversity, and work–family balance at the participating research institutions. These chapters provide detailed summaries of interventions at the environmental level that have brought positive impacts to academia, while identifying areas for continued efforts.

The environmental influences part also focuses on how organizational climate is rooted in values, beliefs and assumptions, and demonstrates some of the ways women navigate and respond to these influences. Monica Adya and Sangeeta Parashar discuss socio-cultural and contextual nuances contributing to different patterns of entrenchment and participation of immigrant Indian- and native-born women in the IT workforce in the US. They found that women born in India demonstrate a functional engagement with STEM careers due to socio-economic and cultural benefits, but those dissatisfied with such careers could feel entrapped due to the same factors. In contrast, native-born women indicate greater career fluidity which, while impacting their long-term participation in IT careers, may lead to greater labor force participation overall. Cynthia K. Riemenschneider and Deborah J. Armstrong address the ups and downs of being a woman academic studying and living out gender within the IT field within the United States. They reflect on their careers in terms of publishing, leadership, and service by looking at issues faced and suggestions for others from the lessons they learned. Their reflexive analysis stresses the importance of intentionality, perseverance, and community. These two chapters show how societal and organizational expectations and policies can improve pathways for women in technology, yet their careers can often hit environmental obstacles and challenges as they attempt to persist and advance.

Part II: Identity Influences

The effect of environmental influences can be varied among the individual identity constructs. Individual identity refers to demographic characteristics (such as ethnicity, age, sexual orientation, disability, socio-economic class, motherhood status, etc.) and how these intersect with gender identity. This part of the book explores how identity influences can facilitate or discourage interest in an IT career.

In the first chapter of this part, Eileen Trauth explores life history interviews conducted with women IT professionals in the United States to illustrate how a woman's intersectional identity interacts with societal and individual influences to enhance or inhibit her participation in the IT sector. More specifically, she applies a framework developed in Trauth and Quesenberry (2006) to explain how a woman's intersectional identity influences her exposure to, experience of, and response to gender biases and barriers. The chapter considers three identities: motherhood, race/ethnicity, and sexual orientation, and demonstrates how an understanding of intersectional identity is crucial to advancing our knowledge about gender barriers and how they affect women in the IT field.

The identity influences part continues with two chapters that consider intersectionality of ethnicity, race, and gender. Lynette Yarger, Courtney Smith, and Adanna Nedd examine the relationships between diversity, equity, and inclusion in the technology workforce and the building of artificial intelligence (AI) hiring systems that discriminate. They discuss two threads of argumentation. The first uses Hill Collins' Four Domains of Power as an organizing framework for presenting factors that contribute to the underrepresentation of women, Black, and Latinx workers in the American technology industry. The second thread discusses how underrepresentation in the IT workforce perpetuates biases in the building and impacts of AI hiring systems. Curtis C. Cain describes BLKGENIUS, which is both a research project to understand the dynamic experiences of Black men in IT against a societal backdrop of barriers that inhibit participation, and an intervention mechanism to highlight successes. His chapter provides a background on the history of Blacks in the United States, Black men in IT, post-secondary education, and the IT workforce. BLKGENIUS stands to make a tangible, positive impact on Black men by providing them the ability to interact with mentors who may share similar lived experiences.

The intersectionality of gender and several other identities is also explored in this part of the book. Eric Patridge uses a retrospective lens to highlight cultural factors affecting lesbian, gay, bisexual, transgender, queer, and allied (LGBTQA+) communities, focusing on a community-based intervention that advances such individuals in the STEM disciplines. He describes how a non-profit organization, oSTEM (Out in STEM), was created to help advance LGBTQA+ people in STEM. The oSTEM leadership has cultivated a community of skilled and diverse people, building bridges across identities while celebrating differences. The community includes both students and professionals from the United States, Canada, the United Kingdom, and beyond. Eleanor T. Loiacono and Shiya Cao explore the intersectionality of gender and disability relevant to IT accommodations and employment. They investigate individuals' experiences and differences in receiving IT accommodations as an organizational diversity intervention that helps employees with disabilities integrate into the workplace. Their work provides the reader with a better understanding of individual differences in the accommodation process, and how to empower women with disabilities in the workplace. Manju K. Ahuja reflects on her career journey and the context and circumstances surrounding the genesis of her conceptual paper on gender in IT (Ahuja, 2002) and the research it generated. She focuses in detail on the research related to motherhood and the specific issue of work–life balance (WLB) and its relationship to career persistence and advancement of women in IT. She concludes with a discussion of the caregiving crisis as the pandemic gave rise to the Work from Home phenomenon, transforming the nature of work, and how it has affected women's employment status. The chapters in the identity influences part of the book show how the inter-connected nature of social categorizations (such as gender, race, sexual orientation, disability, etc.) can create overlapping and interdependent systems of opportunities and challenges to participation in technology fields.

Part III: Individual Influences

Individual influences refer to the role of significant people (such as parents, teachers, role models/mentors, social networks) and life and work interventions in one's interest in a tech-nical career. Individual influences can also involve personal characteristics (such as personal agency, self-efficacy, and empowerment). Personal influences and experiences often help to mitigate negative societal influences. Personal characteristics motivate entry into technical careers and retention within them by utilizing resistance and coping methods. This part of the book explores the ways that individual influences can facilitate or discourage interest in an IT career.

Many interventions that focus on individual influences are frequently aimed at young girls and recruitment pathways. Two chapters in the book provide in-depth examples of such interventions. Tricia Massey, Jenine Beekhuyzen, and Sue Nielsen describe the Tech Girls Movement Foundation, which champions Australian schoolgirls using hands-on learning to transform their futures and encourage equity in the technology industry. They offer an over-view of the origins of Techgirls and highlight some of the integral work being done to expand girls' opportunities to engage with STEM. Their work demonstrates how role modeling and mentorship in multifaceted realms are central to notions of identity construction. Moreover, in her chapter, Roli Varma focuses on organizational efforts to increase the proportion and experiences of women in IT education and employment. She comments on the work of the National Center for Women in Information Technology (NCWIT) from her perspective as a gender scholar, a Social Science Advisory Board member, and one who focuses on areas of particular importance for women of color. The examples provide empirical evidence that intentional interventions can have a positive influence on individuals, yet these approaches should be data-driven and managed in order to maintain sustainable success.

Retention interventions are critical components of individual influences. As such, two chapters explore novel organizational approaches. Hala Annabi reviews limitations inherent in IT interventions aimed at retaining women due to their grounding in essentialist views of women and ignorance of individual differences. She draws on ten years of research to detail how women use individual coping methods to address barriers in the workplace and the impact they experience having to devise and practice these methods. Mari W. Buche introduces an innovative technique called job crafting to help develop and retain diversity in the IT work-force. Job crafting is an approach used to help employees modify their jobs to improve overall work engagement rather than focusing exclusively on the negative aspects of workplace challenges and frustrations. With this intervention, women IT professionals can assess their tasks, professional relationships, and cognitive impressions, identifying ways to improve their experiences in the workplace.

The individual influences part also focuses on personal characteristics. Florence M. Chee, Todd Suomela, Bettina Berendt, and Geoffrey Martin Rockwell examine ethical issues that emerge when conducting research projects that rely on data scraped from online sources (e.g., social media sites, blogs, and forums). Their work focuses on online harassment and hostility, and in particular, the case of Gamergate. Gamergate was an online harassment campaign that promoted sexism and anti-progressivism in video game culture. Their chapter explores two ethical questions: (1) should social media authors be considered research subjects; and (2) how Ethics of Care can support researchers in their exposure to toxic material. Jeria Quesenberry draws from a 15-year longitudinal study of computer science students at Carnegie Mellon University to show how personal efficacy was a driving motivator for women to pursue a technical degree. Interviews with students show that individual characteristics were often the most important factors influencing decisions. Yet there was a growth in the importance women placed on self-efficacy in their decision making. K.D. Joshi picks up this thread by critically examining the role of technical and nontechnical IT self-efficacy in narrowing the IT gender gap. She illustrates how the key sources of self-efficacy that are essential for building confidence are working against women. She concludes with a call to action – if we want to reverse this trend, it is necessary to examine the role of those who are engaged in exclusion.

CONCLUSION

Earlier we posed a somewhat rhetorical question, *why should we care about the "gender problem" in STEM?* As we explained, there are many economic, equity, and moral answers to this question. Perhaps the more challenging question to answer is *how do we address the "gender problem" in STEM?* Despite many years of research, interventions, and funding, we still have yet to solve the problem. Thus, if we are serious about increasing the number of women in technological fields then all of the stakeholders – men, women, scholars, educators, managers, and policy makers – must be part of the solution. But in order to do this we need to share our knowledge about the barriers and best practices for reducing them (Trauth, 2010).

Our intention in this introduction is to set the stage for the chapters that follow in this book, which problematize the important questions and offer informed approaches and solutions. K.D. Joshi (Chapter 22) sums up our motivation for editing this book:

> The accelerating pace with which technology is influencing our work and lives means that the future will be shaped and controlled by people who know how to design and build technology. To ensure that the future of work and life is not decided for women in advance, women need to actively participate in designing and building future systems, algorithms, and technologies.

To address the gender gap in STEM we need to develop a better understanding of socio-cultural factors that serve as barriers to and facilitators of the recruitment and retention of women and other gender minorities in the STEM field. We intend for this book to further our understanding of the influences on people: the external societal and organizational environments; one's intersectional identity; and individual factors that help determine and increase broader participation.

The work ahead promises to be both challenging and necessary. As Hala Annabi (Chapter 18) so critically points out:

> Interventions must be aimed at addressing the barriers women experience rather than trying to fix the women to assimilate to the IT culture. Making the women assimilate is not only immoral, but it also defeats the purpose of bringing a diversity of perspectives to the workplace.

Our authors highlight intervention projects to raise awareness about gender diversity, promote understanding about gender dynamics, and question assumptions about gender, masculinity, femininity, and STEM. In doing so, this collection furthers theoretical developments to enhance our understanding of the gender imbalance. It also presents interventions that flow from research to empower women, and promote meaningful diversity, equity, and inclusion efforts in STEM.

NOTES

1. For a period of time, Eileen served as an Associate Dean for Diversity, Outreach and International Engagement.
2. Jeria currently serves as the Associate Dean of Faculty in the Dietrich College of Humanities and Social Sciences.
3. See Trauth (1999) for the example of Ireland, a country that grew its economy relatively quickly, in large part, through IT and innovation.
4. ENIAC stands for: Electronic Numerical Integrator and Computer.
5. For earlier work on recruitment and retention interventions see Burger et al. (2007); Cohoon and Aspray (2006); and Trajkovski (2006).
6. See *Information Systems Journal* (2012). Special Issue, Women and IT: Increasing the representation of women in information and communication technology: Research on interventions, *22*, 5.
7. Lesbian, gay, bisexual, transgender or queer.
8. For more discussion of these courses see Trauth et al. (2007); and Trauth and Booth (2014).
9. *The Data Base for Advances in Information Systems*: Trauth and Niederman (2006); *Information Systems Journal*: von Hellens et al. (2012); Trauth et al. (2018); *Information Technology & People*: Adam et al. (2002); Cushman and McLean (2008); *Information, Communication & Ethics in Society*: Urquhart and Underhill-Sem (2009); *Journal of the Association for Information Systems*: Bailey et al. (forthcoming).
10. See Chapter 2, and Trauth and Connolly (2021) for a more detailed explanation of this theory.

REFERENCES

AAUW. (2000). *Tech-Savvy: Educating Girls in the New Computer Age*. Washington, DC: AAUW Educational Foundation.

Adam, A. (2005). *Gender, Ethics and Information Technology*. New York: Palgrave-Macmillan.

Adam, A., Howcroft, D. and Richardson, H. (Eds.) (2002). Special issue on gender and information systems. *Information Technology & People*, *15*, 2.

Adya, M. and Kaiser, K. (2005). Early determinants of women in the IT workforce: A model of girls' career choices. *Information Technology & People*, *18*, 3, 230–59.

Ahuja, M. (2002). Women in the information technology profession: A literature review, synthesis and research agenda. *European Journal of Information Systems*, *11*, 1, 20–34.

Annabi, H. and Lebovitz, S. (2018). Improving the retention of women in the IT workforce: An investigation of gender diversity interventions in the USA. *Information Systems Journal*, *28*, 6, 1049–81.

Ashcraft, C., McLain, B., and Eger, E. (2016). *Women in Tech: The Facts*. Boulder, CO: National Center for Women and Information Technology.

Bailey, A., Carter, M., Thatcher, J., Urquhart, C., and Windeler, J. (Eds.) (forthcoming). Technology and social inclusion: Building a dialectic on the role of technology in inclusion and exclusion from societies, organizations, economies, and academe. *Journal of the Association for Information Systems.*

Bell, D. (1973). *The Coming of Post-industrial Society: A Venture in Social Forecasting*. New York: Basic Books.

Blickenstaff, C.J. (2005). Women and science careers: Leaky pipeline or gender filter? *Gender and Education, 17*, 4, 369–86.

Burger, C.J., Creamer, E.G., and Meszaros, P.S. (Eds.) (2007). *Reconfiguring the Firewall: Recruiting Women to Information Technology across Cultures and Continents*. Wellesley, MA: A.K. Peters, Ltd.

Chow, T. and Charles, M. (2019). An inegalitarian paradox: On the uneven gendering of computing work around the world. In C. Frieze and J. Quesenberry (Eds.), *Cracking the Digital Ceiling: Women in Computing Around the World* (pp. 25–45). Cambridge: Cambridge University Press.

Clayton, K., Beekhuyzen, J., and Nielsen, S. (2012). Now I know what ICT can do for me! *Information Systems Journal, 22*, 5, 375–90.

Cohoon, J.M. and Aspray, W. (Eds.). (2006). *Women and Information Technology: Research on Underrepresentation*. Cambridge, MA: The MIT Press.

Connell, R.W. (2005). *Masculinities*. Berkeley, CA: University of California Press.

Cooper, J. and Weaver, K.D. (2003). *Gender and Computers: Understanding the Digital Divide*. Mahwah, NJ: Lawrence Erlbaum Associates.

Corbett, C. and Hill, C. (2015). *Solving the Equation: The Variables for Women's Success in Engineering and Computing*. Washington, DC: AAUW.

Craig, A. (2016). Theorizing about gender and computing interventions through an evaluation framework. *Information Systems Journal, 26*, 6, 585–611.

Cushman, M. and McLean, R. (Eds.) (2008). Special issue on living and functioning in the e-society: Issues of inclusion and exclusion. *Information Technology & People, 21*, 3.

DuBow, W. and Gonzalez, J.J. (2020). *NCWIT Scorecard: The Status of Women in Technology*. Boulder, CO: National Center for Women and Information Technology.

Florida, R. (2002). *The Rise of the Creative Class: And How It's Transforming Work, Leisure, Community, and Everyday Life*. New York: Basic Books.

Fox, M.F., Johnson, D.G., and Rosser, S.V. (Eds.) (2006). *Women, Gender, and Technology*. Urbana and Chicago: University of Illinois Press.

Frieze, C. and Quesenberry, J.L. (Eds.) (2020). *Cracking the Digital Ceiling: Global Views of Women in Computing*. Cambridge: Cambridge University Press:

Frieze, C. and Quesenberry, J.L. (2019). How computer science at Carnegie Mellon University is attracting and retaining women. *Communications of the ACM, 62*, 2, 23–26.

Frieze, C. and Quesenberry, J.L. (2015). *Kicking Butt in Computer Science: Women and Computing at Carnegie Mellon University*. Indianapolis, IN: Dog Ear Publishing.

Glass, J., Sassler, S., Levitte, Y., and Michelmore, K. (2013). What's so special about STEM? A comparison of women's retention in STEM and professional occupations. *Social Forces, 92*, 723–56.

Glover, J. (2000). *Women and Scientific Employment*. London: Macmillan Press Ltd.

Gorbacheva, E., Beekhuyzen, J., vom Brocke, J., and Becker, J. (2019). Directions for research on the gender imbalance in the IT profession. *European Journal of Information Systems, 28*, 1, 43–67.

Harding, S. (1986). *The Science Question in Feminism*. Ithaca, NY: Cornell University Press.

Hewlett, S.A., Luce, C.B., Servon, L.J., Sherbin, L., Shiller, P., Sosnovich, E., and Sumberg, K. (2008). *The Athena Factor: Reversing the Brain Drain in Science, Engineering and Technology*. Boston, MA: Harvard Business Review Research Report.

Hill, C. and Corbett, C. (2015). *Why So Few? Women in Science, Technology, Engineering and Mathematics*. Washington, DC: AAUW.

Howcroft, D. and Trauth, E.M. (2008). The implications of a critical agenda in gender and IS research. *Information Systems Journal, 18*, 2, 185–202.

Huyer, S. (2019). A global perspective on women in information technology: Perspectives from the "UNESCO Science Report 2015: Towards 2030." In C. Frieze and J. Quesenberry (Eds.), *Cracking the Digital Ceiling: Women in Computing Around the World* (pp. 46–60). Cambridge: Cambridge University Press.

Information Technology Association of America (ITAA). (2005). *Untapped Talent: Diversity, Competition, and America's High Tech Future – Executive Summary*. Washington, DC: ITAA.

Klawe, M., Whitney, T., and Simard, C. (2009). Women in computing – take 2. *Communications of the ACM, 52*, 2, 68–76.

Lerman, N.E., Oldenziel, R., and Mohun, A.P. (Eds.) (2003). *Gender & Technology: A Reader*. Baltimore: The Johns Hopkins University Press.

Margolis, J. and Fisher, A. (2002*). Unlocking the Clubhouse: Women in Computing*. Cambridge, MA: The MIT Press.

National Academy of Sciences. (2006). *Beyond Bias and Barriers: Fulfilling the Potential of Women in Academic Science and Engineering*. Washington, DC: National Academy of Sciences.

Niederman, F., Ferratt, T., and Trauth, E.M. (2016). On the co-evolution of information technology and information systems personnel. *The Data Base for Advances in Information Systems, 47*, 1, 29–50.

Panko, R.R. (2008). IT employment prospects: Beyond the dotcom bubble. *European Journal of Information Systems, 17*, 182–97.

Panteli, N. (2012). A community of practice view of intervention programmes: The case of women returning to IT. *Information Systems Journal, 22*, 5, 391–405.

Pew Research Center (2018). Women and men in STEM often at odds over workplace equity. Retrieved from   www.pewresearch.org/social-trends/2018/01/09/women-and-men-in-stem-often-at-odds-over -workplace-equity/ps_2018-01-09_stem_a-09/.

Porat, M. (1977). *The Information Economy: Definition and Measurement*. Washington, DC: Office of Telecommunications.

Quesenberry, J. and Trauth, E.M. (2012). The (dis)placement of women in the IT workforce: An investigation of individual career values and organizational interventions. *Information Systems Journal, 22*, 6, 457–73.

Ridley, G. and Young, J. (2012). Theoretical approaches to gender and IT: Examining some Australian evidence. *Information Systems Journal, 22*, 5, 355–73.

Salminen-Karlsson, M. (2016). *The FESTA Handbook of Organizational Change: Implementing Gender Equality in Higher Education and Research Institutions*. Uppsala, Sweden: Uppsala University.

Schiebinger, L. (1999). *Has Feminism Changed Science?* Cambridge, MA: Harvard University Press.

Schultze, U. (2000). A confessional account of an ethnography about knowledge work. *MIS Quarterly, 24*, 1, 3–41.

Sigacheva, E. (2018). Quantifying the gender gap in technology. Entelo Blog. Retrieved from: https:// blog.entelo.com/entelo-women-in-tech-report.

Smith, D. (2009). The imperative of diversity for institutional viability: Building capacity for a pluralistic society. Presentation at The Pennsylvania State University (University Park, PA, March 27).

Sorensen, K.H., Faulkner, W., and Rommes, E. (2011). *Technologies of Inclusion: Gender in the Information Society*. Trondheim, Norway: Tapir Academic Press.

Tattersall, A., Keogh, C., Richardson, H.J., and Adam, A. (2006). Women and the IT workplace in North West England. In Trauth, E.M. (Ed.), *Encyclopedia of Gender and Information Technology* (pp. 1252–57), Hershey, PA: Idea Group Publishing.

Trajkovski, G. (Ed.). (2006). *Diversity in Information Technology Education: Issues and Controversies*. Hershey, PA: Information Science Publishing.

Trauth, E.M. (2019). *Why Innovations Fail without a Gender Dimension*. Webinar presentation to LIV_IN: Responsible Innovation in Smart Homes and Smart Health Virtual Summit (June 11–12). www.youtube.com/watch?v=9hs1ZpOy-NI.

Trauth, E.M. (2017). A research agenda for social inclusion in information systems. *The Data Base for Advances in Information Systems, 48*, 2, 9–20.

Trauth, E.M. (2013). The role of theory in gender and information systems research. *Information & Organization, 23*, 277–93.

Trauth, E.M. (2011). What can we learn from gender research? Seven lessons for business research methods. *Electronic Journal of Business Research Methods, 9*, 1, 1–9.

Trauth, E.M. (2011). Rethinking gender and MIS for the twenty-first century. In R. Galliers and W. Currie (Eds.), *The Oxford Handbook on MIS* (pp. 560–85). Oxford, UK: Oxford University Press.

Trauth, E.M. (2010). Forward. In A. Cater-Steel and E. Cater (eds.), *Women in Engineering, Science and Technology: Education and Career Challenges* (pp. xviii–xix). Hershey, PA: IGI Global.

Trauth, E.M. (2000). *The Culture of an Information Economy: Influences and Impacts in the Republic of Ireland*. Dordrecht, The Netherlands: Kluwer Academic Publishers.

Trauth, E.M. (1999). Leapfrogging an IT labor force: Multinational and indigenous perspectives. *Journal of Global Information Management*, 7, 2, 22–32.

Trauth, E.M. and Booth, K. (2014). Reflections on blended learning. In Carroll, J. (Ed.), *Innovative Practices in Teaching Information Sciences and Technology* (pp. 207–79). London: Springer-Verlag.

Trauth, E.M. and Connolly, R. (2021). Investigating the nature of change in factors affecting gender equity in the IT Sector: A longitudinal study of women in Ireland. *MIS Quarterly*, 45, 4, 2055–100.

Trauth, E.M., Huang, H., Morgan, A., Quesenberry, J., and Yeo, B.J.K. (2006). Investigating the existence and value of diversity in the global IT workforce: An analytical framework. In F. Niederman and T. Ferratt (Eds.), *IT Workers: Human Capital Issues in a Knowledge-Based Environment* (331–60). Information Age Publishing.

Trauth, E.M., Johnson, R.N., Morgan, A., Huang, H., and Quesenberry, J. (2007). Diversity education and identity development in an information technology course. *New Directions for Teaching and Learning, Special Issue on the Scholarship of Multicultural Teaching and Learning*, 111, 81–87.

Trauth, E.M., Joshi, K.D., and Kvasny, L. (Eds.) (2018). Special issue on social inclusion. *Information Systems Journal*, 28, 6.

Trauth, E.M. and Niederman, F. (Eds.) (2006). Special issue on achieving diversity in the IT workforce. *The Data Base for Advances in Information Systems*, 37, 4.

Trauth, E.M. and Quesenberry, J.L. (2006). Are women an underserved community in the information technology profession? *Proceedings of the International Conference on Information Systems* (Milwaukee, WI, December).

Trauth, E.M., Quesenberry, J.L., Huang, H., and McKnight, S. (2008a). Linking economic development and workforce diversity through action research. *Proceedings of the 2008 ACM SIGMIS Conference on Computer Personnel Research* (Charlottesville, VA).

Trauth, E.M., Quesenberry, J., and Yeo, B. (2008b). Environmental influences on gender in the IT workforce. *The Data Base for Advances in Information Systems*, 39, 1, 8–32.

Urquhart, C. and Underhill-Sem, Y. (2009). Special issue on ICTs and social inclusion. *Journal of Information, Communication and Ethics in Society*, 7, 2/3.

UNESCO. (2020). *Artificial Intelligence and Gender Equality: Key Findings of UNESCO's Global Dialogue*. www.unesco.org.

United States Census Bureau. (2020). *Annual Estimates of the Resident Population by Sex, Race, and Hispanic Origin*. www.census.gov/newsroom/press-kits/2020/population-estimates-detailed.html.

Valian, V. (1999). *Why So Slow? The Advancement of Women*. Cambridge, MA: The MIT Press.

Von Hellens, L., Trauth, E.M., and Fisher, J. (Eds.) (2012). Special issue – Increasing the representation of women in the information technology professions: Research on interventions. *Information Systems Journal*, 22, 5, 343–53.

Wajcman, J. (1991). *Feminism Confronts Technology*. University Park, PA: The Pennsylvania State University Press.

Wajcman, J. (2006). Forward. In E.M. Trauth (Ed.), *Encyclopedia of Gender and IT* (pp. xii–xiii). Hershey, PA: Idea Group Publishing.

Wajcman, J. (1998). *Managing Like A Man: Women and Men in Corporate Management*. University Park, PA: The Pennsylvania State University Press.

Wyer, M., Barbercheck, M., Geisman, D., Öztürk, H.Ö., and Wayne, M. (2001). *Women, Science, and Technology: A Reader in Feminist Science Studies*. New York: Routledge.

Zorn, I., Maass, S., Rommes, E., Schirmer, C., and Schelhowe, H. (Eds.) (2007). *Gender Designs: Construction and Deconstruction of Information Society Technology*. Wiesbaden, Germany: Springer Fachmedien GmbH.

Zweben, S. and Bizot, B. (2020). 2020 Taulbee Survey bachelor's and doctoral degree production growth continues but new student enrollment shows declines. *Computing Research Association*, 33, 5, 2–68.

2. An overview of the individual differences theory of gender and IT

Eileen M. Trauth

INTRODUCTION

A problem that intrigued me from my earliest days of conducting research to examine enablers and barriers to women's participation in STEM and information technology (IT) careers was that women were being viewed as a homogeneous population, as though the same "solutions" would apply to all women. That was not my lived experience as a woman, a woman in IT, or a woman conducting gender and IT research. I began to work on gender issues in the IT field in 2000 on a sabbatical in Australia, in the second half of my career. Prior to that I had spent ten years engaged in IT skills research in the United States, followed by a decade of global IT research where I examined the influence of socio-cultural factors on the IT sectors of countries. This experience of teaching and conducting research in several countries taught me that the variability of societal contexts was an important component of research about the IT field and its workforce. I also witnessed considerable variation in gender roles and the position of women in the IT sectors across countries. And even before I was conducting gender research, I knew that my gendered experience in the IT field was different from that of other women. I was middle-aged, a white American, a lesbian in a long-term but not yet legal relationship, and did not have children. These identity features influenced the trajectory of my career and how I experienced gender biases and barriers, and responded to them. My career trajectory was also different from the path laid out for me by the local context of my youth. I grew up as the middle child in a Midwest, middle class, Catholic family during the 1950s and 1960s when women's life options were very limited. Women typically worked as teachers, as nurses or other health care workers, as secretaries or as hairdressers. They often left the workforce after marriage to raise a family. So how did I come to obtain a PhD in a STEM discipline and become a college professor? Clearly there were other influences beyond societal and identity factors. In my case, my nuclear family – father, mother, and six sisters – were highly influential in helping me to counteract some of the societal messages of the time. So, when I embarked upon gender research, I brought with me these perspectives about influences on women that introduced variability. I knew that interventions based on a monolithic view of women did not provide the answer I sought. This viewpoint was reinforced by one of the first gender and technology authors whose work I read:

> Although studies do find evidence of differences between the sexes, the variation within the sexes is more important than the differences between them. (Wajcman, 1991, 157)

This is what has driven my research and theorizing about gender and IT. I began by focusing on how to understand within-women variation in factors that can explain the gender gap in the IT field. I later expanded my focus to include groups of men who are gender minorities. Just as there is not a single monolithic masculinity or femininity, there is not a single set of factors working in a single way to impact women and men.

In searching for explanations for the underrepresentation of women and girls in the IT field[1] I sought to maintain a balance between environmental and individual factors. I investigated societal influences while also taking into account individual variation in influences on girls and women. It was the desire to articulate this balancing act that led me to propose the *individual differences theory of gender and IT* (IDTGIT). The premise of this theory is that women are not a monolithic group, all experiencing the same barriers in the same ways, and therefore all needing the same interventions. Rather, the reasons for (and solution to) women's underrepresentation in the IT field are to be found in the interaction of societal factors that influence women as a group, and individual factors that influence particular women in varied ways. We cannot truly address the gender gap if we fail to take both sources of influence into account.

My theory also grew out of a desire to unpack this notion of "gender difference" in thinking about the gender imbalance in the IT field. How different *in ways that matter*, are men and women, or other marginalized people who are "different" in an observable way (e.g., based on ethnicity or disability)? On the one hand, a man and a woman might look different on the outside, but may be very similar in their technology orientation. On the other hand, two men might look similar on the outside but be totally different in their relationship to technology. The questions are: When and how should difference matter? and What kind of difference are we talking about?

I was motivated to develop this theory for a few reasons. One was theoretical. I did not see sufficient theorizing about gender underrepresentation in the IT field that took into account what I considered to be all the relevant factors. Another motivation was practice oriented. I wanted to help address the barriers confronted by underrepresented and underserved people in their day-to-day education and workplace. I wanted to contribute to interventions that would address the equity and inclusion issues I had witnessed and experienced in my career. I wanted to help shift the discussion away from diversity being a "problem" for underrepresented people to solve, to being the responsibility of majority groups.

The purpose of this chapter is to take the reader on a journey of understanding about the theory, its evolution over 25 years, and how it has been employed to study the gender gap problem. My goal is for the reader to have a better understanding of the factors that I found to have enhanced or inhibited women's participation in the IT field. In the next section I situate this theory within the gender and technology literature. I then explain the theory's constructs and illustrate how they have guided my and others' research. I conclude by reflecting on the continuing evolution of this theory.

BACKGROUND

In the Introduction chapter of this book, we made a case for addressing the gender gap in technology fields based on the demands of our global digital economy even as workforce statistics show a decline in women's participation. Addressing this problem involves other problems. First, is the knowledge problem: understanding the biases and barriers. But to do that, there is the theory problem: determining how to think about the gender gap. This means deciding what knowledge to seek, and how to organize, filter and evaluate the data. A theory is a lens guiding what you expect to see, what you ignore, and how you interpret what you see. I found that in order to address the knowledge problem I first needed to address the theory problem. The danger is that in the absence of a theoretically informed understanding of the issues, well-intentioned people might resort to "fix the woman" interventions based on gender stereotypes.

Too often theory is disparaged as something abstract, with no connection to the real world. Instead, theory should be thought of as an "intellectual tool toward creating, formalizing, transferring, and applying knowledge" (Niederman, 2021, p. 120). My theory arose from an interest in understanding how some women were able to resist biases and barriers while other women succumbed to them. I sought the intellectual tools to challenge the notion that women who work with technology are exceptions to some universal gender norm.

Gender-Group Level Explanations for the Gender Gap in IT

My initial foray into the gender and IT literature produced what appeared to be two theoretical extremes: gender essentialism and social shaping of gender. But upon closer examination, I saw that they both take a similar perspective on gender issues – looking at the problem and the solution from a group level of analysis. As explained elsewhere, an essentialist explanation for the gender gap assumes the existence of fixed and opposed feminine and masculine natures (see Trauth, 2002; Trauth and Quesenberry, 2007; Wajcman, 1991) that are rooted in biological differences between males and females. Essentialism assumes that observed differences between men and women with respect to technology use or careers are determined by biology: inherent, group-level differences that are based upon bio-psychological characteristics. An essentialist explanation of the gender imbalance could be summed up as: "Women are under-represented because it is not in their nature to work with IT."

Essentialist explanations can be challenged on several grounds. First, the fact that there are any women IT users and professionals challenges the assumption that technology is alien to women's nature. Likewise, the fact that some men are not inclined to work with technology challenges the assumption that it is an integral part of men's nature. Second, historical and cross-cultural variation in gender role differentiation related to technology use and careers demonstrates that biological sex differences cannot account for gender differences in the IT workforce (Marini, 1990). Finally, essentialism assumes an immutable and sex-based gender binary, something that has been challenged by psychologists and feminists alike. Psychologists such as Gill et al. (1987) called for measures of gender in personality scales that allowed researchers to examine both similarities and differences among men and women. Barnett and Rivers (2004) described the damage done to careers and relationships by stereotypes about the differences between women and men. Feminist literature has long challenged the notion of a gender binary, characterizing gender, instead, as socially constructed and fluid (e.g., Butler, 1990; Heinamaa, 1997; Lucal, 1999). With respect to technology, in particular, Adam et al. (2001, 2002) argued that the essentialist theory's narrow focus on psychology, alone, comes at the expense of examining context.

A response to essentialism put forth by feminist and social studies of science scholars is that human outcomes are not determined by biological factors. Rather, social factors shape one's relationship to both gender and technology. The social shaping of technology as masculine looks to social construction theory (Berger and Luckmann, 1966) rather than biological and psychological theories. It asserts that society has constructed a fundamental incompatibility between feminine identity and technology (e.g., Cockburn, 1983, 1988; Cockburn and

Ormrod, 1993; Wajcman, 1991). Hence, a social constructivist explanation of the gender imbalance could be characterized as: "Women are underrepresented because they have been socialized away from technology."

Like essentialism, social constructionist explanations assert that men as a group are different from women as a group in ways that matter for IT, but for sociological rather than biological reasons. Likewise, a critique of this explanation can be found in counter examples. The definitions of masculine and feminine domains are not uniform but vary across cultures. What is considered "men's work" and "women's work" has also changed as a function of time and circumstances. Typewriters used to be masculine machines when men were secretaries; factory work was women's work during World War II when men were in the military.

This group-level theorizing is reflected in IT research on gender that was conducted in the 1990s and early 2000s (e.g., Adam et al., 1994; Balka and Smith, 2000; Joshi et al., 2003; Nielsen et al., 2000; Spender, 1995; Star, 1995; Webster, 1996). In a critical literature analysis of papers published between 1992 and 2012 (Trauth, 2013), I categorized gender and IT work according to whether or not a gender theory was explicitly used. I found that a considerable number of publications looked at surface-level differences in measures of career choice, career progression, technology adoption, etc. and attributed them to assumed essential differences between men and women, or to a fundamental incompatibility between the social construction of feminine and masculine identities. Observed gender differences in email use in one study concluded that men and women differ, inherently, in their use of email, which led to a recommendation that men and women should receive different training for new media (Gefen and Straub, 1997). In another study, observed gender differences in use of a new software package concluded that men and women process information and make decisions about technology in very different ways, which led to the recommendation that new technology be implemented differently for men than for women, with productivity being emphasized for men, and social support for women (Venkatesh and Morris, 2000).

Within-Gender Group Level Explanations for the Gender Gap in IT

The issue that I and others have with both of these theoretical viewpoints is that they look at women as a uniform group when trying to understand under-representation in technology fields. The alternative view is to look for variations among women, to look for differences between those women who take up and persist in technical careers, and those who don't. In response, challenges to a theory of monolithic social shaping of gender have emerged. Sociologists Risman and Davis (2013) urged their field to move beyond normative models of gender, and present a framework that includes attention to the differences and similarities among men and women as individuals. An early critic of universalist gender grouping from a feminist perspective, Barrett (1987) observed that contemporary feminism assumed:

> ... a framework in which the categories of "women" and "men" are themselves seen as relatively unproblematic: the focus is on the difference *between* them. [But] the whole matter looks very different if we attempt to "deconstruct" these categories, examining critically the forms of difference that exist *within* them rather than between them ... [R]ecognizing the idea of difference within the category of women, is radically challenging to conventional feminist arguments. (p. 29)

Other feminists argued that overgeneralizations about women's problems tended to represent only the perspective of privileged women: white, western, middle-class, and heterosexual (e.g., hooks, 1981). This critique has led to the emergence of feminist standpoint literature, which champions the perspectives of diverse, oppressed groups of women (e.g., Harding, 2004; Hartstock, 1983, 1998; Heckman, 1997) particularly women of color (e.g., Collins, 1989, 2009). Challenges come from masculinity studies as well (Connell, 1987, 2005; Halberstam, 1998), which explores alternatives to hegemonic masculinity.

Starting in the 2000s, gender and IT researchers began to offer alternatives to prevailing explanations for the gender gap that were based upon comparisons with men (e.g., Adya and Kaiser, 2005; Ahuja, 2002; Cohoon, 2001; Townsend, 2002; Trauth, 2002; von Hellens et al., 2001). This work focused on factors affecting girls and women. The broadening of research methodology beyond quantitative and positivist to include qualitative, interpretive and critical approaches further enabled more nuanced investigations of the gender gap (Howcroft and Trauth, 2005, 2008; Trauth and Howcroft, 2006a). Consideration was given to the role of societal factors in shaping educational options, career choices, and career progression. A body of research emerged that focused on influences – at the individual level – on girls and women that could serve to mitigate societal biases and barriers. It delved into educational and workplace interventions to address selection of and persistence within IT careers. Factors affecting girls' choice of IT careers came to include influential people such as parents, teachers, counselors, role models, and mentors.

But to me the picture was still incomplete. What we could learn from research about role models and mentors, alone, did not fully answer questions such as: "How is it that some girls and women can resist societal messages and barriers that they encounter while others do not?" and "Why do some girls and women encounter few barriers while others encounter many?" I recognized that: not all women encounter the same messages, stereotypes, and barriers; they process and react to them differently; and some women are able to rise above barriers. All of this suggested that something else was also going on. It was also necessary to include the influence of gender intersectionality. In addition to Black feminist standpoint theory, other types of intersectionality needed to be recognized. For example, Stepulevage (2001) argued that assumptions about young women's developing identity within the context of technology education needed to take into account that gender technology relations are situated within heterosexual social norms. Clearly, multiple meanings of gender, such as those reflected in LGBTQ[2] individuals, or different understanding of gender roles in different cultures needed to be accommodated. These reflections led me to determine that a more fine-grained examination of individual factors was needed. With the incorporation of intersectional identity, I felt that the picture was complete and the IDTGIT was born.

Rather than grouping all women together when trying to explain the gender imbalance, this theory focuses on variation among women in their exposure to, experience of and response to both societal and individual factors that can affect IT career choice (Trauth and Quesenberry, 2006). It incorporates both societal-level and individual-level explanations by focusing on individual variation in response to societal-level gender influences. It posits that the answer to why some women persist in the IT field while others do not can be found in the combined influence of internal and external factors that influence an individual's personal development and subsequent IT career decisions. That is, not all women in a particular society will receive the same messages about gender and IT. But even in situations where they do, both the interpretation of these messages and the response to them will vary as a result of individual factors such as role models, and one's intersectional identity. Thus, the IDTGIT searches for the causes of gender underrepresentation by examining the factors that account for the varied ways that individuals internalize and respond to gender messages. It seeks to understand the sources of individual agency that enable some women to overcome systemic negative influences.

According to this theory, then, the explanation for varied responses to societal influences lies in the combination of identity, individual and environmental influences. Women and men are individuals possessing different technical talents and inclinations, who respond to the social shaping of gender in varied ways. These messages are sometimes communicated differently to subcultures in society (e.g., based on race, ethnicity, sexuality, etc.). Further, significant people and experiences in one's life produce a range of responses to those uniform messages. For example, not all women of a certain age group respond in the same way to commonly received messages. In summary, the IDTGIT asserts that the participation level of women in IT can best be explained by the interaction of: the social shaping of gender and IT in a particular (socio-cultural) context, the intersectionality of gender with other identity characteristics (e.g., ethnicity, age, parenthood), selective (societal and institutional) reinforcement of individual IT inclinations (interest and capability), the influence of significant others in an individual's life/career, and individual responses to generalized societal influences.

Addressing the "Theory Problem"

The articulation of the IDTGIT also addressed some theory problems. I and others saw a lack of theoretical treatment of forces at work that accounted for the gender gap. Much of the gender and IT research that is still being published is under theorized (Gallivan, 2013; Mennega and de Villiers, 2021; Oreglia and Srinivasan, 2016; Trauth, 2006b, 2013).[3] As Gregor (2006) explains, a theory is an attempt to understand a phenomenon. However, I saw studies claiming to be gender research using no gender theories at all – to inform the design of the research or to interpret results. When a phenomenon is first recognized, it is understandable that *pre-theoretical* research would be undertaken, which compiles statistics to document it (e.g., Camp, 1997; Klawe and Leveson, 1995; Montanelli and Mamrak, 1976; Pearl, et al., 1990). Yet despite the decades of evidence for the gender gap in STEM, gender research without an underlying gender theory is still being undertaken in 2022. This research either presents statistics with no theory or, if it has one, it is a theory about technology design, diffusion, use, etc. To interpret results the researchers employ a theory-in-use about gender that is often based upon their implicit gender stereotypes and biases.[4] The problem with this *implicit-theoretical* research is that it is difficult to discuss, challenge, or test. Finally, the IDTGIT addresses what I considered to be *insufficient-theoretical* research, which looks at either group-level (societal) explanations for the gender gap or at individual (psychological) explanations alone.

Another issue with much of the gender and IT research continues to be its tendency to explain the gender gap by theorizing gender as "gender difference." Some of this research is demographic research and not social inclusion research (Trauth, 2017b, p. 13). Its goal is to identify differences between men and women with respect to engagement with technology. This research becomes problematic if it purports to explain the gender gap by looking at differences between men and women. Niederman (2021) and Niederman and March (2019) argue that the IT field has an over reliance on variance research, which might explain why so much gender research has been conducted as between-genders research.

INDIVIDUAL DIFFERENCES THEORY OF GENDER AND IT (IDTGIT)

The *individual differences theory of gender and IT* (Trauth, 2006b, 2002; Trauth and Connolly, 2021; Trauth et al., 2004) explains women's participation (or nonparticipation) in the IT field by considering gender relations at two different levels of analysis. The societal level is concerned with the origins and operations of gender group influences which all women/girls in a given environment would encounter. The individual level focuses on differences among women in terms of how they respond to societal influences. The theory proposes that variation is due to differences in the ways women are exposed to, experience, and respond to gender messages, biases, and barriers related to IT careers (Trauth and Quesenberry, 2006). These differences in the degree to which girls and women encounter gender group biases, and the ways in which they respond to and internalize them results from different demographic traits, personal traits, and influential people in their lives. Hence, this theory takes into account both individual factors and group (organizational/societal) factors. It also concentrates on differences *within* rather than *between* genders in providing reasons for the gender gap in IT.

I approached my investigation of differences among women by starting at the end of the process: interviewing women who had overcome barriers to working as IT professionals. I asked them to reflect on their educational and occupational journeys in order to uncover variation in factors that enhanced or inhibited their careers. The broad, brush strokes of the theory arose during my Australian/New Zealand study (2000–2001) and was subsequently articulated in Trauth (2002). It was in this first gender study that I recognized different societal influences across the United States, Australia, and New Zealand. I also interviewed women working in Australia and New Zealand who were originally from a number of countries, thereby revealing even more within-gender variation in factors. This, then emergent theory, was elaborated upon in two subsequent studies: in the United States (2002–07); and in Ireland (2003–12).[5] For these two studies an extensive coding scheme, based on the Australian/New Zealand data analysis categories, was developed. As we[6] developed the coding guide, the dimensions of within-gender variation began to come into greater focus. In addition, an interview guide (see Appendix 2A.2), based on this theory enabled the interviewer to further draw out the factors of individual variation in subsequent studies. I chose an interpretive epistemology so that I could explore social and psychological processes whereby overt and internalized constraints hold women back, but also the strategies used by women who overcame barriers and how they did so.

The first articulation of the theory was at an ACM SIGMIS conference (Trauth et al., 2004). It put forward three constructs or sets of factors that could explain the gender imbalance in the IT field by showing influences on women's decisions to enter and persist in this career (see Table 2.1). The *environmental influences* construct consists of four sub-constructs related to the geographic region in which an individual lives: *cultural influences* which includes national, regional or organizational attitudes about women and about women and IT; *economic influences* such as the cost of living and availability of IT work; *societal infrastructure influences* such as the availability of childcare facilities; and *policy influences* such as laws about gender discrimination, and policies about maternity leave. The second construct, *identity,* is comprised of *personal demographics* items such as ethnicity or age, and *IT identity* items about the type of IT work in which one currently or will engage. The *individual influences*

Table 2.1   *Constructs of Individual Differences Theory of Gender and IT (IDTGIT)*[7]

Construct	Sub-construct	Examples
Environmental Influences	Economic influences	Cost of living, unemployment rate
		Information sector employment
	Cultural influences	Attitudes, norms and values about: women, women working, women working in IT
		Gendered workplace climate
		Gendered educational climate
	Societal infrastructure influences	Availability of childcare
		Transportation available
		Utilities (electricity, telecom) available
		Stores, doctors, etc. hours of operation
	Policy influences	National: Laws about gender discrimination
		Maternity/paternity leave
		Tax laws
		Organizational: discrimination, etc. policies
Identity Influences	Personal demographics	Age, sexuality, gender identity, race, ethnicity, socio-economic status (SES), marital status, parenthood, geography/nationality, educational background, religion
	IT identity	Type of IT work (e.g., software development, system design)
		Industry of work
		Career progression
Individual Influences	Personal influences	Significant people (parents, counselors, mentors, role models, sponsors, peer networks)
		Significant life experiences
	Personal characteristics	Personality traits, interests, abilities, self-efficacy, agency, work ethic, motivation, assertiveness

construct includes *personal characteristics* items such personality traits and abilities, and *personal influences* items such as mentors, role models, and significant life experiences.

These three constructs characterize factors that operate to inhibit or enhance a woman's participation in the IT field. We explained the variation in the process whereby this occurs in the following way (Trauth and Quesenberry, 2006). *Exposure* refers to the extent to which a woman or girl encounters gendered messages or barriers. *Experience* refers to the degree to which she internalizes them. *Response* refers to what she does as a result of encountering them. My various research studies demonstrate how the constructs of the theory inform data collection, guide data analysis, and contribute to a better understanding of the gender gap. They are discussed below.[8]

Environmental Influences

The *environmental influences* construct refers to contextual factors in society, institutions, and organizations which affect girls' and women's decisions to enter and remain in the IT profession. I articulated environmental influences using a conceptual framework developed in an earlier, ethnographic study of the evolution of the IT sector in Ireland (Trauth, 2000). This framework has four components: culture, economy, infrastructure, and policy. The diversity of participants in my studies provided ample opportunity to consider socio-cultural factors. I conducted interviews in four countries: Australia, New Zealand, Ireland, and the United States.

A number of the respondents were born and raised in a country other than one of these four, which provided additional environmental insights. Finally, my international lecturing and teaching in still other countries aided further refinement of this construct.[9] Below, I provide examples from papers that focused on the role of one or more of these factors in facilitating or discouraging a woman's interest and perseverance in an IT career. In two papers I looked at all four factors: Trauth (2020) and Trauth and Connolly (2021).

Culture

Culture was the most frequently examined of the environmental subconstructs. This is because gender is experienced through culture. A culture – whether national, regional or organizational – reflects the values, beliefs, biases and behaviors regarding gender and gender relations. It communicates the normative understanding of masculinity, femininity, and overall gender roles. Culture reveals attitudes about women in general, women working in the paid labor force, and women working in technical fields.

I explored cultural influences at multiple levels: national, regional, and organizational. I asked the women to reflect on experiences and attitudes about women in general, women working outside the home, and attitudes about women working in technical fields such as IT. We supplemented the interviews with information about gender dynamics in the region or country in which interviews were conducted. But it is important to note that because norms and values about gender vary across cultures, its role in enhancing or inhibiting participation in STEM fields like IT will vary. For example, women who were born and raised in Australia spoke about societal messages that women should not be programmers. But women who came from India received a different message. In India it was acceptable for women to become engineers. I received a similar response from women who were raised in a communist regime (Trauth et al., 2008a). In addition to direct societal messages about women entering a technical career were varying messages that would influence a woman's participation in it: attitudes about maternity, childcare, parental care, and working outside the home.

Insight into cultural influences was particularly noteworthy when countries were at inflection points, when significant transformations were underway, as was the case in Ireland. The longitudinal study of women in Ireland's IT sector against the backdrop of the evolution of the Irish information economy revealed significant cultural changes (Trauth and Connolly, 2021). During the period studied (1970–2019) women's role in Irish society changed dramatically as the country moved from a rural, traditional, isolationist society to a modern global economy. A similarly significant transformation in gender relations was underway in 2008 when I taught a graduate gender and IT course in South Africa (Trauth, 2020). Both men and women students noted the significant cultural change occurring in post-Apartheid South Africa at that time. We discussed the effects of a quota system for employment in all professions, which related not just to race and ethnicity, but also to gender. Women raised concerns about being perceived as simply filling a quota and not qualified for their IT positions. Other discussions revealed cultural influences on women's IT careers that were particular to certain subcultures in South Africa. For example, one woman student who worked for a multinational IT company voiced the challenge she would encounter were she to be managing a man who was from her tribe.

The interviews I conducted in the United States revealed regional cultural influences on women. I studied three specific regions: eastern Massachusetts, central North Carolina, and central Pennsylvania. Each of these regional cultures was shown to exert different influences on women with respect to participation in the IT field (Trauth et al., 2008b). For example, women in eastern Massachusetts were influenced by the overall value placed on diversity in the region as well as the abundance of women role models resulting from the size of the high technology sector there. In contrast, women in North Carolina were experiencing the transition from a predominantly rural culture with defined gender roles to a diversified economy accompanied by new cultural norms about women's role in society and the labor force. The regional culture of central Pennsylvania was rooted in the gendered occupations of the railroad, coal mining and agriculture. But, like North Carolina, it too was undergoing a cultural shift that was accompanying a more diversified economy and employment opportunities for women. Within each of these regions, further variation occurred in the messages and barriers women experienced as a function of such identity factors as race, ethnicity, and socio-economic status (SES).

In addition to national and regional cultural influences, interviewees spoke about organizational culture, which consists of shared assumptions, values and beliefs that govern behavior in the workplace (Quesenberry and Trauth, 2012; Trauth et al., 2009). They commented on gender discrimination, both conscious and unconscious, stated and unstated. A commonly cited example of organizational culture influence was exclusion from men's networks. Women also noted barriers in the form of stereotypes about women's technical aptitude, their ability to balance work and motherhood, and their commitment to their career. Another aspect of organizational culture is educational culture. Women recalled their experiences as students, of being advised away from technical courses, and feeling like they were invading a "man's space" by studying IT in college.

Economy

Across my different studies, I saw a close association between economic factors and cultural attitudes about gender. I also found cultural attitudes adapting to economic changes. Interviews explored themes about the unemployment rate in the nation/region, the availability of IT work and its importance in a nation/region, and the cost of living. These all have a direct or indirect influence on women's IT careers.

With respect to the overall economy, in the early years of Ireland's IT sector (i.e., 1970s and 1980s), when jobs in the country were scarce, a married woman who worked in the paid economy was viewed taking a job away from a man who needed to support a family (Trauth, 1995, 2000). In contrast, beginning in the mid 1990s when the overall economy was robust – due in large part to the IT sector – there was a very different attitude toward women working. It was becoming the norm for women to work outside the home, including working in the IT sector. Irish women students commented that their parents encouraged both their daughters and their sons to go into IT careers (Trauth and Connolly, 2021).

The size of the information economy, that is, the amount of IT employment available in the country or region, is another aspect of economic influence. This was in evidence in the three regions of the United States that I studied. During Boston's high-tech boom of the 1990s the IT workforce made up approximately 4 percent of the overall labor force. During that same time period, IT workers constituted 3.6 percent of the overall labor force in North Carolina, and 2.5 percent in central Pennsylvania (Trauth et al., 2008b). Just like in Ireland, the larger the IT employment sector was in a region, the more normalized it was for women to work in this sector.

Finally, the cost of living was an economic factor influencing women's opportunities in the IT field. In high cost of living regions such as Boston, Massachusetts there was a greater need for two income households. Hence, it was more normal in heterosexual couples to have the woman partner working outside the home.

Societal infrastructure

Societal infrastructure encompasses basic physical and organizational structures and facilities in a society that have a direct or indirect effect on women's engagement with IT. Physical infrastructure includes such elements as telecommunications, transportation, electricity, and water. Relevant organizational infrastructures would include items such as educational and childcare facilities. In countries with insufficient electricity and telecommunications capability, people must travel to internet centers to access computing. This is particularly problematic for women who might experience transportation, education and literacy barriers, or whose domestic duties keep them close to home (Akpan-Obong, 2009; Hilbert, 2011; Rathgeber and Adera, 2000).

The societal infrastructure themes that arose in my research related to childcare, educational infrastructure, and the hours that stores etc. were open. One aspect of societal infrastructure that emerged in the research conducted in every country was childcare facilities. This had an indirect effect on women when inaccessible or unaffordable childcare was a barrier to women having careers outside the home. There was also variation in terms of this influence. In Australia, for example, parents receive state subsidy to offset childcare costs. In contrast, in the United States, childcare is privately funded. Educational infrastructure emerged as a direct influence on women's technology careers when all girls' secondary schools did not offer the mathematics and science courses needed to pursue a STEM degree in university (e.g., Trauth et al., 2003).

Another societal infrastructure theme that arose was the operating hours of stores, doctors' offices, etc. that were based upon the assumption of a stay-at-home-wife. This was an indirect barrier when stores did not have evening or weekend hours. Indeed, when I lived in Ireland in 1990, I had to adjust my work schedule in order to leave campus in time to shop at the grocery store before it closed. Women also spoke about having to leave work for parent-teacher meetings because they were held during the day. Further, women noted that when they went for these meetings, there would be only one chair; it was assumed that the father wouldn't be present because he'd be at work (Trauth and Connolly, 2021).

Policy

Policy refers to both governmental and organizational regulations. In my research, I found that policies influenced women's recruitment and retention in a variety of ways. Some policies encouraged women to enter STEM careers. An example is the Irish policy of funding both men and women students to study IT, which was a part of the country's overall economic development plan (Trauth and Connolly, 2021). Other policies are specific to gender, though not specific to women in STEM. Examples would be gender equality legislation and antidiscrimination laws related to members of the LGBTQ community.

Some policies have not been enablers but rather barriers to women pursuing careers of any kind, including IT careers. An example of this would be tax regimes that favor having a single wage earner in the family, thereby making it less cost effective for a wife to work outside the home. Another example was the "marriage bar" that existed in Ireland until the mid 1970s. It required women to leave their government jobs upon marriage, and many private sector organizations implemented a similar policy.[10]

Parental leave policies exemplify variation in national policies related to gender. Irish women receive 26 weeks' paid and 16 weeks' unpaid maternity leave, whereas Irish men qualify for two weeks' paid paternity leave. Women in Australia are entitled to 18 weeks of paid maternity leave whereas women in New Zealand receive 22 weeks. In both of these countries, fathers qualify for two weeks' paid leave. In contrast to these countries, there is no uniform parental leave policy in the United States. At the federal level, there is guaranteed, but unpaid, family leave of 12 weeks. Of the three states involved in my American study, only Massachusetts provides paid maternity leave (Trauth, 2020). Examples of organizational policies would include gender equality plans, gender antidiscrimination policies, parental leave policies, and childcare policies.

Identity Influences[11]

The *environmental influences* construct of the theory speaks to group-level, societal and organizational factors affecting women's IT careers. The other two constructs – *identity* and *individual influences* – explain the individual variation among women with respect to how these environmental influences have their effect. *Identity* refers to one or a combination of demographic characteristics that reveal the outside world's imposition on the construction of self, and the effect this has on recruitment into and retention within STEM fields such as IT.

In my interviews, the following demographic information was collected: gender, sexual orientation,[12] race, ethnicity, geography/nationality, SES,[13] marital status, parenthood status, age, family of origin, educational background, and religion. I explored ways in which these demographic characteristics interact with societal gender roles, stereotypes, and other factors to influence technology career decisions. In the interviews I invited women to talk about how their identity characteristics interacted with their gender to affect their personal development, as well as their career path.

Another aspect of the *identity* construct is IT identity. This takes into account aspects of a woman's career trajectory including: the industry in which she worked, the type of IT work she did, where she was situated on a technical–behavioral continuum, her career level, and her career progression. The reason for collecting this data was to take into account the effects on a woman's career of working in different aspects of the IT field. For example, a woman working in "hard" disciplines such as computer engineering would likely encounter different barriers than a woman working in a "softer" aspect of the IT field, such as information systems management. Further, I looked at the work sector: primary IT sector (i.e., engaged in the development of information processing hardware, software and services) versus secondary IT sector (i.e., applying IT tools in another sector such as health care, education, government). Our Irish data revealed differences in the work culture and acceptance of women IT professionals across work sectors (Trauth and Connolly, 2021, p. 2070).

Social exclusion as a result of gender, ethnicity and class inequality is one of the most pressing challenges associated with the development of a diverse IT workforce in the United States. Early research into the underrepresentation of women in the IT workforce and educational preparation for IT typically viewed women as homogeneous, thereby blinding the research to variation that exists among women. During the early 2000s when I began my interviews with women in the United States, the intersectionality of ethnicity and gender was not widely considered. Hence, insight into how the experiences of women of color might differ from those of their white colleagues was not investigated. Nor was there much thought given to how women of color negotiate the intersecting challenges associated with being underrepresented in terms of both ethnicity and gender. In response, I collaborated on an investigation of the intersection of gender, ethnicity and class identities that shape the experiences of Black women IT students and workers in the United States (Kvasny et al., 2009). The results showed that the experiences of Black women in the United States vary not only in contrast to white women, but also as a function of their own personal histories (e.g., some have a history of slavery and segregation while others, such as those from the Caribbean or more recently from Africa do not), and personal agency.

The theme of challenges associated with managing motherhood and technical careers arose in every country where I conducted interviews or gave lectures. We examined the challenges of work–life balance that the women raised, and investigated the different ways in which they accommodated motherhood and their IT careers (Quesenberry and Trauth, 2005; Quesenberry et al., 2006; Trauth and Connolly, 2021; Trauth and Quesenberry, 2005). In addition to the actual challenges, women also spoke about stereotypes imposed on them once they became mothers. These were stereotypes about mothers' long-term commitment to their careers.

The identities most commonly thought of in conjunction with the STEM/IT field have been based on gender, race and ethnicity. But more recently, others such as LGBTQ identity have been investigated. We explored within-gender variation in the IT careers of women who were this double minority (Trauth and Booth, 2013). These women spoke about a more complex engagement with and internalization of gender norms as they relate to women in technical careers. And their range of responses to these influences demonstrated yet another type of within gender-group variation.

In the original conceptualization of this theory "gender" was synonymous with "woman." Indeed, the first three field studies using this theory and one dissertation (Quesenberry and Trauth, 2012) focused exclusively on women and the ways in which individual identity characteristics affected girls' choice of an IT career and women's persistence within it. But the intersectionality of women with other identity characteristics such as race, SES, and LGBTQ identity served as a bridge to making intersectional identity more salient in my gender and IT research. Over time, a broader understanding of gender intersectionality has been employed in research studies that I and others have conducted (e.g., Joshi et al., 2017). Consequently, the scope of research using the IDTGIT has broadened beyond issues related only to girls and women to consider other gender minorities. Issues about race, ethnicity and SES related to women; factors related to men of color and LGBTQ men; and persons with disabilities have been examined (e.g., Annabi and Locke, 2019; Cain and Trauth, 2017; Kreps, 2010; Kvasny et al., 2009; McGee, 2018). The theory has also been applied to studies of IT use that included both men and women (Booth and Trauth, 2016; Morgan et al., 2015).

In 2007 I collaborated in a multi-year study of gender and IT career choice, which focused on race and ethnicity as key components, among both men and women. And unlike the previous studies, it was primarily a quantitative study with follow up focus groups at the end. We were interested in the influence of race, ethnicity and SES on gender stereotypes about IT career choice in the United States. We conducted this more fine-grained analysis of university students in order to consider the influence of gender-ethnic intersectionality on gender stereotypes about IT skills and knowledge. We looked at stereotypes held by women themselves as well as stereotypes imposed by men on women (Trauth et al., 2016a). We were interested in understanding what impact gender-ethnic intersectionality (Black, white, Latinx) had on men's and women's gender stereotypes. We learned that gender stereotypes held by a particular gender group (i.e., men or women) varied by ethnicity. Not surprisingly, men tended to rate all of the skills as more masculine than did the women. While technical IT skills (e.g., programming) were generally stereotyped as masculine by both men and women in each gender-ethnic group, non-technical IT skills (e.g., customer relationships) revealed both within-gender and within-ethnicity variation. For example, women rated these non-technical skills as less masculine than did the men, and Black men ranked skills as more masculine than did Latino or white men. Our findings also revealed that ethnic identity is a more salient predictor of career decision-making self-efficacy for people of color (both men and women) than for whites. Further, women of color may have the skills and abilities to be successful in the IT field but may not believe that they will be accepted into it because of the dual barriers of gender and ethnicity. These women's self-efficacy was being undermined by the combination of ethnicity and gender-role stereotypes. As a result, they may forego IT careers, thinking that this field is not truly open to them (Joshi et al., 2013).

This study was followed in 2012 by another collaborative study, this time of Black men as gender minorities in the IT field (Fuller et al., 2015; Joshi et al., 2016; Kvasny et al., 2016). Through life history interviews we explored career pathways of African-American men who were enrolled in an IT degree program at four historically Black colleges and universities (HBCU).[14] Some of the men were interviewed twice: while they were in the early years of their university education and then again towards the end or just after completing their degrees. For these respondents, identity as Black men was a dominant influence on their lives and IT interests. It was the backdrop against which they experienced environmental influences. They were confronted with racial stereotypes embedded in American culture. Many of them had to contend with lower quality public schools resulting from racial segregation in their neighborhoods. This served to reinforce a deficit model to explain Black men's underrepresentation in STEM. It assumed these men were lacking in knowledge, experience and other individual factors rather than pointing to systemic structural barriers embedded in the educational infrastructure and the workplace. SES disparities in American society, related to race and ethnicity, only exacerbated the situation. I conducted my interviews in 2014 and 2015, just before and just after the emergence of the Black Lives Matter movement for racial reckoning in the United States. It was a powerful experience to listen to these men speak about how they had learned at a young age to navigate a sometimes hostile and dangerous world.

In 2012, I also embarked upon a collaborative survey-based study of another gender minority in the IT field – men and women with disabilities (Trauth, 2017a; Trauth et al., 2014, 2015). Participants were military service members and veterans with acquired disabilities who had returned to university to embark on a post-military career. These individuals represented the intersectionality of gender and disability. We found variable influence of disability identity. For some, strong interest and technical self-efficacy overrode concerns they might have had about disability being a barrier. But there was another group of individuals for whom disability was a factor. They did not intend to pursue IT careers because of perceived barriers related to a combination of self-efficacy and disability.

Across my gender studies, other identity factors added nuance and context to respondents' life stories. For example, while age could be the primary factor in a study, in my work it served to contextualize respondents' stories. This was particularly the case in my longitudinal study of women in Ireland (Trauth and Connolly, 2021). Women's ages and, hence, the time period of their education and employment were mapped against societal changes in general and with respect to women. This picture, in turn, formed the backdrop for comparison with the evolution of the IT sector occurring in Ireland over a 50-year period. In other studies, age enabled me to contextualize a woman's experience. This meant situating a woman's gendered experiences within a particular time period in a particular societal setting.

Likewise, educational background also added nuance. This aspect of identity included the degrees earned, type of secondary education received, and exposure to STEM courses prerequisite to university study. While educational background was never the primary focus of attention in interviews, a detailed examination of the Black women's life histories revealed that, regardless of their ages, kinds of secondary schools attended, or where they were located, their educational achievement was what made the difference in their entry into an IT career (Kvasny et al., 2009). Whereas the white women in my interviews exhibited a range of intelligence and academic achievement markers in secondary school, *all of the Black women* had high marks, graduated at the top of their classes, or were enrolled in honors programs. Hence, educational background was a significant factor explaining how these women were able to resist identity-based biases and barriers coming from the environment.

Individual Influences

The other source of individual variation among women with respect to how they are affected by environmental influences is represented in the *individual influences* construct. It encompasses the effect of significant people and experiences, along with personal characteristics. Individual influences inhibit or enhance engagement with technology careers by mitigating societal messages, biases, stereotypes, and barriers. Personal influences refer to individuals and experiences, which influence IT careers. Personal characteristics are aspects of an individual's personality that play a role in attraction to and persistence within a technical career.

Personal influences

Significant people
Parents, teachers, guidance counselors, and friends can make a difference at key points in a student's academic career through role modeling, mentoring, advising and peer support. They can affect her awareness and choice of an IT career, and her persistence within it. In the IT workplace co-workers, mentors, sponsors and advocates can help a woman or gender minority not only survive in their career but thrive in it. These influences can be proactive: people who help a woman navigate her career. They can also be reactive: educational interventions put in place to address issues and barriers. And they can be negative: disempowering advising about working with technology.

Women noted that role models, mentors and other forms of support kept them from internalizing limitations that society may have tried to impose on them. But not always. In my US study, Joanne[15] received negative messages. Despite being a top student in her high school class, she and other African-American students were not encouraged to go on to university; they were advised that they would not succeed. Likewise, in my Australian study, the negative influence of significant others was on display. I was told the story about a young woman who was having difficulty in a database course. Her father's advice was to "stick it out" until the end of the term. If she still was unhappy, she could change majors. On the other hand, when her roommate was experiencing difficulties in her IT course, her mother's response was: "I told you girls weren't cut out for this kind of work!"

Not surprisingly, this young woman did not pursue an IT career.

The American interviewees spoke about the role of mentors and peer support in career development (Trauth et al., 2009). Brandy's mentor drew upon her own experiences in offering career encouragement. Lu appreciated the objective perspective on situations provided by her mentor. Kirsten's mentor suggested books to read. She felt that her mentor was taking a particular interest in her career development. The women had both men and women mentors. Christine's mentor was truly interested in understanding her career goals. She credited him with helping her to translate what she had been thinking of as a "job" into viewing it as a "career."

My longitudinal study of women in Ireland's IT sector (Trauth and Connolly, 2021) demonstrated the dynamic nature of significant people's influence over time. Women interviewed in 1990 said their parents generally supported working in this new IT field because it offered economic stability, but they had little engagement with their daughters about how to do so. In contrast, the parents of women interviewed in later time periods were actively involved in helping their daughters overcome educational obstacles to an IT career. Teachers and guidance counselors similarly changed from reinforcing gender barriers to enabling resistance to them. In 1990 school personnel reinforced traditional gender roles: motherhood, teaching and nursing. However, over time school personnel supported girls' IT career interest. Sally's guidance counselor invited an engineer to speak to students at her all-girls' school. The nuns at Maire's school allowed her to take technical courses at the nearby boys' school. Role models for the oldest informants were stay-at-home mothers. But a wider range of role models evolved from increasing numbers of mothers and older sisters working in IT jobs.

The importance of significant people was also evidenced in our study of IT career pathways of Black men. They acknowledged both emotional and technical support coming from friends, family, and IT professionals (Fuller et al., 2015):

> [My family] actually told me, "If you're not going [to college] I don't know what you're going to do but you got to get out of here; you have got to find something that you are going to do." I knew I didn't want to go to trade school or work at McDonalds or anything like that so I said, "Hey the next option is going to college." That's what I did. [Charles]

However, many fewer received advice specific to pursuing an IT major. Only a limited number of men had friends or family members who were in the IT field or knew about the profession enough to advise them to look into that major.

We mapped the presence or absence of significant others to Pierre Bourdieu's (1986, 2002) concept of capital: material and symbolic resources at one's disposal (Joshi et al., 2016). In particular, we focused on social capital (social networks that improve one's social standing) and technical capital (specific skills a person develops through engagement with computing). One young man attributed his resilience when encountering setbacks in his IT classes to his parents. Another man was inspired by his brother's story.

> My older brother … He's a mechanical engineer now … He received his GED[16] in prison. He didn't even finish high school … To see him come out and try to do the things he has accomplished, was very inspiring.

Still another credited his mother with problem-solving skills that helped him succeed in IT:

She bought me like a toolbox and she always said, "If things break down, you fix them" … I guess technology feature was just the understanding of how things came together because she always made sure … I had to figure it out myself.

Role models and mentors are sometimes conflated. While one individual can play both roles, they are different. A role model is someone others look to as an inspirational ideal. They validate one's identity, personality traits, or life circumstances. Women role models are needed to counterbalance an IT field so dominated by men. It provides girls with affirming images of who can be part of the IT profession. On the one hand, they need to see examples of assertive women, and those who balance work and family. Exposing girls to positive role models in IT helps to dispel myths and stereotypes that computing is a man's profession. Hence, role models typically possess similar identity characteristics. Mentors, sponsors and advocates, on the other hand, provide career advice and help a girl, woman or other gender minority navigate educational and career challenges. The absence of mentoring can lead to isolation, exclusion from peer networks, and affect retention. Mentoring can help minorities in the workplace to overcome a hegemonic masculine culture and/or a hostile climate. Mentors don't need to possess similar identity characteristics. So, women and Black men, for example, might have mentors who are white men.

"Why I learned to fly": social networks in the IT field

In my interviews I also explored the influence of peer networks dominated by men in the IT workplace (Morgan et al., 2004; Morgan and Trauth, 2006). In sociological terms these networks are an expression of homophily – a tendency for people to seek out or be attracted to those who are similar to themselves. In lay terms it means the "Old Boys' Club" – an informal network of men in a position of power and privilege in an organization, who share resources and information in order to help each other gain advantage and opportunities. This cohesive group of men lunch together, go out after work, and generally support each other in the workplace. Particularly germane to the IT field, these networks are also a way to develop new technical skills needed for career advancement. It is in these less formal settings that men get to know and trust each other, and establish personal relationships. Thus, to the extent that women are excluded, social networks become barriers to rather than facilitators of career advancement.

Women were affected by and related to these social networks in a variety of ways. Organizations and networks officially closed to women are becoming increasingly rare in the United States. What is more typical are networks that are *unofficially* closed to women. So, how did the women navigate them? One group of women were successful in breaking the barriers to membership. They found it relatively easy to relate to men's networks due to an interest in sports, having brothers, or through some other way in which they had experience interacting with men and stereotypically masculine activities. Alternatively, they proactively sought ways to connect with the men in the network. They took up a new hobby, developed an interest in a particular sport, or in some other way created common interests with the men. The most extreme example of this was offered by Sharon, who enrolled in flying lessons to be able to join in on lunch conversations with her men colleagues who had pilot licenses!

The other group of women were not part of the men's network. Some said they did not see the value in such networks, or they had family obligations that kept them from after-work activities. But others, like Ivanna, spoke of the isolation she felt working with men in her workplace who shared information about new technologies among themselves, but not with her. And they excluded her from outside-work gatherings for socialization and bonding. Some women who were excluded started or joined a women's network. The women I interviewed found workplace mentoring programs to be a particularly important antidote to the "Old Boys' Club."

Significant experiences
Another aspect of personal influences is significant experiences that had an impact on a woman's IT career. For some women it was their early exposure to computing, observing people working with technology, or playing computer games. For others it was their educational experience: attending a single-sex high school with its absence of negative gendered peer pressure; taking STEM classes; or playing sports and learning that it is acceptable to be assertive and competitive. Insight into women's educational experiences was captured in two ways. In addition to discussing a woman's educational experiences, I also interviewed women IT academics in all three studies: Australia/New Zealand, Ireland, and the United States.

Women also spoke about experiences in their personal lives. Several informants learned at a young age that they needed to be able to support themselves. One mentioned a father who passed away, leaving her mother to be the family bread winner. Another witnessed her father losing his job at a factory and never again being able to support his family. A third reacted to the economic dependence of her mother; she did not want to be dependent on a man.

What had a profound influence on Clare's life and career was having her leg amputated below the knee when she was eight years old, as a result of a traffic accident. In her early 1970s school, all students with "disabilities" were grouped together. Her parents had a constant fight with the school's attempt to place Clare in classes with developmentally challenged students. Thirty years later, this experience had left her with a heightened sensitivity to bullying.

A Black man we interviewed was also affected by a physical disability. He originally went to university on a football scholarship with the goal of playing professional football. But he became paralyzed as an innocent victim of a drive-by shooting. He described the day his father and brother came to the hospital to tell him he would never walk again. He needed to choose a new path; they helped him to rethink his future plans to include IT.

Personal characteristics
The personal characteristics factor refers to inner resources such as personality, interests, abilities, work ethic, personal agency, and resilience, which play a role in mitigating environmental and identity influences. A woman can draw upon these to counter or succumb to societal messages and biases about a gendered technical field. They also influence conscious or unconscious adoption of gender roles from the larger society. These personal characteristics influence the extent to which a girl or woman accepts implicit and explicit norms about behaviors, roles and careers that are considered to be feminine or masculine, and internalized expectations regarding what one should be good at – her self-esteem.

Skills, abilities, and interests
Skills and abilities are arguably more closely tied to IT careers for girls and women than they are for boys and men. This is because it is not only the *reality* of a girl's/woman's technical abilities that is influential. It is also the *perception* of those abilities held by the individual, herself, and by the wider society. Indeed, a theme that was raised by women in all of my

studies was the perception that IT is a masculine field and that women are underrepresented because they are not as qualified as men to be there. Informants repeatedly told stories of being confronted with stereotypes of women as not technical and, therefore, they needed to prove themselves. Likewise, girls have felt the need to be exceptional in order to perceive themselves as capable of entering and succeeding in a technological field. Indeed, in one of my earliest studies of gender in the IT field (Trauth, 2000), a man informant offered his insight about the gender imbalance: an average man can go into IT but a woman needs to be exceptional to do so. Several women characterized the situation as having to do better just to be seen as equal. Hence, the topic of skills and abilities, or technical self-efficacy isn't just about objective measures of skills and abilities. It is also very much about perceived skills and abilities – on the part of both the woman and those who observe her.

What is behind the gender imbalance that is often portrayed as a lack of *interest* on the part of women in STEM fields is actually gender stereotypes about expertise that influence expectations and performance. Internalized norms about what someone should be good at influence the way peers and teachers react to an individual, how she is expected to perform, and the way she responds to difficulties in school. All of these factors influence personal self-efficacy regarding a skill. This has been found repeatedly in the literature across time and geographical location.

In the early 1980s when personal computers and computing courses were first being introduced in some high schools, we surveyed both boys and girls in a high achieving, high socio-economic, college-preparatory high school about gender stereotypes applied to computing (Kwan et al., 1985). Students had little exposure to computers and most had not yet taken the new computing course. We were interested in the extent to which gender stereotypes about mathematics were being transferred to this new domain of computing. Because we surveyed the students in their mathematics courses, we also had access to mathematics and computing grade distributions by gender. We found that attitudes about computing followed traditional gender stereotypes that existed in STEM fields: women were less logical than men, and more afraid of computers. We also compared self-perceptions of mathematical ability against actual grade distributions. We found that the boys overestimated their mathematical achievement, while the girls underestimated theirs.

These findings have been reinforced in studies conducted since that time, and in a range of educational venues in several different countries. Henwood (2000) conducted case studies of gender and IT expertise among university students in the United Kingdom. In these courses the women received better marks than the men. However, the women underestimated their technical competence and men overestimated theirs. Further, comparisons by classroom observers of the "expert" who was seen to be helping other students with perceptions held by peers revealed that women "experts" went unrecognized by their peers. Similarly, Katz et al. (2006) studied first-year computer science students in the United States both before and after completion of a three-course sequence. She found that retention rates for women students with grades that were above average and very good (i.e., B) but not excellent (i.e., A) were less likely to stay in the computer science degree program than men students with the same grades. What these studies show is that technical self-efficacy is not so much about what one actually knows or one's ability as it is about what peers and significant others say and do, and how someone internalizes it all.

We explored perceptions about skills and abilities in two studies of group-level influences – gender and disability stereotypes – as they relate to the IT profession. We found that perceptions, alone, are not sufficient to explain the gender imbalance in the IT profession. Rather, additional personal characteristics such as self-efficacy, gender role incongruity[17] and interest were also influential. Our study of factors influencing American university students' interest in an IT career (Joshi et al., 2010, 2013; Trauth et al., 2016a) examined their gender stereotypes about technical careers, gender role incongruity, their own technical self-efficacy, and their IT career intentions. We found that students enrolled in an IT major who had high technical self-efficacy, overcame gender role incongruity and expressed an intention to pursue an IT career. Those with lower technical self-efficacy were more susceptible to gender role incongruity insofar as they were less inclined to pursue an IT career.

The other study revealed that technical self-efficacy, alone, is not sufficient to attract and retain women in technology fields. They also need to have an interest in the field, as demonstrated in our study of gender stereotypes, disability identity, IT self-efficacy, and IT career choice. Whereas the first study was of university students already enrolled in an IT course, this study was of military personnel and veterans who had acquired disabilities during their service and were now returning to school and choosing a new career after leaving the military (Trauth, 2017a; Trauth et al., 2014, 2015). One group of respondents had a strong intention to pursue an IT career, had high technical self-efficacy, and were interested in the IT field. This group was overwhelmingly men. However, another group was the source of new insight into factors influencing IT career choice. It was comprised of those with a low intention to pursue an IT career even though they had the capability to do so. They had high levels of technical self-efficacy, but with less exposure to the field – having taken fewer IT courses and having less IT experience – it is reasonable to assume that their attitudes about the IT profession were based more on societal images than actual experience of it. The individuals in this group also did not perceive disability as a barrier to pursuing an IT career. What caught our attention was that this group contained the largest percentage of women. While they were *capable* of being successful in an IT career, they were not *interested* in part, we argued, because of gendered societal messages.

Personal agency: "I'm not a feminist, but ..."
In addition to technical self-efficacy, abilities, and interest, another set of personal characteristics that were found to influence IT career persistence relates to personal agency. This theme arose in my interviews during discussions of feminism. During my interviews in Australia and New Zealand, I was struck by the frequency of allusions to feminist ideals made by participants as a way to express their views about gender issues in the workplace. A feminist consciousness means that a woman identifies with feminist thought or theories. This would be a recognition that although men and women are inherently of equal worth, most societies to various degrees privilege men as a group. Hence, there is a need for a social movement, feminism, to achieve greater equality between men and women. What intrigued me was the number of times that a woman would begin with "I'm not a feminist, but ...", and then proceed to make comments about the absence of or need for gender equality, discrimination issues, an uneven playing field, or other references to a feminist consciousness.

I was curious about why women who recognized feminist issues did not want to be labeled "feminist." Hence, in my two subsequent gender and IT research projects – in the United States and in Ireland – I incorporated into the interview guide the question: *Do you consider yourself to be a feminist, why or why not?* In the course of answering this question the women offered their definitions of feminism. I received similarly paradoxical responses. The irony of women whose personal agency enabled them to overcome gender barriers, yet being reluctant to associate that agency with the feminist label led to a deeper consideration of why this might be the case. My graduate students and I uncovered four groups of women.

One group said they were not feminists. They seemed unaware of or else denied the existence of gender discrimination in the IT workplace. They thought inequalities applied equally to both men and women. These women also strived to assimilate into the majority masculine culture, and accepted their domination as if it were justified. They offered varied reasons for adopting this stance. Some women believed gender issues no longer exist. They took gender progress for granted. Still others expressed concern about being viewed as a victim, something they ascribed to feminism. Interestingly, even when these women gave evidence of differential, negative treatment because of their gender, they were reluctant to identify as a feminist. For example, Francie talked about gender discrimination in college, in her internship, in her previous place of employment. But when asked if she was a feminist, she said "no" and explained that:

> I've always gone on the theory that I didn't want to be treated differently than any of the males, so I don't know that there should be special considerations for females, in the IT industry.

Several of the women saw a conflict between feminism and feminine gender identity. To them feminism conjured up images of angry women looking for trouble; being radical, aggressive and activist. Other anti-feminists did not believe in equal relationships: women with a family should be the ones working part time or from home.

The second group of women were ambivalent about feminism. They agreed with, and even espoused, feminist ideals but rejected the label. They were aware of gender inequities, but did not feel comfortable challenging the system. They were not sure if they were feminists:

> I don't really understand the definition of feminist. I hate chauvinism but I don't know if that makes me a feminist. Because sometimes I do expect men to treat me like a lady. So, would that make me not a feminist? I don't know. I have no idea. I agree with a lot of things that feminist ... things that I read about I agree with. [Jill]

Or they wanted to achieve on her own merits, not by playing "the female card", something that they associated with feminism.

The women who embraced the feminist label accepted sexist-related explanations for discriminatory events in the workplace. They characterized themselves as feminist and recognized the need to fight the system for equality. This acknowledgement emerged from study or life experiences. Siobhann, an Irish respondent, felt that certain things are expected of women automatically, such as: care-giving to parents, taking care of a sick child, and attending parent-teacher meetings. A woman's role is taken for granted. Linda didn't consider herself to be a feminist in the 1970s because she believed in meritocracy, something she viewed as being at odds with feminism. While she experienced meritocracy in high school, college and in her family, she became a feminist in the workplace. She learned that when she was doing the same job as a man, the contribution was seen by customers as unequal. The man was perceived as playing a larger role than he actually did. Other women saw feminism as a belief that equality is only fair; there should not be the unequal pay for equal work.

I consider myself a feminist because of my life. I took a really tremendous risk when I got divorced. And I fought the system. I didn't win, I fought the system in the early 80s and I feel like in this field that I'm in, I've always been fighting the system in some way or another. So, I think by my actions rather than … My life is kind of a testament to that ... It's not because I go to marches, do you know what I mean? It's not, it's because I've been fighting for my right as a woman for a long time. So, if that encompasses part of the definition of that then, yes [I'm a feminist]. [Sue]

The fourth group of women embraced feminist ideals, and were committed to a non-sexist world, but did not use a gender lens to interpret their experiences. These women considered themselves to be post-feminist. If they had a negative experience in the workplace, they did not immediately interpret it as due to their gender. They were aware that they had benefited from feminism but they did not focus on it. Some, especially younger women, had no sense of limitations. They expect equality in the workplace and in their personal lives. They thought the term "feminism" was passé and that the overly aggressive revolutionaries of the 1970s and 1980s were no longer needed.

These interviews revealed a conundrum: members of an underrepresented group being reluctant to affiliate with a movement directed at redressing this imbalance. Women lacking a feminist consciousness tended to interpret differential treatment in individual rather than group terms, and to attribute their marginalized positions to personal rather than structural factors. The variability of feminist consciousness sheds light on the different ways that women employ personal agency to respond to gender bias and barriers in the workplace.

How women cope: the angry, the accepting and the oblivious
Another personal characteristic that emerged from my interviews is about resilience: the various ways women IT professionals coped with negative educational and workplace experiences. The barriers these women discussed include: power displays, a raised bar for women, needing to prove oneself, a hostile work environment, a hostile educational environment, sexual harassment, and work–life balance challenges. Coping responses are perceptions, cognitions and behaviors that serve as a protective shield to mediate the impact that working in a masculine-dominated profession can have on a woman. Some women were action-focused. They wanted to change their situation. Their coping was directed at altering the work environment. They sought to define the problem, generate alternatives, and take action toward resolution. At the other end of the continuum were women who took an emotion-focused approach, to regulate their own emotions. These women attempted to modify their perceptions and, at times, engage in self-deception so as to avoid acknowledging certain facts. In between were women who did a little of both. Of course, like all the other factors in this theory, coping methods vary as a function of other factors: other personal characteristics such as personality; ones' identity (e.g., age, motherhood, or the part of the IT field in which she worked), how long she had been employed in the organization, and her management level.

A framework that emerged from the Australian study resulted from an offhand remark that I made when I was reporting back to colleagues about my interviews.[18] In one meeting as I was summarizing some of my observations from a round of interviews in several different Australian cities, I was asked how the women coped with being a minority in this man-dominated profession, I replied, "I would classify the women as the angry, the accepting and the oblivious."

Below is an overview of coping methods employed by the women in my studies (Kase and Trauth, 2003a, 2003b; Trauth et al., 2003).

The Angry. These women were conscious of an uneven playing field at work, were unhappy about it, and were willing to talk about it. They cited examples of a "raised bar" for proving themselves. They also spoke about uninvited sexual advances. Denise, an academic, talked about sexual innuendo and lies circulating about her sexual conduct with co-workers; she finally left that university. These women were activists. They questioned the inconsistencies and contradictions of the gender imbalance. They recognized institutional barriers and expressed the need to be strong and fight the system. They were often proactive, championing gender equality. They questioned the man-dominated workplace, and were unafraid to make people uncomfortable by raising issues about women not being supported for career advancement. They recognized institutional barriers and worked to alter and resolve discrimination.

The Accepting. The women who represented this second type of coping response were aware of the uneven playing field, but accepted it as a fact of life in the IT industry. Some even acknowledged their status as "second class citizens" in their workplaces. To them, this was simply the way things were. They also accepted that a mother working in IT had two full-time jobs. Cynthia said if she had had a daughter, she would discourage her from becoming an engineer because it was simply too difficult managing domestic responsibilities while at the same time trying to keep up with a rapidly changing field. Due to perceived impossible time constraints, these were women who felt they had to choose between scaled-back professional aspirations and remaining childless. They accepted that gender discrimination was an integral part of the IT workplace. Their coping response was to try to accommodate to their situation without being overwhelmed by it.

The Oblivious. This third group of women appeared more or less unfazed by being a woman in a man-dominated field. They didn't see themselves as operating on an uneven playing field. They were generally at the top of their class in high school and university. They were extreme minorities as STEM students and in their careers. These women tended not to focus on the fact that they were women, saying generally that gender was not an issue. Nonetheless, instances of an uneven playing field often crept into the interviews. Laura viewed herself as a person, not a woman. When she had her two children, she said she never expected any special treatment; she kept her motherhood to herself. As long as she kept up her workload, no one ever complained that she was a woman or a mother. In late twentieth-century language, she was the prototypical "supermom." These women were at the other end of the continuum from the angry activists; they appeared to be oblivious of gender discrimination in the workplace. They seemed to selectively ignore problematic observations and experiences.

CONCLUSION

This chapter provides an in-depth overview of the individual differences theory of gender and IT. I explain how and why I developed this theory, describe its dimensions, and illustrate its application with examples from my own research. My purpose in writing this chapter is to assist other researchers who are interested in gender and social inclusion research by providing detail about this theory and how it can be used.

It is only recently that researcher standpoint has become an acceptable component of gender and IT research (e.g., Kvasny et al., 2005). This means bringing oneself into the research. As some of the chapters in this book reveal, such reflexivity can make a significant contribution to understanding. I explained in the beginning of this chapter how my own career journey into the IT field was influenced by environmental, identity and individual factors. Likewise, they played a significant role in my theory development. Seeking an answer to the question of how I came to be who I am, helped me to identify the combination of societal and individual influences affecting women and gender minorities in STEM. My collaborative projects with colleagues possessing identities that are different from mine, reinforced the value of taking researcher standpoint into account.

When I began my gender research in 2000, my objective was new theory development. That is, I sought empirical data to confirm my argument for a theoretical perspective on the gender imbalance that took into account *both* group-level *and* individual-level factors. But over time the theory has evolved. While it originally focused only on women, it came to be used in research about other gender minorities in STEM along with greater emphasis being placed on intersectional identities (e.g., Cain and Trauth, 2017). As a result, this theory is now viewed as a more comprehensive framework for social inclusion research (Gorbacheva et al., 2019; Trauth, 2006a, 2017b; Trauth and Howcroft, 2006b; Trauth et al., 2018).

The theory continues to evolve. In Trauth and Connolly (2021) it was extended to include a dynamic dimension. The more I employed this theory, the more interested I became in understanding the *actual process* by which the factors represented in the theory's constructs had their influence. In my longitudinal study of women in the Irish IT sector, I had that opportunity. Through interviews conducted with Irish women at four different time periods, we were able to discern not just *what* the environmental and individual influencing factors were, but the *changes* in these influences over time. The resulting *dynamic individual differences theory of gender and IT* not only articulates the societal, organizational and individual influences on gender in the IT field. It now has another dimension that characterizes *how* these influences occur (p. 2065). They can be direct or indirect, they can be abrupt or incremental, they can act in conjunction with other factors. Some factors aren't consistently influential; sometimes they appear as impacts, reflecting changes occurring elsewhere. This theoretical evolution is consistent with Niederman's definition of theory as part of a multistep process for systematically refining knowledge of the world (2021).

The IDTGIT enables investigation of the variety of issues that confront women and gender minorities in their technology career journeys. The findings that I present in this chapter document the theory's claim that it is not fruitful to make large-scale generalizations about causes of underrepresentation. Rather, we need to explore the ways in which individuals possessing a particular gender, intersecting with other identity features, encounter and respond to societal and organizational biases, barriers and behaviors. It is only then that we can implement meaningful interventions.

Some interventions need to target the majority population, to create greater awareness about the gender gap, and stimulate attitude change about the underlying reasons for it, so as to change behaviors. An example would be cultural change. Cultural interventions might include questioning a domestic imbalance that requires women in heterosexual couples to have two full time jobs – one inside the home and one outside the home – while men have only one. Since gender roles are tightly interwoven into a culture, these interventions may challenge definitions of masculine and feminine work. But as opportunities for IT careers grow, the economic incentive to address these cultural challenges should also increase.

I have contributed to cultural change in two ways. One was through developing and teaching two undergraduate courses as educational interventions. The first was a diversity course, of which gender was a part. The second was a course on gender and STEM. My goal with both courses was to offer students a counternarrative by critically examining societal assumptions and stereotypes about women, gender minorities and STEM (Trauth and Booth, 2014; Trauth et al., 2007). The other intervention effort was a collaboration with my playwright sister, Suzanne Trauth, to write a play, *iDream*.[19] The purpose was to communicate to an audience beyond the academy themes from my research about gender barriers confronting girls and women (Trauth et al., 2012, 2016b, 2019; Trauth and Trauth, 2019).

Other interventions are directed at ameliorating the effects of biases, stereotypes, and barriers that interfere with the pursuit of STEM careers by women and gender minorities. These interventions need to be situated in both the educational setting and the workplace (e.g., Annabi and Lebovitz, 2018). Just as the constructs of the IDTGIT can be used to understand influences on this population, they can also be used to guide interventions (e.g., Ridley and Young, 2012). For example, the intervention needs of individuals who experience barriers related to their identities would differ from the intervention needs of those who have not had the benefit of role models or mentors. Intervention needs and their enactment also vary over time. The dynamic extension of the IDTGIT takes into account these changing circumstances which require similar changes in interventions.

NOTES

1. IT is used as an overarching term that includes information systems, computer science, informatics, information science, etc.
2. LGBTQ is a label used to designate the community of individuals comprising the sexuality and gender identity community. It stands for individuals who identify as lesbian, gay, bisexual, transgender or queer.
3. The issue of theory extends beyond gender and IT research to the broader context of social inclusion research as well.
4. See Trauth (2013) for examples of implicit-theoretical gender and IT research.
5. See Appendix 2A.1 for a complete list of research studies involving this theory.
6. Three colleagues who, as graduate students, worked with me on the coding and data analysis of the interview transcripts: Dr. Allison Morgan Bryant; Dr. Haiyan Huang; and Dr. Jeria Quesenberry.
7. An earlier version of this table appeared in Trauth et al. (2016a).
8. See Trauth and Connolly (2021) for a more complete list of publications using IDTGIT.
9. See Appendix 2A.1.
10. Civil Service (Employment of Married Women) Act 1973, Irish Statute Book, Retrieved from: www.irishstatutebook.ie/eli/1973/act/17/enacted/en/htmlet.
11. See Chapter 10 in this book for a detailed examination of how I conducted research involving the identity construct.
12. This wasn't explicitly asked but if respondents signaled lesbian or bisexual identity (no one signaled transgender or gender queer), it was recorded and used in analyses.
13. This wasn't explicitly asked but was inferred through responses to questions about parents' work background.
14. HBCUs are higher education institutions whose principal mission is the education of Black Americans.
15. These names are pseudonyms given to respondents to ensure anonymity.
16. General Educational Development (GED) is a high-school equivalency diploma.
17. Occurs when there is a perceived conflict between the role expectations of a particular gender and the requirements of a social role, in this case, an IT professional.
18. My field study was part of a larger study funded by an Australian Research Council grant held by Dr. Liisa von Hellens and Dr. Sue Nielsen.
19. See iDreamThePlay.com.

REFERENCES

Adam, A., Emms, J., Green, E., and Owen, J. (1994). *Women, Work and Computerization: Breaking Old Boundaries – Building New Forms.* Amsterdam: North Holland.

Adam, A., Howcroft, D., and Richardson, H. (2001). Absent friends? The gender dimension in IS research. In N.J. Russo, B. Fitzgerald, and J.L. DeGross (Eds.), *Realigning Research and Practice in Information Systems Development: The Social and Organizational Perspective* (pp. 333–52). Boston, MA: Kluwer Academic Publisher.

Adam, A., Howcroft, D., and Richardson, H. (2002). Guest editorial. *Information Technology & People, Special Issue on Gender and Information Systems, 15,* 2, 94–97.

Adya, M. and Kaiser, K. (2005). Early determinants of women in the IT workforce: A model of girls' career choices. *Information Technology & People, 18,* 3, 230–59.

Ahuja, M. (2002). Women in the information technology profession: A literature review, synthesis and research agenda. *European Journal of Information Systems, 11,* 20–34.

Akpan-Obong, P.I. (2009). *Information and Communication Technologies in Nigeria.* New York: Peter Lang.

Annabi, H. and Lebovitz, S. (2018). Improving the retention of women in the IT workforce: An investigation of gender diversity interventions in the USA. *Information Systems Journal, 28,* 6, 1049–81.

Annabi, H. and Locke, J. (2019). A theoretical framework for investigating the context for creating employment success in information technology for individuals with autism. *Journal of Management & Organization, 25,* 4, 499–515.

Balka, E. and Smith, R. (2000). *Women, Work and Computerization: Charting A Course to the Future.* Boston: Kluwer Academic Publishers.

Barnett, R. and Rivers, C. (2004). Same Difference: How Gender Myths Are Hurting Our Relationships, Our Children, and Our Jobs. New York: Basic Books.

Barrett, M. (1987). The concept of "difference." *Feminist Review, 26,* 29–41.

Berger, P.L. and Luckmann, T. (1966). *The Social Construction of Reality: A Treatise in the Sociology of Knowledge.* New York: Doubleday.

Booth, K. and Trauth, E.M. (2016). Do this, not that: How teens make decisions about contradictory health information on social media. *Proceedings of the 22nd Americas Conference on Information Systems* (San Diego, CA).

Bourdieu, P. (1986). The forms of capital. In J. Richardson (Ed.) *Handbook of Theory and Research for the Sociology of Education* (pp. 241–58). New York: Greenwood.

Bourdieu, P. (2002). *The Social Structure of the Economy.* New York: Polity.

Butler, J. (1990). *Gender Trouble: Feminism and the Subversion of Identity.* New York: Routledge.

Cain, C.C. and Trauth, E.M. (2017). Black men in IT: Theorizing an autoethnography of a Black man's journey into IT within the United States of America. *The Data Base for Advances in Information Systems, 48,* 2, 35–51.

Camp, T. (1997). The incredible shrinking pipeline. *Communications of the ACM, 40,* 10, 103–10.

Cockburn, C. (1983). *Brothers: Male Dominance and Technological Change.* London: Pluto Press.

Cockburn, C. (1988). *Machinery of Dominance: Women, Men and Technical Know-how.* Boston: Northeastern University Press.

Cockburn, C. and Ormrod, S. (1993). *Gender and Technology in the Making.* London: Sage.

Cohoon, J.M. (2001). Toward improving female retention in the computer science major. *Communications of the ACM, 44,* 5, 108–14.

Collins, P.H. (2009). *Black Feminist Thought.* New York: Routledge.

Collins, P.H. (1989). The social construction of Black feminist thought. *Signs: Journal of Women in Culture and Society, 14,* 4, 745–73.

Connell, R.W. (1987). *Gender and Power: Society, the Person and Sexual Politics.* Stanford, CA: Stanford University Press.

Connell, R.W. (2005). *Masculinities.* Second Edition. Berkeley, CA: University of California Press.

Fuller, K., Kvasny, L., Trauth, E.M., and Joshi, K.D. (2015). Understanding career choice of African American men majoring in information technology. *Proceedings of the ACM SIGMIS Computers and People Research Conference* (Newport Beach, CA, June).

Gallivan, M. (2013). A structured review of IS research on gender and IT. *Proceedings of the ACM SIGMIS Computers and People Research Conference* (Cincinnati, OH, May).

Gefen, D. and Straub, D. (1997). Gender differences in the perception and use of email: An extension to the technology acceptance model. *MIS Quarterly*, *21*, 4, 389–400.

Gill, S., Stockard, J., Johnson, M., and Williams, S. (1987). Measuring gender differences: The expressive dimension and critique of androgyny scales. *Sex Roles*, *17*, 7/8, 375–400.

Gorbacheva, E., Beekhuyzen, J., vom Brocke, J., and Becker, J. (2019). Directions for research on the gender imbalance in the IT profession. *European Journal of Information Systems*, *48*, 1, 43–67.

Gregor, S. (2006). The nature of theory in information systems. *MIS Quarterly*, *30*, 3, 611–42.

Halberstam, J. (1998). *Female Masculinity*. Durham, NC: Duke University Press.

Harding, S. (Ed.) (2004). *The Feminist Standpoint Theory Reader*. New York: Routledge.

Hartsock, N.C.M. (1998). *The Feminist Standpoint Revisited and Other Essays*. Boulder, CO: Westview Press.

Hartsock, N. (1983). *Money, Sex, and Power*. New York: Longman.

Heckman, S. (1997). Truth and method: Feminist standpoint theory revisited. *Signs: Journal of Women in Culture and Society*, *22*, 2, 341–65.

Heinamaa, S. (1997). What is a woman? Butler and Beauvoir on the foundations of the sexual difference. *Hypatia: A Journal of Feminist Philosophy*, *12*, 1, 20–39.

Henwood, F. (2000). From the woman question in technology to the technology question in feminism. *The European Journal of Women's Studies*, *7*, 209–27.

Hilbert, M. (2011). Digital gender divide or technologically empowered women in developing countries? A typical case of lies, damned lies, and statistics. *Women's Studies International Forum*, *34*, 479–89.

hooks, b. (1981). *Feminist Theory: From Margin to Center*. Boston: South End Press.

Howcroft, D. and Trauth, E.M. (2005). Choosing critical IS research. In D. Howcroft and E.M. Trauth (Eds.), *Handbook of Information Systems Research: Critical Perspectives on Information Systems Design, Development and Implementation* (pp. 1–15). Cheltenham, UK: Edward Elgar Publishing.

Howcroft, D. and Trauth, E.M. (2008). The implications of a critical agenda in gender and IS research. *Information Systems Journal*, *18*, 2, 185–202.

Joshi, K.D., Kvasny, L., McPherson, S., Trauth, E., Kulturel-Konak, S., and Mahar, J. (2010). Choosing IT as a career: Exploring the role of self-efficacy and perceived importance of IT skills. *Proceedings of the International Conference on Information Systems* (St. Louis, MO, December).

Joshi, K.D., Kvasny, L., Unnikrishnan, P., and Trauth, E.M. (2016). How do Black men succeed in IT careers: The effects of capital. *Proceedings of 49th Hawaii International Conference on Systems Science* (Koloa, HI, January).

Joshi, K.D., Schmidt, N.L., and Kuhn, K.M. (2003). Is the information systems profession gendered? Characterization of IS professionals and IS careers. *Proceedings of the ACM SIGMIS CPR Conference* (Philadelphia, PA, April).

Joshi, K.D., Trauth, E., Kvasny, L., Morgan, A.J., and Payton, F.C. (2017), Making Black lives matter in the information technology profession: Issues, perspectives, and a call for action. *The Data Base for Advances in Information Systems*, *48*, 2, 21–34.

Joshi, K.D., Trauth, E., Kvasny, L., and McPherson, S. (2013). Exploring the differences among IT majors and non-majors: Modeling the effects of gender role congruity, individual identity, and IT self-efficacy on IT career choices. *Proceedings of the Thirty-Fourth International Conference on Information Systems* (Milan, Italy, December).

Kase, S.E. and Trauth, E.M. (2003a). Assimilation, accommodation and activism: How women in the IT workplace cope. *Proceedings of the Information Resource Management Association International Conference* (Philadelphia, PA, May).

Kase, S.E. and Trauth, E.M. (2003b). Toward a model of women in the IT Workplace. *Proceedings of the Americas Conference on Information Systems* (Tampa, FL, August).

Katz, S., Allbritton, D., Aronis, J., Wilson, C., and Soffa, M.L. (2006). Gender, achievement, and persistence in an undergraduate computer science program. *The Data Base for Advances in Information Systems*, *37*, 4, 42–57.

Klawe, M. and Leveson, N. (1995). Women in computing: Where are we now? *Communications of the ACM*, *38*, 1, 29–35.

Kreps, D. (2010). Introducing eco-masculinities: How a masculine discursive subject approach to the individual differences theory of gender and IT impacts an environmental informatics project. *Proceedings of the Americas Conference on Information Systems* (Lima, Peru, August).

Kvasny, L., Greenhill, A., and Trauth, E.M. (2005). Giving voice to feminist projects in management information systems research. *International Journal of Technology and Human Interaction, 1,* 1, 1–18.

Kvasny, L., Trauth, E., and Joshi, K.D. (2016). The role of HBCUs in preparing African American males for careers in information technology. In C.B.W. Prince and R.L. Ford (Eds.), *Setting a New Agenda for Student Engagement and Retention in Historically Black Colleges and Universities* (pp. 234–50). Hershey, PA: Information Science Reference.

Kvasny, L., Trauth, E.M., and Morgan, A. (2009). Power relations in IT education and work: The intersectionality of gender, race and class. *Journal of Information, Communication and Ethics in Society, 7,* 2/3, 96–118.

Kwan, S.K., Trauth, E.M., and Driehaus, K.C. (1985). Gender differences and computing: Students' assessment of societal influences. *Education and Computing, 1,* 3, 187–94.

Lucal, B. (1999). What it means to be gendered me: Life on the boundaries of a dichotomous gender system. *Gender & Society, 13,* 6, 781–97.

Marini, M. (1990). Sex and gender: What do we know? *Sociological Forum, 5,* 1, 95–120.

McGee, K. (2018). The influence of gender, and race/ethnicity on advancement in information technology (IT). *Information and Organization, 28,* 1–36.

Mennega, N. and de Villiers, C. (2021). A quarter century of gender and information systems research: the role of theory in investigating the gender imbalance. *Gender, Technology and Development, 25,* 1, 112–30.

Montanelli, Jr., R.G. and Mamrak, S.A. (1976). The status of women and minorities in academic computer science. *Communications of the ACM, 19,* 10, 578–81.

Morgan, A.J., Myers, Y., and Trauth, E.M. (2015). Consumer demographics and internet based health information search in the United States: The intersectionality of gender, race and class. *International Journal of E-Health and Medical Communications, 6,* 1, 58–72.

Morgan, A.J., Quesenberry, J.L., and Trauth, E.M. (2004). Exploring the importance of social networks in the IT workforce: Experiences with the "Boy's Club." *Proceedings of the Americas Conference on Information Systems* (New York, August).

Morgan, A. and Trauth, E.M. (2006). Women and social capital networks in the IT workforce. In E.M. Trauth (Ed.), *Encyclopedia of Gender and Information Technology* (pp. 1245–51). Hershey, PA: Idea Group Publishing.

Niederman, F. (2021). The philosopher's corner – A minimalist view of theory: Why this promises advancement for the IS discipline. *The Data Base for Advances in Information Systems, 52,* 4, 119–30.

Niederman, F. and March, S. (2019). Broadening the conceptualization of theory in the information systems discipline: A meta-theory approach. *The Data Base for Advances in Information Systems, 50,* 2, 18–44.

Nielsen, S., von Hellens, L. and Wong, S. (2000). *The Game of Social Constructs: We're Going to Win IT!* Brisbane, Australia: Griffith University, School of Computing and Information Technology.

Oreglia, E., and Srinivasan, J. (2016). ICT, intermediaries and the transformation of gendered power structures. *MIS Quarterly, 40,* 2, 501–10.

Pearl, A., Pollack, M.E., Riskin, E., Thomas, B., and Wu, A. (1990). Becoming a computer scientist: A report by the ACM Committee on the Status of Women in Computing Science. *Communications of the ACM, 33,* 11, 48–57.

Quesenberry, J.L., Trauth, E.M., and Morgan, A. (2006). Understanding the "mommy tracks": A framework for analyzing work–family issues in the IT workforce. *Information Resources Management Journal, 19,* 2, 37–53.

Quesenberry, J.L. and Trauth, E.M. (2005). The role of ubiquitous computing in maintaining work-life balance: Perspectives from women in the IT workforce. In C. Sorensen, Y. Youngjin, K. Lyytinen, and J.I. DeGross (Eds.), *Designing Ubiquitous Information Environments: Socio-Technical Issues and Challenges* (pp. 43–55). New York: Springer.

Quesenberry, J. and Trauth, E.M. (2012). The (dis)placement of women in the IT workforce: An investigation of individual career values and organizational interventions. *Information Systems Journal, 22*, 6, 457–73.

Rathgeber, E.M. and Adera, E.O. (Eds.) (2000). *Gender and the Information Revolution in Africa.* Ottowa, Canada: International Development Research Centre.

Ridley, G. and Young, J. (2012). Theoretical approaches to gender and IT: Examining some Australian evidence. *Information Systems Journal, 22*, 5, 355–73.

Risman, B.J. and Davis, G. (2013). From sex roles to gender structure. *Current Sociology, 61*, 733–55.

Spender, D. (1995). *Nattering on the Net: Women, Power, and Cyberspace.* North Melbourne, Australia: Spinifex Press Pty Ltd.

Star, S.L. (1995). *The Cultures of Computing.* Oxford: Blackwell Publishers.

Stepulevage, L. (2001). Gender/technology relations: Complicating the gender binary. *Gender and Education, 13*, 3, 325–38.

Townsend, G.C. (2002). People who make a difference: Mentors and role models. *ACM SIGCSE Bulletin, 34*, 2, 57–61.

Trauth, E.M. (2020). Field studies of women in Europe, North America, Africa, and Asia-Pacific: A theoretical explanation for the gender imbalance in information technology. In C. Frieze and J.L. Quesenberry (Eds.), *Cracking the Digital Ceiling: Global Views of Women in Computing* (pp. 61–72). Cambridge, United Kingdom, Cambridge University Press.

Trauth E. (2017a). Breaking down their own stereotypes to give veterans more career opportunities. *The Conversation*, May 22. [http://theconversation.com/breaking-down-their-own-stereotypes-to-give-veterans-more-career-opportunities-76965].

Trauth, E. (2017b). A research agenda for social inclusion in information systems. *The Data Base for Advances in Information Systems, 48*, 2, 9–20.

Trauth, E.M. (2013). The role of theory in gender and information systems research. *Information and Organization, 23*, 4, 277–93.

Trauth, E.M. (2006a). An agenda for research on gender diversity in the global information economy. In E.M. Trauth (Ed.), *Encyclopedia of Gender and Information Technology* (xxix–xxiii). Hershey, PA: Idea Group Publishing.

Trauth, E.M. (2006b). Theorizing gender and information technology research. In E.M. Trauth (Ed.), *Encyclopedia of Gender and Information Technology* (pp. 1154–59). Hershey, PA: Idea Group Publishing.

Trauth, E.M. (2002). Odd girl out: An individual differences perspective on women in the IT profession. *Information Technology & People, 15*, 2, 98–118.

Trauth, E.M. (2000). *The Culture of an Information Economy: Influences and Impacts in the Republic of Ireland.* Dordrecht, The Netherlands: Kluwer Academic Publishers.

Trauth, E.M. (1995). Women in Ireland's information industry: Voices from inside. *Eire-Ireland, 30*, 3, 133–50.

Trauth, E.M., Avital, M., Kendall, J., Kendall, K.E., and Boland, R. (2012). Out of the box and onto the stage: Enacting information systems research through theatre. *Proceedings of the Thirty-Third International Conference on Information Systems* (Orlando, FL, December).

Trauth, E.M. and Booth, K.M. (2013). How do gender minorities navigate the IS workplace? Voices of lesbian and bisexual women. *Proceedings of the Nineteenth Americas Conference on Information Systems* (Chicago, IL, August).

Trauth, E.M. and Booth, K. (2014). Reflections on blended learning. In J. Carroll (Ed.), *Innovative Practices in Teaching Information Sciences and Technology* (pp. 207–79). London: Springer-Verlag.

Trauth, E., Bryant, A.J., Cain, C., Potter, L.E., Quesenberry, J.L., Trauth, S., and van Slyke, C. (2019). Addressing social inclusion in the IS field through theatre. *Proceedings of the ACM SIGMIS Computers and People Research Conference* (Nashville, TN, June).

Trauth, E.M., Cain, C., Joshi, K.D., Kvasny, L. and Booth, K. (2016a). The influence of gender-ethnic intersectionality on gender stereotypes about IT skills and knowledge. *The Data Base for Advances in Information Systems, 47*, 3, 9–39.

Trauth, E.M. and Connolly, R. (2021). Investigating the nature of change in factors affecting gender equity in the IT Sector: A longitudinal study of women in Ireland. *MIS Quarterly, 45*, 4, 2055–100.

Trauth, E.M. and Howcroft, D. (2006a). Critical empirical research in IS: An example of gender and IT. *Information Technology and People, Special Issue on Critical Research in Information Systems, 19,* 3, 272–92.

Trauth, E.M. and Howcroft, D. (2006b). Social inclusion and the information systems field: Why now? In E.M. Trauth, D. Howcroft, T. Butler, T., Fitzgerald, B. and DeGross, J. (Eds.), *Social Inclusion: Societal and Organizational Implications for Information Systems* (pp. 3–12). New York: Springer.

Trauth, E.M., Johnson, R.N., Morgan, A., Huang, H., and Quesenberry, J. (2007). Diversity education and identity development in an information technology course. *New Directions for Teaching and Learning, 111,* 81–87.

Trauth, E.M., Joshi, K.D., Graham, K., and Nithithanatchinnapat, B. (2015). An exploratory study of identity and IT career choice for military service members and veterans with disabilities. *Proceedings of the 21st Americas Conference on Information Systems* (Puerto Rico, August).

Trauth, E.M., Joshi, K.D., and Graham, K. (2014). Modeling IT career choice for the differently abled: Military personnel and veterans with disabilities. *Proceedings of the 20th Americas Conference on Information Systems* (Savannah, GA, August).

Trauth, E.M., Joshi, K.D., and Kvasny, L. (2018). Editorial. *Information Systems Journal, Special Issue on Social Inclusion, 28,* 6, 989–94.

Trauth, E., Keifer-Boyd, K., and Trauth, S. (2016b). *iDream*: Addressing the gender imbalance in STEM through research-informed theatre for social change. *Journal of American Drama and Theatre, Special issue, Alt Inq: Scientific Research and Inquiry in American Theatre, 28,* 2.

Trauth, E.M., Nielsen, S.H., and von Hellens, L.A. (2003). Explaining the IT gender gap: Australian stories for the new millennium. *Australian Computer Society Journal of Research and Practice in IT 35, 1,* 7–20.

Trauth, E.M. and Quesenberry, J.L. (2005). Individual inequality: Women's responses in the IT profession. *Proceedings of the Women, Work and IT Forum* (Brisbane, Queensland, Australia, June).

Trauth, E.M. and Quesenberry, J.L. (2006). Are women an underserved community in the information technology profession? *Proceedings of the International Conference on Information Systems* (Milwaukee, WI, December).

Trauth, E.M. and Quesenberry, J.L. (2007). Gender and the information technology workforce: Issues of theory and practice. In P. Yoong and S. Huff (Eds.), *Managing IT Professionals in the Internet Age* (pp. 18–36). Hershey, PA: Idea Group Publishing.

Trauth, E.M., Quesenberry, J.L., and Huang, H. (2008a). A multicultural analysis of factors influencing career choice for women in the information technology workforce. *Journal of Global Information Management, 16,* 4, 1–23.

Trauth, E.M., Quesenberry, J.L., and Huang, H. (2009). Retaining women in the US IT workforce: Theorizing the influence of organizational factors. *European Journal of Information Systems, 18,* 476–97.

Trauth, E.M., Quesenberry, J.L., and Morgan, A.J. (2004). Understanding the underrepresentation of women in IT: Toward a theory of individual differences. *Proceedings of the ACM SIGMIS Computer Personnel Research Conference* (Tucson, AZ, April).

Trauth, E.M., Quesenberry, J., and Yeo, B. (2008b). Environmental influences on gender in the IT workforce. *The Data Base for Advances in Information Systems, 39,* 1, 8–32.

Trauth, E. and Trauth, S. (2019). Addressing the gender gap in STEM through theatre. In O.A. Pilkington (Ed.), *Lab Lit: Exploring Literary and Cultural Representations of Science* (pp. 185–96). Lanham, MD: Lexington Books.

Venkatesh, V. and Morris, M. (2000). Why don't men ever stop to ask for directions? Gender, social influence, and their role in technology acceptance and usage behavior. *MIS Quarterly, 24,* 1, 115–39.

Von Hellens, L. and Nielsen, S. (2001). Australian women in IT. *Communications of the ACM, 44,* 7, 46–52.

Wajcman, J. (1991). *Feminism Confronts Technology.* University Park, PA: The Pennsylvania State University Press.

Webster, J. (1996). *Shaping Women's Work: Gender, Employment and Information Technology.* London: Longman.

APPENDIX 2A.1: RESEARCH ACTIVITIES TO DEVELOP AND REFINE INDIVIDUAL DIFFERENCES THEORY OF GENDER AND IT

Phase I: Theory Articulation

Field studies of factors influencing gender and the information technology profession carried out in the United States, Ireland, Australia and New Zealand served to identify and refine the constructs and subconstructs of the IDTGIT. Over 200 life-history interviews with women IT professionals sought to identify individual responses to socio-cultural factors that inhibit or encourage women's participation in the IT profession. Face-to-face, open-ended interviews with women IT practitioners and academics, averaging 90 minutes were guided by the constructs of the IDTGIT. (See Appendix 2A.2.) Strategic, convenience sampling techniques were used to generate a representative sample in each study. In the case of the American study the sample was limited to three geographical regions – Massachusetts, North Carolina, and Pennsylvania – and women academics in the IT field. The interviews were recorded and transcribed in order to facilitate coding and analysis. The interviewees were assigned pseudonyms to guarantee confidentiality. Themes that emerged from the Australian/New Zealand interviews were used to code subsequent interviews. Analysis of the interviews was supplemented by participant observation notes about the women and their socio-cultural environment as well as by literature about the culture of the regions/countries in which the interviews were conducted.

- Australia/New Zealand: 2000–2001 (Australia Research Council) – 31 interviews. Part of a larger ARC grant held by Dr. Liisa von Hellens and Dr. Sue Nielsen.
- USA: 2002–2007 (National Science Foundation, #0204246) – 123 interviews.
- Ireland: 63 interviews. Qualitative longitudinal study of women in IT in Ireland against the backdrop of the evolution of the Irish IT sector
  1989–1990 (Fulbright Foundation)
  2003–2006 (Science Foundation Ireland)
  2010–2012 (with Dr. Regina Connolly).

Phase II: Theory Extension

The population was expanded to apply the theory to underrepresented men.

- USA: 2007–2013 (National Science Foundation, #0733747) – survey of 4,046 white, Black and Latinx men and women college students about the interaction of gender stereotypes, technical self-efficacy and IT career choice, with Dr. K.D. Joshi and Dr. Lynette Kvasny Yarger.
- USA: 2012–2017 (National Science Foundation, #1232344) – 100 interviews with Black men about the influence of gender-race intersectionality on IT career interest, with Dr. K.D. Joshi and Dr. Lynette Kvasny Yarger.
- USA: 2012–2018 (National Science Foundation, #1245124) – survey of 699 men and women military personnel and veterans with acquired disabilities about the influence of technical self-efficacy, gender stereotypes, and disability identity on IT career decisions, with Dr. K.D. Joshi.

Phase III: Theory Validation

Validation of the theory and its constructs was sought through member checking focus groups and presentations to verify the findings rang true to audiences who were representative of those studied. In addition, numerous peer-reviewed conference presentations and publications verified the credibility of the theory's constructs among the academic community. Finally, feedback was received from international lecturing and teaching that was based on the theory. In these ways, globally varied audiences (both students and other researchers) were able to validate that the theory resonated with their experiences. Finally, a play based on the theory's constructs was written and performed for the general public 11 times, followed by audience talkback sessions, about the themes present in the playscript and characters.

- Lectures on gender and IT research findings (2000–2019) in several European countries, South Africa, New Zealand, Australia and throughout the United States.
- Gender and IT courses taught: Finland (2007); South Africa (2008); Austria (Fulbright Foundation, 2008); USA (Pennsylvania State University, 2005–2017).
- Writing and staging performances of *iDream*, an original play with characters and story line based on findings of prior gender studies (National Science Foundation, #1039546, 2010–2019), with Suzanne Trauth.

APPENDIX 2A.2: INTERVIEW GUIDE

1. Self
   Name/alias
   Date/location of interview
   Age
   Race/ethnicity/nationality
   Family status – married, partner, children, other care giving responsibilities
   Family of origin – birth location, parents' & siblings' education & work
   Sexual orientation/gender identity (not explicitly asked, followed up if mentioned by informant)

2. IT education: type, institution, degrees
   Secondary school, university degree
   Educational background/experiences related to gender

3. Path into IT work
   First exposure to IT & computer games
   Choice of IT career, how decision made

4. IT Work History
   Job title, description, industry
   Type of IT work done

5. Career progression/experiences regarding gender
    Special characteristics of IT work that bring unique challenges/barriers to women
    Are you successful? elaboration, reasons
    How important is work in your life?
    Work–life balance issues

6. Significant influences on (IT) career (events, people, things read, family)
    Mentors, role models, advocates, peer support/professional networks
    Experience with societal influences
    Role of family, education, society, other significant people
    Other significant activities in life (hobbies, etc.)

7. Influence of personal characteristics & skills on (IT) career
    Sports & being competitive
    Other skills & personality traits (both positive and negative)
    Coping methods
    Networks & support groups
    Are you a feminist? Why/why not? Your definition of feminism
    Do you use a gender lens to interpret your experiences?

8. (Academic interviews)
    Messages & dynamics about women in IT present in educational climate
    How academic IT women are treated? Same as in industry?

9. Society, Gender & IT
    Is IT a masculine domain? Are masculine traits – being logical, aggressive – required?
    How, why? How do women navigate it?
    Women in IT different from other women? Are you different, in what ways, why?
    Does a woman have to be different from the feminine stereotype to be successful?
    Is there a conflict between gender identity and IT identity?

10. National/regional attitudes about women, women/mothers working, women working IT,
    women having jobs vs. careers
    Societal influences/barriers affecting gender and IT: in general, personal experience
    Importance of IT to a nation's/region's economy
    Has the position of women (women in IT) in your country/region changed with the growth
    of IT work?

11. Recommendations for:
    Society
    Government policy
    IT profession
    Education institutions

Note: Permission to use this interview guide is conditioned upon inclusion of the following statement: "These interview questions are used with permission of Dr. Eileen M. Trauth."

PART I

ENVIRONMENTAL
INFLUENCES

3. Invisible but ubiquitous: leveraging ICTs for development in gendered systems of exclusion – Nigeria and Cameroon

Patience Akpan-Obong

INTRODUCTION

The first indigenously owned commercial bank in Africa, the United Bank for Africa (UBA), was established in Nigeria in 1949. Sometime in the late 1970s, it created an advertisement that included the following words: "Wise men bank with UBA." It became a popular slogan on radio, television and on huge billboards. People sang along gleefully when the commercial came on the air. Then in 1984, for the first time, a woman became the chair of the bank's board of directors. The print version of the advert was edited to include "and women too," inserted between "men" and "bank" (with a visible proofreader's insertion mark). The broadcast script had a similar revision: "Wise men bank with UBA – and women too bank with UBA ..." (*Daily Trust*, 2018). In part, it required the appointment of a woman as the chair of UBA's board of directors for the bank to acknowledge the presence of women among its customers. But when it finally came to this realization and reworked its key advertisement slogan, it did so with an explicit portrayal of its female customers as afterthoughts.

The UBA marketing campaign illustrates the general attitude to women's position in public life. Although women have been integral to national socio-economic processes, acknowledgment of their roles in development theories and policy often seems like an insertion point. This echoes early theories of development some of which have carried over to current discourse on information and communication technologies (ICTs) for socio-economic development (ICTD). Though pervasive, the practice of including women in development theorization and policy making only as an afterthought contrasts with the enduring tradition of women's participation in these processes. Oftentimes, that participation involves the deliberate articulation of strategies by women to navigate persistent institutional and structural barriers. The purpose of this chapter is to examine how women in two neighboring African countries, Nigeria and Cameroon, navigate these barriers to utilize ICTs as tools for socio-economic empowerment.

Building on previous research (Tabuwe et al., 2013; Akpan-Obong, 2009), I examine the specific strategies through which women have leveraged ICTs to achieve personal economic goals. These strategies have evolved from the early days of girls as receptionists in cybercafés, mobile payphone operators and vendors of airtime credit (recharge cards) on roadsides to their current global presence as digital entrepreneurs, computer science and information systems professors, software developers, job creators and CEOs of major ICT corporations in 2021. The chapter is anchored on the empowerment framework approach of ICT for development (ICTD). This is the current nexus of development theory and practice as driven by assumptions that ICTs can be deployed for socio-economic development. The empowerment framework focuses "on individual agency and opportunity structure in achieving development outcomes"

(Chib, 2015: p. 7). It is useful for explanations of women's interactions with ICTs to create socio-economic empowerment and satisfaction.

Two factors inform the focus on Nigeria and Cameroon. First, Nigeria is the largest English-speaking country in West Africa. It is also the most populous and demographically diverse in Africa. Though bilingual (English and French), Cameroon (in Central Africa) is the largest French-speaking country in Africa. Both countries share geographical boundaries and British colonial antecedents. It is expected that women's ICTD experiences in the two countries will provide insights into the ways that other African women navigate. Second, given my sustained ICTD research interest in Nigeria and Cameroon, I have a trove of evidence-based primary data to add value to this volume. My experiences in Nigeria and Cameroon also explain the deliberate framing of this chapter on *resilience*. For far too long, ICTD research on Africa has focused on what is absent (low ICT diffusion and adoption, barriers to ICT access, etc.), rather than on what is present (how Africans navigate the terrains of "absence").

Development is used in this chapter to refer to the socio-economic process deliberately geared toward the satisfaction of the basic needs of the greater number of people, especially the most vulnerable (Akpan-Obong and Parmentier, 2009; Brohman, 1996). Information and communication technologies are devices and applications that support "activities involving information (such as) gathering, processing, storing and presenting data" (Gokhe, 2000: p. 1). They include networks, networking components, digital applications, and systems that facilitate communication and interactions for efficiency and productivity (Pratt, 2021). Other concepts, especially those drawn from the discourse on gender and development, are defined/explained as they occur in the text.

The chapter is arranged in four sections including this introduction. The next section is an overview of the evolution of early theories and assumptions about the role of women in the development process. In the third section, I discuss the initial policy-making processes at the continental level and how they set the stage for the utilization of ICTs as strategic tools for economic development in Africa. Also in this section, I describe how different categories of women in Nigeria and Cameroon have interacted with ICTs to achieve personal economic goals and empowerment despite institutional and structural barriers. Some of these barriers are discussed if only to reinforce the significance of the modest accomplishments of women who have persisted despite various institutional and socio-cultural challenges. I summarize and identify the key contributions of the chapter in the Conclusion.

BACKGROUND

Gender/Women in/and Development

In many African societies, women are fully responsible for the wellbeing of the family in significant ways. They grow the food that the family consumes. They are often involved in trading (from basic household items to capital-intensive products such as clothing items). Often, mothers are the first contact for children or other members of the family when they need food, clothes or school supplies. However, "development" in the definition of Western development "experts" and agencies often focuses on the extractive value of development thus ignoring women's contribution to national development goals from a basic livelihood standpoint (Waring, 1997).

It is perhaps the dissonance in the definition of development that created (and continues to create) the uncertainties about the participation of women/gender in the development project. This goes back to the peak of the modernization theory of development[1] and associated development policies and projects of the 1960s when women were "in" development (WID). The late 1970s brought on the era of women "and" development (WAD). The concept eventually evolved into Gender and Development (GAD) by the 1990s, coinciding with the rise of Third Wave Feminism, and occupied theoretical and policy space for many years. As explained below, the issue was not simply about naming/labeling conventions ("gender" or "women"; "in" or "and"). Rather, each concept was founded on specific theoretical frameworks with delineated development priorities and paralleled the challenges about how to situate women in the development process.

The WID was anchored on modernization theories of development which "dominated mainstream thinking on international development from the 1950s into the 1970s" (Rathgeber, 1990: p. 490). Modernization theories prescribed specific paths to development. It gave rise to liberal feminism[2] which focused on change from within the institutional systems of states. The argument was that women's rights were achievable through the integration of women's concerns in the core processes of economic, political and social development at the national levels particularly as prescribed by international aid agencies (Rathgeber, 1990; Jaquette, 2017). Most WID projects focused on the expansion of women's access to resources within established institutions on the premise that "gender relations will change of themselves as women become full economic partners in development" (Rathgeber, 1990: p. 492). However, WID was criticized for failing to address the structural "causes of women's subordination" (Jaquette, 2017: p. 247). By acknowledging pre-existing participation of women as agents rather than passive subjects of development policies and scholarship, WAD was proposed as a better path. The argument was that before formal theories or policies, women were active in the process of development, and therefore did not need to be "integrated" in development especially in the manner in which the process reinforced structures of patriarchy and dependency (Rathgeber, 1990).

In some ways, WID was the feminist version of the modernization theory of development. It was therefore not surprising that WAD would be rooted in dependency theories (which emerged as a critique of modernization theory). As advanced by Latin American dependency theorists (e.g. Frank, 1967; Dos Santos, 1998), underdevelopment was the direct outcome of the global inequalities created by the unequal relations of dependency between the Global North and Global South. Specifically, the same process that created wealth in the core countries resulted in underdevelopment in countries in the periphery. For instance, the colonial extraction of resources and raw materials, as well as the financial and military dependency of periphery countries on the Global North contributed significantly to underdevelopment (Dos Santos, 1998; Frank, 1967). Similarly, in failing to disrupt or at least interrogate the patriarchal structure[3] of gender relations that perpetuates men's privilege and power over women, WID reinforced the unequal socioeconomic conditions of women in the South.

But WAD itself was considered insufficient to explain the role of women in development and was quickly overtaken by the arguably more expansive "Gender" and Development. This version of theorization about women "and" or "in" development originated from socialist feminism,[4] drawing on Marxist explanations of women's oppression as a function of the social construction of production and reproduction (or class and gender). Its proponents insisted that GAD was a more holistic approach to understanding the role of women in development (Rathgeber, 1990; and Jaquette 2017). The argument was that women's oppression and economic inequalities were caused by the tripartite of patriarchy, production/reproduction, and gender, the latter defined as "the social construction of gender and the assignment of specific roles, responsibilities, and expectations to women and to men" [Rathgeber, 1990: p. 494]). But it went beyond the fundamental livelihood issues of gender inequities in educational and employment opportunities; it included various ways in which women are oppressed *as women*. Though GAD showed a broader understanding of the condition of women, it was criticized for ignoring the very basic issues that confront women living in poverty in non-Western societies. Indeed, it was dismissed for its Western-centrism and the assumptions that concerns about gender-based inequalities such as reproductive rights were the same as those of women in developing countries (Mohanty, 1988; Singh, 2007). In casting the social construction of gender as a contest between men and women, GAD was criticized for failing to recognize that men of color can be allies of women of color in their shared race-class oppression. Many Third World feminists argued that it was easy for Western feminists to fight for reproductive rights and control over their bodies when they did not have to worry about basic "bread and butter" issues such as food, shelter, and clothing.

Gender/Women in/and ICTD

Perhaps the different versions of women's participation in the development process are less about the absence of consensus on "the" best explanation but more on how to tackle a core concern from multiple perspectives. Advocacy for change from within (WID) does not negate efforts to interrogate the institutional structures that reinforce the oppression of women (WAD) nor do the efforts of GAD feminists who challenge the social production and reproduction of gender. It is interesting however to observe how contestations on the theorization of women/gender in/and development and the symbiotic relationship between the different phases of development theories and feminist theories continued into the 1990s as the advances in ICTs paved the path for a new phase of development theorizing. The promises of the new technologies generated a new strategy of development that focused on ICTs simultaneously as integral components and goals of development. This resonated with the integration of radio, newspapers, and cinemas as indicators of development in the 1950s and 1960s (UNESCO, 1955). The ICTD field has spawned journals and research centers in universities around the world as well as national policies and programmatic priorities. It has also revived the role of women in development and their place in the digital revolution that emerged toward the end of the twentieth century.

Technology has always been associated with development but the role of women in the design, production or usage of technologies was often ignored. This is despite the interdependence of gender and technology as each other's co-creators (Fox, Johnson and Rosser, 2006). For instance, as noted by Asiedu (2012), women were excluded from the implementation of agricultural technologies as core development strategies in some African countries in the 1960s even though they made up 80 percent of the farmers. Also, indigenous agricultural knowledge systems on which women depended for their livelihoods were relegated to the background when mechanization took center stage thus doubly relegating the role of women. The exclusion of women at the dawn of new technologies or revolutions is a historical pattern, according to Spender (1995) who argues that the arrival of new technologies always disadvantages women during the early adoption stages. She presents a historical account of the negative impact of revolutions on women by superseding their existing knowledge systems and status.

For instance, before the invention of the printing press, women (especially nuns sequestered in abbeys) were the scripters/producers and keepers of social, political and religious records. Women's specialization in these roles conferred on them political empowerment and social relevance.

The print revolution ended the manuscript era and scripting skills became obsolete. With a wider diffusion of information and knowledge through printed scripts, women began to create new opportunities for knowledge. Then came restrictions on women's literacy beyond a certain age. Those who learned to read books in secret and appeared too knowledgeable were accused of witchcraft. Even the French Revolution that promised liberty, fraternity, and equality excluded women from the privileges of citizenship. Based on this historical pattern, Spender predicted that women would fare worse at the dawn of the ICT revolution before things got better even as:

> Computers have not always been seen as the province of the male. On the contrary, when computers first made their appearance in the business world, it was considered perfectly appropriate to place them in female hands. After all, they had keyboards: and women were the ones with typing skills and keyboard experience. (Spender, 1995: p. 166)

However, as soon as computers assumed "identity" as technologies rather than enhanced typewriters, men went in for the "power grab" (according to Spender) and women were swiftly edged out. As the technologies moved further into the public domain, they inextricably assumed the identity of masculinity. This was inevitable given their origins in the "deep academia and military-industrial complex" (Spender, 1995: p. 166).

Admittedly, much has changed since Spender's 1995 book. As the digital revolution matures (demonstrated by the global diffusion of ICTs), girls and women are participating actively as users, designers, producers, and industry leaders. Technologies generally, and ICTs specifically, are no longer considered to be exclusively male tools and preoccupations, at least in theory. Still, the reality is that the digital economy is overwhelmingly masculine. For instance, the technical design industry is "dominated by men (and) are generally designed by professionals employed by powerful companies" who fill design careers disproportionately with men (O'Donnell and Sweetman, 2018: p. 221). Also, the gender digital divide (the percentage of women who have access to ICTs relative to men) persists in dissonance with the expectations of ICT to empower women and support the development processes. Indeed, ICTs lie at the intersections of the discourses about women/gender and/in development, and ICT for Development (ICTD) in ways that crystallize issues in both areas (development and ICT) to generate an integrated field of theory, research, and policy known as women/gender and ICTD. There is an emphasis on the centrality of women/gender (as agents) and ICTs (as enabling tools) of socio-economic development. Nevertheless, the problems that attended earlier phases of theorizations on women in the development process have migrated to ICTD.

First, the general concern about women's access to ICTs resonates with modernization and liberal feminist theories of development. These theories advocated access to established institutions and structures as a precondition for women's empowerment. In the case of ICTD, the argument is that women need access to enable a transformative harnessing of ICTs for socio-economic and political development. This generates the following intertwined areas of concern for women and ICTD: policies and regulations, connectivity and infrastructure, and ICT services and usage – all of which "share a common goal of addressing development issues and encouraging socio-economic growth in the global south" (Asiedu, 2012: p. 1187). These areas mirror WID's assumptions about the expansion of opportunities for women as the goal of development. But "access" (to the technologies) by itself negates the prospects of participation as designers, producers and decision makers in the ICT industry. Also, emphasis on access as both a problem and end-goal outside systemic issues echoes the technology transfer component of modernization theory and liberal feminism/WID. The notion was that development would occur with the right influx of inputs (e.g., capital, technology, expertise) regardless of endogenous constraints.

Initial ICTD research stressed the importance of access as a strategic imperative. Developing countries such as Nigeria turned to the importation of ICTs as tools to revolutionize their economies. In the era of ICTs, past failures of the technology transfer approach were forgotten. Once again, technology assumed certain neutrality stripped of the ideologies of its design and production site with the expectation of universal outcomes when adopted in other geopolitical and socioeconomic contexts.

As occurred in the heyday of modernization and feminist theories of development, it was not long before ICTD discussions moved from the "bread and butter" issues of ICTs as tools for economic growth to problems of inequalities, knowledge/content creation, and social issues such as violence against women and other concerns for girls and women around the world. There is also a broader discussion of the potentials for empowerment and opportunities inherent in the digital revolution. However, for "ICTs to be as empowering as they possibly can be, more online content needs to challenge gender biases and fill in the gaps in history, offering an alternative account, or 'herstory'" (O'Donnell and Sweetman, 2018: p. 221). Indeed, the development of local content is considered crucial to participation. It explains why community participation has become a recurring theme in recent literature on gender and technology (Asiedu, 2012).

In the introduction to a special edition of *Gender and Development* journal, editors O'Donnell and Sweetman (2018) suggest that a useful way to understand the impact of ICTs on gender is to consider the two elements of ICT: information and communication. As communication, the technologies facilitate contact with others and information across time and space but do they substantially bridge the gaps between people? As information, one must examine whether ICT-created content supports social justice, human development and equalities or if it reinforces gender inequities (O'Donnell and Sweetman, 2018). This interrogation of the claims of ICTs for empowerment has raised new scholarship in ICTs and gender focusing on gender-based violence (GBV), cyberbullying and harassment, and women's agency in everyday decisions in ways that resist societal efforts to prevent their exercise of "transformative agency at the household level" (Hussain and Amin, 2018: p. 220). This transformative agency includes women's capacity to utilize online platforms to find romance and intimate relations (Philip, 2018), and as a source of education for women who lack access to formal education or continuous learning programs (Zelezny-Green, 2018). Other empowering utilization of ICTs includes leisure entertainment for women in ways that challenge a narrow view of development as "production and economic development … exclusive of the totality of the human life." O'Donnell and Sweetman (2018) add that ICTs provide relief for women, especially where movement and opportunities are restricted. Social justice activists have utilized ICT resources such as apps and hashtags (e.g., #MeToo, #IWillGoOut) to lobby for women-friendly legal reforms. These functionalities provide complexities to ICT usage and a "long distance away from seeing ICTs for development as a question of enabling access to devices" (O'Donnell and Sweetman 2018: p. 226).

A 2012 survey of 163 Facebook users indicated that most women use social networks to (re-) connect with family and friends and form new relationships. The authors of the research concluded that "from an economic or political standpoint, social networking sites are not the tools of liberation that African women may have been waiting for. Ironically, only they can determine to what ends they will utilize the technologies already available to them" (Akpan-Obong and Aquah-Braden, 2012: Abstract). That has changed considerably over the years. Social media networks have now become veritable "storefronts" for small business owners many of whom are women. Many studies describe the plethora of ways that women utilize different ICTs for various income-generating activities beyond social connections (Nyoh, 2021; Abdelmonem and Galan, 2017; Abraham, 2014; El-Neshawy, 2014).

Women have indeed come a long way in their engagement with ICTs in their various permutations as devices, processes and outcomes. These interactions exceed the basic issues of economic empowerment to the point where there are female-specific apps such as those that detect the menstrual cycle. While this trajectory, echoing Maslow's hierarchy of needs (McLeod, 2020), can be more valuable and useful in understanding the utility of ICTs for women in the Global North, most women in the South are still at the level of basic needs (how ICTs can help to secure the everyday needs). For most women in the South, the starting point for understanding how best ICTs can serve their needs is at the state level: government policies, priorities and the socio-cultural institutions that encourage or discourage women's access and participation in the ICT sector. This requires an unapologetic return to the much-maligned WID and focus on change from within. This approach is premised on the centrality of the state in resource-poor societies where the state runs schools and most public services and utilities and is the largest employer of labor. It also provides foundational technologies (such as electricity) and creates an environment conducive for individuals to leverage ICTs toward the achievement of personal economic goals. The pivotal role of the state in developing economies is indicated in the responses to Question 50e in the Spring 2019 Global Attitudes by the Pew Research Center: *Generally, the state is run for the benefit of all the people.* Most of the respondents in the 34-country survey who chose "Completely Agree" and "Mostly Agree" came from developing economies (Pew Research Center, 2019). While the various aspects of ICTs were already being used in many African countries, usage accelerated following a formal acknowledgment of the potentials for ICTs to drive socio-economic development and an agreement to create the enabling national policies and institutions in 1996 at a conference in Ethiopia.[5] That process is discussed in the next section.

The Emergence of the African Information Society and ICTs for Development

The adoption of ICTs by countries in Sub-Saharan Africa has undergone various stages of development in the last 20 years. Much of the impetus began with the adoption of the African Information Society Initiative (AISI) Framework to Build Africa's Information and Communication Infrastructure at a conference in Addis Ababa, Ethiopia organized by the United Nations Economic Commission on Africa (UNECA) in 1996. At the conference, ministerial-level officials from 48 African countries agreed to adopt "national e-strategies (to complement) their development efforts" by harnessing ICTs as tools to achieve national and regional development goals (United Nations E-Government Survey 2018: p. 134). The immediate outcome of the Framework was the elimination of restrictive policies and regulations in the telecommunication sector in member countries. This cleared the path for participation by private telecommunication providers and the gradual disengagement of government agencies from the sector.

The focus on telecommunications (defined narrowly as telephony) prevailed in much of the 1990s though countries approached the expansion of their telecommunications network differently. It was not until the turn of the century that considerations were given to ICTs (broadly defined to encompass all forms of telecommunication, computer technology, connectivity infrastructure, the Internet, and digital communication) as tools for socio-economic development. When Eritrea, the last country in Africa to connect to the Internet did so in 2000, the region had become fully integrated with the emerging networked society. Indeed, the development in Eritrea, a country that gained independence from Ethiopia in 1991, became possible only when its public telecommunications operator built an international gateway with funding from the Leland Project, a USAID initiative, as one of the final stages of its independence from Ethiopia (United Nations E-Government Survey 2018). That underscored the role of the state in the adoption of any technology on a national scale either through enabling policies or by providing the start-up infrastructure for the private sector investments. If the fundamentals of technology diffusion rest on the state, it is logical to expect that growth in the sector will benefit those included in the related policy-making processes. As already demonstrated, women are often ignored in these processes even if they eventually become integrated as an afterthought. This was the case in Nigeria and Cameroon during the initial implementation of the recommendations of the AISI Framework. In Nigeria, three specific policy documents accelerated the diffusion of telecommunications and ICTs: the National Telecommunication Policy, Information Technology (IT) Policy of 2001, and the ICT4D Plan of 2010. Cameroon connected to the Internet in 1997 and the following year, it established the Telecommunications Regulating Board.

For Nigeria, the National IT Policy was the first deliberate attempt by the government to prioritize and develop the information technology sector in the country. The 2010 Plan was aimed at promoting the utilization of ICTs to "achieve the government's Seven Point Agenda and Vision 20: 2020, thereby deploying ICT to achieve Nigeria's Millennial Development Goals, NEPAD development initiatives and the World Summit on Information Society's plan of action" (NITDA, 2021). There were consultations and contributions by "stakeholders" in the processes leading to the formulation of these policies but the participants in consultative forums and workshops were almost entirely men. One of the few women or possibly the only woman involved in the process reported that "as soon as the technology became widely known and utilized, the 'boys moved in'" and shoved her aside (Personal interview, 2018).

It was therefore not a coincidence that the document that emerged from that process, the IT Policy, barely acknowledged the role of women in the diffusion and utilization of ICTs for development. Indeed, going by the 60-page document, it appeared that women were not central to any ICT-led development. The policy thus ignored the pivotal participation of women in the informal economy as traders and farmers. Overall, the formulation and language of the policy were in sync with the patriarchal culture of the Nigerian society that traditionally ignored women during discussions of "important" issues. The policy also reinforced the near-universal assumption that women did not do technology. This was particularly so in a society where most women were overtly or implicitly directed to "pink careers" such as teaching and nursing. Secondly, there persisted an assumption that linguistically, "men" represent "men and women" and specific references to women (or gender) were superfluous.

The policy documents reflected the "gender blindness" that prevailed in mainstream policy making in the country.

In Cameroon, the Telecommunications Regulating Board (TRB) was mandated to "regulate, control and monitor the activities of users and operators in the telecommunications sector … and ensure respect of the principle of equity in the treatment of users by all telecommunications enterprises" (Baaboh, 2021: para. 7). One of its first acts was the authorization in 2000 of two private (foreign-owned) mobile telephone companies to operate in the country thus expanding the telecommunications landscape in the country. As the notion of telecommunications expanded to capture ICT so was the responsibility of the TRB to include the development of initiatives "geared towards creating a vibrant digital economy for both women and girls" (Nsaidzedze, 2020: p. 3). Also, in 2012, the National Agency for Information Communication and Technologies (NAICT/ANTIC) was created to promote ICT policies and regulate, control and monitor "activities related to the security of electronic communication networks and information systems." It was also tasked with the development and implementation of national ICT strategies and to "identify the common needs of public services in software and computer equipment" (Nsaidzedze, 2020: p. 4). The 2011 National Gender Policy and the Digital Vision of 2020 were later established to address the gender digital divide. Like previous policies, they too lacked sufficient implementation strategies, and policy makers, in particular, failed to support digital rights and access for women and girls (Nsaidzedze, 2020).

Nevertheless, shortcomings in policy and strategies have not deterred Nigerian and Cameroonian women from utilizing ICTs to achieve personal growth and empowerment. This was the conclusion from observations and interviews with different groups of women in the two countries between 2008 and 2019. As summarized in the next section, these women have also made significant contributions toward the achievement of national socio-economic goals.

WOMEN DOING ICT: PERSPECTIVES FROM WOMEN IN CAMEROON AND NIGERIA[6]

The Callbox Operators of Cameroon

One hot morning in December 2008, I walked from my hotel in Doula, Cameroon to Avenue de l'Indépendance, a major street in the city. It was there, I was told, I would find "ordinary people" engaging with ICTs. My first contact on the street was Monique,[7] a woman in her mid-20s sitting on a low plastic stool on the side of the road. In front of Monique were a mid-sized plastic table and two plastic stools on the other side of the table. There was a small flip phone on the table and a notebook beside it. A few minutes later, a young woman stopped by, spoke to Monique in a local language, and gave a piece of paper to her. On the paper was a phone number written in blue ink. Monique dialed the number, set a timer on a small rectangular device, placed it on the table, and handed the phone to the customer who began a conversation. When the conversation ended, the customer returned the phone to Monique who told her the cost of the call. The young woman gave her some francs (the Cameroonian currency) and left. Soon, another customer arrived. The day of a Callbox (fixed wireless and mobile public payphones) operator continued as the hot December sun made its way across the sky in the largest city in Cameroon. In between attending to customers, Monique told her story in halting English after we realized that her English was better than my French.[8]

Monique was a single mother who previously worked at a pre-school center. The incentive to work there was the tuition reduction for her three-year-old daughter. However, the income was too low to meet their basic needs. Then someone told her about the "Callbox" business. She borrowed money from a relative and bought the cellphone and required furniture and started the business:

> Now, I can send my daughter to a good nursery school. I can buy nice dresses for her. I can pay rent. We eat well. This is good business.

It was good business relative to what was available at the time. As Nzepa and Keuchantkeu (2007) found from their research, this business had already generated more than 10,000 jobs between 2002 and 2004 in Cameroon. This explained the speed with which it flourished in the country's city centers. Around Monique on both sides of the road were other "good businesses" though the more capital-intensive sections of the ICT roadside enterprise were run by men.

Laid out on the sidewalk were electric adaptors connected to several cellphones. Power generators hummed in the background. Some stood by silently but ready to be powered up when customers who needed to charge their phones came by. This business was conducted by men of various ages who called out to passersby in English and French. Unlike the Callbox business, the phone-charging operation required a generator, numerous adaptors, extension cables and plug-ins to accommodate multiple phones simultaneously. In most cases, the men worked for others to whom they remitted a fixed amount of money at the end of each day. The business was also more rigorous because of the steep competition to attract customers. While the Callbox operators could pick any street corner – the quieter the better – to operate, the phone chargers needed busy streets especially around bus stops and other areas with heavy foot traffic to attract enough business to pay their investors or employers at the end of the day – and have some money left for themselves.

The Callbox women and their male colleagues in the roadside ICT enterprise were leveraging the same resource scarcity: access to cellphones and electricity. Yet, the solutions were gendered. Women did what women have always done – work as telephone operators even if not for big telephone companies or in the basement of large corporations. Men did the more technical operations – the generators, the adaptors and hustle for customers. In a few years though, these would change as more women navigated their way into the higher echelons of the ICT industry.

The Umbrella Women of Nigeria

In Calabar, a southeastern Nigerian city just across the Cameroon border from Doula, young women were also in the same business as Monique. Here, they were known as the "Umbrella People" because they sat under large umbrellas emblazoned with logos of cellphone service providers. They too had the same business equipment as the women in Doula: three plastic stools, a plastic table, a cellphone, and a timer. Missing were the phone-charging stations because that business had not traveled across the border. Often, people who had phones in 2008 but had no access to electricity would power up wherever there was electricity and an unused outlet in public places.

In 2012, I met Ekaete, one of the Umbrella Women in Calabar. She got into the business after ending a previous job on the other side of town so she could work closer to home. On average, the 21-year-old woman made N300 (about 60 US cents) a day in sales and was paid N4,000 (about $8 US) a month by the owner of the business.[9] She had finished high school and her goal was to go to a catering school because "I love catering, cooking, housekeeping and all those things." She pictured her life beyond selling phone recharge cards but admitted that in the interim, she was making a living from it as well as from the occasional customer who paid to use her cellphone.[10]

Mary had been in the business as an Umbrella Woman for two years before she advanced to the next level of the business: selling of "bricks." A brick, she explained, was a pack of 100 recharge cards (for airtime credit); the sale of bricks was the first level to becoming a big-time wholesaler. Women like Ekaete would purchase the recharge cards from Mary and retail to customers. Mary was selling for another person (a man) who bought the cards directly from the cellphone services providers. Mary's goal in December 2012 was to generate enough income from the bricks to afford the capital to purchase directly from the cellphone services providers. "I will make more money if I am a distributor," she said while completing a transaction with an Umbrella Woman who came to buy half a brick from her. As more people now owned phones, the demand for recharge cards outstripped the need for pay-phone services. With more than 98 percent of mobile phone usage on the pre-paid plan, recharge cards were in constant demand. But even that alone was insufficient to meet the basic needs of the women and so they added snacks (biscuits, groundnuts, candies, etc.) and fruits to the business. Ostensibly, the mobile pay-phone business and trade in recharge cards did not look like much but it was a step up from the initial revenue-generating interactions with ICT by women.

Women's entrance into the ICT and livelihood terrain in Nigeria began in cyber cafés and telecenters. As noted by Spender (1995), when computers first got into the public domain, they were considered as advanced typewriters and therefore machines for women in the typing pool of organizations and government offices. This is captured by Shetterly (2016) in *Hidden Figures*, a best-selling non-fiction book about how the intersections of race and gender contributed to a failure to acknowledge the pivotal role of four Black women in America's initial space exploration. In the story, women in the typing pool were described as "computers" and occupied low-level positions while the men were engaged in the more serious business of preparation to send Americans to outer space. As soon as computers became "technologies," they occupied male spaces while typewriters and typing skills remained in the predominantly female typing pools. To fill the gap in the demand for computer and digital communication services, cybercafés and telecenters began to pop up in major cities in Nigeria and Cameroon at the turn of the new century.

Women's typing skills became an asset once again especially for younger women most of whom had learned typing as preparations for secretarial careers. They worked long hours for low wages and were constantly harassed by male customers. As the cost of ICTs declined and more people had direct access to the Internet, businesses in the public access points declined. Also, these places were quickly gaining infamy as hubs for crimes: young women hooked up with men for sex and money while young men transformed themselves into "Nigerian princes" to run the notorious e-mail scams that trapped foreign seekers of unearned fortunes. Given the notoriety of the cybercafes, the Callbox and Umbrella businesses became an improvement though with unique security and safety challenges. Most of the women reported that male customers routinely "confused" their ICT business with prostitution.

The next level, particularly for Cameroonian women in the ICT terrain, was a direct outcome of the accelerated twin processes of globalization of production and services and rapid advances in the technologies. As multinational corporations began to outsource their services (especially customer services and sales), call centers mushroomed in various parts of the developing world beginning in Southeastern Asian countries such as India (Friedman, 2012) as cost-serving tactics. The phenomenon was (still is) so widespread that it became a major segment in the 2009 Academy Award Winner for Best Picture, *Slumdog Millionaire* (Boyle, 2008). Women in Cameroon were some of the earliest in Africa to get into the call-center business. In a study of pink-collar ICT jobs conducted in Cameroon by Tabuwe et al. (2013), it was found that call centers had quickly become "a highly feminized sub-sector" as employers (both Cameroonian and MNCs) preferred female workers because their "voices were more pleasant." They were also inexpensive as entry requirements to answer incoming calls and/or initiate outgoing sales calls were low. Many female job seekers flocked to this industry while waiting for better-paying positions or admission to universities. In Nigeria, there were fewer opportunities to work for MNCs but young women found similar customer service jobs with mobile phone operators who showed a preference for younger women.[11] Soon, a new pool of pink-collar jobs had emerged in the ICT sector.

Transformative ICT Users and Participants: From Pink-Collar Workers to CEOs

The early aspirations of the Moniques and pink-collar ICT workers of Cameroon and the Ekaetes and Marys of Nigeria mirror a distant pre-take-off past (the second stage of Rostow's five stages of economic development). In the past ten years, the ICTD sector has changed in unimaginable ways. The early attraction of young women to learn computer skills to combat idleness during school vacations blossomed into an influx of women in computer science and information degree programs. Outside academia, girls and women have developed creative ways to leverage ICTs to create wealth, employment and empowerment for themselves and other women in their society and spheres of influence. From selling recharge cards on street corners in Doula and Calabar, women are now selling clothing and accessories online and setting up technology-enabled recycling businesses. In the early days of ICTD, women sought loans from men for start-ups in the mobile payphone business but now women are giving microloans to rural women and providing them with digital literacy and skills training to start and succeed in small-scale businesses. From paying a few francs to make a phone call, women are using cellphones as a necessary tool in their businesses as traders, farmers, hairdressers, and domestic workers to connect with customers and clients.

In Nigeria, women are active in different sectors of the Nigerian ICT ecosystem. The high point was in 2011 when a woman, Omobola Johnson, was appointed as the first minister of the newly established Ministry of Communication Technology by the Goodluck Administration. Funke Opeke is the founder of Mainstreet Technologies, the developer of Main One Cable, a leading provider of telecommunication and network services in the West African region. She previously worked with Verizon Communications in the United States. The country manager for Google in Nigeria is also a woman, Juliet Ehimuan-Chiazor, dubbed in the Nigerian media as "Queen of Technology." She previously worked with Microsoft, UK. Olatorera Oniru owns an online clothing store, Dressmeoutlet.com, and was recognized in 2016 by Forbes "30 under 30" as one of the Most Promising Entrepreneurs in Africa. Eno-Obong Essien, widely recognized for her business achievements, was one of the pioneers of the vehicle-tracking industry in Africa and the first Nigerian woman to enter the sector. She founded Rheydolence Limited, a leading provider of GPS vehicle tracking systems, in 2007. She remains the only licensed female and youngest CEO in the Nigerian vehicle-tracking industry. Mary Uduma, a ubiquitous presence in the Nigerian IT industry, was the founding president of the Nigerian Internet Registration Association and fought hard to secure a top-level domain name for Nigeria. Currently, the managing director of Jaeno Digital Solutions, Uduma played an active role in the development of the Internet in Nigeria. Damilola Anwo-Ade is the founder of CodeIT, a platform that mentors the next generation of coders, including young women. She advocates strongly for girls in STEM education. Carolyn Seaman started Girls' Voices to serve as a platform for girls to share their stories and inspire other girls on the use of technology and digital media. Martha Omoekpen Alade is the founder of Women in Technology in Nigeria (WITIN), described on its website as "one of the World's Largest and Leading Community of Women in Tech … dedicated to the advancement of women and girls." Among several projects to promote the acquisition of digital skills for girls and active participation in the technology sector the organization:

> Showcases the amazing women in the tech ecosystem as a role model for girls while working closely with educators to bridge the leaky pipeline in STEAM [Science Technology Engineering Arts and Math]. (WITIN, 2021)

Similar engagement with ICT by women is observed in Cameroon where women are pushing the frontiers of the ICT landscape in revolutionary ways. Janet Fofang is the director of Girls in Tech – Cameroon and founder of NextGen Center in Yaoundé where girls are trained in robotics, STEM and coding. She also trains women on how to use basic IT tools to earn income to improve the living conditions of their families:

> We've seen technology prove to every one of us that it's a good method to move away from poverty. And we all know that in these parts, women always remain at the poverty line. (Lazareva, 2018)

Danielle Akini is a digital entrepreneur, "a computer analyst and developer who runs a company that trains thousands of children to code" (Bonny, 2020: para. 2). She thrived despite discrimination from men whom she said acted as if the digital environment was "dedicated to men, and they thought that a woman was not capable of doing anything in the field of technology."

Women in both countries have demonstrated extraordinary capability in the ICT sector. Most have dedicated their lives to raising the next generation of Nigerian and Cameroonian girls and women as "ICTpreneurs," digital experts able to compete with their male counterparts and the rest of the world in utilizing ICTs to achieve personal, communal and national socio-economic goals. These achievements occur even within the context of persisting barriers to more transformative participation by women and the exclusion of their agency in many national policies.

Persisting Barriers to Women's Participation in ICT Beyond Access

Despite the progress made by women in ICTD practice, the digital gender divide persists because of institutional and socio-cultural barriers that go beyond technological access. As noted in the Cameroon ICT gender policy "more women than men still face barriers including language, digital illiteracy, limited time ... due to family responsibilities" but the implementation does not always match policy statements and recommendations (Nsaidzedze, 2020: p. 3). The more obvious barriers are costs and skills both of which are manifestations of the intersection of patriarchy, social-cultural norms about women in development, and technology that circumscribes women's participation in the overall development process. Much has already been written about these issues and there is no intention to repeat them here. However, the discussion of how women in Cameroon and Nigeria have navigated the ICT terrain will be incomplete without an examination of some of the obstacles encountered in the journey to create value and meaning through ICT usage.

Institutional and policy barriers

National ICT-related policies routinely exclude women. This is demonstrated by the absence of inclusive language in policy documents and the lack of gender-disaggregated statistics in other socio-economic sectors. I have already addressed how language in policy documents excludes women and subsumes their concerns by using "men" as an umbrella term for "men and women." In this section, I highlight how this invisibility of women in national policies including those in the ICT sector is reinforced by the absence of gender-disaggregated national statistics. Many African countries are notorious for their poor recording-keeping practices even as they claim to adhere to international norms on gender-specific national data. Indeed, a Women's Online Rights Report by the UN Women Foundation shows that only 24 countries in Africa and Asia collect sex-disaggregated data on internet access. Cameroon and Nigeria are not among those countries.

To comply with the Beijing Platform for Action from the Fourth World Conference on Women in 1995, the Nigerian National Bureau of Statistics publishes a Statistical Report of Women and Men (National Bureau of Statistics, 2021). However, it covers only the conventional data (birth and mortality rates, school enrollment, etc.) and its most recent report is from 2016. The Nigerian Communications Commission (NCC), the official regulatory agency for the telecommunications industry in Nigeria, monitors and tracks activities in the telecommunications industry. It collects data and analyzes them to identify "industry trends (in) services, tariffs, operators, technology, subscribers, issues of competition and dominance ... (and) areas where regulatory intervention would be needed" (NCC, 2003: para. 1). Accordingly, the Commission keeps extensive data on a myriad of telecom-related activities such as teledensity, internet subscriber data, broadband penetration, network type, market share by technology, porting activities by network operators, and time-of-day (peak or off-peak period) tariffs. There is no disaggregation by gender. Without this information, it is unclear how the NCC identifies "areas where regulatory intervention might be needed," such as strategies to acknowledge women's participation and/or expand access to more women.

The Nigerian Information Technology Development Agency (NITDA) implements the Nigerian Information Technology Policy of 2000 and the general development of the ICT sector. The authority of the agency was enhanced by the National Information Technology Development Act of 2007. Its mission was expanded to include a framework for the planning, research, development, standardization, application, coordination, monitoring, evaluation and regulation of Information Technology practices, activities and systems in Nigeria (NITDA, 2021). Despite this expansive role in the country's IT sector, NITDA does not collect data on general ICT usage or women's participation in the IT sector.

Cameroon collects some ICT data for the general population but does not disaggregate by gender. However, some researchers and external agencies such as the Women's Rights Online, HootSuite Digital Report and Research ICT Africa (RIA) have used interviews, surveys and other instruments to determine gender ICT participation data. In one study, the RIA used focus groups from different parts of the continent to create data capturing the degree of access to ICT by women and men (Milek, Stork and Gillwald, 2011). International agencies such as the International Telecommunications Union (ITU) use data reported by national regulatory agencies. Reports from African countries are not always consistent, have missing years or indicators. For instance, of the several ICT indicators that the ITU reports annually for all countries, there is only one indicator for Nigeria: mobile phone ownership disaggregated by gender. It shows that of the 41 percent of the population who own mobile phones, 32 percent were women and 49 percent were men. The report is for 2017 and remains the most recent on the ITU database. The most recent report for Cameroon is also from 2017. Of the 23 percent of the total population who used the Internet in 2017, 19 percent were female and 27 percent were male. Noting that Cameroon does not collect sex-disaggregated ICT data, the Women's Rights Online reports that its independent survey in 2015 showed that 36 percent of women were internet users, compared to 45 percent men. In 2020, however, internet penetration for Cameroon was recorded at 30 percent of the population by Data Reportal (2021), a data aggregator website. It is not clear why the numbers dropped so drastically from 2015 to 2017 (in the ITU numbers). These differences underscore the need for accurate and consistent mechanisms for tracking official data at the national level and organizing them by gender. Data aggregator sites and social media management platforms such as HootSuite also gather extensive ICT data for different countries and disaggregate many of the indicators by gender. The gender ICT columns for Nigeria[12] and Cameroon are blank.

Gender-disaggregated ICT data are critical to making "policies that address the specific needs of women (in ways that take) into account specific experiences of men and women" (World Wide Web Foundation, 2020: p. 6). The United Nations Conference on Trade and Development (UNCTAD) affirms the importance of gender-disaggregated data generally and specifically ICT statistics because of the broad-based impact on the technologies and utilization in all aspects of society. Also, as technology and gender are both social constructs, "social attitudes and norms influence the relationship between the two" such that ICT policies, strategies and projects are not gender-neutral but affect men and women differently (UNCTAD, 2014: p. 3). The absence of gender-specific ICT data in Nigeria and Cameroon perpetuates the invisibility of women's participation in the ICT sector while further excluding them from more transformative engagement with ICTs. It is obvious that what is not counted does not matter (Azcona and Valero, 2018) and not prioritized in the policy-making process:

> [T]he paucity of sex-disaggregated ICT data, particularly from developing countries, makes it difficult, if not impossible, to make the case to policymakers for their consideration of gender-related issues in ICT policies, plans and strategies. The lack of adequate data resulting from the scarcity of gender statistics affects policy and its implementation. Indeed, the dearth of gender-specific data available to policymakers is reflected in the absence of gender awareness in ICT and ICT-related policies and in the undertaking of costly gender-related initiatives based on insufficient evidence. (UNCTAD, 2014: p. 3)

Socio-cultural barriers

Related to the absence of institutional and policy priorities on intentional inclusion of women's participation in the ICT sector, perhaps more insidious barriers occur in the social-cultural and personal sphere most of which are connected with structures of patriarchy. It is not an accident that countries such as Nigeria and Cameroon where the gender digital gap is extensive are also those with significant gender inequities anchored on customs and traditions. For sure, these countries do not have laws that specifically prohibit women from engaging in the ICT sector. Indeed, Cameroon's National Gender Policy of 2011 is aimed at closing the gender digital divide. Yet, unwritten rules, social norms and cues persist. Even the notion of women as ICT experts continues to receive pushback. In northern Nigeria, for instance, 55 percent of husbands and 61 percent of fathers surveyed by the Centre for Information Technology and Development (CITAD) said they do not want their wives and daughters, respectively, to use the Internet (Equal Access International, 2021). Ultimately, the women (those wives and daughters) internalize the concept that these technologies are not for them. This spills over to the classrooms as girls are socialized into non-STEM "female disciplines." Also, the perception (often supported by research) that women engage in ICT activities for social purposes (communication and social networking) leads to the dismissal of women's engagement in ICT as a productive, that is, income-generating, behavior. Nsaidzedze (2020) cites a report that indicated that in Cameroon, "more women and girls dominate digital spaces when it comes to social communication with family members and friends, entertainment, lifestyles, among other leisure activities" (p. 3). A 2015 Pew Survey also seemed to indicate women are more likely to engage in social activities when online than men. For instance, in that study, 94 percent of women sent e-mails when they were online while 88 percent of men surveyed during the same period did. Conversely, 82 percent of men (and 75 percent of women) were involved in "research product/service."

The negative notion about women's relationship with ICTs gained currency in Nigeria on several fronts starting from the rollout of mobile phones in the country in 2001. The initial cellphones were basic and featured voice and limited texting functionalities but were quite expensive and affordable only by the very wealthy. Women who had cellphones in those early days faced two assumptions. First, given that men as a class are wealthier than women everywhere but especially in developing countries, it was assumed that any woman who had a cellphone did so because a man (husband, boyfriend, etc.) purchased it for her. Second, given the basic functionalities of the initial set of phones, they were perceived as a tool for women to "gossip" and chatter about unimportant issues. Many Nigerian movies (commonly known as Nollywood) promoted these assumptions by portraying young women (especially those in the universities) as obsessed with cellphones so much that they would have sex with older rich men in exchange for a cellphone. Soon, these sex-for-phones narratives were spilling out into real life to the point that young women's social status became associated with the types of cellphones they had and the types of men they attracted. The more socially "sophisticated" category of young women became known as "BB babes" (after Blackberry Bold cellphone, the top of the line in cellphone technology at one point). The boundaries between Nollywood and life began to blur.

This early reputation with regard to cellphones and the women who used them influenced general attitudes toward the participation of women in the broader ICT landscape. In concert with the patriarchal infantilization of women, it justified the limitation of women's access to ICT as demonstrated by men who would not allow their wives and daughters to go online.

Those who did were exposed to hostile online environments like those that I witnessed in newsgroups and bulletin boards, the 1990s precursors to contemporary social media. One of those groups was a political discussion forum for Nigerians in the diaspora. Single women who participated in the discussions were dismissed as "angry and frustrated" and told that they were spending time in the public space because they had no men in their lives or were using the forum to seek men. Married women were often asked to "go and cook for your husband and don't concern yourself with politics;" some were accused of being too "out of control" for their husbands to rein them into order. Soon, the forum became an all-male arena as most women unsubscribed over time. Those who remained no longer participated in the discussions. This silencing of women's voices continues on social media, although there is a new generation of fierce and feisty African women who are staking their territories in cyberspace by blogging, tweeting, telling their stories, and breaking out of the veils of invisibility.

CONCLUSION

A significant portion of the chapter has been a historical overview of theories, policies, and early interactions of Nigerian and Cameroonian women with ICTs. This is deliberate to high-light how women in these countries (and in the developing world generally) have been active participants in the various processes of development over time even as national policies seem to ignore their role. The exclusion of women from the development process is certainly not new. Attempts by feminist scholars of development to address this problem got tangled on conceptualizations of women or gender as objects of or partners in development, agents or victims. Those debates remain unresolved in the current era of gender and ICTD. National policies also mirror these uncertainties: to mainstream women, set them outside the context of development, or be more inclusive in response to pressures from feminist activists.

Regardless of the theoretical interplay in academia and national policies that trend toward the exclusion of women or minimization of their contributions, women in Nigeria and Cameroon have created spaces for themselves in the ICT sector. From working in cyber-cafés or as mobile payphone operators to selling airtime credits on roadsides, women have leveraged ICTs to achieve personal economic goals. They have emerged as ICT educators, entrepreneurs, industry leaders and have created jobs and opportunities for other women. This chapter focuses on these contributions and the specific ways that women have interacted with ICTs historically. It showcases how women, circumscribed by the structures and institutions of patriarchy, customs, and traditions, have persisted in confronting and engaging with the system in progressive and empowering ways. It demonstrates that while masculine linguistic narratives and the absence of gendered ICT statistics perpetuate women's invisibility and the misperception that "women do not do" ICT, women in these two countries indeed "do ICT" at all levels of society and complexity.

The advertisement slogan referenced at the beginning of this chapter serves as a metaphor for women's persistence even if they are insertion points in national socio-economic pro-cesses. When Nigerian women were excluded from a bank's marketing campaign, they contin-ued to engage with the country's economic and productive sectors. When they were eventually inserted into the bank's advertisement materials in an unabashed afterthought, they continued to dominate the Nigerian informal economic sector as traders, farmers, craftswomen and entrepreneurs. Recent TV advertisements by the same bank now feature women prominently in various gendered scenarios (shopping for clothes, hanging out with friends and engaging in other "women's activities") enabled by the bank's financial products. This gradual recognition of the contribution of women to the economy parallels the trajectory of development theories including the current ICTD. After the initial blind spot about women's participation in the socio-economic processes, eventually, it seems theory and policy have caught up to integrate women and their economic inputs into the broader development landscape. This is inevitable as women, invisible in policies and national ICT statistics, remain ubiquitous in the spaces that they create for themselves often for basic survival.

NOTES

1. The modernization theory of development is a body of theories that prevailed during much of the 1950s and 1960s as international development agencies and policy makers scrambled for the best way to bring development to newly decolonized African countries and keep them from turning to the Soviet Union. It was rooted on the five stages of growth advanced by Rostow (1959). The stages are traditional society, transition to take-off, take-off, drive to maturity, age of high consumption. The core assumption was that with the right inputs (such as capital and technology), all societies can evolve from the first stage until the last and that would signify "modernization."
2. Liberal feminism is based on the assumptions that the subordination of women results from customary and legal constraints that block women's access and success in the public arena. Gender inequality therefore arises from the denial to women of the rights exercised by men. To achieve gender equality therefore, reforms must occur from within the legal and institutional structures of the state (McCann and Kim, 2010).
3. Sylvia Walby in her 1990 book, *Theorizing Patriarchy*, broke down the structure of patriarchy into six constituent parts: paid employment, household production, culture, sexuality, violence and the state. She argued that it is in these areas that gender inequalities and socially engendered male privilege and dominance occur.
4. Socialist feminism is rooted in Marxism but rather than focusing on class as the only source of inequality, it argues that women's oppression stems from primarily from systems of capitalism and patriarchy. To achieve gender equality, it advocates for the structural transformation of society to eliminate class and gender barriers (McCann and Kim, 2010).
5. This was the African Information Society Initiative (AISI) Framework Build Africa's Information and Communication Infrastructure held in July 1996 in Addis Ababa, Ethiopia on the auspices of the United Nations Economic Commission on Africa (United Nations E-Government Survey 2018: p. 134).
6. The data were generated from interviews with women in Nigeria and Cameroon engaged with ICTs at every level, participant observations, stories in the media and information on websites of various organizations involved in the promotion of ICT use by girls and women. Some of the findings have appeared in the following peer-reviewed journal articles: (1) Alozie, Nicholas and Akpan-Obong, Patience (2016). The digital gender divide: confronting obstacles to women's development in Africa. *Development Policy Review*, 35(2), pp. 137–60; (2) Tabuwe, M.E., Muluh, H.Z., Tanjong, E., Akpan-Obong, P. et al. (2013). Gendering technologies: Women in Cameroon's pink-collar ICT work. *International Journal of Management and Information Systems*, 17(4), pp. 213–22.
7. The names used for Umbrella and Callbox women in this section are not their real names.
8. Cameroon has two official languages, English and French, reflecting its colonial British and French colonial heritage. Doula is the largest commercial city in the country and bilingual in the two formal languages, as well as several indigenous languages and dialects.
9. These were the exchange rates in April 2021 but the US$ equivalent would have been slightly higher in 2012.
10. I wrote a story about Ekaete (not her real name) for publication in a Nigerian newspaper, *Saturday Punch*, January 5, 2013. A reader responded to the article by paying for Ekaete to go to catering school. The entire article, "Tell me, how did you succeed?" became part of a collection of essays in

Akpan-Obong, P.I. (2013). *Letters to Nigeria: Journal of an African Woman in America.* Charlotte, NC: Createspace, pp. 324–26.

11.   Some job vacancy announcements in the newspapers explicitly ask for women usually between the ages of 18 and 25. Hiring preferences are often given to unmarried women with no children.

12.   The site posts two financial indicators by gender for Nigeria: credit card use and online transactions for 2020. The financial institutions collect this information and the data aggregator must have included did as part of ICT data because of the technology involved in those processes.

REFERENCES

Abdelmonem, A. and Galan, S. (2017). Action-Oriented responses to sexual harassment in Egypt: The cases of HarassMap and WenDo. *Journal of Middle East Women's Studies*, *13* (1), pp. 154–67.

Abraham, K.B. (2014). Sex, respect and freedom from shame: Zambian women create space for social change through social networking. In Buskens, I. and Webb, A. (Eds.) (2014). *Women and ICT in Africa and the Middle East: Changing Selves, Changing Societies*. London, UK: Zed Books, pp. 195–207.

Akpan-Obong, P.I. (2013). Tell me, how did you succeed? In *Letters to Nigeria: Journal of an African Woman in America.* Charlotte, NC: Createspace, pp. 324–26.

Akpan-Obong, P. (2009). *Information and Communication Technologies in Nigeria: Prospects and Challenges for Development*. New York: Peter D. Lang Publishing, Inc.

Akpan-Obong, P. and Aquah-Braden (2012). "What's a married woman doing on Facebook?" Gender, cultural narratives and social networking in Sub-Saharan Africa. Presentation at the Southwest Social Sciences Association, San Diego, California, April 4–7, 2012.

Akpan-Obong, P. and Parmentier, M.J.C. (2009). Linkages and connections: A framework for research in information and communication technologies, regional integration, and development. *Review of Policy Research*, *26*, pp. 289–309. https://doi.org/10.1111/j.1541-1338.2009.00383.x.

Alozie, N. and Akpan-Obong, P. (2016). The digital gender divide: Confronting obstacles to women's development in Africa. *Development Policy Review*, *35*(2), pp. 137–60.

Asiedu, C. (2012). Information communication technologies for gender and development. *Information, Communication and Society*, *15*(8), pp. 1186–1216, DOI: 10.1080/1369118X.2011.610467.

Azcona, G. and Valero, S. D. (2018). Making women and girls visible: Gender data gaps and why they matter. www.unwomen.org/en/digital-library/publications/2018/12/issue-brief-making-women-and-girls-visible.

Baaboh, F.H. (2021). Telecommunications law in Cameroon. www.hg.org/legal-articles/telecommunications-law-in-cameroon-7157.

Bonny, A. (2020). Cameroon: Women seek their rights in unequal system. *Anadolu Agency.* www.aa.com.tr/en/africa/cameroon-women-seek-their-rights-in-unequal-system/1758538.

Boyle, D. Director (2008). *Slumdog Millionaire*. Celador Films and Film4 Productions.

Brohman, J. (1996). *Popular Development: Rethinking the Theory and Practice of Development*. Malden, MA: Blackwell Publishers Ltd.

Chib, A. (2015). Research on the impact of the information society in the Global South: An introduction to SIRCA. In Chib, A.; May, J. and Barrantes, R. (Eds.) (2015). *Impact of Information Society Research in the Global South*. International Development Research Center. DOI 10.1007/978-981-287-381-1_1.

Daily Trust Newspaper (2018). Daily Ads of old. https://dailytrust.com/great-ads-of-old.

Data Reportal (2021). Digital in Cameroon. https://datareportal.com/digital-in-cameroon.

Dos Santos, T. (1998). The structure of dependence. In M.A. Seligson and J.T. Passé-Smith (Eds.), *Development and Underdevelopment: The Political Economy of Global Inequality* (pp. 251–61). Boulder, CO: Lynne Reiner.

El-Neshawy, S. (2014). Transforming relationships and co-creating new realities: Landownership, gender and ICT in Egypt. In Buskens, I. and Webb, A. (Eds.) (2014). *Women and ICT in Africa and the Middle East: Changing Selves, Changing Societies*. London, UK: Zed Books, pp. 275–87.

Equal Access International. (2021). Tech4Families: Addressing the gender digital divide. www.equalaccess.org/our-work/projects/closing-the-gender-digital-divide-tech4families/.

Fox, M.F., Johnson, D.G. and Rosser, S.V. (Eds.) (2006). *Women, Gender, and Technology*. Chicago: University of Illinois Press.

Frank, A.G. (1967). *Capitalism and Underdevelopment in Latin America: Historical Studies of Chile and Brazil*. New York: Monthly Review Press.

Friedman, T.L. (2012). *The Lexus and the Olive Tree: Understanding Globalization*. Picador.

Gokhe, M. (2000). Concept of ICT. www.hzu.edu.in/csit/IV.1_information_and_communication _technology.pdf.

Hussain, F. and Amin, S.N. Amin (2018) "I don't care about their reactions": Agency and ICTs in women's empowerment in Afghanistan. *Gender and Development*, 26(2), pp. 249–65, DOI: 10.1080/13552074.2018.1475924.

Lazareva, I. (2018). From robots to girl power, getting Cameroon's women into work. Thomson Reuters Foundation. www.reuters.com/places/africa/article/us-africa-education-women/from-robots-to-girl-power-getting-cameroons-women-into-work-idUSKBN1KK00U.

McCann, Carole and Seung-Kyung Kim (Eds.) (2010). *Feminist Theory Reader: Local and Global Perspectives*. New York, NY: Routledge.

McLeod, S.A. (2020, March 20). Maslow's hierarchy of needs. *Simply Psychology*. www .simplypsychology.org/maslow.html.

Milek, A., Stork, C. and Gillwald, A. (2011). Engendering communication: A perspective on ICT access and usage in Africa. *Info*, 13(3), pp. 125–41. https://doi.org/10.1108/14636691111131493.

Mohanty, C. (1988). Under Western eyes: Feminist scholarship and colonial discourses. *Feminist Review*, 30, pp. 61–88.

National Bureau of Statistics (2021). Statistical report on women and men in Nigeria. https://nigerianstat .gov.ng/elibrary.

National Communications Commission, NCC (2021). Industry statistics. www.ncc.gov.ng/statistics -reports/industry-overview.

National Information Technology Development (2021). NITDA Act. https://nitda.gov.ng/nitda-act/.

Nsaidzedze, S.B. (2020). Digital gender divide and political participation of women in Cameroon. Kafu Policy Institute, Denis and Lenora Foretia Foundation. www.nkafu.org.

Nyoh, I. (2021). African Women to Drive Digitalization and Transformation: AfChix and Internet Society Renew Partnership. www.internetsociety.org/blog/2021/05/african-women-to-drive-digitalization -and-transformation-afchix-and-internet-society-renew-partnership/.

Nzepa, O. and Keutchankeu, R.T. (2007). Cameroon telecommunications sector performance review. www.researchictafrica.net/new/images/uploads/ RIA_SPR_Cameroon_07.pdf.

O'Donnell, A. and Sweetman, C. (2018). Introduction: Gender, development and ICTs. *Gender and Development*, 26(2), pp. 217–29, DOI:10.1080/13552074.2018.1489952.

Pew Research Center (2019). Spring 2019 Global Attitudes Topline. www.pewresearch.org/wp-content/ uploads/2019/11/Pew-Research-Center-Value-of-Europe-Topline-for-Release-FINAL.pdf.

Philip, S. (2018). Youth and ICTs in a "new" India: Exploring changing gendered online relationships among young urban men and women. *Gender and Development*, 26(2), pp. 313–24, DOI: 10.1080/13552074.2018.1473231.

Pratt, M. (2021). Definition: ICT (information and communications technology, or technologies). https://searchcio.techtarget.com/definition/ICT-information-and-communications-technology-or -technologies.

Rathgeber, E.M. (1990). WID, WAD, GAD: Trends in Research and Practice. *The Journal of Developing Areas*, 24, pp. 489–502.

Rostow, W.W. (1959). The stages of economic growth. *The Economic History Review*, 12(1), new series, pp. 1–16. doi:10.2307/2591077.

Shetterly, M.L. (2016). *Hidden Figures: The American Dream and the Untold Story of the Black Women Who Helped Win the Space Race*. William Morrow and Company.

Singh, S. (2007). Deconstructing "gender and development" for "identities of women." *International Journal of Social Welfare*, 16(2), pp. 100–109. DOI: 10.1111/j.1468-2397.2006.00454.x.

Spender, D. (1995). *Nattering on the Net: Women, Power and Cyberspace*. Spinifex Press.

Tabuwe, M.E., Muluh, H.Z., Tanjong, E., Akpan-Obong, P., Sikali, L., Ngongban, A., Itegboje, A.O., Samake, K.D., and Mbarika, V. (2013). Gendering technologies: Women in Cameroon's pink-collar

ICT work. *International Journal of Management and Information Systems (IJMIS)*, *17*(4), pp. 213–22. https://doi.org/10.19030/ijmis.v17i4.8097.

United Nations Conference on Trade and Development (2014). Measuring ICT and gender: An Assessment. https://unctad.org/system/files/official-document/webdtlstict2014d1_en.pdf.

United Nations E-Government Survey (2018). https://publicadministration.un.org/egovkb/en-us/Reports /UN-E-Government-Survey-2018.

UNESCO (1955). *Research paper no. 9.*

Waring, M. (1997). The invisibility of women's work. *Canadian Women Studies/Les Cahiers de la Femme Canadienne*, *17*(2), pp. 31–8.

Women in Tech in Technology in Nigeria, WITIN (2021). About Us. https://wit.ng/about/.

Women's Rights Online (2021). Report Card: Cameroon – Measuring Progress, Driving Action. http:// womensrightsonlinecmr.internetwithoutborders.org/en/assets/wf_cameroon_final2.pdf.

World Wide Web Foundation (2020). Women's Rights Online: Closing the digital gender gap for a more equal world. Web Foundation.

Zelezny-Green, R. (2018). "Now I want to use it to learn more": Using mobile phones to further the educational rights of the girl child in Kenya. *Gender and Development*, *26*(2), pp. 299–311, DOI: 10.1080/13552074.2018.1473226.

4. The gender gap in information systems service organizations in Korea: a contextual hierarchy perspective

Gyeung-min Kim, Namjae Cho and Hee-Sun Kim

INTRODUCTION

Despite observed increases over recent decades in women's participation in the economic, and political dimensions of activities in the Republic of Korea, it still ranks 108th out of 153 countries in the global gender gap index 2020 announced by the World Economic Forum (WEF, 2021). In addition, the latest government report shows that the role of women in executive and administrative activities is still limited (KMOGF, 2018). This is also the case with the information systems (IS) industry in Korea, which is considered to be a new genre of business, creating more expected opportunities for female workers.

Although the size of the IS industry has grown, with a shortage of highly qualified IS professionals, women in the Korean IS workforce are largely under-represented. This applies in particular to the middle to upper levels of organizational hierarchy. The general notion is that workload pressure and disruptive work–family balance are the main factors in relation to female under-representation (KMOGF, 2018).

Kim and Cho (2018) investigated the IS gender gap problem in Korea more systematically from the perspective of the IS industry's project-based work practices, based on interview data collected in 2010. They analyzed how the IS industry's project-based work practices influence the problem of female under-representation. An IS project is typically characterized as challenging because it has to accomplish a project goal while also managing the limited resources allocated to the project. The characteristics of the work lives of IS project members are described as a high-level of time pressure, a disrupted work–family balance, and negotiations concerning scope, tasks, and responsibilities (Lindgren and Packendorff, 2006). According to Kim and Cho's study results, male bonding after work, group culture, and disrupted work–family balance were major factors accounting for female under-representation in that part of the IS industry that is engaged in project-based work practices. This situation was considered to be slightly worse in big IS service companies, than in small to medium-sized organizations.

By including more interviews with female employees in IS service firms in Korea, this chapter aims to extend the previous research by Kim and Cho. The chapter aims to investigate the following research questions:

1. How have project-based work practices in the Korean IS service industry evolved in terms of the Korean understanding of masculinity/femininity in work lives since the 2010 interviews.
2. In relation to its encompassing industrial and social contexts, how has the nature of the IS service business created opportunities for and constraints on the careers of female professionals since the 2010 interviews?

The recent spread of cloud computing has enabled organizations to deliver IS by putting technology components of cloud service providers together (Sebastian et al., 2017) instead of outsider-led (i.e., IS service companies) project-based IS development (Krancher et al., 2018). However, due to regulations and compliance issues, a project-based approach is extensively used especially by organizations in the IS service industry of which the major business is to provide client organizations with the development and management services related to IS (Humble and Molesky, 2011; COBIT, 2007).

Organizational practice is a socially constructed outcome emerging from a complex inter-action between the organization and outer entities situated in the whole economy and society. Business practices are influenced by the environmental differences among countries in terms of cultural, political, economic, and legal systems (Hills, 2006). Since the last interviews performed in 2010 (Kim and Cho, 2018), the industrial and social contexts of the IS service businesses have changed. For example, in 2013, a regulation forbidding big IS service firms from being hired to carry out public IS projects was introduced. In 2018, the Korean govern-ment enacted a 52-hour work system law: workers are prohibited from working more than 52 hours per week in companies with over 300 employees. Although Korean society has been deeply rooted in Confucian ethics – men in public, women at home – over hundreds of years since the Yi dynasty was established in 1392, the level of consciousness in the sense of equal gender rights has increased since the book titled *Kim Jiyoung, Born 1982* by Cho Nam-Joo was published in 2018. This book deals with inequality among men and women in Korea and shares a present-day view of the collective plight of women. The book has also been nomi-nated as "The 100 Must-Read Books of 2020" by the *New York Times*.

In this chapter, the research questions are explored from a gender-construction perspective using narrative analysis of managers and employees in multiple organizations in Korea. This chapter maintains that a gender perspective conveyed in language, gesture, and all types of symbolic social signs is historically constructed, both socially and culturally shared, and acknowledged by others (Tyler and Cohen, 2008). The chapter also maintains the contex-tual hierarchy perspective (Figure 4.1), which is crucial in understanding the emergence of business practices. In the contextual hierarchy perspective, organizational culture and practice influence task characteristics in the workplace. Organizational culture itself cannot be independent from industrial and market contexts including customer organizations' work practices. The industrial practice is again impacted by the social context, which, in turn, is influenced by the historical context.

In order to investigate how masculinity in project-based IS service firms in Korea has changed since 2010, we first review previous research regarding recent changes in system development environment, the project-based work, gender construction, and the Confucian context. We then discuss the research methodology and interview data, which are followed by data analysis and conclusions.

BACKGROUND

Changes in the System Development Environment

The recent popularity and rapid diffusion of cloud computing in forms of Platform-as-a-Service (PaaS), and Infrastructure-as-a-Service (IaaS) are changing the system development environ-

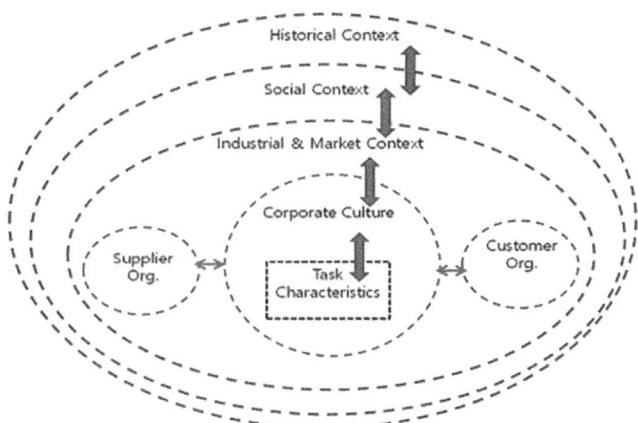

Source: Adapted from Kim and Cho (2018).

Figure 4.1   Contextual hierarchy and business practice

ment (Farnell, 2019). The characteristics of cloud computing are described as easy set-up of development environment, instant deployment of code and rapid elasticity (Krancher et al., 2018). In the cloud environment, Agile/DevOps is used as a development/delivery method to accomplish short lead time and fast feedback. Agile/DevOps provides lean, collaborative and cross-functional software development and deployment methods in the cloud computing environment. While Agile focuses on developing and experimenting with a lean system to accommodate constantly changing user requirements, DevOps focuses on bringing in software engineering experts from software development (the dev) side and operation personnel from IS operation (the ops) side together for continuous system operation and upgrades.

Unlike the traditional methodology consisting of relatively serial tasks such as requirement definitions, analysis, design, development, and deployment, Agile/DevOps is characterized not only as involving operation in the design, development and deployment but also as giving developers operational responsibilities: "representatives from IS operations should attend applicable inceptions, retrospectives, planning meetings, and showcases of project teams. Meanwhile, developers should rotate through operations teams, and representatives from project teams should have regular meetings with the IS operation people" (Humble and Molesky, 2011, p. 7).

While traditional system development takes a project approach, Agile/DevOps takes a service-oriented approach – build, run and change service like a product. In Agile/DevOps, the cost of building and running the service as well as the value of the service to the organization are calculated. The key for Agile/DevOps is to form a cross-functional team to define shared responsibilities and improve collaboration for the overall quality of the service delivered to the customers (Humble and Molesky, 2011).

Software development in the cloud environment is described as collective learning (Krancher and Dibbern, 2015) and self-organizing activities (Humble and Molesky, 2011; Vidgen and Wang, 2009; Schwaber, 2004), rather than outsiders directing the software development teams.

Project Approach in the IS Service Industry

Despite the increasing popularity of cloud computing, the traditional development approach is still widely used, especially for public IS projects where most IS service companies are focused. This is due to regulations and industry standards. Segregation of duties to reduce fraud and errors leads to separation of roles and responsibilities to prevent a single person from compromising the process, thus separating development, operation, and support functions (COBIT, 2007). Often, these tasks are even performed by separate organizations.

A project-based structure is defined in terms of project goals and requirements, the activities of each project phase, deliverables at the end of each phase, resources and time requirements of each phase, and project members' responsibilities (Valacich and DeLuca, 2006). By drafting its members across organizational boundaries both internally and externally, the aim of the project-based structure is to deliver one-off assignments with strict control over time, resources, and quality. The project manager monitors and controls the progress of project members throughout the project life cycle, on the basis of task assignments and deadlines.

Project management is thought to be challenging, in contrast to ordinary work practices because project goals and efficiency need to be pursued simultaneously (Pinto, 1996; Gill, 2002). Progress monitoring to keep the project on schedule and ensuring the quality of project deliverables requires each team leader to evaluate and appraise each team member based on progress reports and pre-designed standards (Hodgson, 2004). Often, changes in task assignments result in changes in personnel and feedback to a team member's supervisor (Murray and Crandall, 2006; Metclaf, 1997).

The performance orientation of modern business organizations along with the popularity of cloud computing treats IS services as a "commodity" or "utility" at a lower price. To reduce these problems, US firms have been offshoring system development and operation for cost reduction associated with workforce, focusing on value-added system analysis and design domestically (Yoo, 2018). However, Korean IS service firms dominated by subsidiaries of large conglomerates are accustomed to providing services to the family of companies within the conglomerate without separating analysis and design from development and operation, which makes it difficult to reduce IS project costs. As a result, Korean IS service firms must cope with IS projects with tighter budgets with lower head counts due to higher costs related to the domestic workforce (Yoo, 2018).

Masculinity in IS Project Management

Bureaucracy has been characterized by rigidity, stability, and inefficiency. In contrast, project management is regarded as "post-bureaucratic", with work organization methods based on efficiency, empowerment, and autonomy (du Gay and Salaman, 1992; Hodgson, 2004). However, project management is also thought to be a bureaucratic system of control based on the principles of visibility, predictability, and accountability (Hodgson, 2004; COBIT, 2007; Humble and Molesky, 2011). Through clear definitions of roles and responsibilities and oversight mechanisms such as deadlines, procedures, standards, and inspection, project management is regarded as a bureaucratic system of control.

The instrumentality (serving as a means of pursuing a goal) and performance orientation (achieving the goal with fewer resources) are thought to be typical masculine business practices. Such masculine perspectives tend to regard stereotypical behavior of women – caring and spending time with minorities – as non-economic, inefficient, and resource-wasting managerial behavior. Historically, the majority of those in management positions tend to be men. Thus, their stereotypical behaviors for success such as devotion to work, winning in a competitive context, controlling subordinates, commitment, and a conquest orientation are set as behavioral exemplars in organizations (Acker, 1992; Gherardi, 1994; Martin, 2001; Kerfoot and Knights, 1993; Knights and Tullberg, 2011). Sustaining male bonding, displaying authoritative expertise, thinking rationally, and taking risks are also considered to be masculine traits (Knights and Kerfoot, 2004; Tyler and Cohen, 2008; Knights and Tullberg, 2011). If women are to reach and maintain leadership positions, they are expected to adopt these masculine traits (Pullen and Knights, 2007; Wajcman, 1998; Knights and Tullberg, 2011). Women's traditional child-bearing and family-caring duties are considered as preventing long working hours and devotion to work that are seen as conforming to masculine behaviors. To achieve career success, female managers are expected to act like men (Gore, 1993).

The notion of masculinity being above femininity in the hierarchical order is thought to be historically constructed and socially and culturally shared. Acker (1992) asserts that "the law, politics, religion, the academy, the state, and the economy, are institutions historically developed by men, currently dominated by men, and symbolically interpreted from the standpoint of men in leading positions, both in the present and historically" (p. 567). Gender hierarchy is present "in the processes, practices, images and ideologies, and distributions of power in the organization" (Acker, 1992, p. 567). Thus, gender hierarchy is deeply institutionalized and difficult to disrupt (Acker, 1992; Martin, 2001).

Confucian Ethics in Korean Cultural Background

Korean culture is deeply rooted in the culture of Confucius. Confucian philosophy was adopted as the formal national mission by the founding king of the Yi Dynasty in 1392. Since then, the philosophy has been maintained and has evolved over five hundred years up to the time of modern Korea. When Confucian philosophy was adopted by the Yi Dynasty, it emphasized strict hierarchy centered on the king or emperor. This hierarchical view of the world was extended into daily life practices by government administrators, the ruling classes of the country, and the dominant gender of the society. Recommended ethical practices included clear role models according to gender. For example, boys and girls were not supposed to be in mixed groups after the age of seven. A woman was supposed to show wholehearted obedience to her father before marriage, to her husband after marriage, and to her sons after her husband's death.

The gender inequality problem in Korea was systematically investigated in a study of 130 female college students with an average age of 22 (Kim and Kim, 2005). The study revealed that certain behaviors are allowed for men, but not for women. Examples of those behaviors included smoking in public and swearing loudly. When women violated codes of behavior, they were treated more negatively than men and felt ashamed and uncomfortable. Typically, parents at home educate daughters more strictly than sons as regards drinking and smoking.

The problem of gender inequality in Korea has been publicized dramatically since the publication of *Kim Jiyoung, Born 1982*. Later, the movie version of this fictional book was released in 2019. Jiyoung had to cope with her father's and later her mother-in-law's subtle unequal treatment toward their daughter and daughter-in-law, respectively. Since inequality among men and women were socially acceptable and prevalent throughout Korean society, she accepted it as an inevitable reality throughout her life. Although Jiyoung was working for a firm as a dual career employee, she had to quit her job in pursuit of her child-caring duties once the couple's child was born. Her obedience to conventions rooted in Confucian ethics, eventually made her give up her career dream and made her mentally ill. Kim Jiyoung's story, shows how women in Korea had to deal with inequality in their everyday lives at home and at work. It shows how Korean women are raised to be obedient (in unequal relationships between male and female, parent and child, old and young) and thus to be "good girls", having to struggle with themselves in silence at home and work. Consequently, some become mentally ill.

Among many episodes in *Kim Jiyoung, Born 1982*, her story shows how Jiyoung struggled when she had to quit her job due to women's child raising responsibilities. Confucian ethics – women at home and men in public – continue to be prevalent in Korean society and still apply to women to different degrees. Career women are considered to be obligated to carry out household chores and to manage their children's growth and even their careers. In this regard, women constantly weigh up the cost and benefits of working. They often end up quitting their job if the costs of working outweigh the benefits.

Research Methods

To address the research questions, first, through a literature review, initial themes to be used in the interview were identified. Second, after interviewing managers in multiple organizations within the IS service sector in 2021, the 2021 results were compared with the 2010 interview results. A literature base allows us to identify certain narratives that have significant potential for understanding gender construction in the project-based IS service industry. Finally, different narratives related to gender construction are extracted by means of thematic analysis. Thematic analysis is a combination of the deductive and inductive approaches in which theoretical themes are formed deductively as a framework to extract specific narratives inductively (Boje, 2001).

We sought stories from the following points of view: organizations' views on female professionals, views on work in association with family life, and views on organizational policy. These views are essential to find out the masculine discourses in the IS service industry. In 2021, three informants – an employee (Lucy), a mid-level manager (Sunny) and a senior manager (Judy) – were added. With the exception of the employee, the informants have extensive experience participating in projects and managing employees in IS service organizations. The data was collected via telephone conversations. Based on the transcribed documentation, we extracted narratives related to masculinities in project-based IS work.

By combining 2021 interview data with 2010 data, we have a total of eight informants. Six of the eight are from IS service organizations which are subsidiaries of large conglomerates. One executive manager, Sue, was from D corporation, which used to be a subsidiary of a large conglomerate, but later became an independent company. Another manager Judy worked in the IS service sector and is currently working with a global IS consulting company as a senior manager. Thus, we used data collected from eight total informants for data analysis. The summary of the informants is shown in Table 4.1. The narratives of three male managers interviewed in 2010 are not quoted in this chapter. However, their narratives provided male managers' points of view and are reflected in the previous study published in (Kim and Cho, 2018).

Table 4.1   *Overview of companies and interviewees*

Year of Interview	Company	Type	Interviewee Name/Position	
2021	DX	subsidiaries of conglomerate	Sunny/Manager	
2021	L	subsidiaries of conglomerate	Lucy/Former employee	
2021	K	global IS consulting firm	Judy/senior manager	
2010	D	independent company (formerly, subsidiary of conglomerate)	Sue/an executive manager	
2010	L	subsidiaries of conglomerate	Suzan/an executive manager	
2010	S	subsidiaries of conglomerate	Sam/Manager	Not quoted here
2010	C	subsidiaries of conglomerate	Tom/Manager	Not quoted here
2010	DX	subsidiaries of conglomerate	Mike/ manager	Not quoted here

NARRATIVES ABOUT GENDER ISSUES IN THE KOREAN IS SERVICE INDUSTRY

In their career paths, employees in the IS service sector can choose to be in the IS project team, operational support, or back-office support such as quality control and technical manual writing. Participating in the IS project team is considered to be demanding, requiring long hours of work, adjusting to group culture, and often working in the client site. To remain on a successful career path to be a project manager or a sales manager, an employee needs to stay in the IS project team. In fact, after a certain point, the career path available to employees is either to become a project manager or a sales manager. Within an IS project team, entry-level employees are initially involved with software development; later they move up to software analysis and design (SAD hereinafter); and they ultimately become a project managers.

Female Ambition

Being a subsidiary of Conglomerate, DX Company has many female-friendly policies such as flexible hours. By law, companies with more than 500 employees or 300 female employees must have a child day-care center within the company. Since implementing female-friendly policies including day-care centers are costly, firms with more financial resources such as a subsidiary of a conglomerate can afford the implementation. Yet, DX company has only one female executive manager. While the female ratio at the entry level is 37 percent, female ratios at assistant manager, manager, middle level and senior level managers are, 30, 24, 20 and 6.4 percent respectively. Sunny, a female manager in DX Company talks about her experiences in job interviews for new employees:

> In the job interviews, most interviewers give higher marks on female candidates than male candidates since the problem-solving skills of the female candidates are more splendid than male candidates. In the recent workshop hosted by HR department, managers expressed their difficulties to assign female workers to important positions. This is based on their experiences in the past with the female workers who use a-day-off vacation all of a sudden to take care of children. As a female manager who raises a child, I cannot deny the managers' concerns. Still in Korea, the child raising responsibility is on female even though the female is working. (Sunny)

The subject of the workshop hosted by the human resources (HR) department was why the number of male employees is as much as 3.4 times more than female employees in software analysis and design (SAD). In the career path, SAD positions are higher paid and more prestigious than software development and operation. Female employees join the IS workforce with high levels of ability and ambition, but their ambition seems to dwindle following marriage and child birth.

This is consistent with the following interviews done in 2010 with Susan, a female executive of L Corporation, an IS service subsidiary in Korea's leading conglomerate (Kim and Cho, 2018, p. 78):

> Female professionals are very capable and show good performance in the company. However, once they reach at a certain point of their career (i.e. the director of a department), they usually quit for child caring. We try to provide several motivations to keep them, but such measures are not yet very helpful in persuading them. (Suzan)

In a 2021 interview, a former employee of L Corporation, Lucy, talks about a similar story about herself:

> L corporation where female workforce is 35 percent of the entire employees, is reputable as a good workplace for women. At the beginning of my career, I was ambitious for my work, but I quit L corporation due to child care; and now work for a different company as a part-timer. After I quit my job, my self-esteem becomes very low. (Lucy)

Marriage and Child Care

To have a successful career in an IS service company, an employee needs to be in the project team. After marriage, female employees tend to prefer back-office support or operation, which guarantees normal working hours and in-office work. Several interviewees pointed out long working hours as industry conditions, and noted that these are very difficult to endure after marriage:

> I would have liked to be in the project team, if I had not had a child. Before marriage, many female employees actually prefer to be in the project team. The work in the project is more dynamic and fun. (Sunny)

> Due to industry culture, the employees in the project team have to work late at night and solve client's unexpected problems when clients make requests. However, child care often requires a short-term vacation, which makes the company difficult to deal with the client's requests. For capable female workers to continue their career, child care issues have to be social and enterprise issues rather than individual issues. (Sunny)

When Lucy was in quality control, she could balance work and family. After she was assigned to SI, everything changed:

> It was impossible for me to continue my career in IS industry unless my parents help me to raise kids. Often times, we do not know what time we can go home. Because we do not know when the work to be done. (Lucy).

Lucy understood that PM or the client organization could not help her to solve the problem. When the problem of raising her child became serious, she finally quit her job:

My daughter came home at 2 pm. Helpers to take care of her got changed frequently. When my daughter became ill with depression and the doctor advised me to find someone to stay with her, I quit my job. I was so fearful and hopeless. I wanted to keep my job desperately but I could not. Now I work as a part timer. (Lucy)

Judy is an executive manager in a business consulting company. She has moved to the current company from an IS service company:

The duration of IS projects involving system development is much longer than other business consulting project. This means the duration of overnight work for an IS project is much longer than for a business consulting project. This becomes a major obstacle for female workers to enter and stay in the IS service sector. (Judy)

Changes in Industry Practices

In IS service companies, after a certain point, becoming a project manager or sales manager represents a successful career path for employees. According to the interview done in 2010, male bonding including drinking sessions at night was necessary to keep a tight relationships with client companies. This was one of the major reasons for female employees to quit their jobs. For example, the following quotes come from 2010 interviews (Kim and Cho, 2018, p. 77):

Once you reach a certain point of your career, the career path available becomes either PM or sales. There used to be female professionals in the sales department, but now they all left. If you made a choice to become a PM, you are expected to keep a close relationship with clients and have lots of drinking sessions with them. Female professionals can't handle this. Thus, their job assignments are limited to certain areas. There are so many problems. Too much … Women can't sustain for a long period. (Sue)

The people who place an order in the client firm, are 40s or over age of 35. They don't want a female PM counterpart. (Sue)

Due to the 52-hour work system law enacted in 2018, the drink culture is disappearing. After hours drinking sessions were considered to be an unofficial part of the job assignment. However, long working hours still remain an industry condition as they were in 2010.

In the 2010 interviews, tight controls with regard to time and budget made it difficult for female employees to work in IS projects. Such problems got worse for public IS projects that used a national IS evaluation system adopted in 1997 for maximum performance of such investment in the public IS projects (Kim and Cho, 2018). In 2010 interviews, the public sector was looking for a company that could do carry out public IT projects with 80 percent of the cost of the original budget. Thus, firms that were part of a conglomerate bid at a much lower rates than the announced budget because they had enough revenue from servicing member firms of the conglomerate.

This problem has been reduced to a certain degrees since the 2013 regulation forbidding big IS firms from participating in public IS projects. However, the budget estimation system based on head counts instead of complexity of user requirements, still makes it difficult for female workers to adopt flexible work hours:

> Since the project needs to maintain a certain number of head-counts, female employees cannot use flexible hours to go home early. The project even maintains a log book that records in and out times for employees. (Sunny)

Organizational Gender Policy

In the 2010 interviews, many companies recognized the importance of female employees and institutional support such as flexible work, telecommuting, and day care. However, such policies were of limited value in practice (Kim and Cho, 2018). Despite such policies being in place over ten years now, female employees still view the support programs as being of limited practical value:

> Although many companies have flexible work hours, it is impossible to use that benefit. The reasons are industry characteristics to work in the client site and handle the client's issues when the client makes requests. Thirty percent of female employees are not returning to the company after child-care-year because they cannot find care givers. Finding child care that operates during the working hours of the company, is difficult. Once the facility is found, a waiting line is actually a few years. If employees' parents cannot help, female employees have to quit the job. (Sunny).

Lucy's previous workplace had a very good policy such as a child-care center, flexible hours and job rotation. However, when she was assigned to a project team, those policies were useless:

> In order to use those policies, client firms as well as society in general must change as well. When I started work as a part-timer, my self-esteem becomes very low. However, I think this is the best for my daughter. (Lucy)

Despite all the programs to support female employees, child care still remains the major reason why female employees in the IS service sector leave work.

Integration of the Narratives

The integration of the narratives helps reveal an overall picture of masculine characteristics in the Korean IS service industry. The analysis of 2021 interview results shows some similarities with the 2010 interviews in that to be on a successful career track, an employee needs to be in the IS project team, work late at night and adhere to the group culture. All the members of the project team need to remain at the project site when system integration and tests are performed as a part of system development. Not being able to be a part of IS project teams due to child care and work family balance excluded the female employees from a successful career path. In 2021, there were still masculine organizational environments characterized by long hours of work, and these play a role in excluding female employees from a successful career path and preventing them from performing to their highest potential. Unlike the 2010 study, the 2021 interview results reveal that late-night drinking sessions to maintain close relationships with clients are disappearing due to the 52-hour work system law enacted in 2018.

The masculine characteristics of the IS service industry are still intensified due to an economic factor, as well as the social and historic factor. From the economic perspective, the problems associated with the subsidiary firm of a conglomerate bidding for public IS projects at much lower rates than the announced budget has been reduced to a degree following the

Table 4.2   *Summary of 2021 interview results compared to 2010*

How masculinity in project-based IS service organizations in Korea has been changed?		
Practices associated with female under-representation	2010	2021
Head-count based budget system	Present	Present
Long working hours	Present	Present
Group culture	Present	Present
Male bonding after work	Present	Disappearing
Gap between the institutional policies and the reality of female professionals	Present	Present
Disrupted work–family balance	Present	Present
Deviation from successful career path	Present	Present

2013 regulation forbidding big SI firms from participating in public IS projects. However, most IS projects still use head counts for budget estimation, which makes it difficult for female workers to benefit from flexible work hours.

From social and historic perspectives, although there has been some improvement, Confucian ethics are still prevalent, and women in modern Korea still have to deal with inequality: men in public affairs and women in domestic affairs at home. In addition, Confucian ethics still separate work concerns from personal needs, which prevents female professionals from speaking out publicly about help with family care, child care, and work–life balance. This can eventually result in mental turmoil for female professionals when they quit their jobs due to their child-raising responsibilities. Recognizing the importance of female professionals, organizations have launched various programs such as flexible work options, telecommuting, and day care to provide support for women in their roles as mothers. But only a few female workers use such programs because of the significant gap between the institutional policies and the realities faced by female professionals. The results of the 2021 interviews compared with the 2010 results are shown in Table 4.2.

CONCLUSION

Despite the growth of the IS industry, women in the Korean IS workforce are largely under-represented especially in the middle to upper levels of organizational hierarchy. The contribution of this research is to analyze the female under-representation problem in the IS industry systematically by investigating the impact of the project-based work practice of the Information Systems service industry on the work lives and the career development of female professionals.

In both 2010 and 2021 interviews, we found that masculine discourses in the IS service industry – head-count based budget systems, more control, long working hours, adhering to group culture and a deviation from successful career path – are continuously repeated and reproduced. Every time a new project is launched in this way, it implies another instance of the masculinization of the work lives of female employees. The consequences of gender stratification in IS projects continue to persist. The results presented in this chapter show that masculine practices influence female employees negatively to give up their ambition and may also result in low self-esteem. The chapter also shows that masculine practices influence the formation of management identity in the IS service sector. HR managers are hesitant to assign female employees to important positions. These results are consistent with previous findings (Martin, 2001; Alvesson and Willmott, 2002).

Although the nature of the IS service industry and competition continuously require maintaining a close relationship with both clients and potential clients for sales negotiations and to achieve success, night drinking with clients is disappearing due to the 52-hour work system law enacted in 2018. The limitation of the interviews carried out in 2021 is that they do not reflect the views of both female and male managers equally. While we interviewed two male managers in 2010, we could not interview male managers in 2021 due to time constraints. However, we could reflect the results of a recent HR workshop in DX Company which are consistent with the views of male managers in 2010. Thus, one limitation is that our findings cannot be generalized beyond the managers reported here. Future research can be carried out to understand masculine practices in the IS service sector more systematically and provide a foundation for further improvements to gender policy.

REFERENCES

Acker, J. (1992). From sex roles to gendered institutions. *Contemporary Sociology*, 21, pp. 565–68.

Acker, S. (1990). Hierarchies, jobs bodies: A theory of gendered organizations. *Gender and Society*, 4(2), 1990, pp. 139–58.

Alvesson, M., and Willmott, H. (2002). Identity regulation as organizational control: Producing the appropriate individual. *Journal of Management Studies*, 39(5), pp. 619–44.

Boje, D. (2001). *Narrative Methods for Organizational and Communication Research*. London: Sage.

Carayon, P., Schoepke, P., Hoonakker, M., Haims, M., and Brunette, M. (2006). Evaluating causes and consequences of turnover intention among IS workers: The development of a questionnaire survey. *Behavior and Information Technology*, 25(5), pp. 381–97.

Collinson, D., and J. Hearn, J. (2005). Men and masculinities in work, organizations, and management. In J. Kimmel, J. Hearn and R.W. Connell (Eds.). *Handbook of Studies on Men and Masculinities* (pp. 289–310). Thousand Oaks: Sage.

COBIT (2007). COBIT 4.1. IT Governance Institute.

du Gay, P., and Salaman, G. (1992). The culture of the customer. *Journal of Management Studies*, 29(5), 615–33.

Farnell, I. (2019). Will there be a dominant IIoT cloud platform? https://il.farnell.com/will-there-be-a -dominant-iiot-cloud-platform.

Gherardi, S. (1994). The gender we think, the gender we do in our everyday organizational lives. *Human Relations*, 47(6), pp. 591–610.

Gill, R. (2002). Cool, creative and egalitarian? Exploring gender in project-based new media work in Europe. *Information, Communication and Society*, 5(1), pp. 70–89.

Gore, J. (1993). *The Struggle for Pedagogies: Critical and Feminist Discourses as Regimes of Truth*. New York: Routledge.

Hodgson, D. (2004). Project work: The legacy of bureaucratic control in the post-bureaucratic organization. *Organization*, 11 (1), pp. 81–100.

Humble, J., and Molesky, J. (2011). Why enterprises must adopt devops to enable continuous delivery. *Cutter IT Journal*, 24(8), pp. 6–12.

Kerfoot, D., and Knights, D. (1993). Management, manipulation and masculinity: From paternalism to corporate strategy in financial services. *Journal of Management Studies*, 30(4), pp. 659–77.

Kerfoot, D., and Knights, D. (1996). The best is yet to come? Searching for embodiment in management. In D. Collinson and J. Hearn (Eds.). *Masculinity and Management*. London: Sage.

Kim, G., and Cho, N. (2108). Masculinity in project-based IS service organizations. *Asian Journal of Information and Communications*, 10(2), pp. 70–84.

Kim, J., and Kim, K. (2005). Preliminary study of psychological function and development of social norms for women in Korean society. *Korean Journal of Woman Psychology*, 10(2), pp. 157–71.

KMGEF (2018). *Report on Rate of Female Executives in 500 Korean Firms*. Korea Ministry of Gender Equality and Family.

Knights, D., and Kerfoot, D. (2004). Between representations and subjectivity: Gender binaries and the politics of organizational transformation. *Gender, Work and Organization*, 11(4), pp. 430–54.

Knights, D., and Tullberg, M. (2011). Managing masculinity/mismanaging the corporation. *Organization*, 19(4), pp. 385–404.

Krancher, O., and Dibbern, J. (2015). *Knowledge in software-maintenance outsourcing projects: Beyond integration of business and technical knowledge*. The 48th Hawaii International Conference on System Sciences, Kauai, HI.

Lindgren, M., and Packendorff, J. (2006). What's new in new forms of organizing? On the construction of gender in project-based work. *Journal of Management Studies*, 43(4), pp. 841–66.

Martin, P. (2001). Mobilizing masculinities: women's experiences of men at work. *Organization*, 8(4), pp. 587–618.

Murray, M., and Crandall, R. (2006). IT offshore outsourcing requires a project management approach. *SAM Advanced Management Journal*, 71(1), pp. 4–12.

Pink, D. (2005). *A Whole New Mind*. New York: Riverhead Books.

Pinto, J. (1996). *Power and Politics in Project Management*. Sylva: PMI.

Pullen, A., and Knights, D. (2007). Editorial: Undoing gender: Organizing and disorganizing performance. *Gender, Work and Organization*, 14(6), 505–11.

Rajah, K., and Jones, L. (2007). Creativity, enterprise and socio-economic change. In K. Rajah (Eds.). *From Complex Creativity: The Pathway to Opportunity Finding* (pp. 21–86). Greenwich University Press.

Schwaber, K. (2004). *Agile Project Management with Scrum*. Redmond: Microsoft Press.

Sebastian, I., Moloney, K., Ross, J., Fonstad, N., Beath, C., and Mocker, M. (2017). How big old companies navigate digital transformation. *MIS Quarterly Executive*, 16(3), pp. 197–213.

Tyler, M., and Cohen, L. (2008). Management in/as comic relief: Queer theory and gender performativity in the office. *Gender, Work and Organization*, 15(2), pp. 113–32.

Valacich, J., and DeLuca, D. (2006). Virtual teams in and out of synchronicity. *Information Technology and People*, 19(4), pp. 323–44.

Vidgen, R., and Wang, X. (2009). Coevolving systems and the organization of agile software development. *Information Systems Research*, 20(3), pp. 355–76.

Wajcman, J. (1998). *Managing Like a Man: Women and Men in Corporate Management*. University Park, PA: Pennsylvania State University Press.

WEF (2021). *Global Gender Gap Report*. World Economic Forum.

Yoo, H. (2018). Opportunities and directions of Korean IT service industry. *SPRI*, January 19.

5.   The FESTA project: doing gender equality work in STEM faculties in Europe

Minna Salminen-Karlsson

INTRODUCTION

This chapter presents the FESTA project, which worked to advance gender equality in STEM departments in seven European countries, by improving the working environment of mainly junior researchers. The aim is to showcase organizational change work in project form to advance women's careers in different national environments. FESTA stands for "Female Empowerment in Science and Technology Academia" and was a gender equality project financed by the European Union (EU) between 2012 and 2016. It included six university partners from Sweden, Denmark, Ireland, Germany, Bulgaria and Turkey and a research institute from Italy.

The prevailing EU ambition at that time was to help women to become professors, by measures directed towards women. Instead, the FESTA project set out to improve the working environment of junior researchers in STEM, in areas where previous research had shown that women had particular disadvantages. Our project team argued that this was a necessary condition for keeping women in the academy. We also argued that projects aimed at helping women were not a sustainable strategy, while a project that aimed at changing organizational structures and cultures was more so. As we expressed it in the proposal:

> Our starting point is the conviction that it is not women who are lacking essential characteristics for being good researchers, but that the academic environment is lacking essential characteristics for fostering the research potential of women – or, rather, anybody who does not fit the traditional norm of for example being available 24 hours, seven days a week, always putting the job first and perceiving research being about publications rather than discoveries.

All this may sound self-evident today, but it was not so in 2011 when we wrote our proposal. At that time the EU perception of the women in STEM problem was just moving from looking at individual women to looking at structures and processes (FESTA being a small part of this shift), to later move into also looking at gender in research content.

Our explicit reason for working with science and technology was that the need was most acute in that area. Our approach of working with the research environment was also fitting in that context. Many STEM environments are very male dominated and, therefore, pose particular problems. By working with the environment, we wanted to improve the situation not only for those staff members who were identified as women, but also others who did not easily adapt to the prevailing masculine context. Our second argument for working with the environment was a reaction against the prevailing discourse where "excellence" is a (never defined) characteristic that is to be praised and rewarded and where the most valoured researcher is the "single genius". In contrast, we wanted to remind people of the fact that most researchers do their teaching and research as parts of teams, in particular in STEM. It is their work that in the end produces many of the research results. O'Connor et al. (2018) find that among women who strive for careers either by neglecting relations in the private sphere or by doing constant balancing between work and other relationships, there are also women who like their work as STEM academics, but rather just do the work because they enjoy it, even if it would not forward their careers. Research has not particularly paid attention to these women. While we wanted to facilitate more women becoming research leaders, we also argued that these women, who may never become leaders, deserve a gender-equal workplace that assures them that their choice of STEM was the right one. And finally, we thought strategically about the resistance that we knew we would meet and launched the project with the motto that a better working environment for women is a better working environment for all.

We were also very clear that we wanted to work at a low institutional level, the daily working environment of junior researchers. On the basis of both research and experience we knew that junior researchers base their decisions for continuing or not continuing their research and applying or not applying for grants and promotions on their everyday experiences. At least this is the case with many female researchers, and everyday experiences are a contributing reason for them to leave the academic setting (Cabay et al., 2018). We also argued for diversity and that for women to bring in diversity, their voices needed to be heard. We saw this being particularly important in STEM disciplines, not only because there is currently a male majority, but also because science and especially engineering have historically been constructed as masculine. It is not only the image of STEM that is masculine, STEM itself is masculine. The teaching of STEM disciplines still carries a ballast from the time when male professors taught male students (Ottemo et al., 2021; Gunter et al., 2021) and STEM research questions are largely formed and evaluated by men (Bautista-Puig et al., 2019). Consequently, the aim of FESTA was to empower the young women who actually have chosen STEM not just to be present in STEM institutions with their female bodies, but also to be able to influence STEM preferably as early as possible and in the long run making careers that would give them influential positions.

The chapter continues as follows. First, as a background, the framework of a European project is explained. The diversity of the European academic scene when it comes to gender equality in the STEM disciplines is exemplified by differences between the FESTA countries. After this background, the presentation of the FESTA project starts with a presentation of the partners and the nine FESTA actions. Then, four of the FESTA actions are described more in detail: (1) Collecting and using institutional data to increase awareness; (2) PhD supervision; (3) informal decision-making processes; and (4) resistance. The chapter ends with general experiences on conducting gender equality work in STEM environments.

BACKGROUND

Projects as Gender Equality Tools in the EU

Since the early 2000s the European Union has financed three to four gender equality projects every year, to start or improve gender equality work in single higher education institutions and at the same time gather knowledge and find best practices. GenPort, a project created to gather material on gender equality in higher education, lists some 50 projects where a number of universities and research institutions in different countries have collaborated, with about 15 of these having a STEM focus.[1] Many of the projects have also produced publicly available reports. There is a wealth of experience – actually enough to get lost in all the material. As so much is produced, much also goes unnoticed. In spite of all this wealth, the EU has not arranged any coordination function to help newcomers to the field to navigate the terrain that has already been mapped.

A European project is not really one project, but a number of different projects in different institutions, where hopefully participants from different countries learn from each other and support each other. Such projects can be complicated to keep together. The EU requires that some documents are produced by the whole consortium, but this can also be done by each member producing their own pieces and then somebody putting it all together. Even if English is the common language used in the projects, participants' proficiency in English varies, which influences interactions in meetings. However, for many partners, particularly those who come from environments with little previous gender equality work, having contacts across Europe is inspiring.

In many of the projects, not least the STEM-related ones, the main actors are still female researchers, who drive the project alongside their own research. Nominally, somebody high up in the university hierarchy needs to sign the agreement with the EU, but the work is done by somebody who, because of the project money, now gets paid for (at least some of) the work she does, but who could do research instead. She may help female researchers' careers in the long run, but it is possible that she is stalling her own career. In addition, being a woman and trying to change a male-dominated organization is not an enviable position.

The composition of the project team (normally it is more than just one person) is crucial for achieving any success. Someone in a key position in the institution needs to be involved, actively and not just passively, to counteract the resistance which, according to our experience, will surface sooner or later (Jordão et al., 2020). In a STEM faculty this person normally needs to be male for two main reasons. First, not all women in STEM departments are accorded the same informal authority as their male colleagues and counterparts, regardless of their official position. For the second, even a woman who actually has a stable and respected position can find it difficult and may actually risk her position if she is perceived as being partial to her own gender. In an environment with such a majority of men as is the case in many STEM departments, it is practically impossible to achieve something without involving men.

Involving men in STEM gender equality teams has different motives than involving women in boards and committees in STEM institutions, and may also meet different obstacles. Finding women to be committee members may be difficult, when they are so few that they would be over-burdened by administrative work, if all decision-making bodies would have equal numbers of women and men. Getting equal numbers of men and women into gender equality work meets totally different difficulties. There are large numbers of men to share the work, but very few men are interested in participating. The effects are also different: while women in male-dominated committees need to be strategic in their talk to be heard (Baxter, 2011), our experience is that men in gender equality committees have quite a lot of voice. While being a member of institutional committees often looks good in a CV, the value of having done gender equality work is more questionable. But still, because in STEM faculties most decision-making positions are normally held by men, and because men in many of the European cultures more readily listen to other men (Fisher and Kinsey, 2014), involving men in gender equality work is crucial.

What Counts as Success

EU projects normally last three to four years. Sustainability – what will happen after the project ends – is an issue that has been properly addressed only in the last few years. It is an issue that is very difficult to address. Three to four years is a short time to change an organization. What counts as success depends on where you start. In a European project, starting points are very different. Institutions in some countries have done gender equality work for years, while it is a novelty in others. Societal gender equality cultures are very different – from very traditional to quite radical. Because the conditions differ so profoundly, what is doable in the institutional context varies greatly, and what counts as success in gender equality work should also vary. There is notable gender fatigue (Kelan, 2009) in some of the countries where gender equality work has been going on for a long time. There are also countries and institutions where gender equality work has been non-existent, and where enthusiasm and hopes can be high at the start of such work. From the perspective of a project partner suffering from gender fatigue, those hopes can seem to be too high, but they can spur the work and achievements.

An important effect of a European project for an institution situated in a very patriarchal society may be just discovering that there are problems that need to be addressed. That may take almost all of the project time. In that case the project may seem unsuccessful when it comes to end, as no improvement may appear to have taken place. However, the long-term effect is very difficult to assess at that point in time. It is possible that the ideas that have been sown will grow in the coming years. Unfortunately, the opposite is also possible: if the project, with the project funding, ends when that insight has developed, there is also a risk that the awareness that has been built up is not taken forward.

The Diversity of the European Scene

Approximately every two years the European Union publishes comprehensive statistics on gender in research and innovation, called She Figures. The figures in this section come from She Figures 2018 (European Commission, 2019), and they give an idea of the diversity of the European higher education scene. All the figures in this section come from there. The countries in the FESTA consortium are used here as an example of the differences.

It is somewhat misleading to group STEM disciplines together when it comes to gender balance. Looking at statistics at the European level there is a gender balance in natural sciences, mathematics and statistics at the PhD graduate level (46 percent women). Engineering, manufacturing and construction is much more problematic (29 percent women) and information and communication technologies even more so (21 percent women). To say that there are few women in STEM is only partly true. Among the FESTA countries Bulgaria and Italy have the best gender balance, while the Scandinavian countries Sweden and Denmark rank lowest. However, Stretenova (2011) argues that while the numbers are good at the PhD level, the glass ceiling is particularly hard in Eastern European countries with traditional gender regimes, as in the FESTA country Bulgaria. Bozzon et al. (2017) show that in the FESTA country Italy neoliberal university governance which has increased instability, precarity and competitiveness in academic careers has affected women to a greater extent than men. Hence, obstacles for women's careers after a PhD look different in different countries, but there are obstacles in most of them, not only because of policies in individual institutions but also because of policies at national levels.

When it comes to career progress, the STEM disciplines are diverse. Natural sciences show a different pattern from that of engineering and technology, at least when comparing She Figures, which give the percentage of women at the professor level in four of the FESTA countries. There is just no clear pattern: in Italy and Denmark the ratio of women professors to women PhD graduates is higher in science than in engineering, while it is the other way around in Sweden and Germany. While Sweden and Germany have the lowest percentage of female PhDs in engineering and technology, they have the highest ratio of female professors to female PhD students in this area. When it comes to science the situation is more mixed, with Italy – where gender equality work largely has been lacking – having the best ratio of female professors compared to female PhD students. All this indicates that (1) there are thus far unexplained differences between European countries when it comes to women's careers in academic science and engineering and (2) STEM is just not one concept, but contains large areas, science and mathematics and technology and engineering with different career possibilities. The "science" in STEM itself is not one thing, either. Biological and environmental sciences have either a female majority or a sizable female minority in all FESTA countries where She Figures data is available and, hence, numerical gender balance problems are restricted to physical sciences and mathematics. Generally, figures on women's underrepresentation depend on how areas are defined and therefore they should be read with forethought.

What makes the situation even more complicated is that being a professor does not mean the same in all European countries. Both status and remuneration differ between countries. For example, it is much more common to be a professor in Italy (25 per cent of male and 11 per cent of female academic staff are professors) than in Germany and Ireland (8 per cent of male and 3 per cent of female academic staff). That indicates that there is tougher competition for those positions in Germany and Ireland and, consequently, they may be invested with more status.

FESTA – ACTIONS AND LESSONS LEARNED

FESTA Partners

The FESTA teams were located in Sweden, Denmark, Ireland, Germany, Italy, Bulgaria and Turkey (which does not belong in the EU but is considered an associated country). Four were large comprehensive universities, where the project worked with the science/technology faculty (Sweden, Denmark, Ireland and Bulgaria), two were large technical universities (Germany and Turkey) and one was a research institute (Italy). The aim had been to found a consortium with partners from all corners of Europe, as that was seen as a strength in an EU proposal. Personal networks were examined and people were contacted until we had representatives from the South (Italy, Turkey), the East (Bulgaria, a post-communist country), Central Europe (Germany), the English-speaking West (Ireland) and the North (Sweden and Denmark).

The Danish and Irish partners were about to start gender equality work, with or without an EU project, and were the original initiators of the project. The Swedish and the German partners had done gender equality work for several years and were contacted to be a resource in the project. The Swedish partner (and the author of this chapter), finally became the coordinator of the project. Turkey had no official gender equality work but had had an equality-minded vice-chancellor, who now as a former vice-chancellor, was the Turkish team leader. Bulgaria had taken part in another EU gender project, but had no gender awareness spread in the organization. In Italy, the EU project meant a possibility to do gender equality work, which otherwise would not have come about. Hence, our starting points were very different.

University cultures and structures are diverse across Europe, not only when it comes to gender equality work, but also more generally. Space does not allow for detailed description of the differences between our countries (one overview on structures can be found in Bennetot et al., 2017). Some that had great impact on our work was the extent of autonomy, for example in regard to appointments: in one country a single department manager had a decisive influence, while in another a decision was made outside the university in the state ministry. In collecting data, we found that some data was too sensitive to be collected and how some data could be interpreted varied between countries. For example, whether salary data could be obtained, and how data on leaves would be interpreted varied by country. When working with decision-making processes, both formal and informal, we soon realised that the formal power of different bodies varied between countries, and how much informal power was in use and the general attitude to transparency varied as well.

We were not only geographically and culturally diverse, but our positions in our institutions also varied. We were STEM researchers, high-level university administrative staff, gender equality officers, and gender researchers. This proved to be important for our work and we realised that we needed all three areas of knowledge: the experiential, hands-on and even emotional knowledge of the STEM researchers; the larger frame of gender research to understand why certain phenomena take place; and the knowledge of the administrative staff and gender equality officers on how issues possibly could or could not be addressed. While a common composition is to have (female) STEM researchers address their own environment or gender researchers address the STEM environment, the practical knowledge of administrative people who work with change processes in their institutions was also important.

Overview of FESTA Actions

As expected, some actions succeeded better than others. Four of the actions are described in more detail in this chapter, but more of what we learnt from the project and the actions can be found in *The FESTA Handbook of Organizational Change* (Salminen-Karlsson, 2016). In the Handbook, all FESTA actions are described and lessons learned are pointed out in bullet point lists. The second part of the book is written as an encouraging "letter" to a newly recruited gender equality specialist, with some advice on how to handle the situation in which she finds herself.

Visibility and transparency became, in practice, our core concepts. We made the equality situation visible for the leadership and all staff with striking statistical data. We made meeting participants aware of the unconscious processes that take place in meetings and which silence some participants. We tried to tease out what the overarching and fuzzy concept of excellence actually entails and how it may relate to gender. We named resistance as resistance. We made the implicit rules of the career game visible for young female researchers. We advocated increased transparency in decision making, formal as well as informal, with hiring as a particularly important case.

Something we did not do was work with work–life balance issues. We were conscious of the fact that these are crucial issues in particular for young female researchers, but we decided to leave them outside the project plan. We did so because our experience is that STEM departments have a strong inclination to attribute women's slower career progression to family issues rather than issues in the organization. We wanted to direct our searchlight into the organization rather than outside it. We decided that FESTA was to be about showing that the working environment itself raised enough hurdles for junior researchers, in particular for women, even without the further problem of care obligations.

Our experience was that the partners with the initially least gender-equal environments got the biggest gains from the project and the added value was less notable among partners who already were relatively gender equal. That difference may not have been mirrored in numbers, but in improved awareness and improved opportunities for female researchers. Improved numbers during a time of the project are not the only indicator of success, as changing the gender balance in STEM is really long-term work. Some partners, however, actually improved their numbers of women in STEM staff within the five-year time frame of the project, which was very inspiring for the fatigued partners whose numbers were higher but not rising any more. Networks and mentoring are often measures for empowering women in STEM, but the increased organizational awareness and improved opportunities could have the same empowering effect for women who previously had not observed the inequalities in their working environments:

1. Collecting and using institutional statistics for increasing awareness of gender equality issues among leadership and staff. Awareness of gender issues is the foundation for organizational gender equality work, and this awareness needs to be based on facts. This action is explained in more detail later in the chapter.
2. Changing institutional hiring processes to become more transparent and less biased. Transparent and gender-equal hiring processes are a cornerstone for creating gender-equal career paths. The partners analyzed the hiring processes of their organizations,[2] identified critical points where biases could influence the process and raised awareness among those responsible for hiring and promotion.
3. Getting more women into boards, committees, and leadership positions. The first step was often to map the current situation and raise awareness about it. The partners also analyzed appointment processes to such positions, which sometimes were quite murky, and suggested improvements. The other side of the coin was to encourage women to apply for such positions and to use their voice once they were there.
4. Improving meeting cultures. This action was based on the research showing that it is complicated for women to make themselves heard in male-dominated meetings (Baxter, 2011). The approach was not gender specific, but aimed at improving meeting cultures so that all participants would be able to participate on an equal footing. Training in conducting meetings was arranged and the partners observed meetings in their organization and gave feedback.
5. Analyzing the informal decision-making processes and making them more transparent. This action was based on the research showing that men more easily are taken in and get an understanding for such processes (O'Keefe and Courtois, 2019). This action is described in more detail later in the chapter.
6. Improving PhD education. This action was based on research and our experiences showing that PhD education discouraged many women from continuing on the academic path. This action is explained in more detail later in the chapter.

7. Initiating career training for young female researchers. This was our only action that only targeted women. It was based on the insight that women and men are differently prepared and get differential support to navigate the academic environment after their PhD (O'Connor et al., 2020). Partners organized career training for junior women to boost their self-confidence and to make them aware of unwritten aspects in advancing a research career. A career training program for facilitators conducting such training was compiled.[3]

8. Trying to change the current perceptions of excellence to such that would allow a more gender-equal working environment. This was our least concrete action. It was based on research results showing that current perceptions of excellent research are based on a masculine model and disadvantage women (van den Brink and Benschop, 2011). The partners interviewed academic staff at different career levels on their perceptions of excellence, analyzed the results in regard to how they would influence the daily working environment and fed the results back to the departments concerned.

9. Analysing the resistance that we would meet, to be able to counteract it better, based on research showing that gender equality measures regularly meet resistance (Mampaey et al., 2020). This action is described in more detail later in the chapter.

A Selection of FESTA Actions

Raising organizational awareness with local data

Numbers are important in STEM. Numbers are convincing. It is through numbers that it is possible to point out that something is amiss when it comes to gender equality in the institution. But there are a lot of potential numbers, and the questions are which of them are useful, which are possible to collect with reasonable effort and which numbers are interesting. Working with STEM faculties, we thought quite a lot about such issues. The number and percentage of women at different career stages is a natural starting point. It shows a clear picture, but it is not very inspiring. In most STEM organizations it is not news that there are more men than women in almost all positions. Everybody already has a notion that there is a gender difference. Those figures can be interpreted as women not choosing STEM. What figures could we use to show that institutional factors treated women and men differently?

Like everything else, this was situational. Partly it depended on which figures were already there: Were staff statistics split in genders, so it would be possible to know the presence of women at different career stages? How about any surveys sent to staff – were the replies gender analyzed? Partly it depended on how difficult it would be to collect statistics. Was the data retrievable from some database? Were there administrative staff who could go to files and make gender compilations? How much time would it take for a project member to do it herself? Partly it depended on what statistics were kept secret – salary statistics, for example, could be assessed in some universities but not in others. And, finally, it depended on ideas about which statistics could be particularly interesting in a certain institutional context. Four partners, Sweden, Denmark, Germany and Italy worked with such considerations. Commonly, interesting statistics were assumed to be: (1) what kind of employment contracts women and men had; (2) salaries; (3) uptake of maternal and paternal leave; (4) success rates for women and men in appointments to different positions, such as lecturer and professor; (5) share of women in leadership positions and committees and boards; and (6) turnover rates for female and male employees. Single institutions were also interested in the gender split in other possible statistics: part-time work due to caring obligations, number of research projects which have a gender aspect, age at first appointment of professorship, internal research resources, PhD students' degree of research activity, distance working, job satisfaction, achievements such as projects led, invited talks, etc.

A great deal could be shown with such figures, but not everything was possible to collect. Preferably a process would be set up, so the same data would be collected recurrently to evaluate and monitor the gender equality situation, but that was even more difficult. All different institutions use different computer systems, they collect different data from their academic staff, and the academic staff and its leaders are not always happy to supply the data, when it means extra work for them. Suggesting a collection of even more data was often not popular. Even gender specification of the existing data could be difficult, when such features were not built into the computer systems. Some partners thought it worthwhile to manually collect some data that was not in computer systems, just to illustrate a certain point to start a discussion, even if that data would not be collected repeatedly.

We learned that rather than present everything we found out in digging up and compiling data, the most effective approach when presenting it to different stakeholders was to concentrate on some central figures, preferably those that were new and revealing. The robustness of data was also an important aspect. Questioning the validity of the facts we presented was common and a natural reaction for someone who never had seen reality that way. For us it was also important that the data were interpreted in relation to institutional structures. For example, by showing how achievements such as number of publications or invited talks differed in terms of gender, we wanted to argue that it depended on women's more limited possibilities, while our listeners could conclude that it depended on women's more limited capacity. We also learned that presenting positive data, when possible, was beneficial, to create a positive atmosphere and a positive identity as a gender equality actor.

PhD supervision

FESTA concentrated on female junior researchers' working conditions, and one of the tasks was to work on PhD supervision. If women's voices are to be heard in STEM faculties, the women need to find their voices as early as possible. The position of PhD students, again, varies considerably between countries, from being salaried faculty members with teaching duties, to working outside the university alongside their studies. The time it takes to get a PhD also varies, from three years up to six or seven, even if the EU has a general agreement (called the Bologna agreement) on three to four years. The percentage of female PhD students also varies, from Denmark's 37 per cent to Italy's and Bulgaria's 53 per cent in science, and from Germany's 19 per cent to Italy's and Bulgaria's 37 per cent in engineering (European Commission, 2019). PhD study is a crucial stage in an academic career, with different and extra considerations for women. They are not fewer in relation to their male peers than they were during their undergraduate studies – actually, in some countries the percentage is higher for PhD than for undergraduate studies – but they are fewer in numbers. They come into an environment with much more interaction with senior researchers who are often men, and many of them become dependent on male supervisors. The environment is also starting to show its competitiveness.

While female PhD students in STEM may get adequate tutoring from their supervisors, they do not often get the implicit sponsoring that their male peers do (O'Connor et al., 2020). O'Connor et al. mention help to navigate the competitive academic environment, assistance in building networks or help to find and enter good positions with recommendations from the supervisor. Career counselling is not a practice in all countries, and this is a particular disadvantage for female PhD students who often do not benefit from informal mentoring. The aim of the FESTA supervision tool, as described below, was to make more visible the wide-reaching impact of a supervisor and the institutional context where supervision happens, and to help supervisors consider aspects that are important from female PhD students' point of view.

In Europe it is quite common that PhD students study outside their native country (Lauchlan, 2019). At the start of their PhD studies they meet a different culture, a different educational system and, even if the studies in themselves are conducted in English, also a new language in their daily lives. A transition period is needed. For female students who are a minority it may be more difficult to make friends at their workplace, and understand the culture. For them, particular persons to whom a PhD student can go to for just a chat can be extremely valuable. Most often that is expected to be the supervisor, but there are also examples of departments which assign another senior faculty member, or a PhD student in the final stages of their studies to be a kind of department mentor during the first weeks. Also, arranging lunches, after-work activities or excursions for all PhD students before certain social patterns have become cemented makes it more likely that these patterns become less exclusive.

We found that female PhD students' stress (Hazell et al., 2020) would be ameliorated by straightforward and honest documentation of tasks and expectations. Senior academics at the department are not always good role models in making time for other things than work in their lives. Unfortunately, having these work-addicted role models transmits an unbalanced work–life to a younger generation – and becomes one of the reasons why young women PhDs choose to leave the academy (Cidlinská, 2019). The "academic housework" (Hejstra et al., 2017) – all the small invisible administrative and social tasks that tend to fall on women rather than men – start to be distributed unequally by gender very early. Such tasks can be writing minutes for meetings, organizing social events for the department, administering projects, mentoring students etc. The division of tasks also concerns teaching duties, if PhD students have any – not only the number of contact hours, but also the kind of courses makes a huge difference to how stressful teaching becomes.

The FESTA project created a "supervision tool",[4] a website with advice to supervisors, PhD students and administration related to different phases of the PhD career. It is based on previous research and on interviews with PhD students, both male and female, in six countries. The tool takes up a number of concerns that can affect any PhD students but especially women, and that can be ameliorated either by supervisors or the departmental administration.

Informal decision making

The importance of "old boys'" networks (Fisher and Kinsay, 2014) and gatekeepers (van den Brink and Benschop, 2014) has been established by research on gender in the academy. There are advantages to informal decision-making processes, for example they are fast. However, as participation in such processes is not monitored, the members of the female minority are more easily left out in the processes, or their viewpoints are not considered. Many formal processes where women participate practically confirm decisions that have been made in informal processes. For those members of formal decision-making boards who have not been part of the informal decision making, it can be difficult to change a decision that other board members already see as practically made. By not connecting informal decision making directly to gender we also discovered that women were not the only minority group who did not take part in the informal decision making. Sometimes foreign researchers had problems, if much of the information, or little but crucial information was given only in the native language of the country. PhD students and junior researchers in general had a poorer understanding of what was going on in their department, and of their possibilities to influence departmental life. As learning about informal decision-making happens informally and even implicitly, it is likely to happen between male senior and junior researchers in homosocial relationships.

The concept of informal decision making was possibly the one where our national differences came out most strikingly: what is meant by informal and how it can be talked about were very different things in the different contexts. For one partner it meant, for example, transparency behind decisions made by individuals who had formal power. Another partner found it difficult to see informal processes at all, and the task became interpreted as an issue of creating a nicer workplace where people were able to talk to each other. Still another partner was quite familiar with the idea that informal decision making often happens in the shadow of formal processes but found it very difficult to influence it, as people had learned to be quite reticent about what they were actually involved in. And one partner felt that women could already navigate the informal negotiations, but needed to be more visible in formal bodies.

Our way to get a better grasp of informal processes was to set up a series of interviews with male and female staff members at all different levels, from PhD students to professors. Because the word informal was problematic, we were quite concrete. We asked about different practices – how people were assigned to positions, how work was assigned and resources distributed, how travel money was obtained, how our interviewees kept themselves informed about what was happening at the department etc. The results varied, not only by country but also by department. However, it was quite common for our interviewees to tell that they really did not know.

Informal decision making is closely tied to departmental communication. It is difficult to influence decisions if you do not know what decisions are about to be taken. It is difficult to put your needs forward if you do not even know that it can be done and how. It seemed that many women were unaware that they were expected to put themselves forward, make their achievements visible and their demands heard, but instead waited for the achievements and demands to be recognized. Regular communication to all employees on opportunities and decisions made became an important point in the actions we suggested. Sometimes it was realized, but those benefiting from informal decision making and some leaders were sometimes quite reluctant to change their communication practices.

Resistance

Not all partners conducted all the actions, but instead chose those that seemed most relevant and achievable in their institutions. However, everyone participated in the resistance action, reporting resistance stories and discussing possible countermeasures. Resistance to gender equality measures takes different forms in different national contexts. In some countries, where supporting gender equality is the officially sanctioned right thing to do, resistance is mostly hidden and may show in inertia; in some other countries it is open and blatant. Resistance to an EU-financed gender equality project can be about resistance to gender equality but it can also relate to a general resistance to the overarching influence of the European Union that is relatively strong in some countries. In those cases, resistance is ultimately directed to EU decision makers who are perceived as trying to impose their ideas and culture in countries which have a very different (and what the EU centrally would call, conservative) understanding of sex and gender. General societal gender ideologies differ widely among different parts of Europe.

For those of us who were new to gender equality work, resistance from women was a surprise. Such resistance is often assumed to be grounded on the worry that making particular measures to enhance gender equality means that women would not be regarded as equal and competent researchers in their own right. This may have been the case, but Britton's (2017) explanation of how gender inequality is not all-pervasive in female STEM researchers' lives, and, hence, often not seen as an overarching problem was also relevant. Our firm approach – that the project was about improving the working conditions of both women and men – mitigated this to some extent, but still, our project was not popular among all the female employees in the departments involved.

We also found men who were positive about our work. Unfortunately, they would not always express it in public, but would, for example, apologize for their brusque male colleagues after a public occasion. We could only hope that when we were presenting in a meeting and were met with harsh comments, there would be men who took to heart what we were saying, even if they did not have the courage to acknowledge it in front of their male colleagues and seniors and remained silent. However, often we did not meet this kind of resistance simply because, if we organized events with voluntary participation, very few people turned up, and a majority of them tended to be women. We learnt that having 15 minutes' talking time in a meeting that people regularly attended could be more valuable than organizing a meeting by ourselves.

We learned that whenever we presented any research that proved some kind of bias in the academy, we had to be able to answer questions about how that research had been conducted, to prove that it was really valid. Interview studies, whether published in peer-reviewed journals or our own interviews from inside the institution, did not carry much credibility. We found that, unfortunately, qualitative research was not that convincing for many people who work in science and technology. This was problematic because if audiences do not believe in research results, it is natural for them to resist measures or recommendations that are based on those results.

In the project we decided to collect "resistance stories" – instances where we had met and had to deal with resistance. This would be done both for our own sake, to put the resistance on paper and become a little distanced from it, and for learning more about resistance and being able to describe it to others. We had the ambition to reflect on the resistance together and find ways to mitigate it, but considering the mitigation we did not always find effective means. It was a little disheartening, but we were also aware that resistance was an indicator that we actually were making change: if there is no resistance, then there is probably very little happening to gender equality in the institution. The resistance stories were used as examples in the online handbook on resistance published by the project.[5]

CONCLUSION

This chapter shows that gender equality work with the aim of changing, not women but structures and cultures, is possible in STEM departments. It presents a project which, to different degrees, succeeded in doing so in a variety of European countries. It gives examples of actions that led to positive outcomes. The description of the action on data collection and presentation stresses that collecting, analyzing and presenting local data is a basic prerequisite for making a STEM institution aware of gender inequality. The presentation of the actions on PhD supervision and informal decision-making processes argues that gender equality work does not always need to focus on gender, but rather institutional cultures and structures that disadvantage all genders, but primarily minority genders in STEM departments. The action on resistance stresses that silent as well as open resistance from both women and men is to be expected when institutional structures and cultures are to be changed to become more gender equal, and, therefore, resistance is actually an indicator that change is taking place. This concluding section presents the argument that STEM institutions have the potential of being forerunners in improving gender equality in the academe.

The skewed gender balance in the science and technology areas has initiated a lot of gender equality work. In the EU, gender balance in the research workforce has been on the agenda for a long time. Many representatives of scientific and technical disciplines are aware that something needs to be done when financers become increasingly observant of the gender composition of research teams. In most European countries science and technology simply have to try to do something about the gender balance. And because they have to do something, structured and conscious gender equality work can take place, and even show the way for other scientific areas.

In social sciences, and maybe particularly in humanities, there is a long tradition of studying and theorising gender. While men and women in social sciences and humanities can be much more knowledgeable about gender issues than men and women in STEM disciplines, this does not mean that subtle gender power mechanisms are not at work. *Knowledge on gender issues does not always equal gender equality measures and achievements.* When there are many women on the lower echelons, the mechanisms that result in men still having it easier to reach professor positions are more invisible, in particular because quite a few women also reach those positions. Because of the many women, the culture is seen as more female friendly. Hence, the need for gender equality work is not always perceived as being so pressing as in a male-dominated environment (Cidlinská, 2019).

In science and engineering the situation is different. The percentage of women is low in most countries. Science and technology are seen as the drivers of development, and as currency in the international competition for status and economy. The women who go to other fields and occupations are often seen as an untapped talent reserve. The rhetoric of using only half of the available talent prompts universities to work to attract more women. Attracting women to academia is not of much use if the universities are not able also to keep those women. Whatever might be said about fairness and women bringing in new viewpoints to research, the argument about more balanced numbers is the one that most often makes university and faculty leaderships initiate gender equality work.

Another reason why STEM faculties may be ones to lead the way in gender equality work is that in particular engineers are doers rather than talkers. Provided that they really see the numerical imbalance as a problem, there is also hope for action. Insofar as the overall problem with imbalanced numbers can be translated into delimited problems that can be addressed with clear actions and measurable results, engineers are often relatively easy to get on board.

While the above may seem encouraging, there are also several constraints. First, gender equality can be threatening. Successful gender equality measures in STEM can threaten foundational biases and misconceptions. Such misconceptions associate science and technology with the "male brain," or believe that being a successful STEM researcher requires something innate that only a few women (Marie Curie and a few more) possess. Gender equality work also threatens the identity of scientists who believe that what they do is pure rationality.

Secondly, not everybody thinks that the numerical balance is a problem. When the numerical balance is seen as a problem, the perception that the roots of the problem lie somewhere else often still prevails. The lack of women in STEM can be perceived as the same kind of issue as other natural phenomena – "fish don't fly" and "women are not interested in STEM." Even if women demonstrate adequate abilities, there is often an assumption that they choose family before career – which, of course, is a nice assumption for those men who want to have a career without being rivalled by women. In some European countries the discourse of innate mental differences between women and men is still very much alive. In other countries, when blaming women's "innate characteristics" is not politically correct, blaming their upbringing can do the same thing – the way upbringing is often used as an explanation for women's presumed disinterest in STEM makes it almost as decisive as an explanation on the basis of genes.

Thirdly, a problem in particular in STEM areas is that there are women who do not want increased visibility to gender, who have learnt to downplay their gender and believe that gender is or at least should be irrelevant. STEM education generally does not equip students to analyze human relations, and many women go through their studies and initial stages of their career without really seeing the implicit power plays that take place, and from which they may be excluded from the beginning (Martin, 2001). In such cases women, encouraged by their male colleagues, can resist gender equality work because they feel that gender equality initiatives downplay their competence by trying to give them special favours.

But let us say that for some reason or another, some influential or engaged men have been convinced that at least part of the problem lies in institutional structures and cultures. There is still a way to go to make that problem visible and manageable. The FESTA teams have met inertia and unwillingness among leaders, but they have also found genuine perplexity about what can be done to improve the numbers. It is here that an institution can really benefit from an EU project that comes in with the resources that are needed to work with the issue. There is a lot of research on gender inequality in STEM institutions, but every institution has its own features. Research greatly facilitates, but cannot fully replace, the analysis of the key issues in each particular institution and, especially, which measures would be both useful and achievable in that institution. Analyses of the local situation are more convincing than ever so many research articles, and what is doable in an institution depends on a number of factors. Of course, both analyses and measures may meet with difficulties. Even if interested and engaged people in good positions understand that there are problems in the organizational environment, it is a different and sometimes disappointing thing to become aware of concrete aspects where women may be disadvantaged. Those aspects often lie well embedded in the core of organizational culture, hierarchies and processes, which means that changing them implies changes in something that people perceive as fundamental. However, to achieve a common understanding of the problems is essential for doing something about them. If that understanding is not relatively widespread, the thought-out actions meet resistance, not only from men who think that women get undue advantages, but also from women, who do not want anybody to think that they would not work and compete on the same terms as their male colleagues.

Even if (male) leaders have a genuine interest in improving the balance, this interest is naturally weighed against other interests, such as keeping on good terms with the majority of the staff. Hence, the support for a gender equality project may remain passive: it is good that there is a gender equality project, and it has an overall blessing of the leadership, but its activities are not actively promoted by them. That the measures to improve the situation need to be doable often means that they must not raise too much resistance. As to gender issues in the working environment, the understanding that a better working environment benefits both women and men may both motivate leaders and mitigate resistance.

The last part of the change work is the measuring of the effects of the actions. It is also a tricky issue. If the project is to continue after the EU funding has ended, it needs to show that it has achieved at least something. Numbers convince STEM people, but gender equality improvements do not always show in numbers, or do so in very long term.

All the above appears to make gender equality work in a STEM faculty a very challenging (and seemingly hopeless at times) endeavour. Still, STEM faculties may be easier to work in for a gender equality project than some other faculties. There is a clear problem in numbers, and there is a tradition of dealing with and acting on problems. If gender equality work is packaged with regard to these traditions, it does have a good chance of succeeding.

NOTES

1.   See www.genderportal.eu/projects.
2.   See www.festa-europa.eu/sites/festa-europa.eu/files/PRINT_Handbook_FESTA.pdf.
3.   See www.festa-europa.eu/sites/festa-europa.eu/files/WP3.1%20FESTA%20Career%20Training%20Programme.pdf.
4.   See www.festatool.eu/.
5.   See www.resge.eu/.

REFERENCES

Bautista-Puig, N., García-Zorita, C., and Mauleón, E. (2019). European Research Council: Excellence and leadership over time from a gender perspective. *Research Evaluation*, *28*(4), 370–82. https://doi.org/10.1093/reseval/rvz023.
Baxter, J. (2011). Survival or success? A critical exploration of the use of "double-voiced discourse" by women business leaders in the UK. *Discourse and Communication*, *5*(3), 231–45. https://doi.org/10.1177/1750481311405590.
Bennetot Pruvot E. and Estermann, T. (2017). *University Autonomy in Europe: The Scorecard 2017*. Brussels: European University Association. http://infosztrajk.hu/wp-content/uploads/2021/03/University-autonomy-in-Europe_2017.pdf.
Bozzon, R., Murgia, A., Poggio, B., and Rapetti, E. (2017). Work–life interferences in the early stages of academic careers: The case of precarious researchers in Italy. *European Educational Research Journal EERJ*, *16*(2–3), 332–51. https://doi.org/10.1177/1474904116669364.
van den Brink, M. and Benschop, Y. (2011). Gender practices in the construction of academic excellence: Sheep with five legs. *Organization*, *19*(4), 507–624. https://doi.org/10.1177/1350508411414293.
van den Brink, M. and Benschop, Y. (2014). Gender in academic networking: The role of gatekeepers in professorial recruitment. *Journal of Management Studies*, *51*(3), 460–92. https://doi.org/10.1111/joms.12060.
Britton, D.M. (2017). Beyond the chilly climate: The salience of gender in women's academic careers. *Gender and Society*, *31*(1), 5–27. https://doi.org/10.1177/0891243216681494.
Cabay, M., Bernstein, B., Rivers, M., and Fabert, N. (2018). Chilly climates, balancing acts, and shifting pathways: What happens to women in STEM doctoral programs. *Social Sciences (Basel)*, *7*(2), 23. https://doi.org/10.3390/socsci7020023.

Cidlinská, K. (2019). How not to scare off women: Different needs of female early-stage researchers in STEM and SSH fields and the implications for support measures. *Higher Education*, *78*(2), 365–88. https://doi.org/10.1007/s10734-018-0347-x.

European Commission. (2019). *She Figures 2018*. Luxembourg: Publications Office of the European Union.

Fisher, V., and Kinsey, S. (2014). Behind closed doors! Homosocial desire and the academic boys club. *Gender in Management*, *29*(1), 44–64. https://doi.org/10.1108/GM-10-2012-0080.

Gunter, K.P., Gullberg, A., and Ahnesjo, I. (2021). "Quite ironic that even I became a natural scientist": Students' imagined identity trajectories in the figured world of higher education biology. *Science Education*, *105*(5), 837–54. https://doi.org/10.1002/sce.21673.

Hazell, C.M., Chapman, L., Valeix, S.F., Roberts, P., Niven, J.E., and Berry, C. (2020). Understanding the mental health of doctoral researchers: A mixed methods systematic review with meta-analysis and meta-synthesis. *Systematic Reviews*, *9*(1), 1–197. https://doi.org/10.1186/s13643-020-01443-1.

Heijstra, T.M., Einarsdóttir, Þ., Pétursdóttir, G.M., and Steinþórsdóttir, F.S. (2017). Testing the concept of academic housework in a European setting: Part of academic career-making or gendered barrier to the top? *European Educational Research Journal EERJ*, *16*(2–3), 200–214. https://doi.org/10.1177/1474904116668884.

Jordão, C., Carvalho, T., Diogo, S. (2020). Implementing gender equality plans through an action-research approach: Challenges and resistances. In M. Au-Yuong-Oliveira and C. Costa (eds), *Proceedings of the 19th European Conference on Research Methodology for Business and Management Studies*. A virtual conference hosted by University of Aveiro, Portugal, June 18–19, 2020, pp. 124–31.

Kelan, E.K. (2009). Gender fatigue: The ideological dilemma of gender neutrality and discrimination in organizations. *Canadian Journal of Administrative Sciences*, *26*(3), 197–210. https://doi.org/10.1002/cjas.106.

Lauchlan, E. (2019). *Nature PhD Survey 2019*. London: Shift-Learning. www.shift-learning.co.uk/case-studies/phd-survey-2019/.

Mampaey, J.L.J., Roos, H., Huisman, J., and Luyckx, J. (2020). The failure of gender equality initiatives in academia: Exploring defensive institutional work in Flemish universities. *Gender and Society*, *34*(3), 467–95. https://doi.org/10.1177/0891243220914521.

Martin, P.Y. (2001). Mobilizing masculinities: Women's experiences of men at work. *Organization*, *8*(4), 587–618. https://doi.org/10.1177/135050840184003.

O'Connor, P., O'Hagan, C., and Gray, B. (2018). Femininities in STEM: Outsiders within. *Work, Employment and Society*, *32*(2), 312–29. https://doi.org/10.1177/0950017017714198.

O'Connor, P., O'Hagan, C., Myers, E.S., Baisner, L., Apostolov, G., Topuzova, I., Saglamer, G., Tan, M. G., and Caglayan, H. (2020). Mentoring and sponsorship in higher education institutions: Men's invisible advantage in STEM? *Higher Education Research and Development*, *39*(4), 764–77. https://doi.org/10.1080/07294360.2019.1686468.

O'Keefe, T., and Courtois, A. (2019). "Not one of the family": Gender and precarious work in the neoliberal university. *Gender, Work, and Organization*, *26*(4), 463–79. https://doi.org/10.1111/gwao.12346.

Ottemo, A., Gonsalves, A.J., and Danielsson, A.T. (2021). (Dis)embodied masculinity and the meaning of (non)style in physics and computer engineering education. *Gender and Education*, *33*(8), 1017–32.

Salminen-Karlsson, M. (2016). *The FESTA Handbook of Organizational Change: Implementing gender equality in higher education and research organizations*. Uppsala University: Personalavdelningen. http://uu.diva-portal.org/smash/get/diva2:1098070/FULLTEXT01.pdf.

Schlamp, S., Gerpott, F.H., and Voelpel, S.C. (2021). Same talk, different reaction? Communication, emergent leadership and gender. *Journal of Managerial Psychology*, *36*(1), 51–74. https://doi.org/10.1108/JMP-01-2019-0062.

Stretenova, N. (2011). Eastern countries' gender and science: Analysis and meta-analysis. *Brussels Economic Review*, *54*(2–3), 177–9.

6. National culture and policy institutionalizing workplace change: supporting women's career progression in STEM through Athena SWAN

Regina Connolly and Ita Richardson

INTRODUCTION

Despite progress over the past decade, the science, technology, engineering and mathematics (STEM) gender gap continues to persist in Europe, with only 6 out of 36 countries having 50 percent or more of scientists and engineers who are women (Eurostat 2019). An example of this is Ireland, where women remain vastly under-represented in the STEM workforce, despite rising demand for STEM professionals and the importance of the sector to the country's success. This chapter describes why macro and micro-level interventions are equally necessary to advance progression of Irish women in the STEM workplace. Macro-level approaches indicate large-scale government intervention, frequently represented through new laws and national policies. On the other hand, micro-level approaches are characterized by small-scale interventions that are specific to a particular organization or institution, one example being introduction of flexible working opportunities for women.

The chapter starts with a short background discussion outlining how gender equality change differs according to country and how COVID-19 has served to illustrate the fragility of gender equality progression. The nuanced nature of that progress is elucidated through an examination of the gap between men and women's workforce participation in European countries, particularly exemplified through the lack of women in CEO positions in leading STEM organizations, such as Intel and Dell. This trend is also reflected in gender difference within academia. For example, in all Irish Higher Education Institutes (HEIs) (2017–18), while 45 percent of academic staff are female, only 35 percent of senior academic staff are women.[1] This is followed by a discussion of several core factors that continue to inhibit that realization of gender equality, including gender stereotypes, gender role congruity perceptions and the lack of flexible working opportunities. Research detailing how culture influences the nature of change is described. This points to the need for a multi-pronged strategy of intervention mechanism, rather than an exclusive reliance on national cultural and economic progressions, which although valuable, are slowly and incrementally achieved. Women's under-representation in higher levels of academia and recognition of the need for such as a dual strategy approach is discussed.

We then evaluate a specific initiative, Athena Scientific Women's Academic Network (SWAN), a framework established by the UK Equality Challenge Unit (now part of Advance HE) which focuses on supporting and transforming gender equality in higher education and research and recognizes the success of HEIs in advancing gender equality goals. Athena SWAN was initially set up to focus on the advancement of academic women's careers in science, technology, engineering, mathematics and medicine (STEMM). It has expanded to include arts, humanities, social sciences, business and law (AHSSBL), staff working in professional, managerial and support roles (PMSS), trans staff and students, and men in particular under-represented disciplines. Each of the Athena SWAN awards, Bronze, Silver and Gold, recognizes the success of HEIs and their constituent departments in advancing gender equality, with HEIs progressing from Bronze to Gold through investigation, data analysis and action plan implementation. In the United Kingdom, Ireland, Australia and parts of the United States, this is being used to address gender inequality in higher education, and has caused significant change in STEM disciplines, amongst others. We discuss the challenges and opportunities associated with this initiative, as well as its implications for policy makers. The chapter concludes by summarizing why a dual focus that encompasses both macro- and micro-level interventions is necessary to advance workplace change that supports women's career progression.

BACKGROUND

Socio-economic development and accompanying modernizations have resulted in profound transformations of society, particularly in relation to the erosion of traditional sex-segregated roles, participation of women in the paid workforce and the mobilization of women in public life. This modernization process has been characterized (Inglehart and Norris, 2003) as consisting of two stages, the first being the move from agrarian to industrialized societies, which has reduced fertility rates and brought more women into the paid labor force and increased education levels, and the second being the move from industrial towards postindustrial societies, which has underpinned significant progression towards gender equality in the workplace and public life.

However, the nature of change has differed and there is clear evidence of variation in countries' gender equality advancement. For example, Nordic countries (Finland, Denmark, Norway, Sweden and Iceland) tend to be at the forefront of progression, as demonstrated through numbers of women participating in the workforce and public life, pay parity, comprehensive childcare provisions, improved education opportunities and higher quality of life. For other countries, progress and the pace of change has been slower. Some of this can be explained through the cultural norms and institutional structures of societies, often reflected through varying degrees of democratization. However, differences still exist among countries with similar levels of democratic development, such as the United States and the United Kingdom (Inglehart and Norris, 2003), and even progressive Nordic countries are observing the development of a gender-specific management gap (Sanandaji, 2018).

This gender difference is more pronounced in the STEM subjects where only 36 percent of graduates in these fields and only two of five scientists and engineers are women.[2] This gender difference is equally present in academia. For example, in Ireland, in December 2020, only 27 percent of full professors were women (December 2020),[3] While figures are not published regarding STEMM professorships nationally, we can state that in the University of Limerick, for example, 40 percent of STEMM academic staff are women, yet only 25 percent of the professoriate is female. This is not atypical across Ireland. Furthermore, given the structure of STEM career policies in academia, the FESTA project[4] (of which one of the authors was a member) established that a significant change is needed in how STEM career development and recruitment is undertaken. Through Athena SWAN implementation, we have been enabled to change the system rather than changing the women. This has been significant.

Gender Equality

Gender equality has been described (Coron, 2020) as a polysemous concept, indicating that it can have multiple meanings. In part, this is because it encompasses multiple dimensions including equal pay, equality of access to positions of responsibility and gender diversity, a fact that may result in organizations and their employees emphasizing one of these dimensions at the expense of others. Within a workplace context, equality discrepancies manifest in multiple ways. In the first instance, women have less access to the labor market (Mandel, 2013) and gender segregation is evident in the enactment of labor (Cech, 2016). They are more likely to experience barriers to their progression (Kalaitzi et al, 2017), which have been likened to a "glass ceiling," to be subject to pay disparities (Eurostat, 2021) and to suffer the effects of work–life imbalance (Muzio and Tomlinson, 2012). Because it is characterized by multidimensionality, it can be difficult to address gender equality in a comprehensive and systematic way. Adding to the problem is the fact that how gender inequality is addressed in an organization is often dependent on the benevolence of senior managers (Woodhams and Lupton, 2006) and their willingness to prioritize the implementation of specific initiatives (Perrier, 2015). Moreover, both the emphasis that they ascribe to addressing gender inequality and the way in which they implement those initiatives will reflect their personal perspectives, something that will affect the importance that they ascribe to specific issues.

Women's Participation and Advancement in the Workplace

It is true that significant steps have been made to increase women's participation in the workplace. Consequently, labor force participation rates among women have increased considerably in most countries over the past decades. In the United States, as of December 2020, women held 50.04 percent of jobs, with the greatest gains in the healthcare and retail industries (Law, 2020). However, the picture is more nuanced than at first glance, as low-paying jobs in these sectors are predominantly held by women. Moreover, these are aggregate figures and labor force participation rates for many groups of women, including women aged between 25 and 54 remains lower than that of men. Added to that is the fact that women are also more likely to work part-time due to caring responsibilities and the high cost of childcare. The number of women working in STEM varies internationally, with the median figure being 38 percent.[5] According to these figures, 48 percent of STEM workers in the United States and 40 percent of those in the United Kingdom are women. However, STEM "occupations" vary greatly when collecting statistics. There is a large majority of women in health-related positions, but if we examine other STEM disciplines, such as computing and engineering, the percentage of women employed is much lower. As an example, Women in Tech in the United Kingdom have reported that only 19 percent of the tech workforce are women.[6]

Across European countries, increases in women's labor force participation are also evident, but the gap between men and women's participation remains more marked, with most countries showing employment rates for women that are approximately 10 percent less than those for men (Catalyst, April 2020). However, similar characteristics apply. For example, a significant portion of women work part time and they also predominate in what tend to be lower paying occupations (Catalyst, April 2020). Exacerbating this difference is the fact that women remain under-represented at senior management levels. For example, in the United States, the number of women running Fortune 500 companies has now increased to 37, but this accounts for only 7.4 percent of these businesses and only three of these are women of color. The figures tell a similar story elsewhere. For instance, in the United Kingdom at present, women comprise only 5 percent of Financial Times Stock Exchange (FTSE) 100 CEOs, representing an increase of only one woman since 2012 (Ohr, 2020). In 2019, just 23 percent of executive committee members in the FTSE 100 were women, and the combined number of women in executive and senior positions reached only 28.6 percent. Furthermore, a FTSE 100 pay gap persists, with the highest paying male CEO earning almost 90 percent more than the highest paid female CEO. In Ireland, the position is equally stark. Census data from the Irish Central Statistics Office (CSO, 2019) show that, in 2019, only 12 percent or one in nine chief executives of large enterprises were women and women constitute 28 percent of senior executive roles compared with 72 percent men. The majority (93 percent) of company chairpersons were men and women constitute only 20 percent of directors of company boards. In short, the data tells a very clear story of entrenched gender inequality. Reports have shown that men rather than women are taking up jobs in technology. And, causes for this are of concern. A Pew Research Centre report in 2017 has stated that 50 percent of women in comparison to 19 percent of men have experienced gender discrimination. For those in computing jobs, 74 percent of women have experienced such discrimination, while 78 percent have experienced it in male-dominated workplaces.[7]

While women's representation in the workplace has certainly improved over the past decades, these figures point to systemic gendered disadvantage, showing that equality remains aspirational. Awareness of this inequality has not been accompanied by adequate action. In 2015, 193 countries became signatories to 17 internationally agreed targets known as Sustainable Development Goals (SDGs). These goals are a blueprint for addressing global challenges that include gender equality, as well as poverty, climate change, environmental degradation, peace and justice. Their target deadline for achieving these goals is 2030. However, the United Nations Gender Forum Index measures progress towards these targets, and its 2020 findings show that it is unlikely that any country will meet those gender commitments within the agreed time frame.

The causes of the workplace gender equality gap and the reasons for its endurance are multi-factorial and the nature of those factors and the challenge that they present vary according to context. This is because societal progression is unique to each country, reflecting the unique structural context and cultural traditions of a society. However, a number of core challenges repeatedly emerge in the literature as barriers to women's advancement in the workplace, one of which is gender stereotypes. As such, this issue is deserving of particular attention.

Gender stereotypes
Gender stereotypes influence how women are treated in relation to employment (Lips, 2013) and contribute to the imbalance of sex distributions within jobs and in seniority of role. For example, research (McKinnon and O'Connell, 2020; Carli et al, 2016) has shown that they contribute to the ongoing discrimination and under-representation of women in STEM professions, influencing not only career choices for women in STEM but also retention of women in these fields. These stereotypes are grounded on deeply entrenched gendered beliefs about abilities and the acceptability of roles played by women and men in society (Banyard, 2010). These beliefs result in category-based traits or attributes being applied to specific gender groups and result in expectations about what behaviors are or are not acceptable for those

groups (Agars, 2004), reflecting inaccurate generalizations that result in bias. As gender is an obvious social category, it can serve as an easily available cue that can initiate stereotypical beliefs (Blair and Banaji, 1996). Such beliefs tend to draw on two categories of attributes – communal and agentic attributes (Eagly and Karau, 2002). The first of these, communal attributes, encompasses those that are frequently associated with women, such as being kind, helpful, nurturing and expressive. The latter, agentic attributes, are typically associated with men. These encompass attributes such as independence, ambition, confidence, assertion, control and dominance.

Role congruity theory (Eagly and Karau, 2002) provides a framework for understanding how stereotypical beliefs create gender bias regarding workplace role acceptability. It proposes that bias results from the congruence between the beliefs of an individual (such as a manager or interviewer) regarding the stereotypical requirements of the job and the stereotypes that they hold about the characteristics of gender groups. The greater the incongruence between both sets of beliefs, that is between the perceived traits associated with the gender group and the stereotype characteristics of the job, the more this will result in gender bias. Moreover, those who are perceived to deviate from gender norms (e.g., by demonstrating agentic attributes) will also experience bias, despite having the necessary skills required for the job.

In practice, this can mean that highly skilled women who are capable of assuming a position of responsibility may experience hiring or promotion discrimination. That discrimination may cause them to be excluded on the basis of *either* the gender stereotypes associated with a particular job, or *because* they actually meet agentic stereotypes associated with that job. The latter (meeting agentic stereotypes) can cause them to be viewed at interview as deviating unacceptably from societal gender norms, resulting in exclusion. This bias, on the basis of a woman displaying attributes that do not align with gender stereotypes, has been described as a bias backlash effect (Rudman et al., 2012; Rudman and Fairchild, 2004) and is particularly problematic as embodying acceptable agentic attributes in addition to meeting competency requirements is clearly an impossible task. Moreover, it highlights the particularly problematic implications of stereotypes being not only descriptive but also prescriptive in nature, in that they propose how women should be (Eagly, 1987). In this way they normatively perpetuate gender bias and discrimination in the workplace.

A meta-analysis of gender stereotypes and bias in workplace decision making by Koch, D'Mello and Sackett (2015) found that gender-role congruity bias was particularly evident with men being rated more favorably than women for male-dominated jobs and that male raters exhibited more of that bias than did women when evaluating applicants for such positions. This is consistent with literature that indicates that men are more likely to rely on gender stereotypes when evaluating appropriateness for job roles (Koenig, Eagly, Mitchell and Ristikari, 2011). However, the study also found that gender-role congruity bias was reduced, not when individuating information was provided, but when information clearly indicated the high competence of those being evaluated. This contains an important implication for women, specifically the critical need for them to focus on communicating their competence for the role effectively and directly, rather than assuming that it will be automatically recognized or emerge during an interview. One final finding with important implications was that when participants had increased feelings that they would be held accountable for their decisions, they believed that those decisions had real-life consequences or were reminded of equity norms in the organization, this resulted in less biased hiring decisions for male-dominated jobs (Koch et al., 2015). This highlights the need to educate those responsible for hiring and promotions regarding the implications of gender equity for women and the organization. It also highlights the need for accountability in hiring decisions.

Flexible working opportunities

One challenge that remains a persistent inhibitor of women's workplace advancement is the lack of flexible working opportunities (Banyard, 2010). Both flexibility in the duration of work and the scheduling of that work are essential requirements for women seeking to balance the demands of work with family commitments (Applebaum et al., 2006). It may account for the higher percentage of women who continue to work on a part-time basis as compared to men. It may also motivate their choice of lower paid and often precarious jobs which have family-friendly policies rather than higher paying jobs with less flexibility, something that has been described by economists as compensating differentials (Becker, 1991).

Interestingly, research conducted in the United States between 2002 and 2008 (Nadler, Voyles Cocke and Lowery, 2016) found that higher educational attainment did not reduce the gender pay gap for women and that men reported greater levels of work scheduling flexibility than was the case for women with similar levels of education. As Budig and England (2001) note, the expectation that women will assume the majority of child-rearing responsibilities has resulted in a motherhood wage penalty, where mothers earn less than other women because having children can cause them to have fewer years of job experience. Using data from the 1982–93 National Longitudinal Survey of Youth, they found that women experience approximately a 7 percent wage penalty for each child that they have. However, this figure is most likely an underestimate and subsequent work by Gangl and Ziefle (2009) using UK, German and US survey data determined that women across these countries experienced a wage penalty that varied between 10 and 18 percent for each child. This "motherhood wage penalty" reinforces gender inequality as women typically assume the greater portion of child-rearing responsibilities which places them at an automatic disadvantage compared to men who do not have to undertake this caring labor to the same extent. It restricts their ability to advance in the workplace, to achieve pay parity with men and places them in a position of greater probability of experiencing gender inequality.

Cultural Values and the Nature of Change

Some have suggested that gender inequalities will diminish in response to economic growth. From this perspective, any differences that exist between men and women in relation to education, workforce opportunities and progression result from human capital differentials emanating from traditional structures and will naturally diminish in parallel with economic progression and market competition (Forsythe, Korzeniewicz and Durrant, 2000). A study by Inglehart (1997) using national-level data from 43 societies included in the 1990 World Values Survey, identified two main dimensions that accounted for more than half of cross-cultural variance involving many different values across a range of domains spanning politics, economic life and sexual behavior. These dimensions tap an axis of variation in relation to "Traditional vs. Secular-rational values" and "Survival vs. Self-expression values." A later study (Inglehart and Baker, 2000) replicating this analysis with data from subsequent surveys elicited the same dimensions of cross-cultural variation, despite the fact that it included an additional 23 countries. The dimension of Traditional/Secular-rational values dimension reflects the contrast between societies in which religion is very important and those in which it is not. The former emphasizes the importance of familial ties and traditional family values. They tend to be characterized by a deference to authority and a rejection of divorce, abortion, euthanasia and suicide. Such societies tend to have a strong nationalistic orientation. Societies on the other end of the axis (i.e., those with secular rational values) demonstrate far weaker and opposite preferences in relation to these specific issues, as manifested through increased willingness to challenge social norms and values that were previously considered to be absolute and a shift towards values characterized by increased tolerance and with greater emphasis on individual autonomy.

The beliefs and values that correlate with the survival/self-expression dimension reflect a shift in emphasis from a foremost focus on economic and physical security, towards an intergenerational shift in emphasis towards self-expression, a more participatory role in politics, subjective well-being and quality of life concerns (Inglehart, 1990, 1997). These self-expression values correlate with higher levels of democracy and align with Maslow's (1970) "Hierarchy of Needs" theory. Recent work by Inglehart, Norris and Welzel (2002) employed both dimensions to provide a cultural map of the world and found that a coherent pattern emerges between rich and poor countries, with the former tending to rank high on both dimensions and poorer countries ranking low on both dimensions. The first dimension, Traditional vs. Secular, is linked to the transition from agrarian society to industrial society and the Survival/Self-expression dimension is linked to the transition from industrial society to a service or knowledge society. These results confirm that economic development effects changes to societal values. For example, gender equality issues are a fundamental component in the intergenerational shift in beliefs and values as captured by the survival/self-expression dimension. As Inglehart et al. (2004) note, the move from survival values to self-expression values incorporates a shift in child-rearing values, where the emphasis on hard work moves towards a progressively increasing emphasis on imagination and tolerance as important values to teach a child.

The case of Ireland
Ireland is a particularly suitable exemplar of how cultural values shift in line with economic development and their influence on gender equality. The position of Irish women has changed dramatically in line with the country's transformation from an impoverished, traditional and peripheral European country to a contemporary economy and global participant. In a relatively short period of time, as a result of concerted efforts at industrial inward investment, Ireland's identity has been redefined from a conservative, isolationist and rural nation into a progressive, outward-looking, and engaged player in the world economy (Bradley and Murphy 1989; Coogan, 1987; Lee, 1989; O'Malley, 1989). The opening up of careers to Irish women took place in tandem with considerable socio-economic change, and the liberalization of women's position in society has remained consistent even despite economic fluctuations.

A granular insight into the nature of this change, specific to women in STEM/IT careers, is provided by Trauth and Connolly (2021). Through their longitudinal examination of women's participation in the Irish information technology (IT) sector, they show that the position of women and their choice of STEM/IT career does not exist in a vacuum, but rather that environmental, identity and individual factors have exerted a collective influence that affected changes in gender relations in the IT sector in that country across five eras. Societal barriers to change came from the culture, economic factors, societal infrastructure, and laws and policies, while institutional barriers included mass media messages about gender roles, an educational system that limited girls' exposure to STEM/IT careers, and the structure of the industry in which women work. They were experienced through influential people, organizations and women's own personal characteristics. Transformation at times manifested as impact, reflecting normative societal changes, and also through the direct influence of laws. It also manifested through the indirect influence of intermediaries. In a personal context, this could be the influence of a parent or teacher, while in an organizational context, indirect influence could be represented through a mentor or person in authority.

Of particular interest is that the study results demonstrate *how* these factors had their effect with respect to women's participation in STEM/IT careers. For example, factors such as new laws resulted in *abrupt change*; factors such as culture demonstrated *incremental change*. This confirms the critical importance of national laws and policies and their potential to catalyze cultural change that can support women's participation and retention in the STEM/IT sector. In that context, the oversight provided by bodies such as the European Commission is particularly important in ensuring systematic implementation of gender equality policies at the national level and stringent progress accountability in that regard. For example, the recent European Union (EU) Gender Equality strategy (2020–25) includes objectives such as challenging gender stereotypes; closing gender gaps in the labor market; achieving equal participation across different sectors of the economy; addressing the gender pay and pension gaps; closing the gender care gap and achieving gender balance in decision-making and in politics.

The relevance of Trauth and Connolly's (2021) study to the current discussion lies in the fact that it indicates that the gender gap in the IT workplace participation can be reduced through a multi-pronged strategy of intervention and mechanism, rather than an exclusive reliance on national cultural and economic evolutions, which are slowly and incrementally achieved. Neither is economic progression any guarantor of gender equality and it is possible for equality advances to regress in the face of changing circumstances (such as COVID-19). This is not to detract from the valuable scaffolding provided by national gender equality policies, but an acknowledgement that change is not assured even in economically advanced contexts. For example, Duflo (2010) points to the sustained or increasing disempowerment of women that persists regardless of economic progress, a fact that is illustrated by the disappointing findings of the 2020 Gender Forum Index. That multi-pronged strategy is a response to the fact that, while the influence of national culture and policy should ideally be reflected at the organizational level in terms of the degree to which workplaces advance women's career progression, this is not always the case. For that reason, if workplace gender equality is to be achieved, macro-level environmental interventions such as laws, policies and strategic national interventions, need to be matched by micro-level systemic initiatives that focus on embedding gender equality best practice through private and public sector organizations. Academia is one example of where this has been recognized.

ADVANCING GENDER EQUALITY IN STEM THROUGH IRISH ACADEMIA

Women are under-represented in higher levels of academia across many countries. For example, in 2017, only 21.7 percent of heads of higher education institutions across Europe were women and although women academics accounted for nearly half of academic positions, they accounted for a minority of senior academics in most of those countries (Catalyst, January

2020). From a pay parity perspective, figures from 2019 (Times Higher Education) show that women academics in the United Kingdom earned on average 15.9 percent less than their male counterparts, with some institutions reporting a gap of up to 20 percent. In Ireland, the pace of change has been remarkably slow, moving from 18 percent of women professors in 2013 to 26 percent in 2019, compared to over half of entry-level lecturing posts being filled by women. Moreover, prior to 2020, no woman had ever been a President in the Irish university sector.

The availability of such data is informative, but attention needs to focus on identifying and addressing the organizational culture and practices that perpetuate such gendered divisions. Barriers to women's advancement in academia are not uniformly experienced and research capturing the voices of women is in short supply. One such study by Harford (2018) examined the perspectives of women professors in Ireland on the professoriate for the period 2010–17. Her research identified four distinctive themes which were that they:

- viewed universities as operating in line with male definitions of merit;
- had strategically decided against taking on senior management roles and responsibilities;
- considered there to be no acknowledgement of caring responsibilities;
- highlighted the importance of validation, selection, and networks of support.

Importantly, national macro-level drivers have the potential to influence necessary micro-level interventions and in Ireland this has been evidenced through some seismic events. The first relates to a high-profile case, Galway University, where a woman academic (Michelle Sheehy-Skeffington), despite her obvious competency, did not attain promotion from lecturer to senior lecturer in an internal promotion round. She took a successful legal challenge against this decision. Her case and other similar cases in this university received significant media attention.

Consequent to this outcome, the Irish Minister for Higher Education commissioned a national review of gender equality in Irish Higher Education Institutions (HEIs)[8] (HEA, 2016), to provide public insight into the issues. Recommendations presented in this report were further clarified in a Gender Action Plan (HEA, 2018). In order to accelerate progress in achieving gender balance within Irish academia, gender actions listed are incentivized by funding and have associated consequences for lack of engagement. These recommendations are closely monitored by the Higher Education Authority (HEA), which is the body with oversight of and funding decisions for Irish third-level institutions. Thus, these "recommendations" can be seen as "rules and regulations," whose express purpose is to motivate change across the HEI sector. These, in turn, drive cultural change at the micro-level. We describe a selection of recommendations from the Gender Action Plan (HEA, 2018) which are of particular interest to this chapter.

Task Force recommendation (Leadership) – Each institution should appoint a Vice-President/ Director for Equality, Diversity and Inclusion: Although different job descriptions are used, there is now an equality, diversity and inclusion appointee in every university and in many other HEIs in Ireland. This person has responsibility for equality in that institution, and is either a member of or responsible to the executive committee. At a national level, these individuals are members of two national groups who focus on equality, diversity and inclusion. These are within the Irish University Association (encompassing the seven member universities) and Higher Education Authority (encompassing seven universities, two technological universities, and 24 other HEIs). Membership of these committees has provided the opportunity to learn from each other, to jointly lobby for diversity and inclusion advances at national government level and the ability to provide input into the development of national policy regarding equality initiatives. In some cases, the post has extended to social inclusion beyond gender equality, to include, for example, human rights.

Task Force recommendation (Recruitment and Promotion Procedures and Practices) – New and additional gender-specific posts ... Should be considered where they would be a proportionate and effective means to achieve rapid and sustainable change: Generally, in Ireland, we use 40 percent as our target for gender equality. Estimating that it could take up to 20 years to achieve an average of 40 percent women at professor level, Ireland's Department of Education and Science, through the HEA, is now providing additional funding 45 Senior Academic Leadership Initiative (SALI) Professorships for women over three years (2019–22). Others who can prove disadvantage due to grounds of equality, as stated in the Irish Equal Status Acts 2000–18,[9] are eligible to apply for these posts. Initially, HEIs must apply to a competitive call to obtain a professorship or senior academic position – the position applied for is dependent on whether the HEI is a university or not, as only universities can employ at professor level. In 2020, 20 such positions were awarded, with 15 and 10 more are planned to be awarded in 2021 and 2022 respectively. The position must then be filled by the HEI through a competitive process following requirements and standards normal to that position.

Task Force recommendations (Athena SWAN charter in Ireland) – HEIs shall apply for an Institutional Bronze award by 2019 and *HEIs should retain the Bronze award until such time as they obtain a Silver award:* Originating in the United Kingdom, the Athena SWAN framework focuses on supporting and transforming gender equality in higher education and research. In Ireland, in 2013, three universities were partners in EU-funded projects, Trinity College Dublin (TCD) was in INTEGER, Institutional Transformation for Effecting Gender Equality in Research,[10] the University of Limerick (UL) was in FESTA, Female Empowerment in Science and Technology Academia,[11] and University College Cork was in GENOVATE, Transforming organizational culture for gender equality in research and innovation.[12] Both INTEGER and FESTA focused on issues for Women in Science, Technology, Engineering and Mathematics (STEM). As a result of their project findings, the TCD project team proposed to the HEA to pilot Athena SWAN. This proposal was supported by the two other university project teams. This proposal was accepted and funded by the HEA with awards first presented in 2015. Through these recommendations, Athena SWAN is the primary means through which gender initiatives are being undertaken in Irish HEIs. Attainment of this award determines eligibility for research funding, and the HEA require that all publicly funded third-level institutions progressively engage with Athena SWAN.

Athena SWAN in Ireland

As mentioned in the Introduction, Athena SWAN initially focused on academic women in STEMM careers, and has, in recent years, expanded to include AHSSBL staff and students. Awards are given at institutional and department[13] level. Attainment of each award is seen as a progression point. To gain an award the institution or department must set up a Self-Assessment Team (SAT)[14] which takes responsibility for the Athena SWAN process. They organize discussion, data collection and analysis, develop the award submission including the action plan. The Bronze, Silver and Gold are progressive, and recognize that HEIs and departments are advancing gender equality. The Bronze award recognizes that a "solid foundation for eliminating gender bias and developing an inclusive culture that values all staff" has been established; the Silver award recognizes "a significant record of activity and achievement in promoting gender equality and in addressing challenges across different disciplines; and the Gold award recognizes "the advancement of gender quality in higher education, encompassing representation, progression and success for all" (Athena SWAN charter, 2021). Once an award is achieved, the SAT oversees the action plan implementation.

By August 2021, in Ireland, 19 institutions, including the seven universities, and 68 departments have been awarded Athena SWAN Bronze awards. In order to attain Athena SWAN accreditation, HEIs must undertake a self-assessment of the current status of gender equality and identify what actions can be taken as a result of this assessment. In many universities the institutional SAT is chaired by the President and requires collation of qualitative and quantitative staff and student data, undertaking institutional surveys, workshops and focus groups and international benchmarking. The HEI must identify weaknesses and challenges from which they develop an achievable SMART (specific, measurable, attainable, relevant and time-bound) action plan, which identifies responsibilities, and whose progress monitored within institutions and departments. Apart from the task force recommendation regarding Athena SWAN, each HEI must submit an annual Gender Action Plan to the HEA who themselves have set up a Centre for Excellence for Gender Equality. This plan contains explicit recommendations and actions in relation to HEI gender equality.

Furthermore, in Ireland, there is also a legal requirement to implement the Irish Human Rights and Equality Commission (IHREC) Act, 2014,[15] whose overarching legal framework requires that all public bodies, including universities, to be cognizant of equality and human rights. Implementing this act requires that the organization undertakes a human rights audit, develops a strategy and produces an annual report outlining progress.

Case Study – University of Limerick

We present this case study as an example of how the implementation of the national macro-level requirements discussed in the previous section have resulted in micro-level policies accelerating cultural gender equality change within an Irish HEI, the University of Limerick (UL). UL has been at the forefront of this change in Ireland. As an EU FESTA project partner, UL was involved in the initial proposal to the HEA to set up Athena SWAN Ireland. Through Richardson (author 2 and Software Engineer), they had membership on the initial national Athena SWAN committee set up by the HEA, and attained an Athena SWAN award in 2015. UL currently holds a Bronze Award (Expanded Charter), the first HEI to renew to this level, has appointed a Director of Human Rights, Equality, Diversity and Inclusion, and has been awarded two SALI Professorships.

History

There is a history of commitment to advancing gender equality in UL, a young Irish university that was founded in 1972. The first postgraduate program offering a graduate diploma in women's studies in Ireland was initiated at UL in 1990 (Byrne, 1992). This meant that an academic base with a specific interest and focus on gender equality issues was in existence, thus ensuring that gender issues were frequently discussed both informally and formally across the university. For example, members of the graduate course team were involved in setting up events for International Women's Days during the 1990s. In 1999, UL became the first Irish university to build an on-campus crèche (childcare centre) for use by staff and students, after receiving a grant from the European Social Fund,[16] following lobbying by parents. Subsequently, they applied for and received funding in collaboration with a local primary school providing "after school" facilities as an extension to this service.

UL's Equal Opportunities Committee was chaired by the Vice-President of Academic Affairs, with representation from interests across the university. In 2005, this committee was instrumental in receiving a grant from Atlantic Philanthropies to advance gender equality and diversity, through which, the first equality officer was employed in the Human Resources (HR) Division, a position that was then mainstreamed to HR Director of Equality and Diversity. The grant also supported the setting up of a women's forum which was in situ for a period of seven years (2005–12). This forum launched the UL mentoring scheme, the aim of which is to "provide a platform through which experiences can be shared on a one-to-one informal basis, thus providing support to people at various stages in their careers" (University of Limerick Mentoring Scheme, 2021). This is now mainstreamed and still running today.

Science Foundation Ireland, the Irish funding agency for STEM, which had been founded in 2000, recognized that few women were applying for funding awards in these disciplines and offered two awards in 2005 and 2006 to examine this problem within HEIs. Successful applicants were UL, TCD and UCC. In UL, a number of actions followed, such as encouraging women to become external examiners for STEM subjects, to join editorial boards of STEM journals and to become active members of program committees of STEM conferences. Subsequently, UL and TCD were awarded EU grants to examine gender equality issues in STEM (FESTA, INTEGER), while UCC were awarded one to examine gender equality issues more broadly. It is therefore unsurprising that, when Athena SWAN was launched in Ireland in 2015, the first recipients of a Bronze Institutional Award were UL and TCD, followed by UCC in 2016. In the remainder of this section, we describe outcomes from the Athena SWAN effort within UL.

Cultural change

Given the base of academic gender equality expertise and associated initiatives as discussed in the previous paragraph, there has always been support for gender equality within UL. However, this support was episodic, undertaken in pockets around the university and led by individuals. Apart from the equal opportunities committee, which was disbanded in the late-2000s, equality efforts were not explicitly led by executive committee.[17] Equality and diversity was within the HR portfolio, rather than being embedded across UL. Thus, regardless of the desire for change across the organization, individuals involved found it difficult to consistently devote the time and focus necessary to further that change. Gender equality work was not recognized as part of an individual's workload, it was not highlighted in promotional CVs, and, in the main, was undertaken as a "voluntary" effort across the organization.

Once Athena SWAN was introduced, with institutional accreditation now a stipulation from the HEA and funding agencies, this changed dramatically. For example, ensuring that gender equality, diversity and inclusion are mainstreamed throughout the organization has now become a presidential responsibility, with Athena SWAN now included in the work of Executive Committee. A UL Self-Assessment Team (SAT) was established, chaired by the President and including representation from executive committee and management teams in UL. Local level SATs have been established in all faculties, schools and departments, and gender equality is included in their meeting agendas. With over 300 people involved in Athena SWAN SATs, including students, it is now recognized as part of an individual's workload including within promotion and recruitment. The focus on gender equality has expanded, and in 2017, UL appointed a special advisor to the President on equality, diversity and inclusion for three years, resulting in the launch of an Equality and Human Rights Strategy (2019–22), which includes Athena SWAN, and taking account of 12 other equality grounds including ethnicity, race disability and social status. In 2021, this position was replaced by the director of human rights, equality, diversity and inclusion. This responsibility includes Athena SWAN, and recognition of equality, particularly gender equality, now manifests across campus on a regular basis. Twenty-five percent of STEMM professors in UL are women. UL proudly hosts a nationally recognized annual International Women's Day Conference, run by the HR Division, opened by the UL President, supported by local companies and regularly over-subscribed. This is a far cry from those celebrations held in a dark corner in UL during the 1990s! A conference to celebrate International Men's Day has been organized annually since 2019.

When companies come to UL faculties, schools and departments with scholarships to support women, particularly in science, technology, engineering and mathematics, there is a greater willingness to promote these scholarships to students, and there is pride in the students' attainment of these. Breast-feeding facilities have been set up, including rooms and availability of fridges, and buildings have been modified to include gender-neutral toilets. When there are specific days to be celebrated such as those recognizing Lesbian, Gay, Bisexual, Transgender, Queer and Intersex (LGBTQI) communities, the relevant flag is raised on the main flagpoles at the entrance to UL, and/or directly outside Plassey House, where the President's and some Vice-Presidents' offices are located. In 2019, UL set up Ireland's first "Rainbow Housing" on-campus initiative for students who wish to live together in a house/apartment that supports the LGBTQI community, ensuring that they have a living space which is affirming to all sexual and gender identities. Additionally, all of the on-campus accommodation is accessible and has rooms which cater for students with disabilities.

National and Institutional Policies

Apart from national rules and regulations implemented within UL, Athena SWAN and the equality and human rights strategy has ensured the development and implementation of new and modified internal policies and procedures. Equality and human rights is on the Corporate Governance Risk Register, which records the risks and opportunities that may affect the delivery of UL strategy, and so, is closely monitored by governing authority.[18] Additionally, equality is reported on biannually to two governing authority sub-committees focusing on staff – the finance, human resources and asset committee – and focusing on students – the access, equality and student affairs committee. When modified or developed, all UL policies must now exclusively state the effect, either positive or negative, which it has on the each of the equality grounds. Recruitment, promotions and progression policies require that every board is balanced with 40 percent gender representation on boards with more than three members, and at least 30 percent when smaller than this. Internal board members must have completed unconscious bias training, and applicants' career gaps due to maternity or other relevant leave are explicitly taken into account. All senior positions require that candidates show *demonstrable experience of leadership in advancing gender equality*. Both internal quality reviews and those by UL's linked providers include equality as one of the topics to be discussed. UL has also introduced a research grant for returning academic carers, for example, from maternity leave and carer's leave, and, in the 2020 round of promotions, 36 percent of those who were promoted were recipients of this grant.

These changes have extended to the student level. For example, students are involved in the development of policies, mainly through student life and the postgraduate students' union, the representative bodies for undergraduate and postgraduate students. As required at the national level, UL is implementing the Department of Education's *Framework for Consent in Higher Education Institutions: Safe, Respectful, Supportive and Positive – Ending Sexual Violence and Harassment in Irish Higher Education Institutions*. The university has recently set up sub-committees to review academic structures for inclusive learning and to promote racial equality and multiculturalism. A gender identity and recognition policy is currently being developed and trans awareness workshops have been delivered to staff and students.

Outcomes

There has been a dramatic cultural shift in relation to gender equality, diversity and inclusion in UL over the past ten years. That shift encompasses how these issues are prioritized and mainstreamed and is predicated on a basis of accountability directly related to funding recognition. The implementation of national requirements has ensured that equality is now a priority action and the responsibility of the governing authority, President and executive committee. It is no longer something that exists on the sidelines – it is now front and center of university policies. As a result of this culture shift, individuals throughout the university are now more aware about what they can and cannot do from an equality perspective. When setting up research discussion panels or organizing research speakers, a conscious effort is now made to ensure that these events are gender balanced. When marketing materials are being developed, individuals consider the images and pictures, ensuring that no grouping is unfairly or unequally represented. These actions are not governed by rules and regulations – rather, there is a management focus on equality issues which has heightened awareness. Furthermore, the addition of training modules, such as mandatory unconscious bias training for those participating in interview panels, does, in fact, cause people to shift their thinking.

We have seen a shift in culture within UL STEM departments due to Athena SWAN implementation. There is a recognition that the recruitment and promotion systems have been set up to support women and men equally, and seen improvement in the numbers of STEM senior female academics at associate professor and full professor level, and both of UL's SALI professors appointed in 2021 have been in STEM disciplines. People think about equality and how it can be actively implemented, and the environment has changed to support early career academics, both men and women. Staff are now afforded the opportunity to be mentored, to discuss their career plans and issues, often relating to being in STEM departments. From the potential student perspective, marketing materials now include cohorts of female students, women from STEM are shown telling their story on web pages and we run workshops for younger women showcasing STEM as an interesting career. For current students, we now bring in women working in STEM to give talks and our external examiner pool is gender balanced. Students have set up a Women in STEM society which is active in, for example, organizing talks and STEM industry visits. Increasingly, individuals are ready to speak out about equality, to talk about the issues and to provide support to those who need it, regardless of their circumstances.

Athena SWAN – Not A Panacea

While Athena SWAN provides an excellent, systematic way of advancing workplace gender equality at the micro-level, it should not be viewed as a cure-all panacea. Firstly, it needs to be enacted in a way that does not unintentionally reinforce inequity. For example, in their examination of the effectiveness and impact of an Athena SWAN program using data gathered from five departments within a UK medical school, Caffrey et al. (2016) reports both positive and negative outcomes. They found that the implementation of Athena SWAN principles was viewed by staff as a predominantly positive initiative that created the necessary social space to address gender inequity and highlighted problematic practices that could then be addressed. However, they also found that gender inequity was unexpectedly reinforced in the program's enactment, with female staff assuming a disproportionate amount of responsibility for Athena SWAN work, which may negatively impact career progression for those women. This points to the fact that unless recognition is given for the workload that is involved in enacting these programs, they can unintentionally reproduce inequality. Their findings also show that staff considered the impact of the program to be weakened by broader institutional practices (such as inconsistent or limited workshare opportunities), national policies (e.g., the need to extend the duration of paid paternity leave) and societal norms (including expectations that women still shoulder more responsibility for childcare within the home), which extend beyond the remit of the program.

A subsequent study by Ovseiko et al. (2017) exploring the experiences and perceptions of staff who participated in Athena SWAN programs in medical science departments in Oxford University provided similar insights. The study participants also viewed Athena SWAN program participation as resulting in positive structural and cultural changes, advancing development of a more supportive work environment which systematically and intentionally sought to eliminate discrimination and bias. Many participants observed that this would not have been achieved without linkage of government funding to achievement of Athena SWAN accreditation. However, they also considered that Athena SWAN was limited in its ability to improve gender equality in the absence of broader institutional and societal changes. Areas highlighted included entrenched power and pay imbalances, enduring lack of work–life balance in academic medicine, concerns regarding the sustainability of positive changes, concerns that achieving the award could become an end in itself, and resentment about perceived positive discrimination (i.e., the perception that women are benefiting from favorable treatment to the detriment of men).

Similar to Caffrey et al.'s (2015) study, these participants emphasized the need for additional structural and cultural changes in the university and society. The work of both Ovseiko et al. (2017) and Caffrey et al. (2015) points to the need for careful management of how gender equality initiatives are enacted at the organizational level. Specifically, they clarify that for sustainable workplace gender equality changes to be achieved, a symbiotic, mutually reinforcing approach where national policies underpin organizational initiatives and the success of those initiatives in turn stimulates further national gender equity policy development is required. Third, they recognize that cultural and societal change is dependent on state and organizational level incentivization to increase participation of men in unpaid family caring work, something that extends beyond the scope of higher education policy and micro-level initiatives. Such intentional incentivization does however achieve critical social change. For example, countries such as Sweden have implemented a progressive 16-month paid parental leave policy that stipulates that the father must take some of that leave and it cannot be transferred to the mother. Moreover, that leave is available up to eight years post-birth. This has resulted in a cultural shift where a more equal sharing of caring responsibilities has now become the norm rather than the exception, resulting in a more gender-equal society. In summary, for national policies and organizational workplace equality initiatives to achieve sustainable cultural change, they need to be structured in such a way as to intentionally incentivize gender equality in relation to unpaid caring work.

CONCLUSION

Over the past decades, national policies have provided a critical gender equality scaffolding for the removal of workplace access barriers. That scaffolding has facilitated women's participation in the workplace generally. However, although some progress has been made, women continue to remain vastly under-represented in the STEM workforce in most European countries despite rising demand from employers for such professionals. This chapter has described why it is necessary for the environmental scaffolding of national policies to be accompanied by micro-level organizational initiatives that focus on systematically removing multi-layered gender inequalities in the workplace in a way that is specific, measurable, accountable and time-bound. Such interventions are critical if employers wish to attract and retain more women in the STEM sector.

The urgent need to implement systematic organizational initiatives has been highlighted by COVID-19 which exposed the gendered impact of lockdown for working women and the fragility of gender equality progress. As Lewis (2020) notes, pandemics affect men and women differently, with school closures and household isolation moving the responsibility of caring to the unpaid economy, causing women's independence to fall as a silent victim of the pandemic. For example, the UK Office for National Statistics (2020) showed that during the first lockdown, women in the United Kingdom assumed the bulk of childcare in households with young children, typically spending at least three hours a day more on daily childcare duties in households, than did men. Further evidence from a survey of 20,000 working mothers showed that 51 percent of mothers were not able to access necessary childcare supports in order to return to work and as a consequence, 67 percent of key workers were forced to reduce their hours because of a lack of access to childcare (PTS, 2020). Similarly, in the United States, a lack of childcare has been shown (Chotiner, 2020) to be an impediment to bringing workers back into the workforce and Philadelphia Federal Reserve Beige Book data shows that during August to September 2020, labor-force participation among women between the ages of 35 and 54 reduced at four times the rate for men of the same age categories.

These figures point to the inadequacy of childcare provision and its implications for women in the workforce. They paint an uncomfortable picture of entrenched gender inequalities in labor force participation and advancement, one with important ramifications for organizations experiencing STEM talent shortages and exacerbated recruiting and retention difficulties, an issue underpinning a growing global workplace crisis (Amaram, 2019). They confirm that national, macro-level policy approaches to support women's career progression must be accompanied by micro-level organizational initiatives that focus specifically on embedding gender equality best practice, systematically throughout the organization. Athena SWAN is one such example of this type of initiative with significant potential to advance and embed gender equality changes. Although predominantly employed in academic contexts, it is of equal value within broader organizational contexts, including the STEM sector. However, the ability of policy and organizational initiatives to effect sustainable workplace gender equality is likely to remain bounded and localized unless they are accompanied by change at the cultural and societal level. Achieving this will require intentional development of gender equality state policies that incentivize men to assume more responsibilities in relation to unpaid family caring work. Countries and organizations that engage in such incentivization strategies are less likely to suffer talent shortages of women STEM professionals and more likely to become destinations of choice for those seeking gender equal workplaces and societies.

NOTES

1. See https://hea.ie/assets/uploads/2021/01/Institutional-Profiles-2017-18-Jan-2021.pdf.
2. European Parliament Briefing Newsletter 7–10 June 2021 – Strasbourg plenary session www .europarl.europa.eu/news/en/agenda/briefing/2021-06-07/20/tackling-the-under-representation-of -women-in-science-and-engineering.
3. See https://hea.ie/assets/uploads/2019/07/Higher-Education-Institutional-Staff-Profiles-by-Gender -2021.pdf.
4. FESTA EU Project (Female Empowerment in Science and Technology Academia (www.festa -europa.eu/).
5. See https://ilostat.ilo.org/how-many-women-work-in-stem/.
6. See https://www.womenintech.co.uk/8-facts-women-tech-industry.
7. See www.cio.com/article/3516012/women-in-tech-statistics-the-hard-truths-of-an-uphill-battle .html.
8. In Ireland, Higher Education Institutions are universities, technological universities, institutes of technology and other higher education colleges.
9. See www.irishstatutebook.ie/eli/2000/act/8/enacted/en/html.
10. See cordis.europa.eu/project/id/266638/reporting.
11. See www.festa-europa.eu/.
12. See cordis.europa.eu/project/id/321378/reporting.
13. Department level awards can be given to departments, schools or faculties as defined within individual HEIs.
14. See www.ecu.ac.uk/wp-content/uploads/2014/06/SAT-Guidance-July-2017-PDF.pdf.
15. See www.ihrec.ie/download/pdf/ihrec_act_2014.pdf.
16. See ec.europa.eu/esf/home.jsp.
17. See www.ul.ie/presidents-office/university-executive-committee.
18. See www.ul.ie/corporatesecretary/governing-authority.

REFERENCES

Agars, M. (2004). Reconsidering the impact of gender stereotypes on the advancement of women in organizations. *Psychology of Women Quarterly*, *28*, 103–11.
Amaram, D.I. (2019). Attracting and retaining women talent in the global labor market: A review, *Journal of Human Resources Management and Labor Studies*, *7*(1), 1–10.
Appelbaum, E., Bailey, T., Berg, P., and Kalleberg, A. (2006), Organizations and the intersection of work and family: a comparative perspective, in Ackroyd, S., Batt, R., Thompson, P. and Tolbert, P.S. (Eds), The Oxford Handbook of Work and Organization, Oxford: Oxford University Press.
Athena SWAN Charter (2021). Advance HE, UK. Retrieved from: www. advance-he. ac. uk/equality -charters/athena-swan-charter.

Banyard, K. (2010). *The Equality Illusion: The Truth about Women and Men Today*. London: Faber and Faber.

Becker, G.S. (1991). *A Treatise on the Family*, Enlarged Edition. Cambridge, Massachusetts: Harvard University Press.

Blair, I.V. and Banaji, M.R. (1996). Automatic and controlled processes in stereotype priming. *Journal of Personality and Social Psychology*, *70*, 1142–63.

Bradley, J. and Murphy, E. (1989). "Ireland, 1992 and the structural funds: An economic perspective," *Studies: An Irish Quarterly Review*, *78*(311), 274–84.

Budig, M. and England, P. (2001). The Wage Penalty for Motherhood. *American Sociological Review*, *66*, 204–25.

Caffrey, L., Wyatt, D., Fudge, N., Mattingley, H., Williamson, C., and McKevitt, C. (2016). Gender equity programmes in academic medicine: A realist evaluation approach to Athena SWAN processes. *British Medical Journal*, Open access. Retrieved from https://bmjopen.bmj.com/content/6/9/e012090.

Carli, L.L., Alawa, L., Lee, Y.A., Zhao, B., and Kim, E. (2016). Stereotypes about gender and science: women # scientists. *Psychology of Women Quarterly*, *40*(2), 244–60.

Catalyst. (April 2020). *Women in the Workforce – Europe: Quick Take*. Retrieved from www. catalyst. org/research/women-in-the-workforce-europe/.

Catalyst. (January 2020). *Women in Academia: Quick Take*. Retrieved from: www. catalyst. org/research/women-in-academia/.

Cech, E.A. (2016). Mechanism or myth? Family plans and the reproduction of occupational gender segregation, *Gender and Society*, *30*(2) 265–88.

Chotiner, I. (2020). Why the pandemic is forcing women out of the workforce, *The New Yorker*, October 23, 2020, Retrieved from: www. newyorker. com/news/q-and-a/why-the-pandemic-is-forcing-women-out-of-the-workforce.

Coogan. T.P. (1987). *The Disillusioned Decades: Ireland 1966–1987*, Dublin: Gill and Macmillan.

Coron, C. (2020). What does "gender equality" mean? Social representations of gender equality in the workplace among French workers, *Equality, Diversity and Inclusion: An International Journal*, *39*(8), 825–47.

Council for the Status of Women. (1981). *Irish Women Speak Out: A Plan of Action*, Dublin: Co-op Books Ltd.

CSO. (2019). Gender Balance in Business Survey. *Central Statistics Office, Ireland*. Retrieved from: www. cso. ie/en/releasesandpublications/er/gbb/genderbalanceinbusinesssurvey2019/.

Duflo, E. (2012). Women empowerment and economic development. *Journal of Economic Literature*, *50*(4), 1051–79.

Eagly, A.H. and Karau, S.J. (2002). Role congruity theory of prejudice toward female leaders. *Psychological Review*, *109*(3), 573–98.

Eagly, A.H. (1987). *Sex Differences in Social Behavior: A Social-role Interpretation*. Hillsdale, NJ: Erlbaum.

Ebrahimji, A. (2020). Female Fortune 500 CEOs reach and all-time high, but it's still a small percentage, *CNN*, May 20, 2020. Retrieved from: https://edition. cnn. com/2020/05/20/us/fortune-500-women-ceos-trnd/index.Html.

EU Gender Equality Strategy. (2020–25). *Striving for a Union of Equality: The Gender Equality Strategy 2020–2025*. Retrieved from: https://ec.europa. eu/commission/presscorner/detail/en/FS_20_370.

Forsythe, N., Korzeniewicz, R., and Durrant, V. (2000). Gender Inequalities and Economic Growth: A Longitudinal Evaluation. *Economic Development and Cultural Change*, *48*, 573–617.

Gangl, M. and Ziefle, A. (2009). Motherhood, labor force behavior, and women's careers: an empirical assessment of the wage penalty for motherhood in Britain, Germany, and the USA. *Demography*, *46*, 341–69.

Global Gender Gap Report. (2020). Retrieved from: www3.weforum.org/docs/WEF_GGGR_2020.pdf.

Harford, J. (2018). The Perspectives of Women Professor on the Professoriate: A Missing Piece in the Narrative on Gender Equality in the University, *Education Sciences*, *8*(2): 50.

HEA. (2016). *HEA National Review of Gender Equality in Irish Higher Education Institutions*. Higher Education Authority Ireland. Retrieved from: https://hea. ie/assets/uploads/2017/06/HEA-National-Review-of-Gender-Equality-in-Irish-Higher-Education-Institutions.pdf.

HEA. (2018). *Task Force Report: Accelerating Gender Equality in Irish Higher Educational Institutes: Gender Action Plan 2018–2020.* Higher Education Authority Ireland. Retrieved from: https://hea.ie/assets/uploads/2018/11/Gender-Equality-Taskforce-Action-Plan-2018-2020.pdf.

Inglehart, R., Basáñez, M., Díez-Medrano, J., Halman, L.C.J.M., and Luijkx, R. (Eds.) (2004). *Human Beliefs and Values: A Cross-Cultural Sourcebook Based on the 1999–2002 Value Surveys.* Siglo XXI.

Inglehart, R. and Norris, P. (2003). *Rising Tide: Gender Equality and Cultural Change around the World*, Cambridge: Cambridge University Press.

Inglehart R., Norris, P., and Welzel, C. (2002). Gender Equality and Democracy. *Comparative Sociology, 1*, 235–65.

Inglehart, R. and Baker, W. (2000). Modernization, Cultural Change, and the Persistence of Traditional Values. *American Sociological Review, 65*, 19–51.

Inglehart, R. (1997). Modernization and Postmodernization: Cultural, Economic, and Political Change in 43 Societies. Princeton, NJ. Princeton University Press.

Inglehart, R. (1990). *Culture Shift in Advanced Industrial Society.* Princeton, NJ: Princeton University Press.

Kalaitzi, S., Czabanowska, K., Fowler-Davis, S., and Brand, H. (2017). Women leadership barriers in healthcare, academia and business, *Equality, Diversity and Inclusion: International Journal, 36*(5), 457–74.

Koch, A.J., D'Mello, S.D., and Sackett, P.R. (2015). A meta-analysis of gender stereotypes and bias in experimental simulations of employment decision making. *Journal of Applied Psychology, 100*(1), 128–61.

Koenig, A.M., Eagly, A.H., Mitchell, A.A., and Ristikari, T. (2011). Are leader stereotypes masculine? A meta-analysis of three research paradigms. *Psychological Bulletin, 137*(4), 616–42.

Law, T. (2020). Women are now the majority of the US Workforce – but working women still face serious challenges, *Time Magazine*, January 16, 2020. Retrieved from: https://time.com/5766787/women-workforce/.

Lee, J.J. (1989). *Ireland 1912–1985: Politics and Society*, Cambridge: Cambridge University Press.

Lewis, H. (2020). The Coronavirus is a disaster for feminism, *The Atlantic*, March 19. Retrieved from www.theatlantic.com/international/archive/2020/03/feminism-womens-rights-coronavirus-covid19/608302/.

Mandel, H. (2013). Up the down staircase: women's upward mobility and the wage penalty for occupational feminization, 1970–2007, *Social Forces, 91*(4) 1183–1207.

Maslow, A.H. (1970). *Motivation and Personality*, Harper and Row, New York.

McKinnon, M. and O'Connell, C. (2020). Perceptions of stereotypes applied to women who publicly communicate their STEM work. *Humanities and Social Sciences Communications, 7*, 160.

Muzio, D. and Tomlinson, J. (2012). Editorial: researching gender, inclusion and diversity in contemporary professions and professional organizations, *Gender, Work and Organization, 19*(5) 455–66.

Nadler, J., Voyles, E., Cocke, H., and Lowery, M. (2016). Gender Disparity in Pay, Work Schedule Autonomy and Job Satisfaction at Higher Education Levels. *North American Journal of Psychology, 18*, 623–42.

O'Malley, E. (1989). *Industry and Economic Development: The Challenge of the Latecomer*, Dublin: Gill and Macmillan.

Office for National Statistics. (2020). Parenting in Lockdown: Coronavirus and the effects on work–life balance, *UK Office for National Statistics*, 22 July 2020. Retrieved from: www. ons. gov. uk/peoplepopulationandcommunity/healthandsocialcare/conditionsanddiseases/articles/parentinginlockdowncoronavirusandtheeffectsonworklifebalance/2020-07-22.

Ovseiko, P., Chapple, A., Edmunds, L., and Ziebland, S. (2017). Advancing gender equality through the Athena SWAN Charter for Women in Science: An exploratory study of women's and men's perceptions. *Health Research Policy and Systems*, 15,12.

O'Connor, P. (2019). An autoethnographic account of a pragmatic inclusionary strategy and tactics as a form of feminist activism, *Equality, Diversity and Inclusion, 38*(8), 825–40.

PTS (2020). The true scale of the crisis facing working mums. *Pregnant then screwed.* Retrieved from https://pregnantthenscrewed.com/the-covid-crisis-effect-on-working-mums/.

Rudman, L.A., Moss-Racusin, C.A., Phelan, J.E., and Nauts, S. (2012). Status incongruity and backlash effects: Defending the gender hierarchy motivates prejudice toward female leaders. *Journal of Experimental Social Psychology*, *48*, 165–79.

Rudman, L.A. and Fairchild, K. (2004). Reactions to counterstereotypic behavior: The role of backlash in cultural stereotype maintenance. *Journal of Personality and Social Psychology*, *87*(2), 157–76.

Sanandaji, N. (2018). *The Nordic Glass Ceiling*. Washington, DC: Cato Institute.

Times Higher Education (2019). Gender pay gap: how much less are women paid at your university? *World University Rankings*. Retrieved from: www.timeshighereducation. com/news/gender-pay-gap-how-much-less-are-women-paid-your-university.

Trauth, E.M. and Connolly, R. (2021). Investigating the nature of change in factors affecting gender equity in the IT sector: A longitudinal study of women in Ireland. *MIS Quarterly*, *45*(4), 2055–2100.

University of Limerick Mentoring Scheme (2021). University of Limerick, Ireland. Retrieved from: www.ul.ie/hr/current-staff/learning-and-development/mentoring-scheme.

Woodhams, C. and Lupton, B. (2006). Gender-based equal opportunities policy and practice in small firms: The impact of HR professionals, *Human Resource Management Journal*, *16*(1) 74–97.

7. Promoting gender equality at two European universities through structural change interventions: the EQUAL-IST project

Elena Gorbacheva and Isabel Ramos

INTRODUCTION

Policies promoting gender equality have existed in the European Union (EU) since the 1950s and for over two decades policy makers have been attempting specifically to address gender inequalities in academic careers (e.g., European Commission, 2019). A diminishing representation of women can be observed as they move up the academic ladder. According to statistics across the 28 EU Member States, in 2016 women constituted over half of graduates at the Bachelor and Master's levels, but only 24 percent of staff at the highest academic positions (such as full professors). In the Science, Technology, Engineering, and Mathematics (STEM) fields, the gap between women and men is wider; in 2016 women formed 36 percent of STEM graduates at the Bachelor and Master's levels and held only 15 percent of the highest academic positions. At the same time, an insufficient supply of STEM skills is forecasted between 2014 and 2025, which, together with a low participation rate of women in STEM, are perceived as "barriers that impede job-rich recovery and growth in the EU" (European Commission, 2019, p. 37). As a result, the EU research and innovation funding programs[1] continuously included calls for project proposals aimed at promoting gender equality in universities and, in particular, in STEM research institutions. In 2014–20 such a program, called *Horizon 2020*,[2] funded a number of projects promoting gender equality in research institutions. One of these projects, called EQUAL-IST (*"Gender Equality Plans for Information Sciences and Technology Research Institutions"*),[3] is discussed in this chapter. The chapter aims to provide insights into the work done and the challenges encountered by the two STEM research institutions participating in the EQUAL-IST project, namely the Departments of Information Systems at the University of Münster (Germany) and the University of Minho (Portugal). Furthermore, the peculiarities and similarities in lessons learned during the project implementation in Germany and Portugal are discussed.

The EQUAL-IST project was aimed at *implementing structural change interventions to enhance gender equality, diversity, and work–family balance* at the participating STEM research institutions. Such *structural change* interventions are specific activities aimed at introducing sustainable change in organizational rules, regulations, processes, and cultures (Sangiuliano, Canali, and Gorbacheva, 2019). They can be contrasted with *preparatory* interventions focused on raising awareness about various aspects of gender equality. The need for *preparatory* interventions prior to the introduction of *structural change* interventions became apparent to the project teams at all participating research institutions right from the project start. Although the project proposal had to be approved prior to its submission by decision makers at the highest university levels (rectors and deans), at the department level many decision makers and staff members were hesitant to acknowledge that there were challenges related to gender equality and did not see value in the project. Such resistance could have been caused by the fact that all participating research institutions were still at the initial stage of setting up gender policies (which was one of the requirements for obtaining project funding). Therefore, at all participating research institutions the lack of awareness could be observed about what constituted gender equality and gender inequality issues beyond direct discrimination, why it was important to tackle these issues (and how), as well as what interventions already existed at the department, faculty, university, and country levels. In order to address this discrepancy between a high-level commitment and a department-level resistance, design and implementation of *preparatory* awareness raising interventions were vital.

Both preparatory and structural change interventions were included in tailored gender equality plans (GEPs), which were designed and implemented at each participating research institution. A GEP is *"a set of actions aiming at: (i) conducting impact assessment, as well as audits of procedures and practices, to identify gender bias; (ii) identifying and implementing innovative strategies to correct any bias; and (iii) setting targets and monitoring progress via indicators"* (European Commission, 2012, p. 13). Within the EQUAL-IST project, the tailored GEPs were designed in a participatory manner involving a wider audience of staff members (both academic and non-academic), students, and decision makers. This participatory approach was supported by an online crowd sourcing platform, called CrowdEquality,[4] which was developed and applied within the project (Gorbacheva, Moumtzi, and Stein, 2019; Gorbacheva and Barann, 2017).

The EQUAL-IST project consortium included seven STEM research institutions from northern, southern, eastern, and central European countries.[5] The project was coordinated by the ViLabs company (Greece),[6] which ensured that the project objectives were achieved within the defined time and budget. The research institutions in Germany and Portugal, which are focused on in this chapter, are of interest, as they represent different cultural environments and different country and organizational policies towards gender equality. These differences resulted in different strategies to intervention design and implementation.

In the next section, the background information is provided. The interventions implemented at both institutions are then detailed. Afterwards, the efforts to ensure sustainability of the initiated interventions are presented. The chapter concludes with the discussion of the project contribution.

BACKGROUND

In this section, the background information is presented, starting with an overview of related work and moving on to project-specific information. Here the rationale behind the participation of the selected research institutions in the project is clarified, the approach followed within the project is introduced, and the details of conducting gender audits are provided.

State-of-the-Art Analysis

The individual differences theory of gender and information technology (IT) chosen as a framework for this book summarizes the factors that could help to "1) explain the under-representation of women in the IT field; and 2) account for those women who overcame barriers and entered the IT field" (Trauth, Cain, Joshi, Kvasny, and Booth, 2016 p. 15). These factors include individual identity, individual influences, and environmental influences (Trauth, 2002; Trauth, Quesenberry, and Huang, 2009). The environmental influences construct, which is addressed in this chapter, refers to, among others, policies facilitating or discouraging women's IT career intentions, choices, persistence, and advancement (e.g., Quesenberry and Trauth, 2012). The chapter thus extends extant literature on this topic (e.g., Quesenberry et al., 2006; Trauth et al., 2009; Trauth et al., 2008; Trauth et al., 2008). Policies that promote gender equality in academic careers can be implemented at the multinational, country, and organizational levels, and ultimately lead to the so-called *structural change* – a sustainable change in organizational rules, regulations, processes, and cultures (Sangiuliano et al., 2019).

At the multinational level, in the EU the promotion of structural change towards gender equality within the academic landscape is reflected in the European Research Area policies (European Commission, 2012a, 2012b). One instrument here is the "She Figures" reports (European Commission, 2019, 2022), which present statistics on gender equality in research and innovation in all EU Member States. The first report was released in 2003 and updated every three years since. Furthermore, in 2006 the European Institute for Gender Equality – the EU body dedicated to gender equality – was founded. One of the institute initiatives was released in 2016: Gender Equality in Academia and Research toolkit.[7] The toolkit is aimed at providing "universities and research organizations with practical advice and tools through all stages of institutional change, from setting up a gender equality plan to evaluating its real impact" (EIGE, 2016). The EU policy direction toward gender equality is also reflected in dedicated funding to enable structural change in research performing organizations. Alone within the Horizon 2020 program 38 such projects received funding,[8] including such EQUAL-IST preceding projects as GenSET[9] (2009–12), FESTA[10] (2012–17), GENDER-NET[11] (2013–16), GenPort[12] (2013–17) etc.

At the country level, structural change can be triggered, for instance, by an educational reform and laws fighting gender discrimination and supporting paternity leave (e.g., Trauth and Connolly, 2021). In Germany, for instance, gender equality laws are implemented at both the federal and state levels. All laws are aimed at achieving equality between women and men, eliminating existing and preventing future discrimination based on gender, in particular discrimination against women, as well as improving work–family compatibility for women and men. The federal equal opportunities act[13] (German: *"Bundesgleichstellungsgesetz"*) came into force in 2001 and applies exclusively to federal agencies and courts. Universities and the private sector, on the other hand, must comply with the state equality regulations developed since 1990s. For example, the law on equality between women and men for the state of North Rhine-Westphalia[14] (German: *"Landesgleichstellungsgesetz"*) obliged the University of Münster to establish an equal opportunities office at the university level and equal opportunities committees at each faculty, which have to be led by equal opportunity officers, as well as regularly issue GEPs and report on their implementation. Further structural change interventions included ensuring gender parity in all commissions, implementation of gender quotas based on a cascade model[15] etc.

At the organizational level, achieving structural change at the participating research institutions was the goal of the EQUAL-IST project. In line with, for instance, DuBow and Ashcraft (2016), the intervention areas addressed within the EQUAL-IST project included human resources practices and management processes, teaching and student services, and institutional communication. The interventions started with a systematic collection of gender disaggregated data, followed by the analysis of causes of gender imbalance and the development and implementation of counter-measures, which were formalized and documented in the GEPs (see the "Approach followed within the EQUAL-IST project" section). Such organizational counter-measures can include adjustments in the recruitment, career advancement, work–life balance, and performance evaluation policies. Further interventions to tackle the "glass ceiling"[16] phenomenon can include offering women networking, mentoring, and training opportunities (e.g., Armstrong et al., 2018). It should be noted that an intervention successful in one context might fail when there is an attempt to implement it in organizations and countries with different cultures and policies (e.g., Gorbacheva et al., 2018). Therefore, within the EQUAL-IST project the one-size-fits-all approach to interventions was avoided and each participating research institution had freedom to develop a tailored GEP.

The Departments of Information Systems at the University of Münster (Germany) and the University of Minho (Portugal)

The University of Münster (hereafter "WWU", German: *Westfälische Wilhelms-Universität Münster*) and the University of Minho (hereafter "UMINHO") were interested in and committed to participating in the EQUAL-IST project, because the under-representation of women at all levels in their Departments of Information Systems ("DIS_WWU" and "DIS_UMINHO" hereafter) was recognized as a challenge that needed to be tackled. In line with the "She Figures" report by the European Commission (2019), such underutilization of female talent in academic careers "is a missed opportunity for Europe's economy and for European society as a whole" (p. 38). Furthermore, as STEM professionals are in great demand but are in short supply in the EU, attracting more female students to the Information Systems (IS) study programs would contribute to addressing this gap.

At DIS_WWU in 2017, the share of women among Bachelor's IS students was 13.2 percent (vs. 46 percent of women among all WWU students). The shares of women were higher in the Master's IS study program (25 percent) and among doctoral researchers (24 percent), although the numbers were low in these categories too. Out of six full professors, there was only one woman (16.7 percent).

At UMINHO in 2018, 50.6 percent of all employees and 44.9 percent of teachers were women. Yet, at the School of Engineering, where DIS_UMINHO is integrated, only 23 percent of teachers and 30 percent of students were women. Furthermore, a worrying percentage of women were unable to complete their engineering studies: the dropout rate was 44 percent among Master's students and 25 percent among doctoral students. In academic careers, at DIS_UMINHO a woman had *never* been appointed to a full professor position. Only two out of seven associate professors were women and only six of the 24 tenure-track positions were occupied by women. The project team at UMINHO realized that the under-representation of women was relevant not only to DIS_UMINHO, but to the entire School of Engineering. Therefore, it was decided to broaden the project scope for UMINHO and to design and implement structural change interventions at the School of Engineering level.

Approach Followed within the EQUAL-IST Project

The approach followed within the EQUAL-IST project consisted of six phases, which can be grouped as follows (Figure 7.1):

- Investigate and analyze: Identification of the existing challenges related to gender equality, diversity, and work–family balance.
- Co-design and implement: Setting the objectives to address the identified challenges; design and implementation of the interventions to achieve these objectives.
- Assess and report: Continuous assessment and refinement of the initiated interventions.

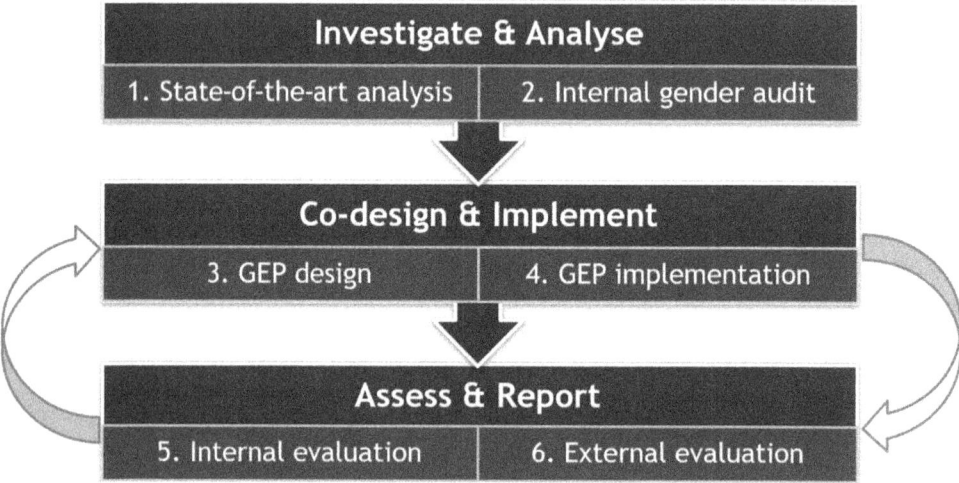

Figure 7.1   Approach followed within the EQUAL-IST project

Research institutions had to follow this high-level approach, as well as adhering to a common timeline and to collaborate with the project external evaluator. At the same time, they had freedom to decide on the exact course of the state-of-the-art analyses and gender audits, on what interventions to include in the tailored GEPs, as well as on how to implement these interventions and to evaluate them internally.

During the first phase, the state-of-the-art analysis of related work was performed to gain insights, which could be valuable for the forthcoming GEP design and implementation. Here the information about GEPs implemented at Departments of Information Systems in the EU, as well as about third-party funded projects aimed at GEP design and implementation, was collected and analyzed. During the second phase, internal gender audits were conducted to investigate current challenges related to gender equality, diversity, and work–family balance faced by students or staff members, and also to collect ideas to address these challenges. The steps performed within the internal gender audits at WWU and UMINHO are presented in the "Internal gender audits" sub-section. Based on the audit outcomes, during the third phase, the GEPs for WWU and UMINHO were designed and in the subsequent fourth phase they were implemented.

The GEPs contained descriptions of the objectives to address the identified challenges, as well as the action plans for the interventions aimed at achieving these objectives. The action plans, in turn, could include information about the intervention goals, target indicators, time-line, resources required, stakeholders responsible, as well as descriptions of the detailed activities that needed to be performed. The fifth and sixth phases were focused on the continuous assessment and reporting on the progress and success of the GEP implementations, which was done both internally (fifth phase) and by an external evaluator appointed for the EQUAL-IST project (sixth phase). The internal assessment was focused on the performed work, while the external assessment was focused on the impact of this work. If the performed work was not in line with the GEPs, corrective actions were discussed and carried out by the responsible stakeholders. The GEPs were implemented in two iterations. Based on the outcomes of the first iteration and the feedback received from the project external evaluator, the initial GEP documents were refined and then implemented further during the second iteration. In addition to these core project phases, the project also included activities dedicated to its management, dissemination of outcomes, and ensuring that the EU ethics requirements were fulfilled.

Internal gender audits
The internal gender audits performed at WWU and UMINHO during the project's second phase included the following steps. To show evidence that women were under-represented at all levels at DIS_WWU and DIS_UMINHO, comprehensive gender-disaggregated statistics for the years 2011–18 were collected for students and staff members. As there was no established practice to collect such statistics, it was time-consuming and involved contacting various university administrative units. These statistics were then analyzed, extended, and used in the subsequent project phases. In particular, the statistics were required to justify the interventions initiated within the project and the need for their continuous implementation after the end of the project. Afterwards different approaches took place at WWU and UMINHO to identify challenges and initial ideas to address them, which are presented in the "Internal gender audit at WWU" and "Internal gender audit at UMINHO" sub-sections.

The identified challenges and the initial ideas to address them were submitted to the WWU- and UMINHO-specific sections of the CrowdEquality online crowdsourcing platform for their open discussion, refinement, and extension. This platform was developed at WWU by a team of eight Bachelor's IS students as part of their project seminar.[17] The students were supervised by a team of academic staff members who acted as stakeholders. The platform was then tested and further improved by the EQUAL-IST project team at WWU. The platform technical basis was the OpenideaL distribution of the open-source Drupal content management system.[18] Within the EQUAL-IST project, the students and staff members from the involved research institutions, as well as external users, had the opportunity to contribute to the platform. All personal data collected on the platform was stored in a secure way on the DIS_WWU server. The platform was maintained by DIS_WWU during the project active phase and was transcoded to a static website at the end of the project to ensure code security and proper maintenance. The platform code (without content) is available on the WWU GitHub repository.[19] By keeping publicly available the anonymized content submitted to the platform and its code, some of the knowledge generated during the project was shared.

Internal gender audit at WWU

At WWU, a comprehensive survey on the status of gender equality was conducted. The survey was informed by the survey on the status of women academics conducted by the Association for Information Systems Women Network, and academic literature on the gender imbalance in the IS field (e.g., Ahuja, 2002; Armstrong and Riemenschneider, 2014; Loiacono, Iyer, Armstrong, Beekhuyzen, and Craig, 2016). Staff members and students at DIS_WWU and the School of Business and Economics (SBE) – the faculty DIS_WWU belongs to – were invited to participate. The survey questions in English and German are still publicly available.[20] The survey was designed in a way that each target group (academic staff members, non-academic staff members, and students) received a set of dedicated questions in addition to the questions relevant for all target groups. Thus, more profound feedback from each target group was obtained. The survey data, together with the collected statistics, were then analyzed and relevant findings were communicated during the subsequent gender audit studies, which included a workshop with six students, five interviews with decision makers, and a focus group with two non-academic staff members.

At a later stage, when further input was collected via the CrowdEquality online crowdsourcing platform, all the information was analyzed by the project team at WWU. The outcomes were then discussed with the stakeholders foreseen to be involved in the forthcoming GEP implementation, including the study coordinator at DIS_WWU, members of the IS student council, and the WWU equal opportunities office. The two main criteria for the idea selection included *priority* and *feasibility*. Priority meant urgency to implement, while feasibility implied that there were, on the one hand, sufficient resources required for the implementation and, on the other hand, it was likely that an intervention would be adopted by DIS_WWU staff members and students and sustain beyond the project runtime. The selected ideas acted as a basis for the interventions included in the GEP.

Internal gender audit at UMINHO

UMINHO at the beginning of the EQUAL-IST project was at a very early stage in what concerns gender equality policies. Despite the impactful research on gender equality carried out predominantly at the Institute of Social Sciences, much of its efforts focused on aspects external to the university. Other schools and institutes at UMINHO also had some activity in the gender area but still at a very initial stage. The School of Economics and Management studies focused on gender issues in the labor market. The School of Law researched various governmental initiatives for gender equality, and the Institute of Education had long studied gendered violence. The initial goal of the EQUAL-IST project team at UMINHO was to engage key organizational actors into the project discussions and interventions. The team implemented a series of workshops and individual interviews to understand how students, teachers, researchers, and administrative personnel understood gender equality, and to elicit the main challenges each of these communities faced. This internal discussion was essential to establish the basis for the GEP.

Twelve interviews were conducted involving various sectors, positions, and functions at the School of Engineering. Two surveys were conducted, one for academic staff members and another for students. The team implemented five workshops, of which four involved specific audiences (academic staff members, administrative staff, researchers, and students), and one final workshop with all participants of the previous workshops. The results of this initial diagnosis were presented to decision makers from most of the university schools and

Table 7.1   *Overview of the intervention objectives at WWU and UMINHO*

Objectives at WWU	Objectives at UMINHO
1. To raise awareness about gender equality aspects and issues	1. To raise awareness about gender equality challenges
2. To increase the share of women among Bachelor students	2. To institutionalize gender equality principles in the mission and strategy
3. To improve work–family balance of academic staff members and students	3. To define structures and services that favor the practice of gender equality and the conciliation of life
4. To enhance inclusion of international students	

research centers. The feedback was registered and later discussed by the project team. This internal gender audit generated the first ideas for the interventions, which were then included in the GEP.

INTERVENTIONS: GENDER EQUALITY PLANS

This section starts with an overview and comparison of the specific objectives, which were defined at WWU and UMINHO to address the challenges revealed during the internal gender audits. These objectives are summarized in Table 7.1. The interventions to achieve these objectives were included in the GEPs for WWU and UMINHO. Examples of *preparatory* interventions implemented at WWU can be found in the Appendix, while the *structural change* interventions and obstacles faced during their implementation at both institutions are described below.

One unifying objective defined both at WWU and UMINHO dealt with *raising awareness about the aspects related to gender equality*. The implemented *preparatory* interventions were all aimed at achieving this objective. At WWU, several interventions were also aimed at raising awareness about the GEP implementation among staff members, students, and a wider audience.

At WWU the second intervention objective was *to increase the share of women among Bachelor students*. The under-representation of women at DIS_WWU at all levels was the main reason for participation in the EQUAL-IST project. During the project internal gender audit, it was revealed that the root problem for it was the low share of women among *Bachelor's* IS students. In order to achieve this objective, both (inter)national marketing activities targeted at a wider audience and local "in-person" activities were implemented. The third objective was *to improve work–family balance of academic staff members and students*. During the internal gender audit, several parents working at DIS_WWU indicated that they experienced difficulties in balancing work and family life. The fourth objective was *to enhance the inclusion of international students*, although it goes beyond the topic of gender equality and deals with the identified challenges related to diversity and inclusion. The results of the internal gender audit showed that international IS students often felt excluded, especially from group work, as German students tended to team up with other German students. This issue was especially relevant for the Master's IS study program, where English is the only language of instruction and the share of international students is especially high. All interventions implemented to achieve this objective were aimed at supporting international IS students and raising awareness of academic staff members at DIS_WWU about the challenges these students faced.

At WWU eight interventions were implemented to achieve the first objective (six *prepara-tory* interventions and two *structural change* interventions). Five interventions achieved the second objective (four *preparatory* interventions and one *structural change* intervention). Four interventions achieved the third objective (three *preparatory* interventions and one *structural change* intervention). Finally, five interventions achieved the fourth objective (three *preparatory* interventions and two *structural change* interventions). Moreover, two further valuable interventions were initiated. A report on all implemented interventions is available on the DIS_WWU website;[21] intervention materials were collected on the WWU cloud storage.[22]

At UMINHO the internal gender audit revealed the lack of institutionalization of gender principles and support, which negatively impacted the success of women. Students and academic staff members mentioned that career and defense committees did not follow the gender parity principle. Participants in the various needs-assessment activities indicated that the assignment of university management leadership to women did not result in greater decision-making capacity. Due to many family obligations traditionally attributed to women, such as the care of children and elderly relatives and the organization of family and social events, women still faced great difficulty in achieving performance levels similar to those of men. Female students mentioned similar responsibilities in the family, making it more chal-lenging to achieve their learning objectives. During the 2008 financial crisis, many of them ended up dropping out of the studies to support their families at home. As a result, the second and third objectives at UMINHO focused on institutionalizing principles and best practices of gender equality.

Structural Change Interventions at WWU

At WWU the following *structural change* interventions were implemented at the department (DIS_WWU), faculty (SBE), and university (WWU) levels. At all levels the interventions were implemented primarily by the EQUAL-IST project leader at WWU in collaboration with the project team at WWU. The interventions at the university level were also supported by the WWU equal opportunities office.

Launching a database of gender equality interventions

The intervention was implemented at the university level and aimed at: (i) structuring and storing online in a secure way information regarding WWU interventions promoting gender equality; (ii) providing an overview of these interventions and disseminating them; and (iii) continuously revising, updating, and extending information about these interventions. During the internal gender audit, a lack of awareness about the goals and content of existing interventions promoting gender equality at WWU was revealed. This has led to a distorted and often negative attitude (among both men and women) towards such interventions. One of the underlying reasons for the inefficient communication and promotion of these interven-tions was a rather unstructured and chaotic way of storing information about them. As this information was collected systematically by the WWU equal opportunities office, it could be possible to develop a database of existing interventions promoting gender equality in study, research, teaching, and career development, which are implemented at WWU and at each of its faculties. One of the members of both the EQUAL-IST project team at WWU and the SBE equal opportunities committee took the lead in the implementation of this intervention. The intervention included development of a conceptual solution for structuring information about the WWU interventions, followed by its discussion and approval by the WWU equal opportunities office. This conceptual solution was then implemented using the Drupal content management system.[23] As a result, the WWU database of gender equality interventions[24] (German: *"Gleichstellungs-ONLINE-Datenbank der WWU Münster"*, GL.ON.DA) was launched in 2018.

The intervention resulted in a stronger impact than foreseen during its initial discussion. The developed online platform has fundamentally changed and improved the formerly rather chaotic processes for collection, storage, revision, update, extension, and dissemination (both internally and externally) of information about the interventions at WWU. These processes were revised, simplified, and automated. The platform target audience was extended in the course of its development and currently includes not only internal staff members and students, but also such external bodies as research funding organizations, the government, etc.

There was no dedicated funding for the platform development, which slowed down the intervention progress. Nevertheless, due to the intrinsic motivation of the involved stakeholders to improve the existing processes, the platform was launched successfully and adopted by the WWU equal opportunities office. The intervention sustainability was ensured due to apparent advantages of the new processes over the previous ones. Furthermore, the WWU equal opportunities office and the system administration at DIS_WWU (both are permanent structures at WWU) took the lead in further content and technical support of the platform. The platform was identified as an important tool in the *WWU Equal Opportunities Future Concept* document and further internal and external reports.

Sensitizing members of appointment committees for tenured positions

This intervention was directed at raising awareness about gender equality among members of appointment committees for tenured positions. It was implemented at the faculty level and aimed at (i) ensuring a fair and transparent appointment procedure at SBE and (ii) sensitizing committee members about the importance of ensuring equal treatment of all candidates and avoiding any form of bias and discrimination. The intervention included intensifying the communication between the SBE equal opportunities officer and committee members. As a result, since 2017, each time a new committee for a tenured position is formed, the SBE equal opportunities officer informs its members (via a set of emails) about existing state and university regulations related to gender equality in recruitment, as well as about the importance of avoiding unconscious bias during the candidate selection.

During the project runtime, the intervention was completed in four appointment committees. In three of them, one at DIS_WWU and two at other SBE departments, female applicants were on the first place, meaning that they received the offer to fill the position. This is an outstanding result, as at the beginning of the EQUAL-IST project women formed only 12 percent of full professors at SBE and at DIS_WWU there was only one woman out of six full professors. WWU equal opportunities office supported this intervention and there were no objections to continuing to implement it after the end of the project, from either the SBE dean or the members of the SBE equal opportunities committee.

Operationalizing the Maternity Protection Act

This intervention was implemented at the faculty level and aimed at supporting and protecting students who have children or are about to become parents. The intervention was motivated by an update in 2018 of the Maternity Protection Act[25] (German: *"Mutterschutzgesetz"*) –

a national law, which, among the other aspects, regulates the duties of the universities in Germany towards students who have children or are about to become parents. A window of opportunity occurred to allocate resources for a contact person at SBE to support such students. Moreover, the examination regulations for all study programs at SBE were extended with information about the rights of such students (e.g., the right to prolong the time for writing a Bachelor's or a Master's thesis). The intervention was triggered by the WWU equal opportunities office and supported by the EQUAL-IST project leader at WWU (acting in her role as the SBE equal opportunities officer). At the WWU level, a coordinating unit "Maternity Protection Act" (German: *"Koordinierungsstelle Mutterschutzgesetz für Studentinnen"*) was established. The appointed at SBE contact person also acted as a connection point between the students and this coordinating unit. Establishing these permanent structures ensures the intervention sustainability. Furthermore, statistics of students who have children or are about to become parents can now be collected. These statistics, however, might not be comprehensive, as students still have the right not to respond to the SBE contact person or the coordinating unit at WWU.

Establishing a new equal opportunities committee

This intervention was implemented at the faculty level and aimed at assembling a new SBE equal opportunities committee with active and motivated members. In order to facilitate the sustainability of the interventions initiated within the EQUAL-IST project, a new SBE equal opportunities committee was formed in 2018. For the first time SBE staff members and students actively volunteered to become committee members. The previous committee, on the contrary, consisted of the members who were assigned to it by the SBE dean mainly to fulfil a legal obligation from the state and university regulations to establish such a committee at each WWU faculty. This time potential committee members were recruited by the EQUAL-IST project leader at WWU (acting in her role as the SBE equal opportunities officer) and those who expressed interest were then elected at a meeting of the SBE council. When forming the committee, the gender balance among its members was ensured. The committee had 15 members, including regular and deputy members from the groups of SBE professors, staff members, and students, in addition to the SBE equal opportunities officer and her two deputies.

Creating gender sensitive marketing materials

This intervention was implemented at the department level and aimed at revising the marketing materials promoting the IS study program applying a gender-sensitive approach. The ultimate goal of the intervention was to ensure that the study program is presented in an attractive and welcoming way for all. The revision was based on the insights from the literature and materials on this topic, which were collected and analyzed.[26] In particular, it was evaluated to determine if there was a balanced representation of men and women in images, gender-sensitive language was used in texts, and gender stereotypes were not transmitted in images or texts. Gender-sensitive language aims at acknowledging gender equality in written and spoken language (EIGE, 2019). This evaluation process was followed during the revision of the brochures and slides promoting the Bachelor's and Master's IS study programs. Recommendations for improvement were communicated to the study coordinator at DIS_WWU and incorporated into the revised versions of the materials. Afterwards the updated brochures were published and disseminated, while the updated slides were used by the study coordinator during promotional events.

In addition, recommendations for gender-sensitive language were for the first time incorporated into the WWU equal opportunity framework document[27] (German: *"Gleich-stellungsrahmenplan der Westfälischen Wilhelms-Universität Münster"*), which is a policy document summarizing WWU regulations related to gender equality. As a result, there is now an official prescription also at the university level to follow gender-sensitive language whenever a new official document is created, including any future marketing materials promoting the IS study program.

Supporting international students

As mentioned above, during the internal gender audit at WWU it was revealed that international IS students often felt excluded, especially from group work. Although this challenge is related to diversity and inclusion and goes beyond the topic of gender equality, it was nevertheless decided to address it within the EQUAL-IST project. Two interventions aimed at connecting local and international IS students were implemented at the department level.

The first intervention included establishing a working group "Mentoring" within the IS student council. Several meetings of the working group took place, where activities were defined and planned. The working group conducted several question-and-answer sessions for first semester Bachelor's IS students, where the students received useful information and could ask questions and share experiences. The events were also aimed at connecting students. Unfortunately, the existence of the working group largely depends on the availability and motivation of the IS student council members. They participate in the council for a limited time and do not receive any payment for their work. Moreover, they have a range of responsibilities in addition to supporting international students. Thus, it was difficult to ensure the sustainability of the "Mentoring" working group.

The second intervention included the development of a new internal guideline recommendation that academic staff members at DIS_WWU assign students to groups within (Master's) IS courses in a random way. Thus, local and international students were brought in contact during group work, which contributed to mitigation of gender and culture prejudices that could exist. The guideline was approved by professors and communicated to academic staff members at DIS_WWU. Anecdotal evidence showed that academic staff members adopted the guideline, although no comprehensive evaluation was conducted.

Structural Change Interventions at UMINHO

At UMINHO the following *structural change* interventions were implemented at the School of Engineering and university levels.

Broadening the scope of the office for inclusion

The intervention was implemented at the university level and aimed at extending the focus of the office for inclusion actions to encompass the analysis and monitoring of gender equality. The office for inclusion had a mission to promote inclusion in the academic context and equal opportunities for all. Over the years, it has supported international students and students with special needs who experience difficulties in adapting to academic life or personal crises. Lesbian, gay, bisexual, or transgender (LGBT) students previously sought out this office to find support for their integration into academic life. Thus, the need to address the topics of gender equality and sexual orientation had already been detected. The new responsibilities of the office for inclusion included the following: (1) collection, aggregation, and dissemination of the information about the situation concerning gender equality at UMINHO on a continuous and regular basis; (2) the organization of the training and other events on gender-sensitive practices and communication strategies; (3) the dissemination of the UMINHO code of conduct for diversity and equality; and (4) support in concrete situations related to gender inequality.

This extension of responsibilities is generally understood as relevant and conducive to greater justice and wellbeing. However, the UMINHO managers raised concerns about the financial effort associated, particularly regarding the increase in human resources needed to carry out the new tasks. This constraint was responsible for the slowness associated with this structural change. Nevertheless, it started with the project and in 2021 was in the final stages of implementation. The agreements established with gender-related governmental bodies such as the National Commission of Citizenship and Gender Equality contributed to it. New legal obligations to report practices to promote gender equality at the national and EU levels were also helpful.

Valuing research on gender equality
The intervention was implemented at the university level and aimed at ensuring that the research on gender equality carried out at all UMINHO schools and institutes is recognized as having the same value as other types of research. One of the challenges encountered by the EQUAL-IST project team at UMINHO was the lack of regular dialogue on gender issues affecting research at UMINHO. Moreover, participants reported that research focusing on gender equality issues in the UMINHO schools and research centers other than the Social Sciences Institute was not seen as having the same value as research on other topics. The dialogue started with the project actions and continues today. A multidisciplinary group was formed during the project and is still regularly consulted. In May 2021, the dialogue about the GEP for UMINHO was resumed, focusing on updating the document produced during the EQUAL-IST project and formalizing it throughout the university. The multidisciplinary group has played a crucial role in identifying the interventions to be included in the updated GEP.

Since the project end in 2019, there has been a growing number of events organized by the rector's office, including involvement in a new gender equality project focused on developing digital skills to support the integration of migrant women. This renewed interest in supporting gender equality projects indicates a culture change. Changing the university culture will inevitably bring new opportunities to address existing gender equality challenges and facilitate structural changes. Another important outcome of the project was bringing together the researchers at UMINHO working on gender studies in various disciplines, from medicine to law through engineering and management to social sciences. Some of these researchers were identified by the project team based on a search in the open repository of articles and doctoral theses at UMINHO. Furthermore, university schools and research centers were called to participate in the multidisciplinary working group committed to pursuing collaborative and cross-disciplinary research on gender equality.

Introducing communication norms

The intervention was implemented at the university level and aimed at aligning the communication of information about study programs with the principles of gender equality. In this regard the project team worked closely with the office for communication, information, and image at UMINHO to define norms and ensure that they are applied in promoting study programs and other marketing endeavors. The aim was to observe linguistic neutrality and promote inclusive marketing messages.

Making gender inequalities transparent

The intervention was implemented at the School of Engineering and aimed at fostering a transparency culture that facilitates pointing out problems and co-designing solutions for student challenges. To achieve this goal, the project team defined interventions in three contexts: (i) definition of partnerships with local entities that carry out regular actions with secondary school students and organization of the annual visits of these students to the School of Engineering; (ii) analysis of the professional experience of female engineers trained at the School of Engineering to better understand the gender challenges they face; and (iii) support for the implementation of gender-sensitive approaches when defining programs and learning dynamics. Regarding the first context, the aim was to demonstrate the potential of women in the engineering disciplines, demystifying some of the stereotypes that still exist. For the second context, a series of events were defined to bring alumni to share their experience about patterns of segregation and discrimination against women in the business world. These events facilitated the definition of research focusing on the particularities of the culture shaping the careers in engineering disciplines to identify ways to address the challenges. Moreover, these events were opportunities for students to hear from inspiring women and help them deconstruct common stereotypes. For the third context, the project team proposed regular training sessions for academic staff members on how to identify and remove gender stereotypes in the information and learning activities they promote. Moreover, the project team proposed creating an observatory of publications and projects about gender equality and diversity as a mechanism to support gender mainstreaming[28] in disciplinary areas and to promote partnerships and exchange of experiences with researchers and teachers.

These proposals have been widely implemented since the end of the project. Only the study of the professional challenges faced by trained engineers has not yet been possible, although several attempts have already been made to initiate it. This delay was caused by the difficulty in obtaining the necessary financial resources.

Project Sustainability

The initiated interventions towards improved gender equality, diversity, and work–family balance need to be sustainable to make a difference. According to Athena Swan, which is a charter used by higher education institutions worldwide to advance the careers of women in STEM,[29] the results of successful interventions become visible only after at least five years following the start of their implementation. The sustainability of interventions depends on the commitment and engagement of decision makers at all university levels, as well as on the availability of resources (human, financial, and infrastructural) to continue the implementation.

Project sustainability plans were developed at both WWU and UMINHO and are discussed in this section. Here the initiated interventions were analyzed, and those which needed or had the potential to continue to be implemented beyond the project runtime, were identified, discussed, selected, and then classified as those relevant for the department, faculty/school, or the entire university. Sustainability plans need to be disseminated, implemented, and later regularly evaluated and refined. It is important to ensure that the stakeholders involved in their implementation remain active and motivated.

Sustainability plan at WWU

At WWU, selected interventions formed the basis for the sustainability plans at the department and faculty levels. At the department level, a new document called *"Plan for Recruitment, Retention, and Advancement of Talented Women, Internationals, and Parents"* was designed for the period 2019–23 and included 17 interventions. The goals of this sustainability plan were to ensure the sustainability of the interventions initiated within the EQUAL-IST project and to improve the processes for personnel recruitment, retention, and advancement. The document has been finalized, approved by the professors, and published on the department website[30] facing no significant resistance. This sustainability plan was foreseen as a living document, which would be reissued every four years. However, after the end of the EQUAL-IST project, no one was assigned to support its implementation on an operative basis, as well as to continuously discuss and monitor the implementation progress.

At the faculty level, after the end of the EQUAL-IST project, an existing document called *"Gender Equality Plan for the University of Münster's School of Business and Economics"* (German: *"Gleichstellungsplan der Wirtschaftswissenschaftlichen Fakultät"*) was updated for the period 2019–22 and published on the faculty website.[31] This policy-planning document is prescribed by the state and university regulations and is obligatory for all faculties at WWU. Unfortunately, the draft version of the document prepared by the EQUAL-IST project leader at WWU (acting in her role as the SBE equal opportunities officer) was not adopted by the new SBE equal opportunities officer and most of the proposed interventions were not included.

In addition to preparing the draft sustainability plan for the faculty, the EQUAL-IST project leader at WWU developed a document summarizing the envisioned tasks and responsibilities of future SBE equal opportunities officers and estimated annual workload for each task. This document was presented to the SBE dean, arguing that the current situation, where the position of the SBE equal opportunities officer does not receive any dedicated funding, needs to be changed, and additional support, resources, and opportunities need to be provided. As a result, certain structural changes related to the position of the SBE equal opportunities officer took place. First, a tenured female professor at SBE took over this role from the EQUAL-IST project leader at WWU. Earlier this role was assigned to SBE staff members who had fixed-term contracts, which hindered the continuity and sustainability of the work carried out. Furthermore, one of the tenured SBE staff members was assigned to support the new SBE equal opportunities officer on an operative basis. Thus, sustainability of the gender equality work at SBE could be ensured. Finally, it should be mentioned that further interventions aimed at enhancing gender equality, diversity, and work–family balance are continuously implemented at the WWU level, independently or with only minor support from the EQUAL-IST project. The WWU Equal Opportunities Office is primarily involved in leading the implementation of these interventions and ensuring their sustainability.

Sustainability plan at UMINHO

At UMINHO, during the three years of the EQUAL-IST project, a considerable effort was made to raise awareness and initiate a dialogue about existing gender equality challenges. Alongside these efforts, other activities were carried out to trigger structural changes, thus ensuring that the GEP is maintained and updated. To guarantee the continuity of the interventions carried out during the project, the UMINHO team ended up abandoning the focus only on DIS_UMINHO, expanding it to the School of Engineering and other university schools and research centers. This broadening of scope resulted in a vibrant experience, as the team identified many researchers and students interested in the topic of gender equality. The internal network that was created has provided a strong guarantee of the success of the changes implemented during the project and of the creation of new interventions.

The project implemented some important permanent changes. The collaboration with the National Commission of Citizenship and Gender Equality was formalized in a protocol signed by the UMINHO rector. To promote transparency regarding gender issues and to disseminate gained insights, an online observatory was established. The UMINHO Gender Equality Network of researchers ensures collaboration in multidisciplinary research projects and events. The annual events to attract girls to STEM study programs were formalized. Modules and classes on gender equality are being integrated into existing courses. Calls for multidisciplinary projects and Master's dissertations on gender equality topics are now issued annually. Project proposals on gender equality topics are submitted annually to local, national, and international funding organizations. Finally, there is regular dissemination of guidelines on gender-sensitive language and practices.

Creating awareness and changing mindsets requires time and effort. Although the EQUAL-IST project was instrumental in triggering a change process, three years is a very short period for the changes to mature and strengthen. Several planned structural changes were not implemented because they collided with cultural barriers and financial constraints. One of them was the implementation of gender parity in career progression and doctoral dissertation panels. Another was the consideration of gender balance among job candidates during recruitment. The final two structural changes that remained unimplemented were the creation of organizational structures to support the balance between professional and family life, as well as better support of women's career advancement in a culture where family care is still an obligation of women. Nevertheless, at the time of writing this chapter, in the fall of 2021, two important initiatives demonstrate that the legacy of the EQUAL-IST project is still alive and producing change. First, driven by new European commission funding rules, the university administration approved the GEP UMINHO to be adopted by all the university units: eight schools and four institutes. At the beginning of 2022, the commission for gender equality at the university will be created, which will be responsible for implementing or promoting the initiatives included in the plan. The second initiative is the initiation of the Master's in Sociology of Gender and Sexuality study program at the Institute of Social Sciences. This Master's will ensure that gender equality will continue to be valued and studied at the university to inform future updates of the GEP.

CONCLUSION

This chapter provides an overview of the interventions to enhance gender equality, diversity, and work–family balance implemented at the two European STEM research institutions and the obstacles and resistance encountered during their implementation. The chapter is aimed at supporting both researchers and practitioners who investigate and address gender imbalance in academic careers in the STEM fields. The practical contribution is to point to the undertaken interventions, the approaches to ensure their sustainability, and the lessons learned. From an academic perspective, the chapter discusses gender equality policies in academic careers and the *environmental influences* construct of the individual differences theory of gender and IT. It addresses the calls to examine this theory further, to conduct comparative studies on gender imbalance in the IT profession in various contexts, and to investigate promising interventions to promote gender equality in IT organizations (Gorbacheva et al., 2018).

The efforts and commitments undertaken to launch, conduct, and complete the EQUAL-IST project show a policy direction toward gender equality at the organizational, country, and multinational levels in the EU. The project was funded by the EU and selected from numerous other proposals submitted by European universities. Prior to the submission of the project proposal to the EU, it was approved by decision makers at the highest university levels (rectors and deans). Furthermore, the driving forces of the project were teams at each participating university who jointly created the project proposal and were motivated to work on the project.

The project revealed differences in the maturity levels of national legislation on gender equality in different European countries, in particular, in Germany and Portugal. The regulations promoting gender equality are well established in Germany, but not in Portugal. Although in Portugal there has been significant progress in recent years, even such basic practices as gender parity in recruitment and assessment panels or encouraging female leadership in academic careers are still not yet rooted in the day-to-day life of universities. Despite this country-level difference at the two universities discussed in the chapter, namely WWU and UMINHO, it is interesting to note that, at both, a lack of awareness as well as resistance were revealed during the project. The project showed that relatively strong gender equality regulations at the country and university levels in Germany did not lead to a smoother implementation of gender equality interventions at the department level. It can be concluded that while gender inequalities might no longer be visible in daily life in the EU, they might continue to exist in indirect forms.

Nevertheless, the EQUAL-IST project provided all participating universities with the resources and opportunities to implement both *preparatory* and *structural change* interventions towards gender equality. *Preparatory* interventions resulted in a positive change in attitudes towards the topic of gender equality or at least in discussions about it, while *structural change* interventions established small, but important sustainable changes in organizational rules, regulations, processes, and cultures. Overall, it can be concluded that the EQUAL-IST project resulted in the raised awareness, increased attention and legitimization, and higher visibility of the topic of gender equality at both WWU and UMINHO, which positively contributed to the sustainability of the project interventions.

NOTES

1. See https://ec.europa.eu/eurostat/cros/content/research-projects-under-framework-programmes-0_ en.
2. See https://ec.europa.eu/programmes/horizon2020/en.
3. See https://equal-ist.eu;https://cordis.europa.eu/project/id/710549;www.uni-muenster.de/forschung az/project/10219?lang=en.
4. See www.crowdequality.eu.
5. See https://equal-ist.eu/partners/.
6. See https://vilabs.eu.
7. See https://eige.europa.eu/gender-mainstreaming/toolkits/gear.
8. See https://cordis.europa.eu/search?q=contenttype%3D%27project%27%20AND%20programme %2Fcode%3D%27H2020-EU.5.b.%27andp=1andnum=100andsrt=/project/contentUpdateDate: decreasing.
9. See https://cordis.europa.eu/project/id/244301.
10. See https://cordis.europa.eu/project/id/287526.
11. See https://cordis.europa.eu/project/id/618124.
12. See https://cordis.europa.eu/project/id/321485.
13. See www.gesetze-im-internet.de/bgleig_2015/index.html.
14. See https://recht.nrw.de/lmi/owa/br_text_anzeigen?v_id=220071121100436242.
15. In the cascade model quotas are based "on the percentage of women at the level immediately below for each type of position" (Wallon, Bendiscioli, and Garfinkel, 2015, p. 2).
16. The "glass ceiling" phenomenon refers to invisible structural barriers that prevent women from advancing.
17. A project seminar is a special teaching format at DIS_WWU, where students work in teams on tasks relevant for research or practice.
18. See www.drupal.org/project/idea.
19. See wiwi-gitlab.uni-muenster.de/equal-ist/crowdequality.
20. See https://bit.ly/3ug5LMG.
21. See www.wi.uni-muenster.de/career/diversity-and-inclusion/report-equal-ist-project.
22. See https://uni-muenster.sciebo.de/s/D1wS3y9rZXRUAWB.
23. See www.drupal.org.
24. See https://glonda.uni-muenster.de.
25. See www.gesetze-im-internet.de/muschg_2018.
26. See https://geschicktgendern.de/; https://koordination-gender.uni-graz.at/de/services/geschlechter sensbile-schreibweise/; https://gb.uni-koeln.de/gender_sensitive_language/index_eng.html; www .philhist.unibe.ch/unibe/portal/fak_historisch/content/e11352/e84118/e566094/Sprach_e_ger.pdf; www.marum.de/Binaries/Binary367/OrientierungshilfeFuerGendergerechteSprache.pdf; https://tu -dresden.de/tu-dresden/chancengleichheit/ressourcen/dateien/gleichstellung/berufungen/leitfaden -geschlechtergerecht-in-sprache-und-bild?lang=de.
27. See www.uni-muenster.de/Gleichstellung/Gleichstellungsrahmenplan.html.
28. "Gender mainstreaming is the (re)organization, improvement, development and evaluation of policy processes, so that a gender equality perspective is incorporated in all policies at all levels and at all stages, by the actors normally involved in policy-making" (Council of Europe, 1998, p. 15).
29. See www.advance-he.ac.uk/equality-charters/athena-swan-charter.
30. See www.wi.uni-muenster.de/career/diversity-and-inclusion/plan-recruitment-retention-and-advan cement-talented-women.
31. See www.wiwi.uni-muenster.de/fakultaet/sites/fakultaet/files/fb_04_gleichstellungsplan_final.pdf.

REFERENCES

Ahuja, M.K. (2002). Women in the information technology profession: A literature review, synthesis and research agenda. *European Journal of Information Systems*, *11*(1), 20–34. https://doi.org/10.1057/palgrave/ejis/3000417.

Armstrong, D.J., and Riemenschneider, C.K. (2014). The barriers facing women in the information technology profession: An exploratory investigation of Ahuja's model. *52nd ACM Conference on Computers and People Research*, 85–96.

Armstrong, D.J., Riemenschneider, C.K., and Giddens, L.G. (2018). The advancement and persistence of women in the information technology profession: An extension of Ahuja's gendered theory of IT career stages. *Information Systems Journal* (July 2016), 1–43. https://doi.org/10.1111/isj.12185.

Council of Europe. (1998). *Conceptual Framework, Methodology and Presentation of Good Practices: Final Report of Activities of the Group of Specialists on Mainstreaming (EG-S-MS)*. Strasbourg.

DuBow, W.M., and Ashcraft, C. (2016). Male allies: Motivations and barriers for participating in diversity initiatives in the technology workplace. *International Journal of Gender, Science and Technology*, *8*(2), 160–80.

EIGE. (2016). *Gender Equality in Academia and Research – GEAR tool*. https://doi.org/10.2839/309020.

EIGE. (2019). Gender Equality Glossary and Thesaurus. Retrieved from https://eige.europa.eu/thesaurus/about.

European Commission. (2012a). *Communication from the Commission to the European Parliament, the Council, the European Economic and Social Committee and the Committee of the Regions on "A Reinforced European Research Area Partnership for Excellence and Growth."* Retrieved from https://eige.europa.eu/sites/default/files/era-communication_en_2012.pdf.

European Commission. (2012b). *Structural change in research institutions: Enhancing excellence, gender equality and efficiency in research and innovation*. Retrieved from http://ec.europa.eu/research/science-society/document_library/pdf_06/structural-changes-final-report_en.pdf.

European Commission. (2019). *She figures 2018*. Retrieved from https://ec.europa.eu/info/publications/she-figures-2018_en.

European Commission. (2022). *She figures 2021*. Retrieved from https://op.europa.eu/en/web/eu-law-and-publications/publication-detail/-/publication/61564e1f-d55e-11eb-895a-01aa75ed71a1.

Gorbacheva, E., Beekhuyzen, J., vom Brocke, J., and Becker, J. (2018). Directions for research on gender imbalance in the IT profession. *European Journal of Information Systems*, *28*(1), 43–67.

Gorbacheva, E., Moumtzi, V., and Stein, A. (2019). An innovative IT-supported approach facilitating co-design of tailored gender equality plans: The CrowdEquality idea crowdsourcing platform. In M. Sangiuliano and A. Cortesi (Eds.), *Institutional Change for Gender Equality in Research*.

Gorbacheva, E., and Barann, B. (2017). IT-Enabled Idea Crowdsourcing – A Mean to Promote Gender Equity in IT Research Institutions. *25th European Conference on Information Systems (ECIS)*, 1281–98. Guimarães, Portugal.

Loiacono, E., Iyer, L.S., Armstrong, D.J., Beekhuyzen, J., and Craig, A. (2016). AIS Women's Network: Advancing Women in IS Academia. *Communications of the Association for Information Systems (CAIS)*, *38*(38).

Quesenberry, J.L., and Trauth, E. (2012). The (dis)placement of women in the IT workforce: An investigation of individual career values and organisational interventions. *Information Systems Journal*, *22*(6), 457–73. https://doi.org/10.1111/j.1365-2575.2012.00416.x.

Quesenberry, J.L., Trauth, E., and Morgan, A.J. (2006). Understanding the "mommy tracks": A framework for analyzing work–family balance in the IT workforce. *Information Resources Management Journal*, *19*(2), 37–53.

Sangiuliano, M., Canali, C., and Gorbacheva, E. (2019). Lessons learned from tailored gender equality plans: classification and analysis of actions implemented within the EQUAL-IST project. *International Conference on Gender Research*.

Trauth, E. (2002). Odd girl out: An individual differences perspective on women in the IT profession. *Information Technology and People*, *15*(2), 98–118. https://doi.org/10.1108/09593840210430552.

Trauth, E., Cain, C.C., Joshi, K.D., Kvasny, L., and Booth, K.M. (2016). The influence of gender-ethnic intersectionality on gender stereotypes about IT skills and knowledge. *The Data Base for Advances in Information Systems*, *47*(3), 9–39.

Trauth, E.M., and Connolly, R. (2021). Investigating the nature of change in factors affecting gender equity in the IT sector: A longitudinal study of women in Ireland. *MIS Quarterly*, *45*(4), 2055–100.

Trauth, E., Quesenberry, J.L., and Huang, H. (2008). A multicultural analysis of factors influencing career choice for women in the information technology workforce. *Journal of Global Information Management*, *16*(4), 1–23. https://doi.org/10.4018/jgim.2008100101.

Trauth, E., Quesenberry, J.L., and Huang, H. (2009). Retaining women in the US IT workforce: theorizing the influence of organizational factors. *European Journal of Information Systems*, *18*(5), 476–97.

Trauth, E., Quesenberry, J.L., and Yeo, B. (2008). Environmental influences on gender in the IT workforce. *DataBase for Advances in Information Systems*, *39*(1), 8–32. https://doi.org/10.1145/1341971.1341975.

Wallon, G., Bendiscioli, S., and Garfinkel, M.S. (2015). *Exploring quotas in academia*. Retrieved from www.embo.org/documents/science_policy/exploring_quotas.pdf.

APPENDIX

Table 7A.1   List of acronyms

DIS_UMINHO	Department of Information Systems at the University of Minho
DIS_WWU	Department of Information Systems at the University of Münster
EQUAL-IST	"Gender Equality Plans for Information Sciences and Technology Research Institutions"
EU	European Union
GEP	Gender Equality Plan
IS	Information Systems
IT	Information Technology
LGBT	Lesbian, gay, bisexual, or transgender
SBE	School of Business and Economics
STEM	Science, Technology, Engineering, and Mathematics
UMINHO	The University of Minho
WWU	The University of Münster (German: *Westfälische Wilhelms-Universität Münster*)

Table 7A.2   Implemented preparatory interventions to achieve the first objective at WWU: to raise awareness about gender equality aspects and issues

Title	Goal(s)	Level of Implementation
Dissemination of the GEP and the implemented interventions: • to academic staff members at a "brown bag meeting" DIS_WWU • to professors DIS_WWU • to the EQUAL-IST project external evaluator DIS_WWU • within the proposals for national and international re-accreditations SBE • to the dean SBE • to the SBE Equal Opportunities Committee SBE • to Decentralized Equal Opportunity Officers WWU.	To disseminate the GEP and the implemented interventions to all target groups.	WWU, SBE, DIS_WWU, external
Support of the photo campaign for the 2018 International Day for the Elimination of Violence against Women.	To support the International Day for the Elimination of Violence against Women.	WWU, SBE
Proposal to adjust existing regulations related to gender equality within the refinement of the Equal Opportunity Framework document.	To improve the existing at WWU regulations related to gender equality.	WWU
Proposal to incorporate the gender equality aspects into the mission statement.	To incorporate the gender equality aspects into the SBE mission statement.	SBE

Title	Goal(s)	Level of Implementation
Awareness raising interventions by the EQUAL-IST project external evaluator.	To raise awareness about the importance of promoting gender equality at DIS_WWU.	DIS_WWU
Keynote speech "Why we Need more Women in IT-Startups" within the "Startup Nights Münster" event.	To raise awareness of the participants of the "Startup Nights Münster" event about the importance of promoting gender equality (referring to the experience gained within the EQUAL-IST project).	external

Table 7A.3   *Implemented preparatory interventions to achieve the second objective at WWU: to increase the share of women among Bachelor students*

Title	Goal(s)	Level of Implementation
Organization and implementation of events promoting the IS study program within the "Hochschultag" annual information day.	To promote the SBE study programs to potential students as an inclusive place welcoming all.	SBE
Organization and implementation of events promoting the IS study program within the "Girls' Day" annual information day.	To promote the IS study program to potential female students as an inclusive place welcoming all.	DIS_WWU
Survey "How did you learn about your study program?" Preparatory intervention: Review of relevant studies.	(i) To understand, how current Bachelor's and Master's IS students, especially female IS students, learned about their study programs; (ii) to identify promising communication channels to promote the IS study program; (iii) to reveal, how current communication channels promoting the IS study program could be improved.	DIS_WWU
Workshop "Why should one want to study Information Systems at the University of Münster?"	To understand, how to make the IS study programs attractive for the best potential students, and especially for qualified and motivated women.	DIS_WWU

Table 7A.4   Implemented preparatory interventions to achieve the third objective at WWU: to improve work–family balance of academic staff members and students

Title	Goal(s)	Level of Implementation
Support in the design and implementation of the survey "Studying with Children."	(i) To identify and analyze the requirements of students who have children or are about to become parents; (ii) to understand their level of awareness about existing (at WWU and beyond) initiatives and opportunities for parents.	WWU
Workshop "Why should one want to work at the University of Münster's Department of Information Systems?"	To understand, how to make DIS_WWU an attractive place to work for the best potential academic staff members, and especially for qualified and motivated women.	DIS_WWU
Improvement in the communication to (potential) staff members of the expectations from them.	To improve the communication of the expectations from staff members at DIS_WWU.	DIS_WWU

Table 7A.5   Implemented preparatory interventions to achieve the fourth objective at WWU: to enhance inclusion of international students

Title	Goal(s)	Level of Implementation
Workshop "Towards higher gender diversity and inclusion of international students in the Information Systems study program at the University of Münster."	To discuss the ongoing and prospective interventions towards enhancing gender diversity and inclusion of international students in the IS study program.	WWU
Organization and implementation of the sessions "How to study successfully in Münster" within the "Master's Orientation Day" information days for Master's IS students.	To provide practical information and recommendations related to studying IS at WWU to those Master's IS students who did not study at WWU before and, in particular, to international IS students.	DIS_WWU
Investigation of opportunities for the involvement of regular international IS students into existing initiatives supporting international exchange students in Münster.	To explore, how regular international IS students could be (further) involved into existing initiatives supporting exchange students in Münster.	DIS_WWU

Table 7A.6   Further implemented preparatory interventions

Title	Goal(s)	Level of Implementation
Appointment of the EQUAL-IST project leader at WWU as (i) the SBE Equal Opportunities Officer and (ii) a member of the WWU Equal Opportunities Committee.	(i) To increase visibility of the EQUAL-IST project and the initiated interventions; (ii) to build alliances for further collaboration.	WWU, SBE

8. Connected and committed? Culture and context in career entrenchment of Indian and native-born women in the United States IT workforce

Monica Adya and Sangeeta Parashar

INTRODUCTION

Janine, a database administrator in a large healthcare organization, was at a crossroads regarding her career. After serving in her current role for over four years, she was looking to move, with a preferred path being in a supervisory project manager position. An opportunity that became available to her involved overseeing a technical team composed primarily of men. But Janine was concerned that should she take this role, she could be setting herself up for failure if her male peers did not respect her. On the other hand, she did not see herself continuing in her current position which required her to routinely update her technical skills. A third option that Janine seriously contemplated was returning to healthcare administration, a job that she had begun before switching to Information Technology (IT). However, the field had changed dramatically enough that she would require additional licenses, training, and possibly education. Janine kept the third option open but knew that the first two were the only viable ones for her.

Radhika, a software engineer in her 40s who worked at a telecommunications organization, was at a similar crossroads. Although she enjoyed her current technical position and preferred to remain in it, the next career move prescribed for her was in a managerial capacity. In fact, she had accepted an opportunity to supervise a team that was developing the next generation of satellite navigation software. But the role of manager was not of particular interest to her. Moreover, all the members of Radhika's team were "young twenty-somethings," leading her to occasionally feel isolated because of her gender and age. As is common with women in the IT field, she struggled to find a woman mentor who could counsel her about future options. But the possibility of leaving her job or changing her career never crossed Radhika's mind "because I've given so much to be where I am today."

Radhika's commitment to her career was different from that of Anita who had an undergraduate degree in business from India and moved to the United States after getting married. With no work experience and a limited skillset, she needed to get a foot in the door. Her husband, an IT consultant, encouraged her to get the necessary technical training and she subsequently joined a health insurance provider as a database developer. When Anita had her first child, both her mother and mother-in-law provided extended childcare, enabling her to continue with her career. However, by the time she had her second child, both mothers' health had deteriorated, requiring Anita and her spouse to manage the demands of a young family on their own. After evaluating their options, the couple determined that Anita quitting her job was the best path forward.

Janine, Radhika, and Anita were participants in a qualitative study that the first author conducted in 2007–08 to understand what drives women to pursue a career in IT, their workplace experiences, and how individual differences stem from the interaction of social, cultural, economic, and psychological factors. Over 30 women professionals working in the IT industry across the United States were recruited through referrals and the author's personal networks; all names included in this chapter are pseudonyms. Half of the participants were born and raised in South Asia, predominantly India, and immigrated to the United States for educational and associational reasons. The other half were native-born (i.e., born and raised in the United States). Interviews ranged from 30 to 45 minutes, were semi-structured (i.e., they began with guided open-ended questions that spontaneously led to follow-up ones), and were audio-recorded and transcribed. Results indicated group differences in perceptions of how the field was gendered, issues of stereotyping and discrimination in the workplace, and long-term commitment to IT careers (Adya, 2008).

As these women shared their stories, one was struck by their divergent paths: career entrenchment and immobility for Janine and Radhika, but career abandonment for Anita. These cameos serve as a springboard to the main goal of this chapter: to explore the social, cultural, economic, and structural factors that lead to variations in career participation among immigrant Asian and native-born women in the IT industry in the United States. We will apply the Model of Career Entrenchment (Carson et al., 1996) to highlight the nature of career entrenchment. Exploring these issues can be used to better understand women's participation in and the challenges they face in IT careers as well as discuss strategies for improved participation based on lessons learned from the two cultural contexts, India and the United States. The chapter draws on the unique cultural perspectives reviewed herein to suggest recommendations for enhancing women's entrenchment in IT careers.

BACKGROUND

A few paradoxes emerge when we compare India and the United States regarding trends in educational attainment, women's labor force participation (LFPR), and the proportion of women in the tech industry. Consider the first. Over the last two decades, women's LFPR in India has dropped from 42.7 percent in 2004–05 to 23.3 percent in 2017–18.[1] However, their share in India's IT and IT Enabled Services (ITES) has steadily increased during the same time period with women now representing 34 percent of the IT workforce in 2017,[2] making it a relatively more gender-integrated industry. This is in direct contrast with trends observed in the United Kingdom and United States where the proportion of women in the IT workforce has remained stagnant or is declining. In the United States, although women comprised approximately half of the workforce in 2018, they held only 25 percent of all tech industry jobs – a reduction since 1980.

One explanation for the observed gender segregation of jobs in the technology industry is the difference in accumulation of human capital – formal education, skills, experience, and training programs – that workers invest in to augment their marketable credentials (Becker, 1957). This often reflects gendered socialization that may influence the selection of courses/concentrations in high school and types of college degrees (England, 2010). Here is where another paradox emerges. In 2014–15, women in India represented 46.8 percent of graduates in engineering and technology careers (NASSCOM, 2022). Yet, for primary school enrollments where government policies have led to greater gender parity, women lag behind in levels of high school and college completion.[3] On the other hand, in the United States, despite women's levels of education converging with and even overtaking that of men, they earned 18 percent of the computer science degrees and 20 percent of engineering degrees (NSF, 2018). These educational choices continue to be reflected in the workforce, at both entry level and in later progression, leading to fewer women and women role models in IT.

Irrespective of geographic context, technological and social processes such as urbanization, industrial expansion, and increasing education have influenced the availability and nature of work, particularly for women (Anker, 1998). In India, because parental and familial influences on career choice are strong from a young age (Ray et al., 2020), children and youth are often steered towards STEM fields, particularly engineering and medicine and now, IT. After economic liberalization in 1991 (Denoon, 1998), the country experienced an IT boom. Thousands of engineering and technical colleges mushroomed in response to the rising global demand, coupled with the need to employ a demographic "youth bulge." A disproportionate number of these colleges are private and oftentimes less regulated because of the government's initial *laissez-faire* approach. Enrollment in these institutions has surpassed that in other disciplines, even as the country has seen a glut of IT graduates compared to employment opportunities over the past decade.

In a poor country with a small formal economic sector such as India, a large proportion of technologists and engineers who also possess good English-language skills, a valuable currency in the global economy, are from the vast middle class which primarily resides in urban areas (Pal, 2003). The IT industry provides opportunities for white-collar jobs with linear career trajectories, relatively high salaries, and sustained job growth.[4] By taking a few training programs in lieu of formal degrees, the field provides quick social mobility and prestige for women who want to break the mold of homemakers and caregivers. Office-based IT jobs also tend to be more work- and family-friendly compared to, for example, civil or mechanical engineering where they may be assigned to fieldwork in settings that might be unsafe or traditional in terms of gender norms (Adya, 2008). Importantly, through co-residence in multi-generational households and/or paying for low cost easily available domestic help, women can manage their productive and reproductive roles.

By contrast, in the United States, an advanced economy with a high level of service sector specialization and differentiation in white-collar occupations, the belief that STEM careers are the only path to socio-economic prestige is less prevalent (Thomas and Watters, 2015; Charles, 2003). There is also a perception of IT being a male-dominated field that primarily entails interaction with computers as opposed to people – a perception bolstered by the media and sustained by various institutions (Papastergiou, 2008). And, as mentioned earlier in this section, the field has become even more gender-segregated, although variations at the occupational level exist (e.g., software programmer jobs being male-dominated versus administrative support being female-dominated). Indeed, good career options exist in other non-technical fields with bureaucratization creating new opportunities for skilled women at the top of the occupational hierarchy. Recent trends indicate a growth in jobs such as pre-school and kindergarten teachers, health aides and nurses, servers, medical assistants, retail workers, etc. that may have functional or symbolic similarities to traditionally domestic activities. Bolstered by organizational adaptations such as part-time flexible scheduling that allows for childbearing and rearing, these jobs often cater to women's "preferences" (Charles, 2003).

Moreover, while correlation does not imply causality, one cannot help but notice the under-performance of 15-year-old US students in PISA (Program of International Student Assessment) math and science literacy exams relative to their peers in other high-income (OECD) nations. The scale on each test ranges from 0 to 1,000.[5] In 2015, American students ranked 24th in science and 40th in math with some striking within-country gender differences in favor of boys. In the case of science scores, the average OECD difference between boys and girls was 4 points, but for the US, it was 7 points. The gender gap in math was wider, both at the OECD average (8 points) and US level (9 points). Expectedly, reading scores trended in the opposite direction in favor of girls (OCED average: 27 points, US: 20 points and 24th position).[6] These trends point to two observations. First, K-12 (kindergarten to 12th grade) education in the United States seems to create comparative disadvantages for girls in math and science – important foundations for technical careers – when contrasted with boys.[7] Second, despite the perceived weakness of US students in STEM disciplines, the nation's global dominance in the technology sector is attributable to skill-based immigration, often through Indian and Chinese students and workers, many of whom are women.

The stories of Radhika, Janine, and Anita, three of several Indian- and native-born women, highlight the interplay of individual and cultural factors which influence one's selection of and long-term participation in IT careers. We will now apply the Model of Career Entrenchment (Carson et al., 1996) to these cases to draw out individual differences among them, and layer it with sociological theories and trends to further explore contextual factors.

CAREER PARTICIPATION AS A REFLECTION OF CAREER ENTRENCHMENT

The previous section highlighted social, economic, and environmental factors that influence women's participation in IT education and career choices (Adya and Kaiser, 2005) and their entrenchment once they enter the IT workforce. Studies across different cultures have highlighted gendered norms, stereotypes, racial and ethnic biases, and economic opportunity, among others to explain variations in career progression among women IT professionals (McGee, 2018; Trauth, Quesenberry, Huang, 2008; Adya, 2008). While most studies have examined women's relationship with IT careers at a single point in time, recent studies such as Trauth and Connelly (2021) have considered the longitudinal interplay between environmental, identity, and individual factors noting that, over time, identity and individual factors are consistently influential in women's engagement with IT while others such as policies, social infrastructure, and economy evolve in impact.

The Model of Career Entrenchment (Carson et al., 1996) notes this "self-perpetuating" nature of career entrenchment and highlights how the passage of time "compounds investments and costs in a self-sustaining manner" (p. 274). The interlacing of financial, emotional, and psychological investments in careers can limit individuals' desire or ability to abandon or change careers. Individuals become career entrenched if they perceive that their financial and psychological investments in their careers are high and few alternative career options are available. Among these entrenched individuals, those dissatisfied with their careers feel entrapped while those satisfied are content and choose to remain career immobile. Career changers are often individuals low on entrenchment because of limited career investments and are dissatisfied with their careers. In contrast, those low on entrenchment but satisfied become voluntary careerists (Carson et al., 1996).

In applying its concepts to the cameos of the three women IT professionals, we rely on the individual differences theory of gender and IT suggesting that commitment, retrenchment, and abandonment are driven by different attitudes, with committed professionals investing immense financial, educational, and psychological resources to remain and succeed in their chosen careers (Fu and Chen, 2015). Their strong ties often stem from job-related factors such as involvement (Blau, 1988), satisfaction, and fit (Goulet and Singh, 2002). Identification with one's career, motivation (Aryee and Tan, 1992), need for achievement, and work ethic (Goulet and Singh, 2002) also enhance commitment. An underlying belief among such individuals is that it is the right career and there is little desire or interest in exploring other options or leaving the existing one (Aryee et al., 1994). A reluctance to abandon the substantial investments made and/or the absence of satisfactory alternatives may sometimes lead to inertia in searching for new opportunities and subsequently, career immobility (Carson et al., 1995). However, when career-entrenched individuals perceive dissatisfaction with their jobs, and find no options ahead of them, they experience career entrapment (Carson et al., 1996) that may eventually lead to career abandonment.

The three women were part of a series of interviews with two groups of women in the US IT workforce – those of Indian descent who were born and raised in the United States and those who immigrated there from South Asia, primarily India. Interviews revealed that the latter group demonstrated greater and early entrenchment in their IT careers than the former. Initial economic investments for immigrant Indian women laid the foundation for an exponentially reinforcing effect on entrenchment over time due to additional social and psychological investments. In contrast, several native-born women in the participant group expressed a desire to move to other careers or abandon IT completely. Since national culture impacts individual behavior (Leidner and Kayworth, 2006), one wonders whether the diverse experiences of these two groups of women can perhaps explain the difference in participation in IT careers between two diverse cultural contexts, India and the United States.

Radhika: Career Entrenched, Contented Immobile: As evident from the cameo at the beginning of the chapter, Radhika's educational and personal inclination towards STEM led to both a financial and emotional investment in her career. Her parents' and sibling's strong encouragement to pursue a science-based profession incentivized her to complete an undergraduate degree in engineering. Radhika subsequently immigrated to the United States from India where she obtained a master's degree in software engineering with an emphasis on satellite technologies and joined a firm that developed software for satellite communications. Her job, at the time of the interview, provided both stability and growth opportunities to move into managerial roles, should she have decided to pursue them.

Radhika was surprised when asked if she had any thoughts of leaving the company or her career. She recognized that as an Indian in the IT industry, she had a positive stereotype that inadvertently helped advance her career. Additionally, the organization valued her education and experience as well as the diversity she brought to the team. As the lone woman in her team, Radhika often felt isolated, but she eventually rectified that by seeking highly supportive women mentors in other units of her organization. Her concern now was that, at some point, age could create another layer of disparity for her. She recognized that even if she wanted to pursue another job, she may struggle to find a similarly satisfying environment. Importantly, as is often the case in India, Radhika was socialized to believe that one "sticks to the chosen career" and she felt no differently.

Janine: Career Entrapped? Growing up in the United States, Janine chanced upon a career in technology at a time when the field had just opened up and opportunities were plentiful and well-compensated. Her prior role as a healthcare administrator had gradually started encompassing database management functions and, along the way, she had received training to support this expansion. When a data management opportunity opened up in a large healthcare company in the Midwest, Janine applied for it and subsequently shifted careers and organization. Now nearly six years later, she missed health administration. She did not see much of a future in her current role, primarily because she understood the need to upskill to a new platform being implemented in her company.

Janine's initial career investment – both financially as well as socially and emotionally – were primarily in the healthcare administrator domain. Her shift to IT was purely opportunistic and, other than the training she received at work, further investment in new skills was a bit daunting. But she recognized that such inaction on her part would likely create roadblocks in her job growth, particularly on the technical side. Janine was concerned that eventually she would be a "misfit" in her role, which had changed skills and expectations faster than she could keep up. Janine had about seven years before her retirement, and she was at that junction where she would lose too much financially if she moved to another organization. But she was also unsure if she had the appropriate credentials to return to her prior career in healthcare administration. When the first author met her, Janine had begun reaching out to colleagues in other parts of the company and in other fields to explore options but she had little clarity on where she would be in a year or two.

Anita: Career Abandonment: Anita, also born and raised in India, received her undergraduate degree in business and subsequently moved to the United States after her marriage. Her husband's advice to build technical skills, especially in light of structural demand for professionals in the field and short training programs, in lieu of formal degrees, served the purpose. She was satisfied working for a prescription management company but the birth of her second child created immense role conflict and led to her leaving her job. With her husband's increased travel over time, she questioned how long she could juggle work and two kids, especially after factoring in the wage differential as well as the cost of childcare.

Anita's investment in her IT career was low, relative to her social and emotional investment in her family. While she would lose her accrued employment benefits and skills, her husband's compensation more than made up for the loss of her income. Today, 14 years later, Anita has still not returned to the workforce even though her children do not require the same level of minding. Initially, after leaving her job, she missed social interactions with her colleagues but, over time, Anita found that "hanging out" with her friends "more than makes up for that loss."

Career Entrenchment Among Indian Women IT Professionals

Both Radhika and Anita were born and raised in India in middle-class families of relatively similar backgrounds. Beyond that, both followed different trajectories to where they stand today, with Radhika remaining firmly entrenched in her career and Anita completely abandoning hers.

Career choice among Indian women

Because gender gaps in education are still quite wide in poor countries such as India, a woman's educational attainment is crucial for her to compete with men for jobs in the formal economy (i.e., jobs taxed and monitored by the government). In such a context, girls are also more likely to pursue STEM careers that offer higher remuneration, economic stability, and social mobility (Stoet and Geary, 2018). Engineering, medicine, and teaching were, and still continue to be, perceived as respected professions in India. As is commonplace in almost all societies, engineering tends to be male-dominated in composition, while medicine is gender-integrated and teaching, female-dominated, with variations at the job level. Girls, along with boys, are often steered into these fields due to strong familial involvement in selection of schooling and career decisions (Ray et al., 2020). Any performance challenges with quantitative skills are perceived as individual and driven by rectifiable environmental circumstances (White, Ruther, and Kahn, 2015), rather than an inherent gender difference in aptitude.

Labor market opportunities or structures influence educational enrollment (Buchmann and Brakewood, 2000), and with the global IT sector boom in the 1990s as well as increasing service sector specialization, new prospects opened up in other fields. IT careers in India are relatively gender-integrated and well-paying, which helps reinforce and sustain women's participation. Within this industry, they have a wider range of options from pure IT careers such as software engineering to ITES such as telemarketing, customer support, and help-desk functions.[8] Whereas some literature on girls' career choice for IT suggests that structural factors such as access to technology at home and school can enhance girls' self-efficacy in such careers, this has not been the case in India (Pegu, 2014). Most homes and schools are still unable to invest in computer technology due to financial constraints, but this lack of access has not affected IT career choices. As noted earlier, despite being a poor country with gender inequities in access to education, in 2014–15, women in India represented nearly half of engineering and technology graduates.

Career experiences of Indian women in IT

Once Indian women have opted for a STEM education and joined IT, it is unusual for them to abandon it. Any investments in education are viewed as purposeful, and deviations from chosen careers not only delay the closure of an important phase of life but also place additional demands on already tight financial resources (Adya, 2008). Indeed, if resources for education are available, they are better utilized if directed towards a tertiary degree, whether in India or abroad, which further enhances one's credentials and market value. Thus, career entrenchment becomes a socially derived attitude that is reinforced through sustained financial and social investments, as evident in the case of Radhika.

For Indian women with specialized IT degrees, immigration to the United States further entrenches them in their careers because they become part of an "occupational niche" or the overrepresentation of foreign-born workers from specific countries in an occupation relative to their percentage in the US labor force (Eckstein and Peri, 2018). Although Asian women are still underrepresented in STEM jobs compared to men, they outnumber native-born women. Occupational niching has been observed, for instance, among Indians and Chinese in IT, Russians in mathematics or Israelis in engineering. It began in the mid-1990s when American companies responded to the market needs of Y2K[9] with a transnational hiring frenzy, and it has been sustained by a continued demand that has not been matched by a local supply.[10] Approximately two-thirds of the three million Indians currently residing in the United

States moved as part of this third wave of migration, and are nicknamed "the IT generation" (Chakravorty et al., 2017).[11] Radhika automatically became a part of this niche because of her rigorous training in both countries, but Anita, responding to the structural demand for IT *after* she immigrated, was less embedded. Indeed, clustering into ethnic enclaves, in this case IT, becomes an important aspect of the assimilation process – whether in the country or in one's career.

Positive stereotypes associated with Indians in IT further drives the process of career entrenchment. Asians in general are viewed as hard-working, and motivated team players because their cultures tend to stress group membership, harmony, and conformity over displays of individualism and self-interest. This perception, however, runs the risk of them being labelled as passive and docile (Adya, 2008). Their advanced degrees and technical skills are highly valued at the workplace, the type of work assigned is proactively aligned to their skills, and opportunities for growth both in technical and managerial roles are made available. In a deadline-oriented IT environment, immigrants who are on temporary work visas or green cards often overextend themselves to prove their competence and to unconsciously live up to these expectations. And, the process becomes self-perpetuating. That is, positive group stereotypes drive performance, and possibly entrenchment for Radhika, but solidifies beliefs regarding innate "Asian" STEM abilities and the "model minority," which have been challenged (Takaki, 1993). At the same time, these pressures of positive stereotyping could be leading to career abandonment for Anita who, in the balance, ultimately chose between work and life. Indeed, cultural labels blind us to the structural factors that create and shape group opportunities to immigrate and achieve socioeconomic mobility in the United States.

Finally, new immigrants tend to be young and in their childbearing years, a phase when they require immense support to achieve work–life balance. However, good-quality childcare in the United States is expensive and often not a company benefit, compared to India where grandparents, other relatives, and nannies step in. This is an important consideration in the IT field that thrives on collaboration, workers with complementary skillsets operate in teams, and tasks may not be confined within the nine to five workday time frame. The pressure to perform in a fast-paced setting, coupled with societal expectations of women as primary caregivers, often rears its head through career abandonment as some women transition out of the workforce after childbirth. Some others persist because of the important, but understudied, phenomenon of trans-nationalized (grand)parent caregiving (Hu, 2018). By providing free and reliable childcare, co-resident or visiting parents help new immigrant mothers stay in the labor force, even if career entrenchment is not guaranteed. Parents visiting their children who have migrated from India are, however, limited to a few months due to immigration policy and restricted visas, and often complicated by issues pertaining to health insurance, language barriers, and cultural adaptation. With the emergence of alternate career paths that allow women to better balance societal demands and personal ambitions, the possibility of opting out of IT careers still remains a possibility, as was the case with Anita.

Indian women in US IT – contented immobiles

Despite exposure to alternate career pathways in the United States, most Indian women IT professionals like Radhika demonstrate career entrenchment. Their financial, social, and psychological investments often begin at a young age with a positive attitude toward STEM education and careers. They garner marketable skills and labor force experience along the way, which translates into better jobs and earnings. These women also benefit from the gender neutrality of the workplace, flexibility provided by their organizations, and the endorsement of their credentials. Reproductive functions such as family involvement and number of dependents does not seem to have any significant effects on career commitment because of the grandparent effect (Goulet and Singh, 2002). As such, Indian women in US-IT are satisfied with their professional roles and experience contented immobility.

One could argue that besides satisfaction, this contented immobility also reflects factors such as perceived wage differentials, competition in non-STEM fields, social networks, or cultural adaptation. The IT/STEM field is associated with smaller wage differentials between native- and foreign-born; pay parity is achieved in less than a decade, even earlier. But switching to non-STEM careers and/or industries (e.g., arts, media, architecture, real estate insurance, law, etc.) is not financially lucrative even for highly educated immigrants because they still earn substantially less than native-born professionals. Indeed, it takes nearly 20 years to reach parity because of the effects of competition and discrimination (Hanson and Slaughter, 2016). Alternatively, an occupational accreditation process in the case of law or real estate may involve relatively high entry costs for recent immigrants, who might also be disadvantaged by their lack of a "nuanced understanding of American culture" or communication styles (Hanson and Slaughter, 2016, p. 485).

Importantly, the selective nature of immigration and niching generates and sustains extensive ethnic networks for Indian immigrants in IT, compared to other fields, which further contributes to their sense of complacency. These transnational networks, motivated by national loyalty and pride, are of vital importance as they provide referrals, friends, work- and employment-related information, advice, and informal mentoring, both in the United States and India (Dossani, 2002). Reentry into another field implies a loss of these informal ties and a reliance on their advanced degree as a mechanism to indicate their competencies. When faced with such a complex interplay of circumstances, contented immobility can set in.

Career Entrenchment Among Native-Born Women

Born and educated in the United States, Janine began her career in a healthcare-related field and moved into IT in response to market opportunities. But her waning interest in updating her skills coupled with the realization that shifting back to her previous job would require additional training, made her feel entrapped in her career. Janine decided to stay on because neither dropping out of the labor force nor career abandonment were viable options.

Career choice among native-born women

Contrary to their Indian peers, factors influencing career choices for native-born women are slightly different, ranging from socialization, weaker societal emphasis on STEM, and the availability of alternate professions, among others. Being a more gender-egalitarian country with a high standard of living, girls do not feel the same singular pressure to pursue STEM/IT education or careers for mobility and economic independence or to learn new technologies for professional growth. Their development of IT competencies may be through "leisure activities" (Gnambs, 2021, p.7), rather than early and sustained exposure to technology at work and home or a deliberate goal to develop skills.

Career preferences also reflect gendered socialization, perceptions, and feelings about one's competence. Many occupations are either defined as prototypically female (requiring dexterity, clerical perception, and nurturance) or male (strenuousness, mechanical ability, and high-order/status interactions), leading to uneven gender representation (Reskin and Roos, 1990). Media portrayals further perpetuate stereotypes of women as teachers and nurses rather than scientists and engineers (Varma, 2002). Because ideas of gender still define the IT field as masculine (Papastergiou, 2015), isolated, and over dependent on technological rather than human interactions (Servon and Hisser, 2011), women may be less likely to recognize their abilities in it. In the process, the field becomes even more male-dominated. These notions, coupled with the underwhelming societal importance given to STEM and the availability of other well-paying non-STEM jobs, contribute to lower career entrenchment for native-born women.

Importantly, as the case of Janine indicated, in some cases women decide on IT careers by choice of opportunities and experiences later in life rather than by choice of early education and sustained training (Adya, 2008). Indeed, whereas families do have an influence on their child's future profession, that of peers, teachers, and college professors appears to be stronger. Christine, an IT professional interviewed by the first author, is a compelling example of such an individual. She served in the military, but was frustrated by the pervasive sexism that she faced there. Her male sibling, a software engineer, suggested a related career but when Christine requested her supervisor for IT-related assignments, she experienced pushback. She eventually left the military, and at the time of the interview, was completing a graduate degree in computer science. Christine seemed elated by her shift in career and skillset because she was "no less than any of those guys."

On the other hand, college professors influenced Rachel (another participant) in her IT degree choice. While enrolled in an undergraduate program in business with a focus on marketing, she took an introductory IT course. Inspired by the content, classroom experience, and application to digital marketing, Rachel added IT as her second major. However, after graduating, she opted for a career in marketing and indicated that her IT background made her a "more effective marketing professional but she [*sic*] just did not see herself in an IT career." Rachel is an example of women in STEM, especially in mathematics, computer science, and engineering, whose underrepresentation reflects not just low enrollments, but also low retention.

Career experiences of native-born women
A significant factor in women's continued participation in the IT workforce lies in their workplace experiences (Ahuja, 2002). A common theme that emerged from the first author's interviews with native-born women was the common perception that they struggled with the negative stereotypes of women in IT and had to work harder to garner respect from their colleagues. Perceptions regarding inherent biases and sexism in the workplace, as well as questioning their credentials and competencies, are their well-founded primary concerns. Indeed, several native-born women, interviewed as part of Adya (2008), expressed frustration about the quality of work assigned to them. They were also gravely concerned about how extensively their clients perpetuated a preference for working with these women's male colleagues.

Yet the stereotypes discussed above and in the previous section are built on the belief that group membership is related to worker qualifications and productivity. According to Reskin and Roos' theory of queuing (1990), labor markets are comprised of a "labor queue" (workers for a particular job are "queued" up with the employer determining the order) and a "job queue" (all jobs available to a worker who then rank them). Whether intentionally or unconsciously, employers often use sex and race-ethnicity as hiring screens to queue candidates not only by their human capital but also by a match between their race-sex combination and the perceived race-sex stereotypes associated with the job tasks, skills, or even personalities (Parashar, 2014). Some examples include "studious or nerdy Asians" who excel in math and science or "ditsy", beautiful, and self-absorbed, loud, feminist White women who don't (Conley, 2013). Native women are also perceived as more vocal about inequities in the workplace as opposed to the more passive approaches stereotypically demonstrated by South Asian women (Adya, 2008). Queuing thus reserves the best jobs for favored groups (Whites, Asian, and men) and relegates other groups (minorities and women) to less desirable ones, with some interactions and exceptions, as with Asian women in IT where race trumps gender.

Moreover, career commitment is significantly and positively related to skill development (Aryee et al., 1994). Compared to immigrant Indian women, native-born women in IT are often more open to roles that require managerial as opposed to technical competence, often when faced with the need to upskill (Quesenberry and Trauth, 2012). Rather than investing in technical training to remain in her current job, Janine actively sought opportunities in IT-related administration and management. She felt that her experience in working with a technical team would benefit her case. But women and minorities face not just a glass ceiling but a "broken rung" on the career ladder that prevents them from moving into senior positions that have greater decision-making power and remuneration. This happens even if they may have expanded their level and range of qualifications and gained experience at lower levels of management.

Finally, having a mentor or being part of a social network is one of the strongest correlates of career commitment (Colarelli and Bishop, 1990). Peers not only help guide career choice but also provide support which seems crucial for gaining a sense of belonging in the workplace. But with fewer women in the IT workforce as well as the phenomenon of occupational niches and ethnic workplace groups described earlier, native-born women tend to lean on peers or mentors outside their organizational units. This further exposes them to alternative careers and pathways and may sometimes lead to career changes. Such a shift is considered acceptable in the United States and is unchallenged by societal norms, even though it is expensive in terms of the purported loss in financial and social investments.

Native-born women in US IT – career changers
Native-born women in the IT workforce demonstrate greater career fluidity compared to their immigrant Indian counterparts. Because the choice of IT careers happens at later stages and, sometimes by happenstance, the financial, social, and psychological investments is lower. Exposure to alternate careers, whether as girls or as women in IT, creates the opportunity for change. Moreover, negative experiences in the workplace due to stereotypes and male dominance can weaken these ties to the career. The threat of professional obsolescence often exerts more pressure on individuals with low professional self-efficacy (Fu and Chen, 2015). This could be a driver for career change or abandonment among women IT professionals in the United States, further adding concerns about the low retention of women in STEM. On the other hand, this labor market fluidity can have a positive impact on women's long-term participation in the overall workforce, perhaps partially explaining the gender parity of women in the US workforce and their declining participation in India.

Increasing Women's Participation in the IT Workforce: Lessons from Cross-Cultural Entrenchment Factors

Career entrenchment has been viewed as an individual attitude shaped by individual invest-ments – financial, social, and psychological – in one's career. Our discussion above extended this emphasis to include macro-level factors that drive individual orientations that manifest as national differences in women's participation in IT. This understanding leads to the following recommendations for increasing women's participation in IT careers.

IT careers need "better marketing"

In a recent opinion piece in the *Wall Street Journal*,[12] Dr. Neil deGrasse Tyson made this appeal in light of the significant distrust of science during the 2020–21 pandemic. Let's consider this in the context of IT careers – they need better marketing. As we look at India where prestige, economic stability, and social mobility are associated with STEM careers, perceptions about IT and STEM careers are no different in the United States. A recent Pew report[13] found that most Americans believe STEM careers pay more, attract the brightest, and are more respected. Yet women with computer/engineering degrees are less likely than their male counterparts to be working in STEM roles. This same report noted that while men and women look for the same things in STEM careers – flexibility, opportunities for advancement, a welcoming workplace – women differ most from men in their desire to have a job focused on helping others. This is where the "marketing" falls short. The force of technology careers on the lives of others is tremendous and unfortunately, under-recognized. If anything, the 2020 pandemic has only served to underscore the many lives and jobs that were saved because of the work of technologists.

For the past several decades in the United States, government incentives and policies, cor-porate investments towards engaging girls in IT, and academic programming have focused on enhancing girls' self-efficacy, exposure, and confidence with technology as a driver for career choice. Indeed, organizations such as *Girls Who Code* have demonstrated both direct (e.g., coding and critical analysis skills) and indirect benefits (e.g., empowerment, bravery) for girls who participate in these programs (Saujani, 2017). These programs and incentives are based on the assumption that early exposure to technology and coding practices will pre-dispose girls to IT careers. Paradoxically, however, over the past two decades, female participation in IT has remained unchanged. Yet, have these programs only furthered technology competency as an act of leisure (Gnambs, 2021) as opposed to deliberately developing career engagement? Perhaps these efforts need to encompass girls' career expectations for service to others. To develop this IT career mindset, the array of career pathways in IT must be promoted as overtly as technological self-efficacy. Too often, technology camps for girls are developed around comfort with reference to coding skills. But the aspects of analysis, design, user engagement, and societal impact are inadequately promoted, furthering the impression that IT is all about programming.

Data-driven insights into mentoring

Genuine mentorship remains crucial in preparing women for prolonged participation in IT careers, irrespective of ethnicity. Indeed, Pew reports suggest that men and women equally view having a mentor as one of the top three ways of progressing in the workplace. Too often, though, the responsibility for mentoring women is placed on other women. But the predominance of males in the US IT workforce creates a gap. Furthermore, decisions regarding the promotion and growth of women often rest with men. And while 30 percent of women in IT experience workplace discrimination, men perceive discrimination against women to a much lesser degree than do women. If a majority of men do not see a problem, how can they develop a more inclusive culture of mentoring for women?

This is where data-driven insights and communications can highlight gender gaps in IT and guide practices that empower all leaders – men and women – to recognize and address barriers faced by women in IT, including demands of caregiving, workplace stereotypes, and the shortage of mentors in IT roles. Workplace climate and professional and social support are crucial to the retention and promotion of women, particularly in male-dominated fields like IT. Understanding the complexity of women's success requires drawing on organizational data to present a dashboard that can raise awareness of equity gaps and enable organizations to bridging these gaps by identifying and addressing factors that promote or hinder women's success in the IT workplace.

Encouraging IT-related career fluidity
Inarguably, IT infuses every aspect of our personal and professional lives. Yet, women in the United States who obtain STEM education often demonstrate greater career fluidity and end up not working in those domains. While a complex interplay of factors contributes to this, corporations should strive to develop pathways so as to draw upon and retain this talent. As such, being deliberate about guiding women to shift from IT to IT-enabled functions is a worthwhile undertaking. Additionally, women continue to value the flexibility afforded for the purposes of work–life balance. Whereas telecommuting options have been possible for a while in the IT sector, in a post-pandemic world, more expansive options for telecommuting are becoming the norm. Continuing to evolve these practices to support greater work–life balance is even more crucial as corporations are becoming increasingly reliant on a stable IT workforce.

CONCLUSION

The year 2020–21 has highlighted both the vulnerability of women's progress in the workplace, and paradoxically, their effectiveness in leading organizations through crisis. Whether in response to social acculturation or through carefully honed practices, women have steered through rough waters with empathy, inspiration, deliberate action, and empowerment. At the same time, as women left the workforce during the pandemic, we watched with alarm as the progress women have made in IT over the past decade dissipated. This is particularly concerning as it has taken significant investments at national and international scale to increase participation of women in IT. Fewer women in the IT workforce means fewer women in leadership and mentor roles. As we build back the engagement of women in the IT workforce, the retention and growth of those who still remain in the workplace is critically important.

With the intention of exploring reasons for differential participation of Indian- and native-born women in the US IT workforce, we have evaluated socio-cultural and contextual nuances contributing to career entrenchment between these two groups. Women's representation in IT is framed as a process that begins early in one's life and sustains over time. Indian-born women engage with STEM careers to gain socio-economic and cultural benefits. Whereas this engagement leads to higher participation and retention in the workforce, for those who become disengaged or dissatisfied over time, perceptions of entrapment can be elevated. In contrast, native-born women demonstrate greater career fluidity which, while impacting their long-term participation in IT careers, lead to greater labor force participation in the long run.

The pipeline for IT builds surely from deliberate career choice but steadily from strong retention practices. Organizations have a distinct opportunity now. Over the past year, most have developed robust practices and infrastructure for remote and hybrid work. Even as the lines between work and personal lives blur, we have seen that productivity sustains over time. Our normative understanding of the intersection between performance and participation must be revisited. A top-down, bottom-line driven mindset often disregards the positive and enduring effect of the inclusive, culture-building leadership style of women. The vulnerabilities that led to women's departure from the IT workforce can be addressed through flexibility and engagement, enabling the progression of women in IT roles.

NOTES

1. See www.strategy-business.com/blog/As-India-advances-womens-workforce-participation-plum mets?gko=762f7.
2. India's Ministry of Electronics and Information Technology.
3. See World Bank, 2018.
4. See https://theprint.in/pageturner/excerpt/indian-it-industry-attracts-more-women-but-many-exit-within-first-5-years-in-the-job/368504/.
5. PISA: https://nces.ed.gov/surveys/pisa/pisa2015/index.asp.
6. See https://nces.ed.gov/surveys/pisa/index.asp.
7. The degree to which performance on PISA standardized tests translates into career selection and success remains under-explored, but it seems evident in the underrepresentation of women in STEM-related fields.
8. The former may still have a higher proportion of men and the latter, women.
9. As described by National Geographic the Y2K bug, a "computer flaw, the so-called 'Millennium Bug,' led to anxiety and the Y2K (Year 2000) scare. When complex computer programs were first written in the 1960s, engineers used a two-digit code for the year, leaving out the '19.' As the year 2000 approached, many believed that the systems would not interpret the '00' correctly, therefore causing a major glitch in the system." See www.nationalgeographic.org/encyclopedia/Y2K-bug/.
10. Between 1997 and 2013, educated Indians with at least an undergraduate degree received almost half of the 125,000 H-1B visas issued annually. A significant proportion of this number are employed in the IT industry.
11. The first wave of early movers came after the Immigration and Nationality Act of 1965 that shifted eligibility to skills and familial relationships. The second wave in the 1980s was based on familial relationships of those already settled in the US (www8.gsb.columbia.edu/articles/chazen-global -insights/singular-population-indian-immigrants-america). The third wave of migration began with Y2K efforts starting mid 1990s as US firms hired IT professionals to address concerns with the millennium bug in light of shortage of local IT talent.
12. See www.wsj.com/articles/neil-degrasse-tyson-on-the-pandemic-year-science-needs-better-market ing-11616106660.
13. See www.pewresearch.org/social-trends/2018/01/09/women-and-men-in-stem-often-at-odds-over-workplace-equity/.

REFERENCES

Adya, M.P. (2008). Women at work: Differences in IT career perceptions and experiences between South Asian and American women. *Human Resource Management*, 47(3), 601–35.

Adya, M. and Kaiser, K.M. (2005). Early determinants of women in the IT workforce: A model of girls' career choices. *Information Technology and People*, 18(3), 230–59.

Ahuja, M.K. (2002). Women in the information technology profession: A literature review, synthesis, and research agenda, *European Journal of Information Systems*, 11(1), 20–34.

Anker, R. (1998). *Gender and Jobs: Sex Segregation of Occupations in the World*, International Labour Office, Geneva.

Aryee, S., Chay, Y.W., and Chew, J. (1994). An investigation of the predictors and outcomes of career commitment in three career stages. *Journal of Vocational Behavior*, 41(1), 1–16.

Aryee, S. and Tan, K. (1992). Antecedents and outcomes of career commitment. *Journal of Vocational Behavior*, 40(3), 288–305.

Becker, G. (1957). *The Economics of Discrimination*, University of Chicago Press, Chicago.

Blau, G.J. (1988). Further exploring the meaning and measurement of career commitment. *Journal of Vocational Behavior*, 32(3), 284–87. https://doi.org/10.1016/0001-8791(88)90020-6.

Buchmann, C. and D. Brakewood. (2000). Labor structures and school enrollments in developing societies. *Comparative Education Review*, 44, 2.

Carson, K.D., Carson, P.P, and Bedeian, A.G. (1995). Development and construct validation of a career entrenchment measure. *Journal of Occupational and Organizational Psychology*, 68(2), 301–20.

Carson, K.D., Carson, P.P, Phillips, J.S., and Roe, C.W. (1996). A career entrenchment model: Theoretical development and empirical outcomes. *Journal of Career Development*, 22(4), 273–86.

Chakravorty, S., Kapur, D., and Singh, N. (2017). *The Other One Percent: Indians in America*. Oxford University Press.

Charles, M. (2003). Deciphering sex segregation: Vertical and horizontal segregation in ten countries. *Acta Sociologica*, 46(2), 267–87.

Cortes, P. and Tessada, J. (2011). Low-skilled immigration and the labor supply of highly skilled women. *American Economic Journal: Applied Economics*, 3(3), 88–123.

Denoon, D.B.H. (1998). Cycles in Indian economic liberalization – 1966–1996. *Comparative Politics*. 31(1), 43–60. https://doi.org/10.2307/422105.

Dossani, R. (2002). Chinese and Indian engineers and their networks in Silicon Valley. Stanford, CA: Stanford University, Asia/Pacific Research Center (March).

England, P. (2010). The gender revolution: Uneven and stalled. *Gender and Society*, 24(2), 149–66.

Fu, J.R. and Chen, J.H.F. (2015). Career commitment of information technology professionals: The investment model perspective. *Information and Management*, 52(5), 537–49. https://doi.org/10.1016/j.im.2015.03.005.

Georgiadou, E., Abu-Hassan, N., Siakis, K.V., Wang, X., Ross, M., and Anandan, P.A. (2009). Women's ICT career choices: Four cross-cultural case studies. *Multicultural Education and Technology Journal*, 3(4), 279–89.

Gnambs, T. (2021). The development of gender differences in information and communication technology (ICT) in middle adolescence. *Computers in Human Behavior*, 114(1), 1–10. https://doi.org/10.1016/j.chb.2020.106533.

Goulet, L.R. and Singh, P. (2002). Career commitment: A reexamination and extension, *Journal of Vocational Behavior*, 61(1), 73–91. https://doi.org/10.1006/jvbe.2001.1844.

Hanson, G. H. and Slaughter, M.J. (2016). High-skilled immigration and the rise of STEM occupations in US employment. Working Paper 22623. Cambridge, Mass.: National Bureau of Economic Research.

Hu, X. (2018). Filling the niche: The role of the parents of immigrants in the United States. *RSF: The Russell Sage Foundation Journal of the Social Sciences*, 4(1): 96–114. DOI: 10.7758/RSF .2018.4.1.06.

Kerr, W.R. and Lincoln, W.F. (2010). The supply side of innovation: H-1B visa reforms and US ethnic invention. *Journal of Labor Economics*, 28(3), 473–508.

Leidner, D.E., and Kayworth, T. (2006). A review of culture in information systems research: Toward a theory of information technology culture conflict. *MIS Quarterly*, 30(2), 357–99.

McGee, K. (2018). The influence of gender, and race/ethnicity on advancement in information technology (IT), *Information and Management*, 28(1), 1–36.

National Association of Software and Service Companies (NASSCOM). (2022). "Facts and Figures." Retrieved from: https://nasscom.in/knowledge-centre/facts-figures.

National Science Foundation (NSF). (2018). "Science and Engineering Indicators 2018." Retreived from: https://nsf.gov/statistics/2018/nsb20181/.

Pal, J. (2003). The developmental promise of information and communication technology in India. *Contemporary South Asia*, 12(1), 102–19.

Papastergiou, M. (2008). Are computer science and information technology still masculine fields? High school students' perceptions and career choices. *Computers and Education*, 51(2), 294–308.

Parashar, S. (2014). "Marginalized by race and place: A multilevel analysis of occupational sex segregation in post-*Apartheid* South Africa." *International Journal of Sociology and Social Policy*, 34(11/12), 747–70.

Pegu, UK (2014). Information and communication technology in higher education in India: Challenges and opportunities. *International Journal of Information and Computation Technology*, 4(5), 513–18.

Peri, G., Shih, K., and Sparber, C. (2015). STEM workers, H-1B visas, and productivity in US cities. *Journal of Labor Economics*, 33(S1), 225–55.

Ramdoss, K. (2012). Job demand, family supportive organizational culture and positive spillover from work to family among employees in the information technology enabled services in India. *International Journal of Business and Social Science*, 3(22), 33–41.

Ray, A., Bala, P.K., Dasgupta, S.A., and Srivastava, A. (2020). Understanding the factors influencing career choices in India: From the students' perspective. *International Journal of Cultural and Business Management*, 20(2), 215–30. DOI: 10.1504/IJICBM.2020.105641.

Reskin, B.F. and Roos, P.A. (1990). *Job Queues, Gender Queues: Explaining Women's Inroads into Male Occupations*. Temple University Press, Philadelphia.

Saujani, R. (2017). *Girls Who Code: Learn to Code and Change the World*. Penguin Random House, NY.

Servon, L.J. and Visser, M.A. (2011). Progress hindered: The retention and advancement of women in science, engineering, and technology careers. *Human Resource Management Journal*, 21(3), 272–84.

Stoet, G. and Geary, D.C. (2018). The gender-equality paradox in science, technology, engineering, and mathematics education. *Psychological Science*, 29(4), 581–93. See https://doi.org/10.1177/0956797617741719.

Takaki, R. (1993). *A Different Mirror: A History of Multicultural America*. Boston: Little, Brown and Company.

Thomas, B. and Watters, J.J. (2015). Perspectives on Australian, Indian, and Malaysian approaches to STEM education. *International Journal of Educational Development*, 45(1), 42–53, https://doi.org/10.1016/j.ijedudev.2015.08.002.

Trauth, E.M. and Connolly, R. (2021). Investigating the nature of change in factors affecting gender equity in the IT sector: A longitudinal study of women in Ireland. *MIS Quarterly*, 45(4), 2055–100.

Trauth, E.M., Quesenberry, J.L., and Huang, H. (2008). A multicultural analysis of factors influencing career choice for women in the information technology workforce. *Journal of Global Information Management*, 16(4), 1–23.

Varma, R. (2002). Women in information technology: A case study of undergraduate students in a minority-serving institution, *Bulletin of Science, Technology, and Society*, 22(4), 274–82.

White, G., Ruther, M., and Kahn, J. (2015). Educational inequality in India: An analysis of gender differences in reading and mathematics, NCAER White Paper, http://drupal-base-s3-drupalshareds3-1qwpjwcnqwwsr.s3.amazonaws.com/ihds/s3fs-public/WhiteRutherKahn.pdf.

9. Thriving as women in IT publishing, leadership, and service: challenges faced and lessons learned

Cynthia K. Riemenschneider and Deborah J. Armstrong

INTRODUCTION

When we were approached to contribute a chapter to this book, our first response was, "WOW – do we have a story to tell!" We have been publishing research on gender and information technology (IT) professionals over 17 years (Allen et al., 2004). This chapter is a reflexive consideration of being women in IT who both research and experience the identified issues. Our focus is on two major cornerstones of academic life – research and service.

The intended audience for this chapter includes doctoral students and junior academics who desire to publish research in gender and IT. Additionally, senior academics who desire to make academia more inclusive are also a valuable audience for this chapter. Furthermore, anyone who is interested in how to be more inclusive and improve diversity, equity, and inclusion (DEI) considerations may also be an audience for this chapter. For example, department heads, associate deans, graduate directors, and others in positions of influence can benefit from reading about the challenges and issues we transparently share in this chapter. One intent for this chapter is to represent a conversation a junior colleague or doctoral student might have with a mentor or senior colleagues at a conference where we transparently describe experiences and knowledge gained from our careers. A second intent for the chapter is for those interested in specific inclusion examples to glean insights and practical tips from the experiences presented.

We recognize that all scholars, no matter their culture, training, or location, face obstacles in terms of their publishing efforts and taking on leadership roles. Since both authors reside in the United States (US), some of the challenges and issues we have faced arise from the US cultural aspects of the IT profession and higher education. Examples include general male domination in the IT profession (Lee et al., 2016), challenges female faculty experience to attain tenure and promotion (Hamlin, 2021), and family responsibilities women in academe juggle throughout their careers (Ward and Wolf-Wendel, 2015). Often universities lag behind industry in offering parental leave for faculty (i.e., no university policy was in place when one author's child was born in 1988). Even though agencies such as the Equal Employment Opportunity Commission (founded in 1965) which oversees discrimination in the workplace and the American Civil Liberties Union (founded in 1920) exist in the United States, parity is not always guaranteed. While the experiences described in this chapter are from the authors' gendered, western cultural perspective, we believe that many of the lessons we have learned are generalizable beyond our immediate sphere.

The goal of this chapter is to share our career experiences, to "pay it forward," and hopefully assist others who are interested in conducting similar research and/or taking on service or leadership roles within academe. This chapter first addresses multiple categories of publishing issues faced, followed by suggestions for others from the lessons we learned. There have been many challenges, publishing learning curves and steep climbs; however, we hope our experiences will benefit others as we share how we have tackled – and continue to tackle – these topics. The challenges that we have faced include criticism from studying the behavioral, cultural, and/or psychological traits typically associated with only one sex, justifying the existence of an IT artifact within our research, and defending why we were studying IT professionals (i.e., what makes them interesting and unique). We discuss these challenges in the chapter and how we addressed them. Then, we discuss leadership and service – to the institution and the profession. We conclude with lessons learned from various leadership roles and experience in the field.

BACKGROUND

Cindy Riemenschneider earned her PhD in management information systems (MIS) in 1997. She has held academic positions at San Jose State University, University of Arkansas, and Baylor University. She became an endowed professor in 2015, served for six-and-a-half years as an Associate Dean for Research and Faculty Development, has chaired multiple dissertation committees, published over 55 peer-reviewed journal articles, numerous conference papers, and received the Association for Information Systems Women's Network (AISWN) Mentoring Award in December 2019. She was recently elected the chair for the ACM special interest group on management information systems (SIGMIS) and will serve until June 30, 2023.

Deb Armstrong earned her PhD in MIS in 2001 and has held academic positions at the University of Arkansas and is currently a full professor at Florida State University chairing multiple dissertations. She has published numerous peer-reviewed journal articles, conference papers, and book chapters, and co-edited a book, *Causal Mapping for Research in Information Technology* in 2005. She served as the AISWN co-chair from 2013–16, and as a senior editor at *Information Technology and People* from 2012–19. She served as the co-editor in chief from 2020–2022 for *The Database for Advances in Information Systems*, the longest continually published MIS journal.

Landscape of the Information Systems Field

While many IS researchers are currently focusing on big data, analytics, robotics, artificial intelligence, and econometrics the need to study behavioral issues of the IT workforce remains paramount. For example, Gallivan (2013) conducted a review of the research on gender in IT covering 20 years. A telling finding Gallivan reported was that "… the concept of 'gender' was only associated with women in the majority of studies" he [Gallivan] reviewed. In fact, most authors assumed that men lacked gender – in short, that being male was the norm, and that gender issues should only arise when women are introduced into the equation" (Gallivan, p. 52).

Trauth (2013) and Mennega and de Villiers (2021) looked at the role of theory in gender and IS research. Trauth (2013) covered the time period from 1992–2013, and Mennega and de Villiers (2021) built on Trauth's work and covered 2012–20. Trauth found that while some research about gender and IS explicitly employed gender theory, other "gender" research only used gender essentialism[1] as a theory-in-use to interpret research findings. Mennega and de Villiers (2021) found that IS research progressed from gender essentialist studies toward gender intersectional studies[2] which opens avenues for exploring issues in today's diverse global IT workforce and the identification of potential interventions.

The landscape of the field with regard to IS academia parallels female representation within the workplace. In the United States, the federal government continues to provide funding to address the lack of gender parity in academia. As announced on the Association for Information Systems (AIS) website[3] on July 2021, a $1 million National Science Foundation (NSF) grant is being led by a team of five IS female academics, and the project is focused on addressing the lack of gender equity within IS academia. The motivations supporting this grant include: only 10 percent of IS full professors are women with only 25 percent of professors overall being women, and within the AIS only 10 percent of the top leadership positions have been held by women and only 10.5 percent of top awards have gone to women. In the announcement, AIS Vice President of Diversity, Equity and Inclusion, Eleanor Loiacono stated that the grant will fund a study that "is designed to foster gender equity through a focus on the identification and elimination of organizational barriers that impede the full participation and advancement of diverse faculty in IS." In addition, an Eos Foundation report[4] released in February 2021, looked at the ten highest-paid core jobs at the nation's top research universities and found that women are notably absent from those top earners (only 24 percent), and even more so for women of color (2.2 percent). These issues persist across the STEM field and are not exclusive to the United States. Falkner et al. (2015) studied female academics in the computer science field and found that perceptions of identity conflict and a lack of belonging to the discipline persisted, even for high-performing individuals. Hamilton et al. (2016) explored the perceptions of international faculty regarding gender equity in the computing discipline. They found a clear need for gender equity programs in academia and the need for more consistent evaluation of the programs and dissemination of gender equity information to faculty.

Considering the culture in the United States around promotion and tenure, some institutions regard the development of a research stream as a major component of justification for tenure and promotion. Even though we have heard of instances where a senior colleague might advise junior faculty to jump from topic to topic or focus on current "hot topics" to ease the path to publication, the requirement to develop one or several intertwined research streams is often emphasized. These illustrative quotes from various university promotion and tenure policies as well as instructions for external reviews guide the predominant culture:

> ... and clear evidence of a promising research stream is required initially for tenure (including national reputation) and for promotion to full professor (including international reputation).

> Tenure-track faculty should demonstrate a clear commitment to scholarly activity by establishing an ongoing program of research, writing, submissions, and publications in one or more active research streams.

> Based on the evaluator's expertise in the area, assess the level of contribution that the candidate has made in the discipline.

> Assess the richness of the candidate's current scholarly or creative agenda and the potential for ongoing successful contributions in the future.

Leadership in IS Programs

Within the IS field professional associations, Gupta, Loiacono, Dutchak, and Thatcher (2019, p. 1882) found a "lack of women representation in AIS leadership positions, events such as panels, SIG officers, SIG board members, and AIS nomination for various awards." This is another area where we have experienced challenges – taking on conference leadership in SIGMIS-CPR, the AIS Women's Network College leadership, and leading women's networking events at the HICSS[5] conference. We have experienced pushback from colleagues regarding the necessity of these types of groups and events but continue to hear positive feedback from the male and female participants. The benefits of taking on these types of leadership roles are many. For the individual they provide exposure which may further one's career through other leadership opportunities, or potential promotion/tenure letter writers. These leadership roles also provide the individual with the opportunity to lean into the community and assist others in developing their own career paths.

Editorial boards play a central role in academia, that of gatekeepers, as they influence the field in pronounced and subtle ways (Bennis and O'Toole, 2005; Braun and Diospatonyi, 2005). For example, looking at 52 journals in the IS field, Burgess, Grimshaw, and Shaw (2017) state:

> Our study shows that, in contrast to the high topic diversity that the contributors to the IS diversity debate identify, diversity of EABs [editorial advisory board] is low for key demographic variables, with male researchers affiliated to US universities dominating boards. (p. 5)

Looking at the editorial boards of the AIS Senior Scholars Basket of Eight Journals, Beath, Chan, Davison, Dennis, and Recker (2021, p. 1) found that the journals "overall have fewer female members than might be reasonably expected" with the College of Senior Scholars 2019 Survey Report indicated that the average ratio of men to women in editorial positions was 3.7 for associate editor (AE) positions and 3.1 for senior editor (SE) positions. In addition, as of 2019, AIS membership was 55.8 percent male, while the Senior Scholar College membership was 82.7 percent male.

Editorial boards can provide advice on journal policy and scope, journal content, attracting new authors, and encouraging special issues and/or submissions. For example, according to one publisher of IS journals, when selecting board members, one should (among other considerations) appoint representatives "from key research institutes" and "existing board members may have suggestions for new members". While on the surface these seem reasonable, how do we expand our perspective if we continue to draw the "gatekeepers" from the same, shallow pool?

REFLEXIVE CONSIDERATIONS AS WOMEN IT ACADEMICS

Publishing Research on Gender and IT: Early Publications and Uphill Battles

Before diving into our experiences in publishing research on gender and IT, we need to note that gender issues such as equality in the IT profession have been studied for over 30 years (Trauth and Connolly, 2021) and yet the implications of gender within the workplace continues to be a focus of study (Serenko and Turel, 2021). It is important to study social inclusion issues in the IT profession for many reasons, including the varying perspectives that diversity can bring to sound and creative decision making. Grounded in seminal pieces on gender in the IT field such as Trauth's (1995) qualitative study of the effect of the developing IT industry in Ireland on women, Trauth's (2002) development of the individual differences theory of women in IT, and Ahuja's (2002) development of a gendered theory of IT career stages, we have added to the body of literature on the implications of gender in IT.

Initially, we published several papers outside of the mainstream information systems (IS) research journals. One reason for this was the composition of our (all women) interdisciplinary research team – two IS researchers (authors of this chapter), one communications researcher (Dr. Myria Allen), and one in public policy (Dr. Margaret Reid). While at times it was challenging to conduct interdisciplinary research, we would not forgo the experience because of the enrichment to our scholarly (and personal) lives – for the new lens to view data, insights, knowledge, exposure to theories, and experience we gained by working with these colleagues outside of our discipline. The communications professor initiated the idea to look at laughter that occurred in focus groups we had conducted with women IT professionals (Allen et al., 2004). Her approach and insights were very enlightening and reinforced the idea that researchers could address challenging societal issues and sensitive topics with empirical data and rigorous analysis.

Within our interdisciplinary research team, our plan was to explore women's perspectives of gender issues in the IT profession from a qualitative standpoint. The use of focus groups was a research design choice that ended up providing rich data from not only the questions asked by the researchers but also from the interaction between the participants in the groups as they responded. This design choice provided detailed descriptions around the topic, while minimizing the time necessary to gather the data (as opposed to conducting individual interviews). However, that design choice also proved to be an obstacle to overcome in terms of convincing reviewers of the legitimacy of the data collected in a group setting. We had to provide extensive details regarding the data-collection procedures such as what is a focus group, what characteristics of the data collection method are important to the context, why we were using this data-collection method for this study, etc. For example, in the Riemenschneider et al. (2006) article (detailed next), two paragraphs were needed to satisfy the review team in terms of justifying the focus group data collection method. Once this hurdle was overcome, several publications resulted from these data collections. One of the strengths of qualitative research is the ability to uncover layers of complexity within a phenomenon, sensitive issue, or societal concern that may make some people uncomfortable. The depth of scholarly understanding with a critical/interpretivist epistemology often exceeds insights gathered from a survey or a-theoretical data analytic solution. This may require a little extra explanation in the publishing process or the education of the editorial team, but the reward of a well-developed theory is worth it.

In the first publication from this data collection, we addressed barriers that women face in the IT work environment (Riemenschneider et al., 2006). From the focus group data, we identified explicit statements to elucidate the barriers women face in the IT workforce using revealed causal mapping.[6] The barriers the women identified in terms of promotion opportunities (both perceived and actual) were linked to their views of family responsibilities, stresses faced within the workplace, job qualities, and the flexibility of their work schedule.

In another study from the focus group data, we identified implicit statements (in addition to the explicit statements) to elucidate the barriers women face in the IT workforce using revealed causal mapping. Three concepts: promotion barriers, work stress, and work schedule flexibility were expressed both implicitly and explicitly by the women. However, ageism and lack of respect were only expressed implicitly. Ageism issues were expressed by the young women and the older women. For instance, in our study, one woman shared the following regarding lack of respect (Allen et al., 2006):

> Actually, I've had experience during analysis on a project where the users internal to the company but outside my department wouldn't even look at me or answer my questions. I'd ask a question and they'd look at my male companion and answer my question to him. This went on all day long. (p. 839)

This finding is still relevant, today. One of the participants in the Trauth and Connolly (2021) study shared her experience with respect:

> Most people I've talked to recently, they kind of don't think that a woman can know … They don't respect you until you start talking about what you know. And then you get the respect. Even the respect is very patronizing, "Oh she is a woman, what does she know?" until you start talking and prove yourself. (p. 2091, Appendix B5)

Other factors that continue to be prevalent gender-relevant issues to study within the IT field include barriers to promotion (Armstrong et al., 2018b), work schedule flexibility (Armstrong et al., 2018b; Serenko and Turel, 2021), work stress (Armstrong et al., 2018b; Riemenschneider and Armstrong, 2021), and personal characteristics (Trauth and Connolly, 2021).

We have repeatedly "stood on the shoulders" of bold scholars such as Eileen Trauth and Manju Ahuja as we have built a career in IT workforce research field. For example, we contributed an empirical investigation of Ahuja's (2002) conceptual model to the 2018 *Information Systems Journal, Special Issue on Social Inclusion* edited by Trauth, Joshi and Yarger (Armstrong et al., 2018b). This study found that social expectations, work–family conflict, a lack of informal networks, and institutional structures influenced women's advancement in an IT career, while social expectations, occupational culture, and institutional structures influenced women's persistence in an IT career. Thus, key influencers of advancement and persistence of women in IT, identified by the study participants, were social expectations regarding gender difference and institutional structures around the job qualities and communication pressures of the IT work environment.

We also, along with other colleagues, conducted a replication of Ahuja, Chudoba, Kacmar, McKnight, and George's (2007) study of turnover intentions for IT "road warriors"[7] with non-road warrior (in-house) IT personnel published in the *AIS Transactions on Replication Research* journal (Armstrong et al., 2018a). Consistent with Ahuja et al. (2007) the relationships between work exhaustion, organizational commitment, and turnover intention were supported, thus confirming their generalizability. Thus, regardless of their "location" (in-house or external), IT professionals experience the core turnover intention antecedents similarly. The influence of work–family conflict on work exhaustion but not on organizational commitment was consistent with the Ahuja et al. (2007) study in both strength and direction. In contrast to Ahuja et al. (2007), the replication study found that fairness of rewards was much more important in the model for in-house IT professionals than job autonomy. Thus, the strong influence of autonomy within the IT road warrior context was not replicable in the in-house IT context. In addition, we found mixed support for the mediation effects of work exhaustion and organizational commitment.

Further benefits from conducting research with an interdisciplinary team were two publications in public policy journals (Reid et al., 2008a; 2008b) and a book chapter (Armstrong et al., 2007). Each of these studies focused on IT professionals working in a state government context. An earlier version of the 2008a paper was presented at the 2006 ACM SIGMIS-CPR[8] Conference and received the Magid Igbaria Outstanding Paper Award (Reid et al., 2006). The feedback from reviewers and the comments received from colleagues at the conference were instrumental in improving this paper and positively impacted the resulting publication.

Publishing research on gender and IT: where is the IT artifact?
One of the most common and recurring issues we faced in studying IT professionals was from reviewers asking, "where is the IT artifact?" How is this IS research (and not organizational behavior/human resources) if there is no IT artifact? Similar to computer science, engineering, and strategy perspectives being a part of IS, organizational behavior and human resource perspectives also are an important part of IS. It is interesting to note that addressing the IT artifact issue is a concern that other researchers have also faced. Most recently, Trauth and Connolly (2021, p. 2061) state, "… the focus is not on the IT artifact, itself, but rather on the people who develop it." We have defended the need to study the IT professionals framed in various ways. For a paper addressing IT personnel working in state governments (Allen et al. 2008), we justified our study by addressing the shortage of new graduates in IT, the decreasing enrollment of students studying IT in universities, and the shortage of IT personnel in state governments. Some of these justifications are still applicable as expressed in a recent study by Serenko and Turel (2021).

We have also repeatedly addressed the related question, "Why is it important to study IT professionals?" One approach to justifying IT professional research was to frame our studies around the significant global demand for IT professionals and the limited number of professionals available to meet that demand. In multiple papers (Reid et al., 2010; Armstrong et al., 2015; Riemenschneider and Armstrong, 2021), we provided extant literature to illustrate the imbalance between the supply and demand for IT professionals, including data from the US Bureau of Labor Statistics, the United Kingdom, Europe, and India. Additionally, we addressed the high costs organizations experience globally to hire and retain IT professionals.

We finally had enough of the question – and attacked it head on. In a recent *MIS Quarterly* paper, we developed the "perceived IS distinctiveness" construct which allowed us to develop a more complete picture of IT workers' perceived distinctiveness including its composition and outcomes. Riemenschneider and Armstrong (2021) found that, "The occurrence of change within the profession, the facets of knowledge needed, and the continuous refinement and adaptation of the knowledge base within a mentally demanding work context are what make the IS profession distinctive from other professions" (p. 1156).

We have also explored IT professionals' exhaustion and turn-away intention (i.e., intention to leave the IT field altogether) and found that perceived workload increased career exhaustion, whereas fairness and control decreased career exhaustion; and career exhaustion strongly influenced turnaway intention via affective commitment (Armstrong et al., 2015). The reviewers of this manuscript (for *MIS Quarterly*) were much more accepting of our "IT artifact" from a career perspective than a job perspective.

Table 9.1   Matrix of challenges by gender

	Men's challenges	Women's challenges
Men's perspective	What men say about their own challenges	What men say about women's challenges
Women's perspective	What women say about men's challenges	What women say about their own challenges

Publishing research on gender and IT: what about the men?
Another pointed question from reviewers we have continually addressed is "Why are you studying women IT professionals – what about the men?"[9] This question directed us to gather additional qualitative data from male managers and female managers in same sex focus groups and ask about their perceptions of the challenges they faced in the IT profession. Our data collection resulted in identifying the challenges in a 2 × 2 matrix, as illustrated in Table 9.1.

Once we collected this rich data, we then faced the task of publishing the findings. Our first attempt at publishing a paper was to use the entire matrix – this was a very optimistic (and unrealistic) endeavor on our part. How does one address all four cells of a matrix of qualitative data within 25 pages? (Hopefully, you are chuckling at our naivete regarding this research conundrum.) After several unsuccessful attempts to include all the data, we determined that we should address a few cells in a single manuscript. In Reid, Allen, Armstrong, and Riemenschneider (2010) we published a paper addressing the cognitive gender gap focusing on the challenges that women faced that their men counterparts do not face as articulated by the male and female IT managers. Reid et al. (2010) applied critical theory "… by examining how societal and organizational values and roles interact or are mirrored in the organizational realities …" as shared by the IT managers (p. 529).

In Riemenschneider, Buche, and Armstrong (2019) we employed the communication theory of identity to frame the challenges men face in the IS workplace as perceived by male and female managers. Gratitude goes to our colleagues in the communications discipline for broadening our lens (repository) of theories we could apply to the IS discipline. This publication was one of the few to tackle IT workforce issues from a men's perspective. Surprisingly, the amount of research dedicated to exploring IT work from a male viewpoint (Fuller et al., 2015; Cain and Trauth, 2017; Joshi et al., 2016; Kvasny et al., 2016) in this male dominated profession has been limited. Perhaps future studies may study social inclusion within the workplace from a broader and/or more nuanced perspective.

It is important to note that we have not shared an exhaustive list of research challenges one may encounter when studying IT workforce topics but have focused on those we have repeatedly encountered. As research in the IS field as a whole, and IT workforce as a subdomain matures, existing methods evolve – and new ones emerge, and journal scope and directions change we will face additional publishing challenges.

Lessons learned in publishing
From our years of experience in publishing gender and IT research, we have learned several lessons; hopefully, others will benefit from the experiences we share here. First, the selection of the journal to which one submits her work is very important. Know which journals to target (e.g., we have published in these journals: *MIS Quarterly, Information Systems Journal, European Journal of Information Systems, The DATABASE for Advances in Information Systems*), because not all IS discipline journals are friendly toward this type of research. In addition, the "friendliness" of a particular journal can fluctuate over time with changes in the

editorial board. On more than one occasion, we have received a desk reject because the journal was not amenable to publishing gender or IT workforce research. This is extremely frustrating; however, do your homework regarding the journal and the editorial team, identify special issues that might be open to the research topic, and be tenacious.

A second lesson is perseverance, perseverance, perseverance. Even if the paper is rejected at more than one journal, take the feedback from the reviewers, modify the paper to address the doable reviewer comments, and submit the paper to another journal. Our philosophy has been, "every paper has a home," you just must keep searching until you find it.

Third, be intentional and thoughtful when considering recommending individuals for the editorial team – senior editor, associate editor, and reviewers for the submitted paper. A significant consideration is to know whether the colleagues you are recommending are supportive of the topic area of the research. Becoming a member of a SIG, and attending special interest conferences, such as SIGMIS-CPR or Informs,[10] offers a great way to build a network of individuals who have a similar research interest, learn about other ongoing research projects, and identify potential reviewers and editors.

Fourth, conducting research in an interdisciplinary team offers many benefits. Colleagues from different disciplines may offer a novel way to look at phenomena, employ various theoretical lenses, and provide guidance in publishing in journals outside the IS discipline. Many universities are currently emphasizing the need to study complex world problems with interdisciplinary teams. Also, having a mixture of academic ranks within the team (e.g., assistant, associate, full) provides opportunities for growth, mentoring, and continual learning for all team members.

Our fifth learned lesson is to only submit research-in-progress to conferences that maintain copyright. Researchers can avoid the expense of paying copyright fees for their own materials published in conference proceedings that they want to use in a journal submission. (*Note:* we had to pay for a graphical illustration we had created and published in the proceedings when we wanted to use it in a journal paper.) Another challenge we have faced is justifying the uniqueness and contribution of the journal paper above and beyond the conference paper. In one situation, we had a multi-page table summarizing extant research; the journal's submission software flagged our manuscript for not having enough "new material." It took multiple emails, phone calls, and substantial justification to show how our work was contributing above the conference proceeding.

A final consideration is the methodology and theory the authors are employing in the research. We suggest that researchers should build a "tool belt" of diverse research perspectives and methods to support their overall research stream. Like a carpenter, electrician, plumber, roofer, or other craftsperson, a "tool belt" is a device worn around one's waist to hold a hammer, pliers, measuring tape, screwdriver, and various tools used by the craftsperson. Over our careers, we have published positivist and interpretivist studies, qualitative and quantitative studies, and have applied diverse theories (including individual differences and critical theory). We have employed partial least squares (PLS), Warp PLS, hierarchical regression, covariance-based structural equation modeling,[11] revealed causal mapping, content analysis, and canonical correlation analysis methods. Having a tool belt full of diverse tools offers greater opportunities for addressing a variety of research questions as well as disseminating your research in a variety of outlets. Admittedly, some journals are more open to positivist, interpretivist, case studies, action research, or critical theory, but not limiting oneself to a single perspective, method, or data analysis technique offers a wider array of research opportunities and publication channels.

Our Leadership Experience in IT

Call for more women in academic leadership

From January 2013 through May 2019, one author served as the Associate Dean for Research and Faculty Development in a school of business. These years were a time of growth, learning the ins and outs of administrative responsibilities, which often reflected similar issues to the findings from the gender in IT research. At the time, she was the only woman serving on the 12-member executive leadership team in the business school. Due to the isolation that she experienced, she approached the Director for the Academy of Teaching and Learning to explain her predicament. Subsequently, they created a women's leadership group, comprised of female department chairs, associate deans, vice provosts and deans from across the university. This group meets once a month to discuss a variety of topics from articles in the *Chronicle of Higher Education* to brainstorming ideas to assist other women in their academic careers. The leadership group worked with the provost's office to offer a workshop entitled "Women's Paths to Promotion" in which a panel of four full professors from various disciplines shared their personal stories. Currently, they are extending their outreach by facilitating additional satellite groups of mid-career women from across the academy who are considering administrative roles.

Another strategy the author found helpful was to join affinity groups pertinent to the role of administrator. The AACSB offers a "Women Administrators in Management Education Affinity Group," which hosts a variety of events, leadership development seminars, and offers networking opportunities with women administrators across the globe. Many states have organizations such as the "Texas Women in Higher Education," that offers conferences, webinars, and events to grow one's network. In these affinity groups, one can ask questions, share ideas, and receive sound suggestions and guidance from others' experiences in an academic leadership environment.

A third approach the author employed was to apply the universal design for learning (UDL) framework (the Center for Universal Design 1997), most often used in teaching, to presentations she was giving in leadership meetings. She used a plethora of tactics to justify an initiative she was advocating to an executive committee or upper administration. Some examples include using props (a picture, stuffed animal, toy) to illustrate a particular mental model (e.g., Why is it important for a male department chair to be aware of the existence and location of a lactation facility within the school of business? Because he is hiring new young female faculty!), telling a story, using humor to demonstrate an opposing view, providing data-driven graphical and quantitative displays, engaging others in a role play, or highlighting supporting research from a prestigious academic journal. Negotiation and persuasion are skills that may be learned and improved with practice.

Finally, she learned to be intentional about selecting and choosing career mentors and psychosocial mentors, including both men and women. Develop relationships with mentors outside your university, within your university, and within your discipline; build a diverse mentor group to expand your lens and assist you in seeing issues from a variety of views. Be patient with yourself as you climb the learning curve to be an administrator.

Apart from a few disciplines (e.g., nursing and K-12 education), higher education/academe/the Ivory Tower has predominantly been and continues to be a male-dominated profession. While writing this book chapter, the *Chronicle of Higher Education* published an article, "Who's Mostly Missing from Among the Highest-Paid Employees at Top Research

Institutions? Women" that referred to the Eos Foundation study "The Power Gap among Top Earners at America's Elite Universities" (February 2021). The academic literature continues to address the power gap in research universities – in which white males continue to hold most leadership positions. Desai, Chugh, and Brief (2014, p. 330) found that "men in traditional marriages – married to women who are not employed – disfavor women in the workplace and are more likely than the average of all married men to make decisions that prevent the advancement of qualified women." While this is not an indictment, it does speak into our lived experiences.

Call for more women in editorial leadership
As an editorial function, reviewing, can be beneficial to one's career on multiple fronts. Reviewing helps you by keeping current with the latest activity in your research area, shows you good/bad communication styles, sparks new research ideas in your topic area, provides positive exposure to more senior scholars (if review is done well and on time). This exposure to senior scholars can aid in terms of potential promotion/tenure letter writers or lead to other service opportunities in the profession. For example, acting as program chair for the SIGMIS-CPR conference led to an invitation to chair the conference. That led to an invitation to guest edit a special issue at *EJIS* (Riemenschneider et al., 2009). As relatively new and unknown scholars in the IT workforce area, we invited Jo Ellen Moore, to join us as a guest editor on the project. Including her in the editorial team provided the gravitas that we could not garner on our own. We also learned from her expertise as an experienced developmental editor. This provided exposure to senior and acknowledged scholars in the field but also taught us the difference between a critical and developmental editor. Over the years we have strived to be reviewers and editors who provide helpful, constructive feedback and encourage authors to push themselves and their work to be the best.

In addition to guest editing a special issue, contributing to a special issue on a topic can be beneficial. This provides the opportunity to get your work in front of scholars, often key gate-keepers in the field and/or your subdomain of study. In recent years there have been several special issues related to gender and IT such as *Information Technology and People* Special Issue: Exclusion, Inclusion and Changing the Face of IS Research (2008) edited by Cushman and McLean; *European Journal of Information Systems* Special Issue: Meeting the Demand for IT Workers: A Call for Research (2009) edited by Riemenschneider, Armstrong, and Moore; *Information Systems Journal* Special Issue: Social Inclusion (2018) edited by Trauth, Joshi, and Kvasny-Yarger. Perhaps even more exciting/encouraging is that at this writing there are at least three relevant open special issue calls in: the *Journal of the Association for Information Systems* Special Issue: Technology and Social Inclusion (in process) edited by Bailey, Carter, Thatcher, Urquhart, and Windeler; *Journal of Information Systems Education* Special Issue: Equality, Diversity, and Inclusion in IS Education (in process); and on a related topic *Management Information Systems Quarterly* Special Issue: Digital Resilience (in process) edited by Boh, Constantinides, Padmanabhan, and Viswanathan. We note these special issues – past and present to highlight opportunities for scholars to make an impact both personally and professionally.

Lessons learned in leadership and service to the profession
The first lesson learned is that as with most aspects of academic life – there are trade-offs to be made in terms of leadership and service. One author accepted an administrative position at her

university, which impacted her availability for potential leadership roles within the profession. During her time as an administrator, she was asked to serve as an associate editor for several journals. Due to the heavy administrative workload, she was not able to accept any of these requests. She was also asked to serve in a leadership role for a professional organization, but again did not have the bandwidth to accept the position. Depending on one's personal goals, trade-offs must be considered.

A benefit of taking on a leadership/service position within the profession increases your visibility and exposure to others in the field. Even if you are relatively young in your career, be willing to take on leadership responsibilities that will increase your exposure to other colleagues in the field. For example, serving in multiple positions at the SIGMIS-CPR conference (doctoral consortium chairs, program committee chairs, conference committee chairs) will lead to multiple other opportunities.

Taking on an editorial role, is not only useful to yourself in terms of understanding the publication process from the inside out, but it can also be useful in supporting others and "paying it forward". One of the authors had a conversation about a paper with a more junior colleague known through one of the women's groups. The author was able to share with the junior colleague some suggestions (regarding the specific journal she edits) to improve the likelihood of obtaining a favorable review team. Paying it forward through serving as a mentor to junior colleagues and helping them "learn the ropes" can be a valuable mechanism for the betterment of the entire community.

CONCLUSION

This chapter is a reflexive consideration of being women in IT who both research and experience the identified issues. Our focus was on two major cornerstones of academic life – research and service. The goal of the chapter was to share our career experiences, to "pay it forward," and hopefully assist others who are interested in conducting similar research and/or taking on service or leadership roles within academe.

Our contribution is that the lessons we have learned in publishing (e.g., editorial team selection, perseverance, target journals, and interdisciplinary teams) could be a springboard to mitigate publishing challenges of the future. Because of recent relevance of the NSF grant to study IS women in academe and the continuous challenges administrators face from the global pandemic, the lessons learned from service as an associate dean provide insight for those IS academicians who desire administrative positions.

These lessons can be applied to societal challenges. For example, Ajjan et al. (2020) studied individuals working from home during the COVID-19 global pandemic. Their findings revealed significant gender group differences with women experiencing less control over time and lower perceived technology usefulness; and these gender differences were even more significant for those who had children living at home. This major societal challenge will persist as a large percentage of the global population remains working from home. As a result, behavioral IS research will continue to be appropriate for exploring complex societal changes.

As there is a cyclical nature to the economy, there is also a cyclical component to discipline maturity. Over time, we have experienced an ebb and flow regarding the job market for tenure track positions within the IS field, and we have also experienced an ebb and flow in the ease/difficulty of publishing IS behavioral research. The current trend is toward a prevalence of articles using econometric models published in premier IS journals, and specific basket journals no longer accepting particular analysis methods/software (e.g., SmartPLS vs EQS vs AMOS), or research designs (e.g., survey). The publishing environment is dynamic, and we cannot predict the future. We know that there will be new analysis methods and tools developed, which no one has yet dreamed of. While behavioral IS research is still needed, the future outlook for publishing behavioral IS research continues to be in flux.

Throughout this chapter we have attempted to highlight the challenges faced and lessons learned from our years of studying and living out gender in the IT/IS workplace. When reflecting on our academic careers the overarching theme seems to be *community*. McMillan and Chavis (1986) identify four elements of community: membership (feeling of belonging), influence (making a difference), reinforcement (fulfillment of needs), and shared emotional connection. We have been fortunate to have become part of the IT workforce community – both giving and receiving the elements of community along our journey. It is in the spirit of community that we continue to work toward a more inclusive worldview.

NOTES

1. Gender essentialism is the belief that a particular trait or person is inherently female and feminine or male and masculine. See www.healthline.com/health/gender-essentialism#definition.
2. Intersectional studies look at the intersection of multiple social categories or groups. In this context, intersectional studies consider individuals as members of multiple [and simultaneously experienced] social categories. For example, everyone has their own experience of ethnicity, culture, and gender and these factors could act individually or interdependently.
3. AIS is the Association for Information Systems professional organization https://aisnet.org/. Announcement accessed https://aisnet.org/news/573422/AIS-members-lead-a-U.S.-National-Science-Foundation-ADVANCE-Grant-ImPACT-IT.htm.
4. The Eos [Greek goddess of dawn] Foundation is a private philanthropic foundation. The Eos Foundation report can be found at www.womenspowergap.org/higher-education/the-power-gap-among-top-earners-at-elite-universities/.
5. HICSS is the Hawaii International Conference on Systems Sciences https://shidler.hawaii.edu/itm/hicss.
6. Revealed causal mapping is a data analysis and presentation method that was used to understand the concepts elicited from the data collection and cast these concepts into appropriate representations. This allows researchers to assess the structure and content of the participants' cognitions. We direct interested readers to Narayanan and Armstrong (2005).
7. "Road warriors" refer to those IT employees who spend most of their work life away from home.
8. SIGMIS-CPR is the special interest group on management information systems within the Association for Computing Machinery (ACM) https://sigmis.org/, that includes an annual conference on computers and people research.
9. Gender (e.g., male versus female) is a descriptive term that refers to characteristic categories that are socially and culturally defined. The term is often used more broadly to indicate a range of identities that do not correspond to established ideas of male and female.
10. Informs is the Institute for Operations Research and the Management Sciences www.informs.org/.
11. PLS is a variance-based structural equation modeling analysis tool using the partial least squares path modeling method. Warp PLS is a variance-based and factor-based structural equation modeling analysis tool using the partial least squares and factor-based methods. Hierarchical regression is a form of multiple regression in which variables are added to the model in steps by blocks or groupings of variables. Covariance based structural equation modeling is an analysis tool that models the observed and predicted covariances between variables, example software includes AMOS and EQS.

REFERENCES

Ahuja, M.K. (2002). Women in the information technology profession: A literature review, synthesis and research agenda. *European Journal of Information Systems, 11*(1), 20–34.

Ahuja, M.K., Chudoba, C., Kacmar, C., McKnight, D.H., and George, J. (2007). IT road warriors: Balancing work–family conflict, job autonomy, and work overload to mitigate turnover intentions. *MIS Quarterly, 30*(1), 1–17.

Allen, M., Armstrong, D.J., Riemenschneider, C.K., and Reid, M. (2008). Factors that impact IT personnel perceived organizational support. *Information and Management, 45*(8), 556–63.

Allen, M.W., Armstrong, D.J., Riemenschneider, C.K., and Reid, M. (2006). Making sense of the barriers women face in the IT work force: Standpoint theory, self-disclosure, and cognitive maps. *Sex Roles: A Journal of Research, 54*(11/12), 831–44.

Allen, M.W., Reid, M., and Riemenschneider, C.K. (2004). The role of laughter when discussing workplace barriers: Women in information technology jobs. *Sex Roles: A Journal of Research, 50*(3/4), 177–89.

Armstrong, D.J., Brooks, J.G., and Riemenschneider, C.K. (2015). Exhaustion from information system career experience: Implications for turn-away intention. *MIS Quarterly, 39*(3), 713–27.

Armstrong, D.J., Riemenschneider, C.K., Buche, M., and Armstrong, K.R. (2018a). Mitigating turnover intentions: Are all IT workers warriors? *AIS Transactions on Replication Research, 4*(1), 1–10.

Armstrong, D.J., Riemenschneider, C.K., and Giddens, L. (2018b). The advancement and persistence of women in the information technology profession: An investigation of Ahuja's model. *Information Systems Journal,* Special Issue on Social Inclusion, *28*(6), 1082–124.

Armstrong, D.J., Reid, M., Riemenschneider, C.K., and Allen, M. (2007). Managing IT employee retention: Challenges for state governments. In *Modern Public IT Systems*, D. Garson (Ed.), Hershey, PA: IGI Publishing, pp. 221–38.

Beath, C., Chan, Y., Davison, R.M., Dennis, A.R., and Recker, J.C. (2021). Editorial board diversity at the basket of eight journals: A report to the college of senior scholars. *Communications of the Association for Information Systems, 48*, 1–15.

Bennis, W.G., and O'Toole, J. (2005). How business schools have lost their way. *Harvard Business Review, 83*(5), 96–104.

Braun, T., and Diószpatonyi, I. (2005). World flash on basic research. *Scientometrics, 62*(3), 297–319.

Burgess, T.F., Grimshaw, P., and Shaw, N.E. (2017). Research Commentary – Diversity of the information systems research field: A journal governance perspective. *Information Systems Research, 28*(1), 5–21.

Cain, C.C., and Trauth, E.M. (2017). Black men in IT: Theorizing an autoethnography of a Black man's journey into IT within the United States of America. *The DATA BASE for Advances in Information Systems, 48*(2), 35–51.

Desai, S.D., Chugh, D., and Brief, A.P. (2014). The implications of marriage structure for men's workplace attitudes, beliefs, and behaviors toward women. *Administrative Science Quarterly, 59*(2), 330–65.

Falkner, K., Szabo, C., Michell, D., Szorenyi, A., and Thyer, S. (2015). Gender gap in academia: perceptions of female computer science academics. In *Proceedings of the 2015 ACM Conference on Innovation and Technology in Computer Science Education (ITiCSE '15)*, ACM, New York, NY, pp. 111–16.

Fuller, K., Kvasny, L., Trauth, E.M., and Joshi, K.D. (2015). Understanding career choice of African American men majoring in information technology. In *Proceedings of the 2015 ACM SIGMIS Conference on Computers and People Research*, ACM, Newport Beach, CA, pp. 41–48.

Gallivan, M. (2013). A structured review of IS research on gender and IT. In *Proceedings of the 2013 ACM SIGMIS Conference on Computers and People Research*, ACM, Cincinnati, OH, pp. 45–56.

Gupta, B., Loiacono, E., Dutchak, I., and Thatcher, J. (2019). A field-based view on gender in the information systems discipline: Preliminary evidence and an agenda for change. *Journal of the Association for Information Systems, 20*(12), 2.

Hamlin, K.A. (2021) Why are there so few women full professors? *Chronicle of Higher Education,* https://community.chronicle.com/news/2519-why-are-there-so-few-women-full-professors?cid=VT EVPMSED1, accessed March 15, 2021.

Hamilton, M., Luxton-Reilly, A., Augar, N., Chiprianov, V., Gutierrez, E.C., Duarte, E.V., Hu, H.H., Ittyipe, S., Pearce, J.L., Oudshoorn, M., and Wong, E. (2016). Gender equity in computing: International faculty perceptions and current practices. In *Proceedings of the 2016 ITiCSE Working Group Reports (ITiCSE '16)*, ACM, New York, NY, pp. 81–102.

Joshi, K.D., Kvasny, L., Unnikrishnan, P., and Trauth, E.M. (2016). How do Black men succeed in IT careers: The effects of capital? In *Proceedings of 49th Hawaii International Conference on Systems Science*, IEEE, Koloa, HI, pp. 4729–38.

Kvasny, L., Trauth, E.M., and Joshi, K. (2016). The Role of HBCUs in preparing African American males for careers in information technology. In *Setting a New Agenda for Student Engagement and Retention in Historically Black Colleges and Universities*, C.B.W. Prince and R.L. Ford (Eds.), Hershey, PA: IGI Global, pp. 234–50.

Lee, P., Stewart, D., and Calugar-Pop, C. (2016). Technology, media, and telecommunications predictions 2016 (predictions for the technology, media and telecommunications (TMT) sectors). Deloitte. Retrieved from www2.deloitte.com/content/dam/Deloitte/global/Documents/Technology-Media-Telecommunications/gx-tmt-prediction-2016-full-report.pdf.

McMillan, D.W., and Chavis, D.M. (1986). Sense of community: A definition and theory. *Journal of Community Psychology*, *14*(1), 6–23.

Mennega, N., and De Villiers, C. (2021). A quarter century of gender and information systems research: The role of theory in investigating the gender imbalance. *Gender, Technology and Development*, *25*(1), 112–30.

Narayanan, V.K., and Armstrong, D.J. (Eds.) (2005). *Causal Mapping for Research in Information Technology*, Hershey, PA: Idea Group Publishing.

Reid, M., Allen, M., Armstrong, D.J., and Riemenschneider, C.K. (2010). Perspectives on challenges facing women in IS: The cognitive gender gap. *European Journal of Information Systems*, *19*(5), 526–39.

Reid, M., Riemenschneider, C.K., Allen, M., and Armstrong, D.J. (2008a). Affective organizational commitment in state government: The case of IT employees. *American Review of Public Administration*, *38*(1), 41–61.

Reid, M., Riemenschneider, C.K., Allen, M., and Armstrong, D.J. (2008b). The role of mentoring and supervisor support for state IT employees' affective commitment. *Review of Public Personnel Administration*, *28*(1), 60–78.

Reid, M., Riemenschneider, C.K., Allen, M., and Armstrong, D. (2006). Affective commitment in the public sector: The case of IT employees. In *Proceedings of the 2006 ACM SIGMIS Conference on Computers and People Research*, ACM, pp. 58–78.

Riemenschneider, C.K., and Armstrong, D.J. (2021). The development of the perceived distinctiveness antecedent of information systems professional identity. *MIS Quarterly*, *45*(3), 1149–86.

Riemenschneider, C.K., Armstrong, D.J., Allen, M.W., and Reid, M.F. (2006). Barriers facing women in the IT work force. *The DATA BASE for Advances in Information Systems*, *37*(4), 58–78.

Riemenschneider, C.K., Armstrong, D.J., and Moore, J.E. (2009). Meeting the demand for IT workers: A call for research. *European Journal of Information Systems*, *18*(5), 458–61.

Riemenschneider, C.K., Buche, M.W., and Armstrong, D.J. (2019). He said, she said: Communication theory of identity and the challenges men face in the information systems workplace. *The DATABASE for Advances in Information Systems*, *50*(3), 85–115.

Serenko, A., and Turel, O. (2021). Why are women underrepresented in the IT industry? The role of explicit and implicit gender identities. *Journal of the Association for Information Systems*, *22*(1), 41–66.

The Center for Universal Design. (1997). *The Principles of Universal Design* (version 2.0). Raleigh, NC: North Carolina State University.

Trauth, E.M. (2013). The role of theory in gender and information systems research. *Information and Organization*, *23*(4), 277–93.

Trauth, E.M. (2002). Odd girl out: An individual differences perspective on women in the IT profession. *Information Technology and People*, *15*, 98–118.

Trauth, E.M (1995). Women in Ireland's information industry: Voices from inside. *Eire-Ireland*, *5*(4), 133–50.

Trauth, E.M., and Connolly, R. (2021). Investigating the nature of change in factors affecting gender equity in the IT sector: A longitudinal study of women in Ireland. *MIS Quarterly*, *45*(4), 2055–100.

Ward, K., and Wolf-Wendel, L. (2015) Academic motherhood: Mid-career perspectives. In *Proceedings of the 2015 Conference for the Association for the Study of Higher Education*, ASHE, Denver, CO, pp. 64.

PART II

IDENTITY INFLUENCES

10. The influence of intersectional identity on women in the IT field

Eileen M. Trauth

INTRODUCTION

The individual differences theory of gender and information technology (IDTGIT), which I developed, posits a balancing act (Trauth, 2002, 2006; Trauth and Connolly, 2021; Trauth et al., 2004).[1] On the one hand, women are confronted with socio-cultural influences that would direct them away from IT careers. On the other hand, these influences are highly variable; they are a function of individual factors that exacerbate or ameliorate societal influences. But how, exactly, does this occur? What emerged from my field studies of women working in the information technology (IT) profession were three themes about individual differences. One was that the social construction of gender and technology is a cause of the gender imbalance. Girls and women are *exposed* to discourses in the external environment about women being nontechnical and technology being masculine. Exposure to this social shaping comes from societal factors such as culture. The second theme that emerged was about the role of individual identity as a source of variation in the extent to which girls/women are affected by these societal messages. My research found that demographics influence both how much one is exposed to and how much one internalizes societal messages and barriers. In other words, one's *experience* of these social shaping factors is influenced by such identity factors as race, ethnicity, age, socio-economic status (SES), marital status, parenthood, sexual orientation, disability, etc. The third theme was about another source of variation in factors influencing girls and women. How one *responds* to societal messages and barriers is related to individual agency, which is influenced by factors such as role models, mentors and family support, and by personal characteristics such as personality, skills, abilities, and interests.

This means that women with different identities often encounter different gender stereotypes, biases and barriers. Further, some women encounter more of these than other women. For example, women coming of age in the 1960s received very different messages about gender roles than a woman who came of age in the 2010s. A woman who has a disability likely encounters different barriers than a woman who does not. A Black woman in the United States might encounter gender stereotypes differently than does a white woman. As these examples illustrate, understanding the influence of identity is a crucial aspect of understanding gender biases and barriers and how they affect women.

But how, specifically, does one's identity help to explain the varied effect of societal forces on girls and women? The purpose of this chapter is to answer this question. I examine identity as a source of variation among women by revisiting transcripts of life history interviews that I conducted with women working in the IT sector in the United States between 2002 and 2006. I do so in order to show how a woman's identity interacts with societal and individual influences to enhance or inhibit her participation in the IT sector. I employ a framework developed in Trauth and Quesenberry (2006), which explains how within-gender variation occurs. Women's intersectional identities result in varied *exposure* to, *experience* of, and *response* to gender biases and barriers. In the Background section I review research that has been conducted on intersectional identity to date and explain this framework in greater detail. The main section of the chapter employs data from my field study to illustrate the effect on women's IT careers of three types of intersectional identity: motherhood; race and ethnicity; and sexual orientation (LGBTQ).[2]

BACKGROUND

People use assumptions about identity – their understanding of gender, race, ethnicity, class, etc. – to organize their world and inform their interactions with others (Kendall, 1998). Some identity characteristics, such as race, ethnicity, and gender, might be primary, fixed and visible in one's life and career. Other characteristics, such as religion and marital status might be secondary. Some people choose to make their sexual orientation invisible, others do not. Identity characteristics such as age and educational background are fluid, changing over time. The important thing is that people have multiple, intersecting identities. Gender intersectionality, then, is one way that within-gender variability comes about (Trauth, 2012, 2013). Crenshaw (1989) coined the term "intersectionality" to capture the influence of identity characteristics in conjunction with gender. It aligns with feminist standpoint theories (Haraway, 1988; Harding, 2004; Hartstock, 1997), which focus on the situated knowledge of marginalized persons. A gender intersectionality perspective adds nuance to studies of gender in relation to STEM. It enables examination of Black women or gay men or transgendered individuals in relation to the IT profession.

Intersectional Identity Influences

There is a growing body of gender and IT literature that focuses on issues associated with multiple, intersecting identities (e.g., Josselson and Harway, 2012; Margolis et al., 2008; Tapia et al., 2004). Probably, the identity factor most often considered is motherhood, which is typically expressed as issues related to childcare and work–life balance. Examples from my data set include: Quesenberry et al. (2006), Quesenberry and Trauth (2005), Trauth and Quesenberry (2005), and Trauth et al. (2009). Examples of others who have investigated these issues include: Adams et al. (1996); Ahuja (2002); Ahuja and Rodhain (2000); Ahuja and Thatcher (2005); Ahuja et al. (2006); Brough et al. (2005); Butts (2013); Greenhaus (1988); Greenhaus and Beutell (1985); Perrons et al. (2006); and Salaff (2002).

Race and ethnicity in relation to technology is another identity characteristic that is currently being examined. While some of this research focuses only on women (e.g., Kvasny, 2006; Kvasny and Trauth, 2003; Kvasny et al., 2009), most of this research focuses on both men and women (e.g., Morgan and Trauth, 2013; Morgan et al., 2015; Payton et al., 2018; Trauth et al., 2016; Windeler and Riemenschneider, 2016; Yarger et al., 2019). In some cases, it focuses just on men (e.g., Cain and Trauth, 2017; Fuller et al., 2014; Joshi et al., 2016; Kvasny et al., 2016).

One aspect of gender intersectional identity that remains largely unexamined is the intersection of gender with sexual orientation and gender identity. Of all the types of identity considered in the literature, research on LGBTQ identity is the least researched. The work of Kreps (2009), Light (Fletcher and Light, 2007; Light ,2007, 2010; Light et al., 2008), and Trauth (Blodgett et al., 2007; Trauth, 2016; and Trauth and Booth, 2013) are the exceptions.

Other identity characteristics that have appeared in the IT literature to a lesser extent include socio-economic status and disability. Kvasny et al. (2009) explored how the intersectionality of race and socio-economic class influenced women's IT careers. Morgan et al. (2015) explored these issues for both men and women in the context of IT use. Recent disability research in the IT literature includes Annabi and Locke (2019); Halonen and Mononen (2014); Loiacono and Ren (2018); Sahasrabudhe and Lockley (2014); and Trauth et al. (2015). However, it typically does not address the intersectionality of gender and disability.[3]

Finally, some of the identity research looks at multiple identities. For example, Annabi and Lebovitz (2018) considered race, ethnicity, nationality, age, and IT identity. McGee (2018) looked at race, ethnicity, educational background, and IT identity. This brief review of identity research in the IT field reveals the need for both more research on specific aspects of identity, and more research on the intersectionality of gender and various identities.

Methodology

The data that I draw upon in this chapter comes from a qualitative field study of 92 IT practitioners who worked in three specific regions of the United States: the northeast region (Boston, Massachusetts); the south (Charlotte-Raleigh-Durham-Chapel Hill, North Carolina); and the mid-Atlantic (Central Pennsylvania, excluding Pittsburgh and Philadelphia). These regions represent differing cultural influences, both urban and rural settings, and differing economies. In addition, I interviewed 31 IT educators who were located all over the United States. For this chapter I revisited these interview transcripts in order to probe more deeply into the influence of three intersectional identities: motherhood, race/ethnicity, and sexual orientation. At the beginning of each interview participants were asked to provide some demographic information including: age, marital/relationship status, parenthood status, race/ethnicity, educational background, and background on family of origin. The interview was conducted as a life history interview in which the woman talked about growing up, her educational history, and her IT work history. Throughout the interviews women invoked their various identities (both prompted and unprompted) when discussing influences on their IT careers.

When intersectionality is applied to this particular context it means that various identities, intersecting with the identity of woman, can shed light on why some women persist in the IT field in the face of environmental gender biases and barriers while others do not. As noted above, societal discourses about gender and technical careers do not affect all girls and women in the same ways. Girls and women in a particular society are often exposed to different forms of bias and encouragement regarding IT careers depending on their identity characteristics. Further, even when women and girls are exposed to the same gender messages, the way they experience these messages can vary. A woman's identity influences how she internalizes social shaping discourses. Finally, girls and women respond to gender messages and biases in different ways based on their individual identities. Therefore, it is this difference in exposure, experience and response that accounts for within-gender variation in women's engagement with STEM and IT careers. The analytical framework (Trauth and Quesenberry, 2006) that I used to examine the dimensions of intersectional identity in this chapter is shown in Table 10.1.

Table 10.1   *Intersectional identity and within-gender variation*

Sources of within-gender variation due to intersectional identity	Definition	Example
Exposure to gendered discourses and barriers	Encountering a societal message	Mass media portrayals of women as primarily homemakers
Experience of gendered discourses and barriers	Internalizing a societal message	Believing that women are not as capable as men in working with technology
Response to gendered discourses and barriers	Coping with, resisting, or acquiescing to a societal message or barrier	Deciding to pursue a career in IT despite societal messaging, because of interest, receiving excellent marks in STEM courses in secondary school and encouragement from teachers and parents

The next section of this chapter examines the influence of identity on women's choice of and persistence within the IT field. "Influence" is operationalized as exposure to, experience of, and response to identity-driven factors. These factors can be positive (facilitators) or negative (barriers). The identity interacts with a woman's gender to produce this effect.

THE INFLUENCE OF INTERSECTIONAL IDENTITY ON IT CAREERS

Below, I consider how each of the three types of identity can evoke barriers to entrance into and persistence within IT careers. I employ interview data to demonstrate how a given barrier might be a factor in one woman's career but not in another's because of their different identities.

Motherhood

In my study, motherhood was the most common identity characteristic affecting women's IT careers. Perhaps because of the demands of an IT career, motherhood seemed out of reach to many of them. Of the 92 IT practitioners I interviewed, 39 had children, 53 did not. However, some of the women without children were still of an age to become mothers, and offered their reflections about motherhood. The most common issue cited by women was the need for work–life balance in order to accommodate family obligations. In Quesenberry et al. (2006) our investigation of the connection between this identity and the IT gender gap revealed that there was not one, single motherhood path that women IT professionals had followed. Rather, the women exhibited a range of ways they absorbed societal messages, and a range of decisions regarding IT career and motherhood.

Exposure to environmental influences about motherhood

My findings showed that motherhood was an influential identity that spanned geographical regions and time frames. The prospect of having children presented women with the societal challenge of reconciling their professional development with attitudes about women working outside the home. All the women were exposed to societal discourses about motherhood as a barrier for women to cope with in a career. During discussions of national, regional, and organizational cultural attitudes towards women and women working outside the home, respondents spoke of cultural messages about women's family obligations taking precedence over professional ones. They invoked the not-so-subtle message that women should assume domestic childcare roles and men should assume professional income-earning roles.

Betty Jean, who does not have children, explained her perception of stereotypical gender roles: women should "date, get married and have kids," with the only acceptable jobs being domestic or traditional feminine occupations. Francie, who also does not have children, observed that:

> Typically, the family obligations take precedence over the professional obligation … I think typically [the societal view] is, that, when the woman has a child that she should stay home and take care of them. The male would be the financial supporter. (Francie)

Nancy, as well, was not a mother. Nevertheless, she spoke about the pressures she felt to have children despite the fact that it conflicted with her own personal desires. In speaking about a traditional view of society, she said:

> A woman's place is to make the home and raise the kids and cook the meals and clean, and all that kind of thing. It really was very clearly what I saw happening in our house [growing up], and what I saw happening on TV … There was always this kind of disconnect, it was like I could not imagine not having kids, and yet it was not like I really wanted to, it was just, that's what you did. (Nancy)

The women also spoke about the difficulties of reconciling their professional development, particularly in job attainment, with that of a socially constructed view of women as primarily mothers. They noted the difficulty they faced in obtaining IT jobs because of such attitudes. For instance, Janet spoke about a co-worker of child-bearing age who was held back by gender stereotypes. During an interview she was told "well you are young, you are going to get married and have children anyway, so there is really no point" in hiring you. When Janet was asked if gender stereotypes in society influenced women's career attainment in the IT industry, she responded:

> I think a woman interviewing a woman would not be worried about her having a baby and leaving, but maybe it is more prevalent to me as coming from a man … I think because women are a little bit more understanding of, maybe the biological clock, or the need to have children, or the desire to have children … I do not think men see that as easily. (Janet)

As an example of this, Sandra was told by one of her supervisors that "you should be home making your children better people," rather than working outside the home.

For these reasons, women who were not mothers felt that they could adjust more easily to the IT workplace culture with its time demands, including long work days and late hours. Francie viewed the ability to work late and long hours as particularly essential during project deadlines and software development because her managers were able to count on her being in the office. Hence, she felt that it was very difficult for women IT professionals to have children and a career. Other women commented on the difficulty that mothers have participating in after-hours networking events. Julie, who didn't have children, explained how she was able to overcome her feelings of being an outsider when she first moved to her job. She adjusted her schedule to more easily spend time networking with co-workers, which eventually led to a heightened sense of community.

Experience of environmental influences about motherhood

Societal messages were complex, difficult to digest, and were processed in different ways by different women. So, while all women in a given culture might have been exposed to the same discourse about motherhood and the barriers it represented to women's careers, these messages were being interpreted differently. Women pointed out that the "motherhood message" – societal stereotypes about a woman being primarily responsible for family with the consequent need for work–life balance – did, indeed, keep some girls from entering the IT field to begin with. However, while women who did not have children may have heard the message, they were able to ignore it.

But even those women who were mothers, internalized the motherhood message in a variety of ways. Some women acquiesced to it. Kimberly took advantage of opportunities to tele-commute in order to have more work–life balance in her role as primary child minder. Helen stressed the importance of working part time while her children were still in primary school.

Other women were influenced by important people in their lives to reinterpret the moth-erhood message. Leah was the primary breadwinner and her husband had primary domestic responsibilities. Candace was a single mother raising her son without joint custody or child support. So, she didn't consider the messages about being taken care of by some man as apply-ing to her. Jada and her husband shared childcare responsibilities equally:

> [My husband who also works in IT] understands the pressures and the demands. We work more closely dealing with [child care] situations. Dealing with a child being sick, he takes half a day off, I take half a day off. We work around our schedules. (Jada)

Similarly, Sheryl explained that she and her husband agreed that if they had children, there would be a "50–50" split on domestic responsibilities. Rose also pointed to her husband's influence in parenting:

> My husband, I don't think, would have married anybody who wanted first and foremost to be an at-home mom. And he has been very supportive … and he's probably been my biggest advocate. … I've even seen a real change in him … He is taking a lot more responsibility for our daughter … I know that he and I are in constant communication … over who is going to be home at what time to make sure [our daughter] was being cared for. (Rose)

In the Indian culture in which Karen was raised, her mother would provide a significant amount of childcare support.

Some women rejected the motherhood message for personal reasons. Samantha revealed a strongly independent personality.

> By the time I reached high school, I was very independent. I really did not see a need for a man to take care of me. I thought my parents were very silly in trying to push me into marriage. (Samantha)

Donna continued on in her career after having children because

> [Having a career] is very important for me … It is important to keep my mind active, to keep chal-lenged, and to like what I do. (Donna)

Response to environmental influences about motherhood

The women IT professionals who were exposed to and accepted the stereotypical motherhood message, exhibited a range of responses. Some of them perceived that they had to make an "either-or" choice between an IT career and motherhood. Those who chose not to have children explained that they had strong professional aspirations and felt that the societal view of women did not fit with their goals. For example, Julia chose to not have children because she did not want to forgo her career ambitions:

> I knew very early on that I didn't want children, because I saw from my mother's own example how much of her aspirations that she put aside ... I saw how much effort went into raising me and my three brothers. And I knew that there were things that I wanted to do [and could not do and also raise children]. I was not willing to give up my hope that one day I would [be extremely successful in my career]. (Julia)

At the opposite end of the continuum were women who chose to remain in the IT labor force as working mothers. Pamela was raised in an environment in which women were expected to be taken care of by a man. Her father refused to pay for her college tuition but paid for her brother's because

> He was going to be a breadwinner someday. (Pamela)

Nevertheless, she resisted those influences and was very clear that she did not want to be in a position where she would have to depend on somebody else.

In the middle were the women who viewed motherhood and career in sequential fashion. They internalized the message that women should take time away from work to stay home with their children, that it was the right thing to do. They might have delayed career entry while their children were young. Or they might have left the labor force for a long period of time when they became mothers. Or they might have delayed having children, intending to leave the IT workforce permanently when they became mothers. While Sue eventually returned to her IT career, she felt the path laid out for her was to get married and then stay home with her children. Elsie left the labor force for ten years. Hence her coworkers at her same career stage were a decade younger. She felt pressured to constantly demonstrate that her prime working years were her 40s and 50s, which brought on another set of challenges.

> I feel like [fellow coworkers] are all going to retire when [they are] 55 because [they have] worked since they were 22. I started at 38. I want to go all the way; I don't want there to be any limits to what I can contribute. I have got to work until I am 65 because I did not start my career at a young age, so I have got to make up [time] ... But I do not want them to know my age, because they will think I am too old. (Elsie)

Race and Ethnicity

Crenshaw (1989) originally articulated intersectionality in terms of gender and race because Black women have historically been largely absent from diversity discourses. Metrics and interventions have tended to focus either on *women* or on *racial/ethnic minorities* leaving the experience of *minority women* to fall between the cracks. In this section I show some of the ways that the intersection of gender and racial identities has shaped the IT careers of Black women.

Exposure to environmental influences about race and ethnicity

The women of color I interviewed had to contend not only with gender bias; they also spoke of being subjected to both overt and covert forms of racial stereotypes and exclusion emanating from cultural and organizational environments. These women encountered low expectations about their intellectual ability and backlash from affirmative action initiatives.[4] The women gave examples from both school and the workplace of people assuming they were accepted or hired because of an institutional desire to fill a "quota" of Black students or workers rather than because of their capabilities.

Cynthia's fellow students thought she had been accepted into their prestigious university because she is a Black woman. It sometimes made her doubt herself.

> I think about it all the time ... People were wondering if I just got in because I was a Black female ... [I]f there is affirmative action then a white person who was qualified won't get in because they have to fill a spot for an African-American person, and "Oh that poor white person" ... And even with this program, the internship that I got there was a lot of competition and it hurts to have the first thought that comes to mind not to be "Oh this is so exciting I got this!" but "Oh were they just trying to make it [easy for me]?" (Cynthia)

Hence, the same initiatives that were meant to motivate greater racial diversity in the IT sector sometimes also served to reinforce stereotypes that were invoked to diminish these women's achievements. The company that hired Peggy for a summer internship seemed only interested in hiring Black students so that they could show proof of affirmative action.

> I spent the whole summer doing absolutely nothing. (Peggy)

Melissa was an honors student in secondary school, and grew up in a middle-class family. Both of her parents had attained post-secondary qualifications. Nevertheless, when she got to college, she encountered racial stereotypes and exclusion from her peers.

> So, you're in physics class, and people are choosing lab partners and you're the last one picked because you're Black and they assume that you're an idiot. (Melissa)

She noted her fellow students' shock when she got the highest marks in the class on the first exam. June was one of three Black women in a class of 50 students in her computer science major. One of her professors (a man) told her that in order to succeed she would need to work harder than the men students. She interpreted this as being about her gender until she talked with white women classmates and learned they had not received that same message from that professor. In the workplace Joanne was subjected to covert stereotypes about her personal life. She was married with two children, but when people looked at her as a Black woman, they saw someone different:

> ... the perception is that ... Black women are probably single with kids ... You're not considered to be stable because you don't have a man at home ... And those types of things are not really said, they're kind of implied. (Joanne)

Megan gave an example of differential assessment of workers' performance based on race and gender. She witnessed a Black woman and a white man who were hired at the same time. The woman had consistently high work output while the man did not.

But I think they tolerated that laziness a lot more with him, because he was a young white male, versus how long they would have tolerated that with anyone else … It makes you feel angry … They babied him a little bit more, and hoped he would come around. (Megan)

Melissa told a story about the senior systems engineer (a white man) who she believed had set her up for failure by giving her incorrect instructions regarding an important meeting with the client. But his advice did not ring true based on her work experience. So, she took it upon herself to be overly prepared, something that stood her in good stead when the client asked the difficult questions that she had prepared herself for, despite the advice of her so-called mentor.

As a manager Allison's gender-race intersectional challenge was encountering two sets of issues. When she was dealing with a white man:

[W]hat I've found is a lot of the professional men in [city where she works], their wives don't work. So, it's difficult for them, their wives are at home and then they go to work and their boss is a woman. (Allison)

However, when she was dealing with a Black man:

[O]ne of the managers who reports to me, he's African American, and we argue because I find he's very laid back and I can't get him to move as fast as I want to. Whenever I speak to him he says, "Listen, Julie." I say, "Roger, I am not your wife!" (Allison)

Experience of environmental influences about race and ethnicity

In general, the Black women in this study were exposed to the same negative societal influences. However, these barriers were experienced differently due to a variety of individual factors. Since these women did not have the benefit of white privilege,[5] their success was due to a combination of personal traits and influential people in their lives. A significant personal trait they possessed was being very intelligent. Every one of the Black women I interviewed could be classified as "smart" using metrics such as class standing in secondary school, being in the top tier of their graduating classes, and being enrolled in honors and gifted programs. This is noteworthy because course grades, class standing, etc. were not part of the recruitment criteria for participation in this study, nor was it characteristic of all the white women I interviewed. In the Black women's view, they had to be intellectually better just to be viewed as equal.

An important institutional influence on some of the women was their attendance at predominately Black colleges (HBCU).[6]

I wanted to be in a little bit more relaxed environment at least for four more years until I had to get out into the real world and deal with the discrimination. (Megan)

For some of the women, supportive professors played a significant role in their perseverance. Cynthia struggled during her first year in college and almost dropped out because when she thought of a career in computer science, she didn't fit the stereotype of a nerdy, white guy.

I didn't see myself in that at all and I didn't see how there could be a way for me to be who I am in this field. I [went to see] the Director of Students and [told her] I was going to switch majors and she's saying, "No you're not!" I also talked to this professor and he is very interested in females and minorities in computer science. (Cynthia)

This professor even paid for her to attend a conference focused on providing support to women and minorities in the computing fields. In contrast, June felt that having had some Black professors would have helped her to navigate the environment, helped her to experience differently the racism she was being exposed to:

> … somebody who could share their experiences with me, which our experience would be Similar … Maybe give me some advice … Role model … Give me advice as to what to expect and overcome adversity in the industry. (June)

Response to environmental influences about race and ethnicity

These women's stories illustrate how race introduces variation in women's exposure to biases and barriers in the societal environment. But even among this group of Black women, there were differences in how these women experienced them. As pointed out above, those who attended historically Black colleges did not encounter the same racial barriers as did women who attended predominantly white institutions.

No doubt there were Black women contemporaries of those I studied who did not remain in the IT field. But my interest was in those who managed to persevere. How did they do it? Beyond their intellectual capabilities, and a supportive educational environment, they drew upon sources of personal agency in responding to barriers. Some of the women received empowering messages growing up that "you can do anything you want." Megan's parents actively prepared her for what she would face as an adult.

> We were told, "A lot of whites don't want to see you succeed, and you're going to meet discrimination and racism. And you need to stand up for what you believe in." (Megan)

Olivia was raised by a single mother who did maintenance work in Trinidad. Because of this she grew up thinking she would need to take care of herself. She didn't grow up thinking there would be a man around to take care of her. Allison's mother was only educated as far as sixth grade. Not having graduated elementary school, her mother was always nervous that Allison wouldn't do well in school. Consequently, she pushed her children. Allison's father, who did some post-secondary course work, just expected her to do well.

Lydia was a student in the 1950s and 1960s when schools in the American south were migrating from being racially segregated to becoming integrated. Because of segregation her school contained the entire spectrum of students – from those who would go into technical jobs after high school to those who would go on to college. So even though no one in her immediate family had gone to college, she was able to observe students older than her who were doing so. This was a strong influence on her.

> I'd never heard [of engineering]. When my chemistry teacher mentioned that to me, I was like, "OK, Lydia, go home and look this up and see what this is," because all I knew about engineers were railroad people. (Lydia)

She was able to envision the kind of future she wanted to have.

> I guess I looked at my childhood as saying, "This is somewhere I know I don't want to be the rest of my life." And I knew that education was the route to not be left in that situation. (Lydia)

To Olivia, coping with racial bias was more of a marathon than a sprint:

Every day I make decisions about, you know, I decide if I'm going to react or not because I can be angry every day and it doesn't help me. I mean I [was recently] pulled over by the cops coming home in the morning coming from the gym and I thought he had absolutely no reason to pull me over ... And this has happened to me a couple of times in this area ... My point is that I feel I have a lot of these little encounters and I'm not clear as to why it's happening ... I try not to get angry every time something like that happens ... [I]t's going to be a judgment call, if I have the energy to deal with it today and in some instances, yes, I will let it go and in some instances, I will speak up. (Olivia)

Sexual Orientation

These interviews were conducted during a time period in American society when there were greater cultural taboos, lack of legal protections, and overall vulnerability associated with an LGBTQ identity. For this reason, I did not originally include sexuality or gender identity as an explicit item in the interview protocol. Instead, if a woman said something in the interview to suggest she identified as lesbian or bisexual (no participants identified as transwomen or queer), I turned off the recorder and asked the woman for permission to pursue this line of questioning and to record her answers. To increase the women's feeling of safety I pointed out that I am also a member of the LGBTQ community.

Exposure to environmental influences about sexuality

In contrast with straight[7] women, the LGBTQ women I interviewed encountered different environmental influences, specifically cultural and policy ones. They were inundated with heteronormative messages, and vulnerable to societal sanctions[8] and legal actions.[9] While the LGBTQ civil rights movement had been underway for over 30 years,[10] widespread governmental and workplace policies were just coming into existence. At the organizational level, anti-discrimination and same-sex domestic partner benefit policies[11] were varied and constituted a patchwork of employment protections. No national policy provided protection against discrimination in housing, accommodations, services, employment, etc. The first state-level recognition of same-sex marriage occurred in 2004,[12] more than a decade before same-sex marriage would become legal nationally. Hence, an LGBTQ woman's IT career was subject to the culture of the locale in which she worked and the company that employed her. Indeed, my own career was, to a large extent, governed by geographical locations in which higher education institutions had workplace cultures and policies that were LGBTQ inclusive.

These environmental factors affected the women's decisions about "coming out" to publicly identify as lesbian or bisexual in the IT workplace. LGBTQ inclusion was sometimes not a real part of the workplace culture but was simply a line in an anti-discrimination statement. Kristen worked in an office with what she called "conservative values," which had only recently adopted gender and sexuality anti-discrimination policies. She believed that management did not "feel ready" to have them, nevertheless they were adopted so as to remain consistent with corporate domestic partner policies. She also cited a transgendered employee working in a back office as evidence that it was easier for her company to employ gender minorities at lower levels who would be isolated from top management.

These women felt exposed to "double jeopardy" – having more than one minority identity – by virtue of being a woman in IT and a sexuality minority. Because of this intersectional identity they were sometimes exposed to different gender norms. Sol's educational and IT workplace experiences were shaped by her gender, age, sexuality, and ethnicity. She felt that being a "multiple minority" caused important people in her career journey – guidance counselors, employers, and co-workers – to question her IT ability. As a lesbian she felt isolated from both men and women in the workplace. The men could not relate to her as a woman and the women could not relate to her because she wasn't heterosexual. She recalled being fearful of accidentally mentioning her partner because her co-workers were uncomfortable with her sexuality.

Isolation as a woman in a man-dominated field became intensified when already-marginalized women possess non-normative sexualities, suggesting that awareness of a woman's sexual orientation affected the types of gendered messages she received. Kristen believed that gender expectations of lesbians varied from those held for straight women. She said men did not tell lesbian employees to act more feminine because they did not believe these women were capable of it. Yet those lesbians who were perceived to be straight would often be instructed to be "softer" both in dress and demeanor.

Experience of environmental influences about sexuality
These women acknowledged receiving some of the same heteronormative messages as straight women early in life about the behaviors and attitudes considered appropriate for women and girls.

> [I] felt pressure to, you know, follow a path, that there was a path set for me and that I was to follow it. That path was to go into business, get married [to a man] … That was the direction I was supposed to go in. (Sue)

These messages positioned IT careers, along with the associated skills, abilities, and workplace behaviors, in opposition to femininity. It was made clear to Kristen that she would have to fight to be on equal footing and be considered a technical expert. Women had to work harder and "prove themselves" within this masculine sphere.

But while the women may have been exposed to similar messages regarding heteronormativity and femininity, they were internalized differently based on personal factors. Much of this variance centered on their process of "coming out" and being "out" in the workplace. Yvette had always felt different from other women because she considered herself to be neither unequal nor "beholden" to a man. Kristen was not "out" in her conservative company. Yvette, on the other hand, refused to be "closeted" after being forced to lie about her sexuality earlier in her career. Lena was a relatively new employee and not "out" at work yet, but she intended to be at a later date. She usually took her time with this decision and did so as a slow process. Also, being "out" to one individual or group at work did not necessarily mean that she was "out" to everyone in the workplace. Glenna, a professor, was "out" to her colleagues and was active within the LGBTQ community, but was not "out" to her students. Similarly, some participants indicated that they were "out" to their peers at work, but not to upper management.

Some of the women talked about not being held to traditional standards of femininity because of their sexual orientation. On the one hand, it might be easier to be a lesbian in IT than a straight woman because, they said, you already don't fit the mold of conventional femininity. It might be easier for a lesbian or bisexual woman to work with married men because the men's wives would not become jealous about long hours at work or business travel. Lesbian women might more easily participate in activities such as golf outings and drinks events. They might not be bound by the same expectations of femininity, and it might be more acceptable for them to exhibit typically masculine behaviors such as assertiveness or ambition.

On the other hand, similar to their heterosexual colleagues, these women experienced a lack of peer inclusion from what they termed the "old boy's club." Being lesbian only added to feelings of isolation by exclusion from professional contacts and support. Glenna noted that it is difficult enough to be a woman in IT, much less a lesbian. Both identities assign a minority status, serving to further alienate these women from the normative straight, white, man.

Response to environmental influences about sexuality

Even though all of the woman described societal factors that imposed a "double minority" status upon them, they varied in their response to these messages and barriers. Their willingness to openly challenge authority was influenced by individual factors such as support from peers, role models, and mentors, as well as their own individual agency. Lena pointed to the influence of the women's liberation movement in counteracting prevailing messages about appropriate behaviors and roles for women. Reading *Ms. Magazine* offered her an alternative worldview by promoting career and self-sufficiency. She said these messages provided her with a different set of options than those she saw modeled elsewhere.

While these feelings of exclusion from men's groups in the workplace were almost unanimous, the ways in which they were handled varied. Ava sought support from peers by forming a lunch group with other lesbians at work. Kirsten found her own mentors in order to gain support from more established figures in her field. Ava recalled the enormous impact of a woman mentor in college. Lena noted that her older sister and brother-in-law helped her apply to college when the rest of her family was unwilling to do so. Yvette associated her love of math with her aunt, who was a math teacher.

Visible and invisible minority status is a response that derives from the duality of being a visible minority as a woman and (possibly) an invisible minority with a non-normative sexual identity. The "invisibility" of sexuality forced these women to confront a decision about "coming out" in the IT workplace and openly identifying as lesbian or bisexual. They noted that a woman could hide the fact that she is lesbian, but cannot hide the fact that she is a woman. While Lena felt that her sexuality, although not explicitly stated, was obvious to others, this invisibility gave rise to a type of decision making and choice unique to sexual identities.

Kristen noted that while her large company had welcoming policies, daily workplace behaviors were still rooted in the conservative values of her geographic region. This perception that behaviors rooted in local attitudes and values of a company transcend the content of corporate policies provides further insight into the variety of ways that lesbian and bisexual women responded as minorities in the IT workplace. Consequently, Sol's response to this double jeopardy was to refrain from discussing her personal life and family the way her heterosexual co-workers freely did.

CONCLUSION

This chapter contributes to the discourse about women's intersectional identities as a source of insight into the gender gap and how to address it. It does so by illustrating some of the ways in which three intersectional identities – motherhood, race/ethnicity, and LGBTQ identity – affect a woman's exposure to, experience of, and response to gender messages, biases and barriers. While this chapter demonstrates how these three multi-layers contribute to a woman's identity, it is important to emphasize that other forms of intersectional identity require similar examination.

As the women in this study have revealed, they represent considerable variation, even when they are living and working in the same culture. While all members of a society may be exposed to a particular influence or barrier at a particular place and time, the ways in which it does or does not take hold result from individual factors. As this chapter has shown, individual identity is a significant individual factor affecting people. Hence, when addressing gender barriers in society, schools, the workplace, or the profession it is important to understand the nature of identity factors, the effect they can have on individuals, and ways to address them. A woman's or a girl's intersectional identity affects her exposure to societal influences. But individual influences can also mitigate or reinforce the identity barriers that come from exposure to and experience of environmental influences. These women's stories demonstrate ways that individual influences can also affect one's response to the environment.

The findings presented in this chapter reinforce the theoretical argument of the IDTGIT that the project of addressing gender barriers to interest about, preparation for, and persistence in an IT career is not a one size fits all proposition. Interventions need to be tailored to the specific issues of specific groups of people in specific contexts. These contexts might be countries, regions of a country or organizations. While childcare might be a universal need of working mothers, its provision varies. Some countries might provide state-sponsored child care; other countries would not. Further, not all of the interventions should be about women. Men also need to become aware, educated, and encouraged to contribute to addressing the gender gap.

In the time since this study was conducted the underlying theory has been extended with the addition of a dynamic dimension (Trauth and Connolly, 2021). The resulting *dynamic individual differences theory of gender and information technology* incorporates seven themes that characterize the nature of change in factors affecting women IT professionals. This extended theory now enables investigation of changes in the ways in which environmental, identity and individual factors influence women's participation in IT careers. As it relates to the identity construct, in particular, the extended theory can now take into account changes in the effect of intersectional identities on women. For example, the influence of LGBTQ identity on IT careers has changed dramatically in recent years in the United States as a result of the legalization of same-sex marriage. But this change is uneven insofar as such legal protections do not extend to all countries. Similarly, the influence of motherhood and racial/ethnic identities on women's IT careers is undergoing change accompanying altered cultural attitudes about members of these groups. What we can learn from this is that identity-based privilege is a continuously moving target.

NOTES

1. See, also, Chapter 2 in this book for a detailed explanation of this theory and the field studies that employed it.
2. LGBTQ stands for lesbian, gay, bisexual, transgender and queer.
3. Chapter 14 in this book is a notable exception.
4. "Affirmative action" in the United States (called "positive discrimination" in some other countries) refers to proactive efforts to improve educational, employment and other opportunities for members of groups that have historically been subjected to discrimination.
5. "White privilege" refers to special rights, advantages or immunities granted to white people, over non-white individuals.

6.   HBCUs are higher education institutions whose principal mission is the education of Black Americans.
7.   That is, heterosexual.
8.   During the period that this research was being carried out I knew women who had lost custody of their children when they came out as lesbian. They also could lose their jobs.
9.   States that had no laws outlawing the refusal of services to LGBTQ people – hotels, etc.
10. 1969 is generally used to mark the beginning of the gay rights movement in America.
11. Such as health care benefits for domestic partners of an employee.
12. Same-sex marriage was legalized in Massachusetts in 2004.

REFERENCES

Adams, G.A., King, L.A., and King, D.W. (1996). Relationships of job and family involvement, family social support, and work–family conflict with job and life satisfaction. *Journal of Applied Psychology*, *81*, 411–20.

Ahuja, M. (2002). Women in the information technology profession: A literature review, synthesis and research agenda. *European Journal of Information Systems*, *11*, 1, 20–34.

Ahuja, M.K., Ogan, C., Herring, S.C., and Robinson, J.C. (2006). Gender and career choice determinants in information systems professionals: A comparison with computer science. In F. Niederman, and T. Ferratt (Eds.), *IT Workers Human Capital Issues in a Knowledge Based Environment* (pp. 277–302). Greenwich, CT: Information Age Publishing.

Ahuja, M. and F. Rodhain (2000). Mentoring relationships, gender and work–family conflict: The case of IT careers. *Proceedings of the Americas Conference on Information Systems* (Long Beach, CA, August).

Ahuja, M. and Thatcher, J. (2005). Moving beyond intentions and toward the theory of trying: Effects of work environment and gender on post-adoption information technology use. *MIS Quarterly*, *29*, 3, 427–59.

Annabi, H., and Lebovitz, S. (2018). Improving the retention of women in the IT workforce: An investigation of gender diversity interventions in the USA. *Information Systems Journal*, *28*, 6, 1049–81.

Annabi, H. and Locke, J. (2019). A theoretical framework for investigating the context for creating employment success in information technology for individuals with autism. *Journal of Management & Organization*, *25*, 4, 499–515.

Blodgett, B.M., Xu, H., and Trauth, E.M. (2007). Lesbian, gay, bisexual and transgender (LGBT) issues in virtual worlds. *The Data Base for Advances in Information Systems*, *38*, 4, 97–99.

Brough, P., O'Driscoll, M.P., and Kalliath, T.J. (2005). The ability of "family friendly" organizational resources to predict work–family conflict and job and family satisfaction. *Stress and Health*, *21*, 4, 223–34.

Butts, M.M., Casper, W.J., and Yang, T.S. (2013). How important are work–family support policies? A meta-analytic investigation of their effects on employee outcomes. *Journal of Applied Psychology*, *98*, 1, 1–25.

Cain, C.C. and Trauth, E.M. (2017). Black men in IT: Theorizing an autoethnography of a Black man's journey into IT within the United States of America. *The Data Base for Advances in Information Systems*, *48*, 2, 35–51.

Crenshaw, K. (1989). Demarginalizing the intersection of race and sex: A Black feminist critique of anti-discrimination doctrine, feminist theory and antiracist politics. *University of Chicago Legal Forum*, *1989*, 1, 139–67.

Fletcher, G. and Light, B. (2007). Going offline: An exploratory cultural artifact analysis of an internet dating site's development trajectories. *International Journal of Information Management*, *27*, 422–31.

Fuller, K., Kvasny, L., Trauth, E.M., and Joshi, K.D. (2015). Understanding career choice of African American men majoring in information technology. *Proceedings of the ACM SIGMIS Computers and People Research Conference* (Newport Beach, CA, June).

Greenhaus, J.H. (1988). The intersection of work and family roles: Individual, interpersonal, and organizational issues. *Journal of Social Behavior & Personality*, *3*, 4, 23–44.

Greenhaus, J.H. and Beutell, N.J. (1985). Sources of conflict between work and family roles. *Academy of Management Review*, *10*, 1, 76–88.

Halonen, R. and Mononen, J. (2014). Educated people with disabilities in the ICT Field. *Proceedings of the 20th Americas Conference on Information Systems* (Savannah, GA, August).

Haraway, D. (1988). Situated knowledges: The science question in feminism and the privilege of partial perspective. *Feminist Studies*, *14*, 579–99.

Harding, S. (Ed). (2004). *The Feminist Standpoint Theory Reader: Intellectual and Political Controversies*. New York: Routledge.

Hartstock, N. (1997). Comments on Hekman's "Truth and method: Feminist standpoint revisited": Truth or justice? *Signs: Journal for Women in Culture and Society*, *22*, 21, 367–74.

Joshi, K.D., Kvasny, L., Unnikrishnan, P., and Trauth, E.M. (2016). How do Black men succeed in IT careers: The effects of capital. *Proceedings of 49th Hawaii International Conference on Systems Science* (Kauai, Hawaii, January).

Josselson, R. and Harway, M. (Eds.). (2012). *Navigating Multiple Identities: Race, Gender, Culture, Nationality, and Roles*. Oxford, UK: Oxford University Press.

Kendall, L. (1998). Meaning and identity in "cyberspace": The performance of gender, class, and race online. *Symbolic Interaction*, *21*, 2, 129–53.

Kreps, D.G. (2009) Performing the discourse of sexuality online. *Proceedings of the Fifteenth Americas Conference on Information Systems* (San Francisco, CA, August).

Kvasny, L. (2006). Let sisters speak: Understanding information technology from the standpoint of the "other." *The Data Base for Advances in Information Systems*, *37*, 4, 13–25.

Kvasny, L. and Trauth, E.M. (2003). The "digital divide" at work and home: Discourses about power and underrepresented groups in the information society. In E. Wynn, M.D. Myers and E.A. Whitley (Eds.), *Global and Organizational Discourse about Information Technology* (pp. 273–91). Boston: Kluwer Academic Publishers.

Kvasny, L., Trauth, E., and Joshi, K.D. (2016). The role of HBCUs in preparing African American males for careers in information technology. In C.B.W. Prince and R.L. Ford (Eds.), *Setting a New Agenda for Student Engagement and Retention in Historically Black Colleges and Universities* (pp. 234–50). Hershey, PA: Information Science Reference.

Kvasny, L., Trauth, E.M., and Morgan, A. (2009). Power relations in IT education and work: The intersectionality of gender, race and class. *Journal of Information, Communication and Ethics in Society*, *7*, 2/3), 96–118.

Light, B. (2007). Introducing masculinity studies to information systems research: The case of Gaydar, *European Journal of Information Systems*, *16*, 658–65.

Light, B. (2010). Lesbian, gay, bi and trans (LGBT) experiences with digital media: Minitrack proposal for social inclusion track. *Proceedings of the Sixteenth Americas Conference on Information Systems* (Lima, Peru, August).

Light, B., Fletcher, G., and Adam, A.E. (2008). Gay men, gaydar and the commodification of difference, *Information Technology and People*, *21*, 3, 300–314.

Loiacono, E.T. and Ren, H. (2018). Building a neurodiverse high-tech workforce. *MIS Quarterly Executive*, *17*, 4, Article 5. Available at: https://aisel.aisnet.org/misqe/vol17/iss4/5.

Margolis, J., Estrella, R., Goode, J., Holme, J.J., and Nao, K. (2008). *Stuck in the Shallow End: Education, Race, and Computing*. Cambridge, MA: The MIT Press.

McGee, K. (2018). The influence of gender, and race/ethnicity on advancement in information technology (IT). *Information and Organization*, *28*, 1–36.

Morgan, A.J., Myers, Y., and Trauth, E.M. (2015). Consumer demographics and internet based health information search in the United States: The intersectionality of gender, race and class. *International Journal of E-Health and Medical Communications*, *6*, 1, 58–72.

Morgan, A. and Trauth, E.M. (2013). Socio-economic influences on health information searching in the USA: The case of diabetes. *Information Technology & People*, *26*, 4, 324–46.

Payton, F.C., Kvasny, L., and Pinter, A. (2018). (Text)mining microaggressions literature: Implications impacting Black computing faculty. *Journal of Negro Education, Innovations in African American Educational Research*, *87*, 3, 217–29.

Perrons, D., Fagan, C., McDowell, L., Ray, K., and Ward, K. (Eds.). (2006). *Gender Divisions and Working Time in the New Economy*. Northampton, MA: Edward Elgar.

Quesenberry, J.L. and Trauth, E.M. (2005). The role of ubiquitous computing in maintaining work-life balance: Perspectives from women in the IT workforce. In C. Sorensen, Y. Youngjin, K. Lyytinen, and J.I. DeGross (Eds.), *Designing Ubiquitous Information Environments: Socio-Technical Issues and Challenges* (pp. 43–55). New York: Springer.

Quesenberry, J.L., Trauth, E.M., and Morgan, A. (2006). Understanding the "mommy tracks": A framework for analyzing work–family issues in the IT workforce. *Information Resources Management Journal, 19,* 2, 37–53.

Richardson, H. and Bennetts, D. (2007). Work-life imbalance of IT workers in the internet age. In P. Yoong and S. Huff (Eds.), *Managing IT Professionals in the Internet Age* (pp. 37–86). Hershey, PA: Idea Group Publishing.

Sahasrabudhe, S. and Lockley, M. (2014). Understanding blind users' accessibility and usability problems in the context of myITlab simulated environment. *Proceedings of the 20th Americas Conference on Information Systems* (Savannah, GA, August).

Salaff, J.W. (2002). Where home is the office: The new form of flexible work. In B. Wellman and C. Haythornthwaite (Eds.), *The Internet in Everyday Life* (pp. 464–95). Oxford, UK: Blackwell Publishers Ltd.

Tapia, A., Kvasny, L., and Trauth, E.M. (2004). Is there a retention gap for women and minorities? The case for moving in versus moving up. In P. Yoong and S. Huff (Eds.), *Managing IT Professionals in the Internet Age* (pp. 143–64). Hershey, PA: Idea Group Publishing.

Trauth, E.M. (2002). Odd girl out: An individual differences perspective on women in the IT profession. *Information Technology & People, 15,* 2, 98–118.

Trauth, E.M. (2006). Theorizing gender and information technology research. In E.M. Trauth (Ed.), *Encyclopedia of Gender and Information Technology* (pp. 1154–59). Hershey, PA: Idea Group Publishing.

Trauth, E.M. (2012). Are there enough seats for women at the IT table? *ACM Inroads, 3,* 4, 49–54.

Trauth, E.M. (2013). The role of theory in gender and information systems research. *Information and Organization, 23,* 4, 277–93.

Trauth, E.M. (2016). Information technology. In N. Naples (Ed.), *The Wiley-Blackwell Encyclopedia of Gender and Sexuality Studies* (pp. 1419–24). Oxford, UK: Blackwell Publishing Ltd.

Trauth, E.M. and Booth, K.M. (2013). How do gender minorities navigate the IS workplace? Voices of lesbian and bisexual women. *Proceedings of the 19th Americas Conference on Information Systems* (Chicago, IL, August).

Trauth, E.M., Cain, C., Joshi, K.D., Kvasny, L. and Booth, K. (2016). The influence of gender-ethnic intersectionality on gender stereotypes about IT skills and knowledge. *The Data Base for Advances in Information Systems, 47,* 3, 9–39.

Trauth, E.M. and Connolly, R. (2021). Investigating the nature of change in factors affecting gender equity in the IT Sector: A longitudinal study of women in Ireland. *MIS Quarterly, 45,* 4, 2055–100.

Trauth, E.M. and Quesenberry, J.L. (2006). Are women an underserved community in the information technology profession? *Proceedings of the International Conference on Information Systems* (Milwaukee, WI, December).

Trauth, E.M. and Quesenberry, J.L. (2005). Individual inequality: Women's responses in the IT profession. *Proceedings of the Women, Work and IT Forum* (Brisbane, Queensland, Australia, June).

Trauth, E.M., Quesenberry, J.L., and Huang, H. (2009). Retaining women in the US IT workforce: Theorizing the influence of organizational factors. *European Journal of Information Systems, 18,* 476–97.

Trauth, E.M., Quesenberry, J.L., and Morgan, A.J. (2004). Understanding the underrepresentation of women in IT: Toward a theory of individual differences. *Proceedings of the ACM SIGMIS Computer Personnel Research Conference* (Tucson, AZ, April).

Trauth, E.M., Joshi, K.D., Graham, K., and Nithithanatchinnapat, B. (2015). An exploratory study of identity and IT career choice for military service members and veterans with disabilities. *Proceedings of the 21st Americas Conference on Information Systems* (Puerto Rico, August).

Windeler, J.B. and Riemenschneider, C.K. (2016). The influence of ethnicity on organizational commitment and merit pay of IT workers: The role of leader support. *Information Systems Journal, 26,* 2, 158–90.

Yarger, L. Payton, F.C., and Neupane, B. (2019). Algorithmic equity in the hiring of underrepresented IT job candidates. *Online Information Review, Special Issue on Social and Cultural Biases in Information, Algorithms, and Systems.* https://doi.org/10.1108/OIR-10-2018-0334.

11. We cannot build equitable artificial intelligence hiring systems without the inclusion of minoritized technology workers

Lynette Yarger, Courtney Smith and Adanna Nedd

INTRODUCTION

> Those of us building these [AI] systems will choose which subreddits and online sources to crawl, which languages to use or ignore, which data sets to remove or accept. Most important[ly], we choose who we apply these algorithms to, and which objectives we optimize for. We choose the labels we create, the data we take in, the methods we use. We choose who we welcome as data scientists and engineers and researchers – and who we do not. There were many possibilities for the design of the technology we built, and we chose this one. We are responsible. (Raji, 2020)

The underrepresentation of women, Black, and Latinx technology workers is a long-standing and systemic problem in the United States. In terms of the overall workforce composition in 2020, the US Bureau of Labor Statistics (2021) reports that 78 percent was white, 6 percent was Asian, and 12 percent was Black. People of Latinx ethnicity, who may be of any race, made up 18 percent of the total labor force. Women represented nearly half (47 percent) of the total labor force. Of those employed in computing and mathematics occupations, 25 percent were women, 66 percent were white, 23 percent were Asian, 9 percent were Black, and 8 percent were Latinx (Bureau of Labor Statistics, 2021). Based on the US Equal Employment Opportunity (EEO-1) reports from 177 companies in Silicon Valley, racial and gender disparities are even starker with less than 8 percent of the technology workforce identifying as either Black or Latinx and only 30 percent identifying as women (Rangarajan, 2018).

The pipeline metaphor is typically used by US universities, government agencies, and corporations to frame this underrepresentation as a lack of diversity among those preparing for technology careers. The pipeline metaphor describes a rigid educational path with a cultural and social focus that keeps many talented women, Black, and Latinx people from achieving careers in technology. In other words, the pipeline "leaks" women and people of color and emphasizes the challenges for the US education system to produce diverse and appropriately skilled workers (Garbee, 2017). The solution, therefore, is to create interventions aimed at increasing the retention and recruitment of underrepresented individuals who will enter the technology workforce at the end of the pipeline.

Despite its widespread use, the pipeline framing presents significant limitations for improving diversity in the technology workforce in the United States. For example, solutions that focus on technical skills development imply that the leakage of people from underrepresented groups is the problem. By focusing on a perceived shortage of diverse talent, US institutions give insufficient attention to the value that comes from a diverse set of perspectives and experiences (Garbee, 2017). This leads to the more troubling pipeline assumption that technology companies will enthusiastically hire and retain the women and minoritized professionals who persist in the pipeline. Empirical evidence suggests that this is not the case. Moreover, the technology sector faces few consequences for its failure to diversify its technical workforce (Rooney and Khorram, 2020). Freada Kapor Klein, the founding partner at the venture capital company Kapor Capital, states: "If you wasted a billion dollars and nowhere near met your target, you wouldn't get your bonus, you wouldn't have a job. And yet there seem to be no consequences. Despite all the words, despite all the money, despite all the platitudes and initiatives, it's hard to say that the companies are really taking it seriously" (Harrison, 2019). Bias and discriminatory practices in the recruitment, hiring, promotion, pay, and workplace climate perpetuates racial, ethnic, and gender disparities in the technology workforce. While hiring inches forward, the workplace culture and practices remain unfair and unjust. These malicious workplace cultures lead to avoidance of computing careers and significant turnover of women and people of color, suppressing any meaningful diversity gains (Rooney and Khorram, 2020).

Some forward-thinking organizations like the National Science Foundation have revised their thinking and now use "pathways" to describe a variety of educational routes that bring individuals to technology careers (Garbee, 2017). Whereas pipelines privilege academic routes to technology careers, pathways offer more ways to identify and tackle issues of recruitment and retention. As artificial intelligence (AI), the building of computer algorithms capable of performing problem solving and decision-making tasks that require human intelligence, becomes increasingly embedded in various US industries and public policy matters, society needs more than just white and Asian men building technology and information systems. Therefore, diversifying the people in and pathways to technology careers is critically important. AI hiring systems is a particularly fruitful area of exploration since it serves as a gatekeeping function for determining entry to the technology workforce.

According to the AI Now Institute (West et al., 2019), the technology industry's long-standing diversity problems and bias in the systems it builds have tended to be considered separately. However, these two problems are inextricably linked and rooted in discriminatory practices in hiring, in the workplace culture that devalues the contributions of women and minoritized workers, and in the building of biased technology systems. The term "minoritized" is used intentionally to express the historical underrepresentation, marginalization, and resilience of technology workers who identify as Black or African American, Hispanic or Latino, American Indian, Alaska Native, Native Hawaiian, and Pacific Islander. This term emphasizes the humanity and power within these racial and ethnic groups while also acknowledging the reality of oppressive systems that shape their lived experiences (Cooper, 2016). The connection between the underrepresentation of women minoritized technology workers and the building of just AI systems is exemplified in the highly publicized case of Timnit Gebru, a prominent AI researcher and co-lead of the Google ethics team who was pushed out of her job in December 2020. Gebru is one of the few Black women in Google's research organization and is highly engaged with diversity and inclusion in AI research and Google. Her firing's impetus was a research paper that she co-authored that called out potential biases in the language models that power Google's search mechanism. Despite Gebru's exceptional research credentials and stature in the AI ethics field, she was told to retract her paper without being allowed to engage in a conversation about this decision. Instead, she was fired for revealing how AI models harm marginalized communities and the importance of the equitable development of models that underpin many of society's most critical automated systems (Schwab, 2021). This example demonstrates why tackling the underrepresentation challenges within workforce diversity also requires addressing inclusion in hiring, workplace culture, and technical system building.

The purpose of this chapter is to examine the relationships between diversity and inclusion in the technology workforce and the building of AI hiring systems that harm and discriminate. In the next section we present Hill Collins (1990) Matrix of Domination as a synthesizing framework for reviewing extant literature that forms the basis for understanding how women, Black, and Latinx workers are systematically excluded from participation in the technology workforce. This framework includes four distinctive but interconnected domains of power: structural, disciplinary, hegemonic, and interpersonal. We then discuss how the absence of minoritized technology workers contributes to the construction of AI systems that discriminate. By understanding imbalances in diversity in the technology workforce and the harm caused by technology systems built without the participation of women, Black, and Latinx technology workers, we can begin to construct career pathways that are more accessible and inclusive.

BACKGROUND

A combination of biases and barriers in recruitment, hiring, performance review, and promotion processes contribute to racial, ethnic, and gender disparities in the US technology workforce. These disparities are persistent and stark and result in a technology sector that does not reflect the US workforce's demographics. Patricia Hill Collins (1990), a Black feminist scholar, describes four distinctive but interconnected domains of power: structural, disciplinary, hegemonic, and interpersonal. These domains of power provide an approach for examining power relations in society and are used as a sensitizing framework for exploring diversity and inclusion in the technology workforce. The concerns and figures presented are unique to the US context but may have applicability to other Western, educated, industrialized, rich, and democratic (WEIRD) societies (Henrich et al., 2010).

Structural Domain of Power

The structural domain of power describes how societal structures are organized to reproduce power relations that oppress historically marginalized groups. Major societal institutions such as law, government, education, and industry work together to grant or deny access to positions of authority and resources. For example, in 2019, for every dollar of income the median white household earned, the median Black household earned just 61 cents, while the median Latinx household earned 74 cents (US Census Bureau, 2020). This racial wealth gap plays a significant role in determining educational preparation for college, social networks, and cultural capital to gain college admission and financial resources. Black and Latinx students are more likely to pay for college by balancing their studies with work to cover the costs of attending college, making it more challenging to maintain a high GPA. These students are also more likely to accumulate excessive debt. Scott-Clayton and Li (2016) report that Black undergraduates owe $7,400 more than white peers at graduation. This Black-white debt gap surges to $25,000 over the next few years due to the higher default rates of Black borrowers.

Many Black and Latinx high-school students, particularly those in economically marginalized communities, are also less likely to apply to top institutions even though they have the academic talent to succeed (Belasco and Trivette, 2015). This decision (i.e., undermatching) can have lasting consequences for students' job opportunities and wages (Muskens et al., 2019; Ovink et al., 2018). The racial wealth gap and undermatching suggest that access to prestigious technology companies that recruit workers from elite universities may be beyond the average Black and Latinx family's reach. Moreover, while 20 percent of all Black computer scientists graduate from historically Black colleges and universities, companies do not look for talent across a representative sample of HBCUs (Vara, 2016). In one example, Google's Howard West pilot program, designed as an immersive training program for Black computer science students to mold Google's ideal software engineer. Few students who participated in the program were offered full-time software engineer roles. However other assumed roles within Google in project and program management (Elias, 2021).

Howard West is just one example of how hiring discrimination hinders the entry of minoritized college graduates into the technology workforce. For decades, researchers have used field experiments to measure discrimination in selection systems used in the US labor market. For instance, Quillian and colleagues (2017) conducted a meta-analysis of 28 field experiments of hiring discrimination against Black or Latinx job seekers. The results of their study indicate that, since 1989, there have been no changes in the level of hiring discrimination against Black and a modest decline in discrimination against Latinx applicants. Blacks received 36 percent fewer callbacks compared to white job applicants, and Latinx applicants received 24 percent fewer callbacks. The study results remain consistent across the applicant's education and gender, study method, occupational groups, and local labor market conditions (Quillian et al., 2017). In a field experiment to study racial discrimination in the labor market, Bertrand and Mullainathan (2004) submitted fictitious resumes for the same job listings. The fictitious resumes were submitted in pairs manipulated to assign a Black or white-sounding name randomly. Resumes with white-sounding names receive 50 percent more callbacks for interviews than resumes with Black sounding names, and callbacks were more responsive to resume quality for white-sounding names than for Black ones. This disparity held consistent across industries.

For jobseekers from underrepresented groups in technology firms, discrimination in hiring practices can take additional forms, including job descriptions that include a long list of required skills that do not match the job, over-reliance on candidates' academic backgrounds, and tricky technical interviews (Xu, 2020). Job descriptions that include an intimidating list of requirements can deter women and minoritized individuals who are suitable for the job from applying. In contrast, people of privilege may apply even though they do not meet all of the qualifications listed in the job ad (Xu, 2020).

After applicants submit their resumes, technology companies frequently invite suitable candidates to complete a skills-based exam. These exams may involve solving logic puzzles or applying software concepts, and have been found to privilege obscure software concepts and specialized knowledge taught at specific universities (Xu, 2020). The skills-based exams may also include content that does not reflect the fundamental skills needed to perform the job. Perplexing programming challenges at leading US technology firms have given rise to a coding interview prep industry that benefits already-privileged candidates who have time and money to spend on preparation (Xu, 2020). In this way, the skills-based exam behaves like standardized testing in college admissions by creating a discriminatory process that disadvantages minoritized job seekers who lack specialized technical knowledge.

Job candidates also encounter interviews with corporate recruiters, hiring managers, and AI hiring platforms. A vast body of literature finds that the traditional interview process serves as a site for unconscious racism and sexism that clouds the human interviewers' judgments and decision making in ways that favor some groups to the detriment of other groups (Knight, 2018). Our human tendencies to connect more easily with people with shared hobbies, opinions, or experiences may lead to hiring processes that favor an applicant's "cultural fit" rather than the applicant's "cultural add" and the objective outcomes that the applicant has achieved. Segrest and colleagues (2006) report that applicants with ethnic names and speaking with Spanish accents were viewed less positively by interviewers. In another field experiment that examined hiring bias against Muslim women who choose to veil while job seeking, Carmichael (2017) found that white women received callbacks at 4.5 times the rate of Muslim women wearing a hijab or headscarf. Discrimination was highest in the occupation with the highest status and the highest qualification necessary, even though firms had problems filling vacancies.

Johnson and colleagues (2016) uncovered an intriguing finding of the applicant pool's diversity and implicit bias. Using a series of experiments that evaluated hiring decision making, the researchers found that "when there was only one woman or minority candidate in a pool of four finalists, their odds of being hired were statistically zero. But when we created a new status quo among the finalist candidates by adding just one more woman or minority candidate, the decision-makers actually considered hiring a woman or minority candidate" (Johnson et al., 2016). It is important to note that these findings do not support "reverse racism," which is the idea that hiring decision making favors underrepresented job candidates over white candidates. Rather, these findings demonstrate that biased hiring decisions can be mitigated by increasing the diversity of qualified candidates selected as finalists.

Racialized disparities in hiring can be attributed to Black and Latinx job seekers having smaller social networks with fewer acquaintances working in technology companies who can refer them to open positions (Xu, 2020) or, in the case of algorithmic hiring systems, less similarity to "the type" of successful hires historically (Xu, 2020). A *New York Times* study found, for example, that Black and Latinx undergraduates who completed computer science or engineering degrees were more likely than their white and Asian counterparts to hold jobs outside the fields of technology and engineering (Bui and Cain Miller, 2016). Even when obtaining jobs in technology and engineering fields, Bui and Cain Miller (2016) describe an employment hierarchy where 10 percent of Black computer science and engineering graduates have lower paying and lower status office support jobs than 5 percent of white graduates and 3 percent of Asians.

Disciplinary Domain of Power

The disciplinary domain manages oppression by governing people from marginalized groups with limited opportunities and stricter rules that sustain dominant groups' power. Disciplinary power is expressed through organizational policies that obscure the effects of racism and sexism that provide the dominant group unearned benefits. This way of ruling relies on bureaucratic hierarchies and techniques of surveillance to manage power relations. Career advancement and promotion are two areas of human resources where disciplinary power is commonly enacted.

Promotion and career advancement are more limited for women and minoritized groups. At the managerial and executive levels, Black and Latinx employees are significantly under-represented. Rangarajan (2018) analyzed data from the EEO-1 reports from 177 Silicon Valley firms and found that the average management team had only 2 percent Black and 4 percent Latinx representation. At the executive level, those percentages are reduced by half. Six companies did not have a single female executive. As representation declines for women and people of color, white men become increasingly over-represented in leadership positions, holding 46 percent of management roles and 59 percent of executive roles.

Disciplinary power most heavily marginalizes the careers of Latina and Black women. For instance, in 2016, ten large technology companies in Silicon Valley did not employ a single Black woman, while three had no Black employees at all (Rangarajan, 2018). Alegria (2019) interviewed women in technology and found that women of color struggled to switch teams or change career tracks within their companies. However, white women in technical roles were offered advancement opportunities to become managers, even if they were not actively seeking those jobs. These advancement opportunities were often based on the gender stere-otype that women had strong communication skills useful for translating between technical teams and management. However, no women of color interviewed during the study were offered managerial positions, even when they applied for the positions. Women of color often lack the role models, influential social networks, and support systems needed to advance their careers (Nordli, 2019). A Kapor Center (2018) report found that only 12 percent of all women in computing roles are Black or Latina. In Silicon Valley companies, representation of Black, Latina, or Native American women drops to 2 percent (Kapor Center, 2018). The lack of rep-resentation of women of color in technology limits access to higher paying and highly visible leadership positions (Kapor Center, 2018).

Women are also paid less than men for performing similar work. The American Association of University Women (AAUW, 2017) reports that the earnings ratio of women to White men by ethnicity is: Asian (87 percent), White (79 percent), Black (63 percent), Hispanic (55 percent). The wage gaps for Black and Latina women are more considerable and closing more slowly. While the wage gaps for white and Asian women are projected to close in 50 and 22 years, *Black and Latina women's wage gaps will not close for 350 and 432 years, respectively.*

Hegemonic Domain of Power

The hegemonic domain links the structural, disciplinary, and interpersonal domains by creating and sustaining cultural beliefs, narratives, and values about race and gender. The significance of the hegemonic domain of power lies in its ability to shape consciousness via manipulating ideas, images, symbols, and ideologies (Hill Collins, 1990). For example, in a survey of women in the technology workforce, Williams et al. (2015) found that 34 percent reported pressure to play stereotypically American feminine roles such as "office mother" or "dutiful daughter." Fifty-three percent reported retaliation for exhibiting stereotypically American masculine behaviors like speaking their minds directly or being outspoken or decisive. The researchers emphasized that social isolation from colleagues was a pattern of bias that applied mainly to Black women and Latinas. Nearly two-thirds of women with children reported that their employer questioned their commitment and competence to the company. Opportunities for challenging projects that develop new skills and knowledge needed for advancement also decreased after having children.

External stressors related to work–life balance present unique constraints for women in the workforce. For example, the Kapor Center reports that women are more likely than men to be single parents and caregivers for elderly parents (Scott et al., 2017). Working mothers are less likely to be rated as competent and less likely to be perceived as committed to work when compared to equally qualified men with children and women without children (Scott et al., 2017; Cuddy, 2004). The "firefighting" work style (i.e., solution focused, fast-paced environment) in technology companies rewards long hours, risk-taking, and travel. On the one hand, working mothers who meet the demands of this work style and prove their competency may be penalized and viewed as bad mothers for their dedication to the job (Benard and Correll, 2010). On the other hand, intense workplace demands can disadvantage women with heavy household management workloads (Williams et al., 2015). Working mothers also hit "the maternal wall" and are less likely to be interviewed, hired, promoted, and paid equally (Scott et al., 2017).

Researchers have also shown that gender bias and stereotypes limit women's progress in careers in technology and early choices not to pursue technology careers. To be seen as equally competent as men, women often must "prove-it-again" by providing more evidence of competence (Eagly and Thomas-Hunter, 2014; Foschi, 2000). Gender stereotypes about femininity often place women on "the tightrope" where they balance between being seen as too feminine to be competent or too masculine to be likable (Cuddy, 2004; Prentice and Carranza, 2002; Williams, 2014). Women working in gender-biased environments where traditionally masculine traits such as being assertive or self-promoting are rewarded may exhibit "the queen bee" syndrome where they distance themselves and refuse to help other women (Prentice and Carranza, 2002).

Interpersonal Domain of Power

The interpersonal domain describes the power relations that shape social interactions between people in marginalized and dominant groups. Interpersonal power is revealed through the misrecognitions, microaggressions, and social isolation that leads people from underrepresented groups to leave the technology field. For example, Rankin and Thomas (2019) describe repeated acts of misrecognition experienced by Black women within the Human-Computer Interaction (HCI) community. Misrecognition occurs when peers and colleagues disregard an individual's presence or confuse a person, on multiple occasions, with other Black women. Consequently, these acts of misrecognition communicate to Black women that they are indistinguishable, devalued, and not entirely welcomed within the HCI community.

Misrecognition and microaggressions contribute not only to the diversity problem of underrepresentation in the technology workforce; they create hostile workplaces where the few women, Black and Latinx employees who are hired feel isolated, unvalued, and excluded (Kapor Center, 2018; Williams et al., 2015). The technology companies that women, Black and Latinx struggled so long to enter can look entirely different once they get inside. In the book *Opting Out: Losing the Potential of America's Young Black Elite*, Beasley (2011) found that Black undergraduates majoring in technology were less likely than white peers to persist in their major when they felt that they were underperforming. Black undergraduates who persisted and completed their degrees were less likely to apply for technical jobs. They were wary about taking jobs in technology companies due to repeated stories about racist workplace climates and the lack of Black employees. Moreover, trust between underrepresented groups and technology companies tends to be low, suggesting that many will not apply even if there are job openings (Xu, 2020).

Women of all races and ethnicities reported isolation in the workplace. Black (42 percent), Latina (38 percent), Asian (37 percent), and white women (32 percent) perceived that co-workers' perceptions of their competence decreased after socializing (Williams et al., 2015). These perceptions of adverse communication outcomes can activate gender and racial stereotyping. When experiencing stereotype threat, the risk and fear of confirming a negative stereotype about one's group, job performance and engagement decline, negative attitudes towards work increase, and turnover intentions rise (Scott et al., 2017).

To further unpack the intersectional impacts of racial and gender stereotypes, women of color report that biases functioned differently for them when socializing with colleagues. For instance, the results of Williams and colleagues' (2015) survey study found that Black women are aware of negative stereotypes about their group (97 percent) and have been personally affected by negative stereotypes (80 percent) (Jones and Shorter-Gooden, 2003). Black women (77 percent) were more likely than white (63 percent), Asian (64 percent), and Latina (65 percent) women to report having to repeatedly prove their competence (Williams et al., 2015). Moreover, Black women attributed the "prove-it-again" bias more to race rather than gender, were twice as likely (56 percent) than women of other races and ethnicities (25 percent) to report being negatively impacted by queen bee behavior and were more likely to report a sense of "bleak isolation" (Williams et al., 2015; Williams and Dempsey, 2014).

In response to these patterns of bias and exclusion, women and minoritized workers leave the technology sector in more significant numbers than their majority counterparts. The Tech Leavers Study by the Kapor Center for Social Impact (Scott et al., 2017) found that women and other underrepresented groups are more likely to experience unfair treatment, stereotyping, and bullying in technology workplaces. These experiences directly drive turnover. Unfairness or mistreatment within the work environment was the most frequently cited (37 percent) reason for leaving (Scott et al., 2017). Minoritized workers were the least satisfied with their jobs, most likely to report experiencing a higher number of negative work experiences, and most likely to leave the company (Kapor Center, 2011). Moreover, employees in technology companies (42 percent) were significantly more likely to leave their company due to unfairness than technical employees in other industries (32 percent) (Scott et al., 2017). This malicious workplace culture, rather than pipeline issues or personal choice, pushes women out of technology jobs (Williams, Phillips, and Hall, 2015). These long-standing attrition problems suggest that technology companies have not adequately addressed discriminatory practices in hiring, retention and promotion processes that create largely homogenous workforces and biased workplace cultures (Harrison, 2019).

BUILDING AI SYSTEMS THAT DISCRIMINATE

Diversifying the technology workforce becomes more urgent with the advancement of AI systems in society. AI systems are fundamentally classification technologies that differentiate, rank, and categorize (West et al., 2019). However, the harmful effects of these classification technologies are not evenly distributed and tend to reproduce existing structures of inequality in society. The issue is that with fewer women and minoritized employees designing and building AI systems, the systems are not being sufficiently trained to recognize the voices, skin colors, and faces of women and people of color. A steady stream of examples in recent years has shown a persistent problem of gender and race-based discrimination perpetuated by AI algorithms (West et al., 2019).

In one example, Link (2020) reports a Stanford University study (Andrews, 2020) of speech-to-text software used by Amazon, IBM, Google, Microsoft, and Apple. Using a data set that included 2000 voice samples from recorded interviews with Black and white speakers, the researchers found that the software misidentified Black speakers' words (35 percent) at nearly double the rate of white speakers (19 percent). The disparity likely occurred because such technologies are based on machine learning systems that rely heavily on databases of English as spoken by white Americans (Andrews, 2020). The harms to minoritized communities are compounded as companies use speech-to-text software to support a growing number of functions such as screening job applicants and facilitating promotion decisions, assisting people with disabilities to browse the web, automating ordering, and transcribing hearings in court.

In another study, Buolamwini and Gebru (2018) showed how facial recognition systems misidentify people of color, especially women of color. Computer vision systems with inferior performance across racial groups can have profound implications. For example, harm may be done when automated skin cancer detection systems perform inconsistently across people of different skin types. Facial recognition systems that misidentify people of color at higher rates could also be harmful in contexts like policing (Buolamwini and Gebru, 2018).

Artificial Intelligence Hiring Bias

Wide-scale harm from opaque decision making, misidentification in facial recognition, and errors in speech-to-text systems converge to become significant risks in AI hiring systems. To improve the speed and efficiency of the screening, interviewing, and hiring processes, US companies are increasingly adopting AI hiring systems. AI hiring systems automate tasks such as sorting through resumes, using disparate sources of data to make predictive matches between job seekers and positions, correcting biases in the language used in job descriptions, and using bots to schedule candidate interviews (Danieli et al., 2016). AI hiring systems allow companies to mesh traditional data sources of employment data like resumes and candidates' performance on computer programming tasks, along with audio, video, text, and social data posts to construct psychological profiles. However, when algorithms use these profiles as a proxy for measuring organizational fit and predicting a candidate's ability to perform the job, people from significantly different cultural backgrounds may be systematically disadvantaged (Yarger et al., 2019).

Companies are also adopting AI hiring systems to reduce human bias and subjectivity in the selection of job candidates (Maurer, 2021). However, Vasconcelos and colleagues (2017) explain that when creating AI hiring algorithms, companies model historical patterns of hiring from data describing "high-performance" employees as a basis for selecting candidates with similar profiles. Consequently, biases rooted in the traditional hiring process are reproduced and encoded in the data used to train the new systems. The AI hiring system will surface applicants who have attributes similar to historically high-performing white male employees, resulting in workplaces that are just as homogeneous as they were before (Miller, 2015). An example of this bias occurred in October 2018 when Amazon scrapped an AI recruiting tool after realizing that this new system was not rating candidates for software developer jobs and other technical posts in a gender-neutral way (Dastin, 2018). Because the computer models were trained to rate applicants by observing patterns in submitted resumes, and most of those resumes came from men, the algorithm was systemically excluding female candidates. What was novel, in this case, was that the discriminatory effects are data driven (Kim, 2017; Mann and O'Neil, 2016).

Biases are also rooted in the complexity of the algorithms and the inability to explain the logic used by the algorithm. As AI tools learn how to efficiently perform tasks in the hiring process usually performed by humans, the predictive and decision-making processes used by machine-learning algorithms remain opaque and unexplainable. Employers face new legal risks because they are not able to explain why and how a particular hiring decision was made by the software (Dineen et al., 2004; Vasconcelos et al., 2017). Moreover, when humans make employment recommendations based on an algorithms' inferences about applicants' age, race, religion, and sex, it is difficult to determine if firms are adhering to US federal laws that protect job applicants against intentional discrimination and adverse impacts on women and minoritized individuals (Vasconcelos et al., 2017). US employers face significant legal risk including violations of state laws governing biometric data and AI video interview data, the Age Discrimination in Employment Act, and the Title VII of the Civil Rights Act of 1964, a federal law that protects employees and applicants against discrimination based on certain specified characteristics such as race, color, national origin, sex, and religion. Companies are advised to conduct periodic bias audits to validate the software to mitigate these legal risks and ensure that the AI hiring systems perform as intended (Maurer, 2021).

A growing number of researchers, government officials, and concerned citizens are publicly voicing concerns about the (un)intended algorithmic bias and the drastic impacts that such bias can have on human lives (O'Neil, 2016; Kilpatrick, 2016). HireVue, a leading vendor in the AI hiring marketplace, has received tremendous scrutiny. HireVue software analyzes speech and facial expressions from video interviews to quantify human personality traits and compute an employability score. During the interviews, HireVue software presents video scenarios that candidates might encounter on the job. Applicants' responses are analyzed in real time by AI, not humans. The software measures tone of voice and clusters of text being used (e.g., "I am dependable and action-oriented") and captures and analyzes facial movements to predict emotions and personality traits. The applicants' interview performance is compared with the performance data (voice, word clusters, and facial expressions) of high-performing current employees. The software then computes a rating score for the job applicant. Companies can then use the rating scores to rank candidates and view portions of the video for high-scoring candidates.

In 2021, the Electronic Privacy and Information Center (EPIC) urged the US Federal Trade Commission to investigate HireVue. At issue is HireVue's use of proprietary video interview software that scans job candidates' faces and voices, analyzes the candidates' speech patterns and facial movements, and predicts the candidates' employability (Harwell, 2019). Candidates are not told their scores, and they don't know the basis for how the algorithm generated their employability scores. And they neither know how their data is being used nor have an opportunity to consent to such uses (Harwell, 2019). Furthermore, because the algorithms are hidden, there is no way of determining if and how the software perpetuates discriminatory hiring practices and biases based on applicants' gender, race, sexual orientation, or neurological differences (Harwell, 2019). The reliability of AI software analysis of video interviews is also questionable. A team of reporters from Bayerischer Rundfunk, a German Public Broadcasting organization, performed several experiments with similar AI video interview software produced by a Munich-based startup.[1] The experiments' results show that simple changes like wearing a hat or glasses, adding pictures and bookcases to the job candidate's background, and changing the brightness and saturation of the camera settings alter the results of the algorithms. The experiments further demonstrate how AI technologies built primarily by white men in a homogenous workplace culture tend to be biased in favor of white male job applicants.

The Convergence of Power in AI and Human Systems

The examples above demonstrate how AI systems may reproduce bias and discrimination when built without considering a broader community of users and social contexts or critically assessing whether the technology offers a beneficial and worthwhile purpose to society (Schwab, 2021; Sullivan, 2021; Crawford, 2016). AI researchers and system builders are not generally trained in ethical reasoning and social sciences. Therefore, they tend to adopt narrow definitions of "bias" and "fairness" that can easily be operationalized through computer algorithms, computational models, and data manipulation (West et al., 2019). Examples of these technical approaches to address human diversity and reduce discriminatory outcomes include looking for a biased distribution of error rates based on a single variable like gender or race (Andrews, 2020), auditing algorithms for accuracy of predictions and classifications (Sullivan, 2021; Raji and Buolamwini, 2019; Buolamwini and Gebru, 2018; Williams et al., 2018; Kim, 2018; Sandivig et al., 2014), and improving the human diversity represented in datasets (Paullada et al., 2020; Buolamwini and Gebru, 2018).

However, this focus on fixing historical systems of human inequality with contemporary technology systems fails to adequately contextualize the human power dynamics that shape the AI system's use and outcomes. One of the most troubling power dynamics is the growing influence of the largest technology firms on AI research and funding. Abdalla and Abdalla (2020) report that "Big Tech" is actively funding academic research "to put forward a socially responsible public image, influence events hosted by and decisions made by funded universities, influence the research questions and plans of individual scientists, and discover receptive academics who can be leveraged." In one example, Schwab (2021) reports that the National Science Foundation teamed with Amazon to create one of the largest research funding pools dedicated to AI ethics studies. In the first pool of awards, men receive nine of the ten grants awarded. All of the grant recipients were white or Asian. This outcome points to a systematic relationship between patterns of exclusion within the field of AI and the industry driving its production.

CONCLUSION

This chapter offers a discourse about how discriminatory human and AI systems reproduce unjust systems that harm women and minoritized people. Specifically, the first half of the chapter uses Hill Collins' Domains of Power as a framework for considering factors that contribute to the underrepresentation of women, Black and Latinx in the technology workforce. These power domains are linked to both historical systems of inequality and contemporary AI technologies. Specifically, AI hiring systems are examined in the second half of the chapter as a potential site for the biased representation, improper classification, and exclusion of women and minoritized individuals in the technology workforce. The legal risks and ethical concerns that AI hiring decisions create for employers are also discussed.

Reflecting upon these power relations raises many important questions about the connections between the lack of diversity in the technology workforce and the discriminatory effects of AI technologies that are built. The AI Now Institute provides a long list of questions for researchers and practitioners (West et al., 2020). The following are the most salient questions related to the power relations discussed in this chapter:

- Who gets to design and build AI software systems?
- What stories do we tell about how AI software systems are designed?
- Whose interests shape the metrics used to assess the discriminatory uses and disparate impacts of the software?
- How does societal discrimination surface in data provenance?
- How are cultural norms and stereotypes numerated and represented at the time of data creation?
- How do we shield AI ethics from Big Tech's influence?
- Should certain AI systems be designed at all?

The diversity problem in technology is fundamentally about the circulation of power and how this creates barriers for women and underrepresented people of color. However, in the case of AI, the stakes are higher and more diffused. Therefore, the scope of future research inquiry and policy responses should consider not only how AI tools can be biased technically but how they are shaped by the social contexts in which they are built and the people that build them (West et al., 2019).

NOTE

1. The study is published as an interactive website at https://web.br.de/interaktiv/ki-bewerbung/en/?mc_cid=4eaa79778candmc_eid=dd869360f3.

REFERENCES

Abdalla, M., and Abdalla, M. (2020). The grey hoodie project: Big tobacco, big tech, and the threat on academic integrity. ArXiv, abs/2009.13676. Retrieved February 25, 2021, from https://arxiv.org/pdf/2009.13676.pdf.

Alegria, S. (n.d.). Translators or engineers: Gendered labor and the glass escalator in tech work. *Gender and Society*, 33(5), 722–45.

American Association of University Women. (2017). Race and the pay gap. AAUW research and data. Retrieved February 25, 2021, from www.aauw.org/resources/research/race-and-the-pay-gap/.

Andrews, E. (2020, March 23). Stanford researchers find that automated speech recognition is more likely to misinterpret Black speakers. Stanford News. Retrieved February 25, 2021, from https://news.stanford.edu/2020/03/23/automated-speech-recognition-less-accurate-blacks/.

Beasley, M. (2011). *Opting Out: Losing the Potential of America's Young Black Elite*. The University of Chicago Press.

Belasco, A.S., and Trivette, M.J. (2015). Aiming low: Estimating the scope and predictors of postsecondary undermatch. *The Journal of Higher Education*, 86(2), 233–63.

Benard, S., and Correll, S. J. (2010). Normative discrimination and the motherhood penalty. *Gender and Society*, 24(5), 616–46.

Bertrand, M., and Mullainathan, S. (2004). Are Emily and Greg more employable than Lakisha and Jamal? A field experiment on labor market discrimination. *American Economic Review*, 94(4), 991–1013.

Bortz, D. (2018, April 18). Can blind hiring improve workplace diversity? Society for Human Resource Management. Retrieved February 25, 2021, from www.shrm.org/hr-today/news/hr-magazine/0418/pages/can-blind-hiring-improve-workplace-diversity.aspx.

Bui, Q., and Cain Miller, C. (2016, February 25). Why tech degrees are not putting more Blacks and hispanics into tech jobs. *New York Times*. Retrieved February 25, 2021, from www.nytimes.com/2016/02/26/upshot/dont-blame-recruiting-pipeline-for-lack-of-diversity-in-tech.html.

Buolamwini, J., and Gebru, T. (2018). Gender shades: Intersectional accuracy disparities in commercial gender classification. *Proceedings of Machine Learning Research*, 81, 1–15.

Bureau of Labor Statistics. (2021, January 1). Labor Force Statistics from the Current Population Survey. US Bureau of Labor Statistics. Retrieved February 25, 2021, from www.bls.gov/cps/cpsaat11.htm.

Carmichael, S.G. (2016, May 26). Study: Employers are less likely to hire a woman who wears a headscarf. *Harvard Business Review*. Retrieved February 25, 2021, from https://hbr.org/2017/05/study-employers-are-less-likely-to-hire-a-woman-who-wears-a-headscarf.

Cooper, J. (2016). A call for a language shift: From covert oppression to overt empowerment, Neag School of Education, University of Connecticut. Retrieved February 25, 2021, from https://education.uconn.edu/2016/12/07/a-call-for-a-language-shift-from-covert-oppression-to-overt-empowerment.

Crawford, K. (2016, June 25). Artificial intelligence's white guy problem. *The New York Times*, Section SR, page 11.

Cuddy, A.J. (2004). When professionals become mothers, warmth doesn't cut the ice. *Journal of Social Issues*, 60, 701–18.

Danieli, O., Hillis, A., and Lucas, M. (2016, October). How to hire with algorithms. *Harvard Business Review*. Retrieved February 25, 2021, from https://hbr.org/2016/10/how-to-hire-with-algorithms.

Dastin, J. (2018, October 10). Amazon scraps secret AI recruiting tool that showed bias against women. Reuters. Retrieved February 25, 2021, from www.reuters.com/article/us-amazon-com-jobs-automation-insight/amazon-scraps-secret-ai-recruiting-tool-that-showed-bias-against-women-id USKCN1MK08G.

Derks, B., Ellemers, N., Van Laar, C., and de Goot, K. (2011). Do sexist organizational cultures create the queen bee syndrome? *Journal of Social Psychology*, 22, 1243–49.

Dineen, B.R., Noe, R.A., and Wang, C. (2004). Perceived fairness of web-based applicant screening procedures: Weighing the rules of justice and the role of individual differences. *Human Resource Management*, 43(2–3), 127–45.

Eagly, A., and Thomas-Hunter, M. (2014). Condoning stereotyping? How awareness of stereotyping prevalence impacts expression of stereotypes. *Journal of Applied Psychology*, 100(2), 343–59.

Elias, J. (2021, February 21). TECH Google's program for Black college students suffered disorganization and culture clashes, former participants say. CNBC. Retrieved August 15, 2021, from www.cnbc.com/2021/02/21/google-howard-west-program-faced-disorganization-culture-clashes.html.

Foschi, M. (2000). Double standards for competence: Theory and research. *Annual Review of Sociology*, 26, 21–8.

Frey, W.H. (2020, July 1). The nation is diversifying even faster than predicted, according to new census data. The Brookings Institute. Retrieved August 15, 2021, from www.brookings.edu/research/new-census-data-shows-the-nation-is-diversifying-even-faster-than-predicted/.

Garbee, E. (2017, October 20). The problem with the "pipeline." *Slate Magazine*. Retrieved August 15, 2021, from https://slate.com/technology/2017/10/the-problem-with-the-pipeline-metaphor-in-stem-education.html.

Harlan, E., and Schnuck, O. (2021). Objective or biased: On the questionable use of artificial intelligence for job applicants. Retrieved February 25, 2021, from https://web.br.de/interaktiv/ki-bewerbung/en/?mc_cid=4eaa79778candmc_eid=dd869360f3.

Harrison, S. (2019, October 1). Five years of tech diversity results and little progress. *Wired Magazine*. Retrieved February 25, 2021, from www.wired.com/story/five-years-tech-diversity-reports-little-progress/.

Harwell, D. (2019, November 6). Rights group files federal complaint against AI-hiring firm hirevue, citing "unfair and deceptive" practices. *The Washington Post*. Retrieved February 25, 2021, from www.washingtonpost.com/technology/2019/11/06/prominent-rights-group-files-federal-complaint-against-ai-hiring-firm-hirevue-citing-unfair-deceptive-practices/.

Henrich, J., Heine, S.J., and Norenzayan, A. (2010). Most People are not WEIRD. *Nature*, 466(7302), 29–9.

Hewitt, S.A., Luce, C.B., and Servon, L.J. (2008). Stopping the exodus of women in science. *Harvard Business Review*, 86(6), 22–4.

Hill Collins, P. (1990). *Black Feminist Thought: Knowledge, Consciousness and the Politics of Empowerment* (Second Edition). NY: Routledge.

Johnson, S., Heckman, D.K., and Chan, E.T. (2016, April 26). If there's only one woman in your candidate pool, there's statistically no chance she'll be hired. *Harvard Business Review*. Retrieved February 25, 2021, from https://hbr.org/2016/04/if-theres-only-one-woman-in-your-candidate-pool-theres-statistically-no-chance-shell-be-hired.

Jones, C., and Shorter-Gooden, K. (2003). *Shifting: The Double Lives of Black Women in America*. NY: Harper Perennial.

Kapor Center. (2011, September). The tilted playing field: Hidden bias in information technology workplaces. Kapor Center: Leveling the Playing Field. Retrieved February 25, 2021, from https://mk0kaporcenter5ld71a.kinstacdn.com/wp-content/uploads/2017/05/tilted_playing_field_lpfi_9_29_11.pdf.

Kapor Center. (2018, August). Data brief: Women and girls of color in computing. Women of color in computing. Retrieved February 25, 2021, from www.wocincomputing.org/wp-content/uploads/2018/08/WOCinComputingDataBrief.pdf.

Kilpatrick, K. (2016). Battling algorithmic bias: How do we ensure algorithms treat us fairly? *Communications of the ACM*, 59(10), 16–17.

Kim, P.T. (2017). Data-driven discrimination at work. *William and Mary Law Review*, 58(3). Retrieved February 25, 2021, from https://scholarship.law.wm.edu/wmlr/vol58/iss3/4/.

Knight, R. (2018, April 19). 7 Practical ways to reduce bias in your hiring process. Society for human resource management. Retrieved February 25, 2021, from www.shrm.org/resourcesandtools/hr-topics/talent-acquisition/pages/7-practical-ways-to-reduce-bias-in-your-hiring-process.aspx.

Link, J. (2020, July 26). Why racial bias haunts speech recognition. BuiltIn. Retrieved February 25, 2021, from https://builtin.com/artificial-intelligence/racial-bias-speech-recognition-systems.

Mann, G., and ONeil, C. (2016). Hiring algorithms are not neutral. *Harvard Business Review*. Retrieved February 25, 2021, from https://hbr.org/2016/12/hiring-algorithms-are-not-neutral.

Maurer, R. (2021, April 7). Use of AI in the Workplace Raises Legal Concerns, Society for Human Resource Management. Retrieved August 15, 2021, from www.shrm.org/resourcesandtools/hr-topics/technology/pages/use-of-ai-in-the-workplace-raises-legal-concerns.aspx.

Miller, C. (2015, June 25). Can an algorithm hire better than a human? *The New York Times*. Retrieved February 25, 2021, from www.nytimes.com/2015/06/26/upshot/can-an-algorithm-hire-better-than-a-human.html.

Muskens, M., Frankenhuis, W.E., and Borghans, L. (2019). Low-income students in higher education: Undermatching predicts decreased satisfaction toward the final stage in college. *Journal of Youth and Adolescence*, 48, 1296–1310. Retrieved February 25, 2021, from https://link.springer.com/article/10.1007/s10964-019-01022-1.

Nordli, B. (2019, July 29). For women of color in tech, it's "hard to grow" without representation. BuiltIn. Retrieved February 25, 2021, from https://builtin.com/women-tech/women-color-tech-inclusion.

O'Connor, C. (2016, March 3). Black woman engineer launches "blind" job match app to take bias out of tech hiring. Forbes. Retrieved February 25, 2021, from www.forbes.com/sites/clareoconnor/2016/03/03/black-woman-engineer-launches-blind-job-match-app-to-take-bias-out-of-tech-hiring/#228f757d2394.

O'Neil, C. (2016). *Weapons of Math Destruction: How Big Data Increases Inequality and Threatens Democracy*. NY: Crown Publishing Group.

Ovink, S., Kalogrides, D., Nanney, M., and Delaney, P. (2018). College match and undermatch: Assessing student preferences, college proximity, and inequality in post-college outcomes. *Research in Higher Education*, 59(5), 553–90.

Paullada, A., Raji, I.D., Bender, E.M., Denton, E., and Hanna, A. (2020, December 9). Data and its (dis)contents: A survey of dataset development and use in machine learning research. arXiv, 2012.05345v1. https://arxiv.org/abs/2012.05345.

Prentice, D.A., and Carranza, E. (2002). What women and men should be, shouldn't be, are allowed to be, and don't have to be: The contents of prescriptive gender stereotypes. *Psychology of Women Quarterly*, 26(4), 269–81.

Quillian, L., Pager, D., Hexel, O., and Midtbøen, A.H. (2017). Meta-analysis of field experiments shows no change in racial discrimination in hiring over time. *Proceedings of the National Academy of Sciences*, 114(41), 10870–75.

Raji, D. (2020, December 10). How our data encodes systematic racism. MIT Technology Review. Retrieved February 25, 2021, from www.technologyreview.com/2020/12/10/1013617/racism-data -science-artificial-intelligence-ai-opinion/.

Raji, I., and Buolamwini, J. (2019). Actionable Auditing: Investigating the Impact of Publicly Naming Biased Performance Results of Commercial AI Products. Conference on Artificial Intelligence, Ethics, and Society. Retrieved February 25, 2021, from www.media.mit.edu/projects/actionable -auditing- coordinated-bias-disclosure-study/publications/.

Rangarajan, S. (2018, June 25). Here's the clearest picture of Silicon Valley's diversity yet: It's bad. But some companies are doing less bad. Reveal, The Center for Investigative Reporting. Retrieved February 25, 2021, from https://revealnews.org/article/heres-the-clearest-picture-of-silicon-valleys -diversity-yet/.

Rankin, Y., and Thomas, J. (2019). Straighten up and fly right: Rethinking intersectionality in HCI research. *IX Interactions*, XXVI.6 (November–December), 64. Retrieved February 25, 2021, from https://interactions.acm.org/archive/view/november-december-2019/straighten-up-and-fly-right.

Rooney, K., and Khorram, J. (2020, June 12). Tech companies say they value diversity, but reports show little change in the last six years. CNBC. Retrieved February 25, 2021, from www.cnbc.com/2020/06/12/six-years-into-diversity-reports-big-tech-has-made-little-progress.html.

Sandvig, C., Hamilton, K., Karahalios, K., and Langbort, C. (2014, May 22). Auditing algorithms: Research methods for detecting discrimination on Internet platforms. 64th Annual Meeting of the International Communication Association, Seattle, WA.

Schwab, K. (2021). "This is bigger than just Timnit": How Google tried to silence a critic and ignited a movement. Fast Company. Retrieved February 25, 2021, from www.fastcompany.com/90608471/timnit-gebru-google-ai-ethics-equitable-tech-movement.

Scott, A., Kapor Klein, F., and Onovakpuri, U. (2017, August 27). Tech leavers study. Kapor Center for Social Impact. Retrieved February 25, 2021, from www.kaporcenter.org/wp-content/uploads/2017/08/TechLeavers2017.pdf.

Scott-Clayton, J., and Li, J. (2016, October 20). Black-white disparity in student loan debt more than triples after graduation. *Economic Studies at Brookings*, 2(3). Retrieved February 25, 2021, from www.brookings.edu/wp-content/uploads/2016/10/es_20161020_scott-clayton_evidence_speaks.pdf.

Segrest, S., Gillespie, P.L., Mayes, B.T., and Ferris, G.R. (2006). Implicit sources of bias in employment interview judgments and decisions. *Organizational Behavior and Human Decision Processes*, 101(2), 152–67. Retrieved February 25, 2021, from www.sciencedirect.com/science/article/abs/pii/S0749597806000690.

Sullivan, M. (2021, February 11). Fighting AI bias needs to be a key part of Biden's civil rights agenda. Fast Company. Retrieved February 25, 2021, from www.fastcompany.com/90599820/fighting-ai-bias -needs-to-be-a-key-part-of-bidens-civil-rights-agenda.

US Bureau of Labor Statistics. (2020, December 1). Labor force characteristics by race and ethnicity, 2019. BLS Report. Retrieved February 25, 2021, from www.bls.gov/opub/reports/race-and-ethnicity/2019/home.htm.

Vara, V. (2016). Why doesn't Silicon Valley hire Black coders? Bloomberg. Retrieved February 25, 2021, from www.bloomberg.com/features/2016-howard-university-coders/.

Vasconcelos, M., Cardonha, C., and Goncalves, B. (2017). Modeling epistemological principles for bias mitigation in AI systems: An illustration in hiring decisions. arXiv, 1711.07111. Retrieved February 25, 2021, from https://arxiv.org/pdf/1711.07111.pdf.

Vespa, J., Medina, L., and Armstrong, D.M. (2020, February 1). Demographic turning points for the United States: Population projections for 2020 to 2060. Current Population Reports, P25-1144,

US Census Bureau, Washington, DC. Retrieved February 25, 2021, from www.census.gov/library/publications/2020/demo/p25-1144.html.

West, S.M., Whittaker, M., and Crawford, K. (2019, April 1). Discriminating systems: Gender, race and power in AI. AI Now Institute. Retrieved February 25, 2021, from https://ainowinstitute.org/research.html.

Williams, B., Brooks, C., and Shmargad, Y. (n.d.). How algorithms discriminate based on data they lack: Challenges, solutions, and policy implications. *Journal of Information Policy*, 8, 78–115.

Williams, J., Phillips, K.W., and Hall, E.V. (2015). Double jeopardy: Bias against women in science. work life law. Retrieved February 25, 2021, from https://worklifelaw.org/publications/Double-Jeopardy-Report_v6_full_web-sm.pdf.

Williams, J.C. (2014). Hacking tech's diversity problem. *Harvard Business Review*, 92(20), 94–100.

Xu, T. (2020, August 31). The deck is stacked against Black women in tech. BuiltIn. Retrieved February 25, 2021, from https://builtin.com/women-tech/black-women-in-tech.

Yarger, L., Payton, F., and Neupane, B. (2019, December 17). Algorithmic equity in the hiring of under-represented IT job candidates. *Online Information Review*, 44(2), 383–95.

12. BLKGENIUS: a social-academic network for combating the underrepresentation of Black men in computing in the United States

Curtis C. Cain

INTRODUCTION

This chapter discusses the BLKGENIUS project and entry pathways for Black[1] men[2] in information technology (IT). I coined the term BLKGENIUS in October 2010 to record the lived experiences, obstacles, trials, tribulations, and successes of my academic journey to a PhD in IT in the United States. In 2020, with the support of a National Science Foundation grant, an expansion of BLKGENIUS was funded, and work began in 2021. The expansion transforms BLKGENIUS from the chronicle of a single individual's journey to an online social-academic website for Black men in computing, especially those pursuing degrees and careers in IT. It assists such students in finding local mentors and role models in the field. Additionally, BLKGENIUS connects Black students with identified Black mentors in computing at companies who are willing to serve as role models to assist students in developing necessary skillsets for the field, as well as assist them with interviewing techniques and résumé review. The chapter begins with a discussion of racial and gender underrepresentation in the United States, followed by background on the history of Blacks in the United States, Black men in IT, postsecondary education, and the IT workforce. Afterward, the BLKGENIUS project is introduced, along with a summary of objectives. Finally, the chapter concludes with a discussion of takeaways and next steps.

In the United States, a country that professes to promote the concept of social mobility, education is critical. Postsecondary education is now an essential criterion for participation in an increasingly competitive technical labor market (Wilkins, 2006). In higher education, the overall representation of all ethnic minorities is steadily increasing. Although still far behind white students in terms of participation, Black and Latinx students have experienced significant increases in postsecondary education participation over the past decade (McGee, 2018). However, the data on growth rates of Black students indicate more of a good news/bad news scenario. Even though the total number of Black students is increasing, the gender gap between Black men and Black women in higher education has grown wider (National Center for Education Statistics, 2003). Whereas Black women are experiencing notable growth in enrollment and graduation, the participation of Black men is declining and is the lowest of all demographic groups (National Center for Education Statistics, 2003).

In the United States, the underrepresentation of Black men in higher education can be broadly explained in terms of three experiences: educational, environmental, and personal. In terms of education, Black men and women have historically been incorrectly deemed academically inferior to whites (as explained in Allen and Epps, 1991; Allen, 1988; McGee, 2018; McGee and Martin, 2011; Palmer et al., 2009; Morris and Monroe, 2009; Museus, 2008). As a result of the history of socioeconomic stratification disproportionately impacting Blacks in the United States, Black students are more likely to attend ill-equipped schools; these schools lack the resources that contribute to providing quality education at the same level white students receive (Codjoe, 2001). According to Joshi et al. (2016), a significant amount of research has examined Black men through the lens of a deficit model, which primarily focuses on systemic and individual failure as a method to explain a lack of success. However, this research approach is insufficient because it does not consider the positive factors that contribute to Black men's success. Framing the issues confronting the Black community in terms of a lack of resources runs the risk of analyzing it from a deficit-based perspective – which does not tell the whole story and is detrimental to student development and confidence – as well as failing to highlight and uplift Black students.

Black students must also contend with various obstacles in their environment, including household makeup, schools, and geographic location. One obstacle occurs when the prevailing racial environment in which they live is a predominantly Black or an ethnically mixed neighborhood. Black students must then seamlessly shift between environments and adjust behaviors depending on their surroundings, a form of adaptation known as *code switching* (Casimir, 2020; Myers, 2020; Saeedi and Richardson, 2019). Code switching allows Black students to adjust and attune their style, demeanor, and overall tenor to an atmosphere that differs from that to which they are accustomed and wherein they believe their authentic self would not be welcomed. This shifting of behaviors tends to intensify when they attend college away from home, surrounded by people with different demographics from the environment of their upbringing.

Lastly, there are personal experiences that influence underrepresentation. Relevant factors include personality and the ability to adapt and thrive in different surroundings, which overlap with code switching. Another aspect of personal factors is the process of building relationships and surrounding oneself with like-minded, positive friends. One's family also contributes personal factors that can either positively or negatively affect their ability to cope with underrepresentation. The combination of educational, environmental, and personal experiences poses a significant hurdle to full participation in society, let alone a career in IT.

BACKGROUND

Slavery and its Aftermath in the United States

Kept in slavery in the American colonies after being abducted from Africa, these people were forced by the European settlers to work for free (and without freedom) during the seventeenth, eighteenth, and half of the nineteenth centuries in farms and industries. Abolitionist movements and the westward expansion of the United States in the mid-nineteenth century ignited a national debate about slavery that ultimately led to the American Civil War. As much as the Union victory led to the freedom of millions of slaves, slavery's legacy continues to shape the history of the United States. Africans, enslaved and free, played a critical role in America's establishment and long-term dominance as a global power, while reaping little to no benefit themselves. Slavery in America is widely believed to have begun in 1619, when the merchant ship, *The White Lion*, sold 20 Africans who were in the British colony of Jamestown, Virginia.

The emancipation

Slavery in the United States ended during the Civil War when, President Lincoln issued an executive order, the Emancipation Proclamation, on September 22, 1862. The executive order, which took effect on January 1, 1863, declared that "slaves within any State, or designated part of a State ... in rebellion ... shall be then, thenceforward, and forever free" (Lincoln, 1863). The Emancipation Proclamation led to the freedom of almost three million Black slaves in the ten states that rebelled against the Union of the United States. Although the Emancipation Proclamation is generally referred to as the moment the slaves were freed, it is a misnomer.

Since the order freed slaves only in the Confederate states, and not those that remained in the Union, President Lincoln used slavery as leverage during the war. It is assumed that Lincoln strategically freed the slaves in hopes that they would join, and fight on behalf of, the Union. As soon as slaves escaped from a Confederate to a Union state, they were deemed to be freed. This resulted in the Confederacy losing most of its enslaved labor, which accounted for a large percentage of overall labor, thereby swaying the war in favor of the Union. Although the Emancipation Proclamation did not completely eliminate slavery in the United States,[3] abolition came soon after the 13th Amendment to the US Constitution was passed in 1865, just after the Civil War ended. No fewer than 180,000 Black troops joined the Union army, around 38,000 of whom died in the war. From the onset of the Civil War, slaves had been fighting for their freedom. The Emancipation Proclamation served as a reaffirmation of their ideology that the Union's fight had to morph into a fight for their freedom and liberty. Therefore, this gave the Union movement the moral high ground in appearing to support the freedom of slaves while also strengthening its military. Despite the Emancipation Proclamation's place among some historical texts of freedom and liberty as an iconic moment on the path to America's ultimate abolition of slavery, its benevolence remains questionable (Curran, 2019).

Reconstruction

The Reconstruction period lasted roughly a decade, beginning immediately after the Civil War in 1865. It ended the Confederate secession, giving Blacks fundamental rights and freedoms. During this period anybody "engaged in the insurrection" against the United States was barred from holding military or civil office and the national government was permitted to correspondingly reduce the states' congressional representation if those states violated or abrogated the prerogative of their citizens to vote (Braxton and Zapernick, 2001). During this period the 13th, 14th, and 15th Amendments were added to the United States' Constitution. The 13th Amendment abolished slavery throughout the United States. The 14th Amendment, ratified in 1868, gave citizenship to all people naturalized or born in the United States, including ex-slaves. It also assured "equal protection of the laws" for all citizens.

The 15th Amendment, ratified in 1870, banned states from discriminating against voters in terms of "color, race, or previous condition of servitude." However, the amendment allowed for states to institute other qualifications for voting (Drexler, 2018). Even though these rules were adhered to, some states instituted poll taxes[4] and literacy tests that discriminated against Blacks. All these amendments granted equal treatment to everyone, but they were not well enforced. For instance, members of the Ku Klux Klan, a white supremacist terrorist hate group, who primarily targeted Blacks at the time, attacked Black citizens when they exercised their right to vote and ran for public office. However, a series of enforcement acts were passed in 1870 and 1871 to bring such violence to an end and empower the President to use military force to protect Blacks.

Jim Crow laws[5]

These were a set of local and state statutory provisions that made racial discrimination constitutionally legal in the United States. Enacted in the late nineteenth and early twentieth centuries, primarily in southern states, these laws sought to restrict the gains afforded to Blacks during Reconstruction. The laws were designed to deny Blacks the opportunity to vote, find a decent job, receive an education, and other opportunities. This lasted until 1965 – *ten decades* after the Civil War ended. The people who tried to disobey Jim Crow laws were often beaten, fined, and imprisoned, and many were killed. When the 13th Amendment abolishing slavery was ratified in 1865, Jim Crow laws, which looked for and employed legally dubious means of circumventing the amendment were actualized. Strict local and state laws known as "Black Codes" dictated how, where, when, and how much former slaves could be paid to work (Tischauser, 2012). The codes spread throughout the South as a legal means of, in essence, re-enslaving Black people by removing their right to vote and dictating where they could live and travel.

At the beginning of 1880s, Jim Crow laws were not adhered to by the big cities in the South. Black Americans found more freedom in the laws there, so more Blacks moved to the cities. However, as years passed, white residents of the cities demanded more laws to reduce opportunities for Blacks (Tischauser, 2012). The laws quickly spread across the states, outpacing the previous ones. Thus, Blacks were prohibited from accessing public parks, theaters, and restaurants. Jim Crow laws flourished up into the twentieth century within a society identified by violence. Lynchings and race riots increased; as many as 25 occurred in 1919. With the spread of Jim Crow laws, education was also affected such that Black students had few opportunities to attend college.

Racial segregation in education and housing

As mentioned, there were three amendments to the Constitution enacting civil rights for Blacks during Reconstruction. Still, the Supreme Court overturned those provisions and other civil rights legislation in a series of judgments from 1873 to 1883. Therefore, segregation throughout the United States was permitted by legislation, which impacted public and private transportation, civic infrastructure and facilities, prisons and armed forces, and educational institutions because many viewed Blacks as second-class people. In 1896, the Supreme Court decided in *H.A. Plessy v. J.H. Ferguson* that racial segregation was permissible under the 14th Amendment of the United States' Constitution (Herring, 2015). In response, a group of activists formed the National Association for the Advancement of Colored People (NAACP)[6] in 1909. The struggle to eradicate racism and racial division in American culture continues. As recently as the mid-twentieth century, the NAACP was focusing exclusively on legal challenges to school segregation.

As part of the segregation movement, cities passed laws prohibiting Blacks from residing in white-dominated neighborhoods. This was unconstitutional because it interfered with people's rights to own property. Despite this, in the 1920s, President Herbert Hoover formed a committee that persuaded local community boards to pass rules preventing low-income families from moving into middle-income neighborhoods (Herring, 2015) – an effort that particularly targeted Black families. Further, housing segregation was enforced through a variety of local ordinances. One such regulation in Richmond, Virginia prohibited couples who could not legally marry from residing in certain neighborhoods. Interracial marriage was prohibited in Virginia until 1967.[7] For example, in Richmond, Virginia, couples were prohibited from residing in a neighborhood where they were not able to be legally married, as was the case for interracial couples living in that state.

Segregation was also practiced in schools. In 1873, a law was passed that mandated so-called "separate but equal" schools for Black students and teachers. The school segregation laws of West Virginia and Kentucky, states that remained in the Union, showed that racism was not confined to the former Confederacy. Segregation in public education was accepted in most cities as it was seen to be necessary for peaceful racial coexistence. This resulted in a parallel education system in the South. The term Historically Black Colleges and Universities (HBCU) represents the institutions of higher education established before the Civil Rights Act of 1964, to serve the Black community (Kvasny et al., 2016).

Discrimination in the workplace

During the Reconstruction period, the number of labor unions in the United States increased. But since they were white-only unions, Black workers started organizing on their own behalf. An early success was in the agricultural workforce. The South's legislatures divided state land into 40-acre farms and offered low-interest loans to Black farmers in response to a petition delivered to Congress pleading for assistance in alleviating the "situation of the Black workers of the Southern States." Later, an urgent national need for labor and combatants during World War II led to advancements for Black workers in business and industry. Moreover, the National War Labor Board (NWLB) noted, "America needs the Negro ... the Negro is necessary for winning the war" (Cassedy, 1997). Thus, it issued a directive eliminating racial wage disparities. Nevertheless, by the close of the twentieth century, the position of Blacks showed that racial inequality continued. Although there has been progress in the workforce, there is still much to be accomplished in many fields with respect to the participation of Blacks – and certainly Black men in IT.

Black Men in IT in the United States Today

The United States Department of the Interior's Office of Civil Rights website (2016) defines diversity as:

> a term that is used broadly to refer to many demographic variables, including, but not limited to, race, religion, color, gender, national origin, disability, sexual orientation, age, education, geographic origin, and skill characteristics. America's diversity has given this country its unique strength, resilience, and richness.

However, statistics show that diversity based on race/color and gender within science, technology, engineering, and mathematics (STEM) is particularly low (Charleston et al., 2014; McClelland and Holland, 2015; Miriti, 2020; Prey and Weaver, 2013; Vardi, 2015; Whitaker and Montgomery, 2014). The capacity to use IT enables individuals to participate fully in society. This is vital, particularly as capabilities, functioning, and well-being in society are also important measures of relative affluence or poverty, along with measures of income alone (Massey, 1996; Wen et al., 2003). Kvasny (2002) argued that IT can be used in ways that promote social inclusion and that technology capabilities and access are integral to inclusion. However, major published surveys show that Blacks have lower rates of home computers and internet access (Perrin and Duggan, 2015). In comparison, 46 percent of Blacks use their phones as their primary source of internet access, compared with 33 percent of whites (Perrin and Duggan, 2015).

The IT workforce is dominated by men who are either white or Asian (Zweben, 2016). While addressing the National Education Association in 2010, President Barack Obama said, "We understand that our nation's prosperity is tied to innovation spurred on by students' engagement in STEM" (NEA, 2010). He continued by saying, "For America to be technologically competitive in the future, our students must become more fluent in complex science and math." If the United States is to meet its need for world-class talent in STEM, it is essential that a diverse population be attracted to engineering and other technical fields (Chubin et al., 2005).

Black men in IT postsecondary education

A variety of approaches have been used to investigate the topic of diversity in computing. One of the prevailing approaches to studying underrepresentation in computing starts at the primary and secondary education levels (K–12), analyzing mathematical success among young Black men in the United States, since mathematics is widely accepted as the foundation for computing (Berry, Thunder, and McClain, 2011; Berry, 2003, 2008; McGlamery and Mitchell, 2000).

In the educational landscape for IT careers in the United States, Blacks are vastly underrepresented. In the IT fields of study, the numbers are low: although Black men represent 6 percent of the US population, they account for just 2.2 percent of those employed in computing occupations (Pew Research Center, 2018). Black students' interest in computer science is no different than that of any other demographic group (Bonner and Bailey, 2006). One approach to this issue focuses on postsecondary education and the overall student experience of Black men, including the influence of university culture and self-efficacy (Bonner and Bailey, 2006; Cuyjet, 2006; Kvasny et al., 2015; Strayhorn, 2013). The issue of underrepresentation is also evident outside the research space. A College Board report (Ericson and Guzdial, 2014) – the entity that administers the Advanced Placement (AP) exams and grants college credit to high school students – indicated that of the 30,000 students who took the AP exam in computer science in 2013, only 3 percent were Black. In fact, 0 percent of Black students took the AP exam for computer science in 11 states, including Mississippi, where Blacks make up 37 percent of the population. Furthermore, less than a third of Black men and women who receive a degree in computing stay in that field (United States Census, 2013).

Black men in the IT workforce

In terms of race, gender, and sexual orientation, a diverse workplace has been shown to be effective in assisting underrepresented groups adapt to a new work environment, department, or division (Adler, 2002; Martin, 2014). Additionally, diverse workplaces tend to have managerial staff that is more aware of and responsive to different cultures, which aids in the organization's perspective and ability to launch new products, create new ideas, assess emerging trends, and develop new marketing plans (Adler, 2002; Martin, 2014).

Blacks, however, are underrepresented in the workforce. Preparing for and adapting to careers in IT where there are few people of color poses a challenge on a personal level. The vast majority of Black people leave science and engineering occupations (NSF, 2008). Moreover, they are disproportionately underrepresented in higher education. When comparing percentages of women to men, studies of underrepresentation in computing have historically focused on women, and assumed that all men are well represented. Some of the barriers that impeded women's adoption of IT careers were stereotypes and erroneous beliefs about their intelligence. Understanding social factors plays a role; however, it is not the only factor in understanding women's underrepresentation (Trauth and Connolly, 2021). For example, a stereotype about men being intellectually superior to women in mathematics affects women entering fields that rely heavily on mathematical concepts. Similarly, understanding social factors plays a significant role in understanding Black men's representation in IT.

The differences in educational, environmental, and personal experiences lend themselves to research that focuses on further understanding how these factors influence IT identity – the feeling of identification with and sense of belonging in the IT field – development and career choice (Carter and Grover, 2015). Not all Black men are the same; there are differences among them that must be evaluated and accounted for. An investigation of these differences may provide useful insight into what attributes successful Black men in IT possess; differences like this, among the same group, are referred to as *within-gender variation* (Kvasny et al., 2009; Trauth et al., 2009; Trauth et al., 2016).

Most importantly, and central to the focus of the BLKGENUIS project, is that cultural diversity has a positive effect on coalescing the many ways people from different cultures think, which can lead to more ideas and solutions (Martin, 2014; Al-Jenaibi, 2011). There are several places from which to launch an investigation into the underrepresentation of Black men in computing, such as primary or postsecondary education. Inquiries can also be started at the workforce level. The lack of diversity in the field is a problem, and there is a disconnect between early exposure by Black students to computer science in primary and secondary education and the number who graduate with a degree in an IT discipline. The unique challenges that Black men face in the field of computing demand further investigation (Kvasny et al., 2015). In the following section, I describe the academic nature of BLKGENIUS and how it can promote interventions to contribute to supporting Black men in IT. As a Black man who attended an HBCU for undergraduate school, transitioned to a Predominately White Institution (PWI) for graduate school, spent time in industry, and is now a faculty member at an HBCU in computing, I have insight into how to engage Black men in computing and strategies to retain them (Cain and Trauth, Forthcoming; 2017, 2016, 2015, 2013, 2012; Cain, 2021, 2012; Cain et al., 2019, 2018).

BLKGENIUS PROJECT

To support academic success, BLKGENIUS, partially supported by a National Science Foundation–funded grant project that began in 2021, identifies and examines the experiences (i.e., barriers and successes) of Black men in computing through qualitative data analysis to: (1) understand the factors that influence their computing success; and (2) determine how their aspirations for postsecondary and career paths in computing are actualized and cultivated. As the principal investigator, I hired two undergraduate students to work on this research with me. As part of the project, I am also mentoring the students. The project's objectives, activities, and benefits for students, practitioners, and society are summarized below in Table 12.1.

Table 12.1   Objectives and benefits of BLKGENIUS

Objectives of BLKGENIUS	Interactive activities for student participants	Benefits for students	Benefits for practitioners	Benefits for society at large
Collect data to identify intrinsic and extrinsic factors of successful Black men in IT	Serve as a research assistant	Gain research experience as an undergraduate	Ability to help recruit more Black men to higher education in IT	Increase the diversity of the IT workforce
Offer a student–business professional mentorship program for Black men	Project-related website and information repository	Build connections with Black men working in IT	Access to new hires	Offer inclusion solutions for underrepresented groups
Provide opportunities to collect data for a long-term research agenda	Semi-annual webinars Symposium of Black Men in IT	Access to a plethora of Black men in the IT fields	Offer internships to Black men in IT programs	
Dissemination of research results			Expand the representation and retention of Black men in IT	

Summary of Objectives

The main objective is to collect data to identify the intrinsic factors (e.g., attention and motivation) and extrinsic factors (e.g., financial support, mentors, motivation, peer interaction, spirituality) that have led to the success of Black men in computing. Pursuing this objective will unearth the critical incidents in Black men's educational pathways, facilitate a deeper understanding of the computing experiences of these students, and inform future educational objectives. Students currently participating in the project come from two HBCUs and two PWIs. The combined undergraduate enrollment across all four participating higher education institutions is over 70,000 students, of which 2 percent are Black students in IT-related majors. The majority of the Black men participating in this project are attending HBCUs.

Another objective is to develop an understanding of how Black men's aspirations for postsecondary computing study are cultivated and actualized. Moreover, analyzing the data will determine which strengths Black men attribute to their computing career decisions. The expected findings may point to youthful computing experiences, familial influences, career counselors, individual interests in computing, undergraduate research programs, unique institutional computing efforts, and other factors. This objective will be met through the intersection of qualitative semi-structured interviews and qualitative artifacts[8] and is expected to reveal the influences of various contextual markers on Black men's computing career aspirations.

Another goal is to expose undergraduate students to research that can propel them to become future scholars and researchers. Scholars should plan to include undergraduate assistants throughout the implementation of their projects and seek to serve as effective research mentors. The undergraduate research assistants working on the BLKGENIUS project will help with data management and analysis, attend conferences to present project results, and provide other technical assistance with the research process. This exposure to ongoing research will involve them in documenting and analyzing the experiences of Black students in computing, teach them how to manage the logistics of a large-scale qualitative research project, help them further develop their own research agendas, cultivate their interest in graduate school, and allow them to reflect on how they have navigated barriers and obstacles.

In addition to interviews, BLKGENIUS engages study participants in three ways: (1) a project-related website (which is under development in 2022); (2) semi-annual webinars for participating students and their institutions that are conducted by the principal investigator and Black men in the IT workforce; and (3) a Symposium of Black Men in computing.

The BLKGENIUS website[9] serves as a repository for information about the project. It includes a mission statement, links to other useful websites, a list of professional resources, and other related network events and opportunities. The website includes testimonials from students concerning their experiences, successes, achievements, etc., in the computing realm separate from the interview and focus group data. Organizations can partner with the various universities to combine efforts in this domain.

The webinars include success stories (i.e., "learning the ropes") and fruitful efforts regarding the achievements of Black men, as well as other pertinent information concerning BLKGENIUS and relevant events. The webinars are designed to bolster further interest and participation in computing among Black men. Each webinar lasts approximately an hour and is devoted to a specific topic in computing. The webinars are conducted in the fall and spring of each year; the principal investigator of BLKGENIUS conducts one webinar each year, and a Black IT scholar at a university or research lab or Black IT professional conducts the other.

Specific goals of the symposia, which occur biennially, include: (1) building a bridge between research and education concerning computing barriers and successes of students who are Black men, (2) influencing the participation of Black men in computing by establishing mentoring partnerships, and (3) identifying, expanding, and building on ideas to improve the transition to computing fields for Black men. Ultimately, the goal of the symposia is to thrust more Black students interested in computing into the computing pathway.

Overall, BLKGENIUS is being used to support computing excellence by connecting Black computing majors across institutions, in professional settings, and secondary schools. The network also functions as a peer mentoring space, with peer guidance serving as a driving force. There are numerous benefits of BLKGENIUS, including keeping students engaged in their computing studies, showcasing students as facilitators at the symposia, sharing tips for navigating the IT workforce terrain, conversing with other Black men about their experiences as racialized beings in computing, exposing students to research on their experiences, and linking protégés with other Black computing students for future networking and collaboration concerning research opportunities, internship possibilities, and IT job vacancies. The project will end in 2026, but the project-related website and semi-annual webinars are intended to make this network sustainable and adaptable over time and serve as a medium for sharing results.

BLKGENIUS will rely heavily on the dissemination of results, including sharing with educators, department chairs, deans, and others, as well as through publications and presentations. The findings from this work should be shared with key stakeholders, including educators, department chairs, and deans, to mitigate recruitment and retention issues in computing.

CONCLUSION

The underrepresentation of Black men in computing in the United States continues to be a problem in the workforce. This chapter provides an overview of how BLKGENIUS can be used as a tool to promote and highlight interventions for Black men. The core takeaway is that unique educational approaches can lead to more informed and effective interventions and programs designed to increase Black men's representation in computing. BLKGENIUS is a novel approach to integrating prior research that emphasizes the importance of mentoring and role modeling in cultivating identity and building confidence. As such, BLKGENIUS stands to make a tangible, positive impact on Black men by providing them the ability to interact with mentors who may share similar lived experiences.

Another key takeaway is the applicability of outreach techniques mentioned in this chapter that can be extended to companies, corporations, and institutions outside academia. Many companies have employee resource groups, which may have meetings and provide a sounding board for Black employees but which may lack the specific matching of an employee to a mentor. Mentoring that is more purposeful and active, rather than passive, may increase retention. Additionally, the more people who feel a strong sense of attachment to BLKGENIUS outcomes, the stronger the network becomes – with documented experiences that lead to success, resulting in an increase in the number of Black students who return to school for graduate studies.

Lastly, sharing outcomes is not a practice that should be limited to faculty in academia. More corporations are posting annual diversity metrics, and unfortunately many of these figures remain stagnant with little improvement. If an adaptation of the proposed network is adopted, organizations would be able to show their commitment to diversity, equality, and inclusion beyond hollow pledges and empty promises. Corporate executives could demonstrate how they are implementing policies and actions to increase their technical workforces' diversity. Practitioners could then modify the implemented models for their organizations.

Regardless of their role, anyone can be a proponent of diversity, equity, equality, and inclusion in the workforce. There is no shortage of places to begin this work. Although this chapter focuses on Black men and targets the work necessary for the computing workforce, the fact remains that having an inclusive workforce is the moral and ethical thing to do.

NOTES

1. Throughout this chapter, "African American" and "Black" are used interchangeably. Both of these constructs are used to denote students of African descent born in and educated in the United States.
2. In this chapter "men" is used as opposed to "males" in an effort to respect gender identity and refrain from addressing people by their anatomy. In using the term "men," this chapter is referring to those who identify as a "man."
3. Because emancipation only applied to states in rebellion. Four "slave states" remained in the Union.
4. Poll taxes were essentially voting fees, which were implemented in southern states as a technique to prevent Black people, who generally could not afford the fee, from participating in elections. Generally, poll taxes had a clause that excluded poor whites from paying if they had an ancestor who voted before the Civil War.
5. Jim Crow laws were legal means of states and localities subverting federal law to enforce racial segregation, usually occurring in the southern United States.
6. The National Association for the Advancement of Colored People https://naacp.org/.
7. *Loving v. Virginia* was a civil rights case, which the United States Supreme Court ruled that laws that banned interracial marriage to be unconstitutional and a violation of the 14th Amendment to the United States Constitution.
8. Qualitative artifacts are digital or physical items that signify a moment in time (e.g., someone's first memory of IT).
9. BLKGENIUS www.blkgenius.org.

REFERENCES

Adler, N.J. (2002). *International Dimensions of Organizational Behaviour*, 4th edition. McGill University, South Western, Thomson Learning, pp. 105–31.

Al-Jenaibi, B. (2011). The scope and impact of workplace diversity in the United Arab Emirates – An initial study. *Journal for Communication and Culture*, *1*(2), 49–81.

Allen, W.R., and Epps, E.G. (1991). *College in Black and White: African American Students in Predominantly White and in Historically Black Public Universities*. SUNY Press.

Allen, W.R. (1988). Improving Black student access and achievement in higher education. *The Review of Higher Education*, *11*(4), 403–16.

Berry III, R.Q., Thunder, K., and McClain, O.L. (2011). Counter narratives: Examining the mathematics and racial identities of Black boys who are successful with school mathematics. *Journal of African American Males in Education*, *2*(1), 10–23.

Berry III, R.Q. (2008). Access to upper-level mathematics: The stories of successful African American middle school boys. *Journal for Research in Mathematics Education*, *39*(5), 464–88.

Berry, J.W. (2003). *Conceptual Approaches to Acculturation*. American Psychological Association.

Bonner, F.A., and Bailey, K.W. (2006). Enhancing the academic climate for African American men. In Cuyjet, M.J. (Ed.). *African American Men in College* (pp. 24–46). San Francisco, CA: Jossey-Bass.

Braxton, G., and Zepernick, M. (2001). Racism and global corporatization. *Peace and Freedom*, *61*(4), 20.

Cain, C.C. (2021). Establishing a research agenda for broadening participation of Black men in computing, informatics, and engineering. *Technology in Society*, *67*. https://doi.org/10.1016/j.techsoc.2021 .101790.

Cain, C.C., Buskey, C., Bryant, A.M., Washington, G., and Burge, L. (2019). *Research Implications of the Tech Exchange: Immersion of Howard University Computer Science and Informatics Students in Silicon Valley*.

Cain, C., Morgan Bryant, A., Buskey, C., and Goel, R. (2018). *The Role of Tech Corporations at Historically Black Colleges and Universities in American STEM Education*.

Cain, C.C., and Trauth, E.M. (2022). The pursuit of tech degrees for Black men in the United States: Belonging and happiness, an individual differences study. *Technology in Society*, *69*, 101835. https://doi.org/10.1016/j.techsoc.2021.101835.

Cain, C.C., and Trauth, E. (2017). Black men in IT: Theorizing an autoethnography of a Black man's journey into IT within the United States of America. *ACM SIGMIS Database: The Database for Advances in Information Systems*, *48*(2), 35–51.

Cain, C.C., and Trauth, E.M. (2016). Black lives matter: The journey of a Black IT scholar. *Proceedings of the ACM SIGMIS Computers and People Research Conference (Washington, DC)*.

Cain, C.C., and Trauth, E.M. (2015). Theorizing the underrepresentation of Black males in information technology (IT), *Proceedings of the 21th Americas Conference on Information Systems (Puerto Rico)*.

Cain, C.C., and Trauth, E.M. (2013, May). Stereotype threat: The case of Black males in the IT profession. In *Proceedings of the 2013 Annual Conference on Computers and People Research* (pp. 57–62). ACM.

Cain, C.C., and Trauth, E.M. (2012). Black males in IT higher education in the USA: The digital divide in the academic pipeline re-visited, *Proceedings of the 18th Americas Conference on Information Systems* (Seattle, WA).

Cain, C.C. (2012). Underrepresented groups in gender and STEM: The case of Black males in CISE, *Proceedings of the ACM SIGMIS Computers and People Research Conference* (Milwaukee, WI).

Carter, M., and Grover, V. (2015). Me, my self, and I (T). *Mis Quarterly*, *39*(4), 931–58.

Casimir, J.N. (2020). *The cost of fitting in: An investigative analysis of race-based code-switching and social exclusion*. Unpublished Senior Honors thesis, University of North Carolina at Chapel Hill, Department of Communication Studies. https://do i.org/10.17615/wmk4-mf65.

Cassedy, J.G. (1997). African Americans and the American labor movement. *PROLOGUE-WASHINGTON*, *29*, 113–21.

Charleston, L.J., George, P.L., Jackson, J.F., Berhanu, J., and Amechi, M.H. (2014). Navigating underrepresented STEM spaces: Experiences of Black women in US computing science higher education programs who actualize success. *Journal of Diversity in Higher Education*, *7*(3), 166.

Chubin, D.E., May, G.S., and Babco, E.L. (2005). Diversifying the engineering workforce. *Journal of Engineering Education, 94*(1), 73–86.

Codjoe, H.M. (2001). Fighting a "public enemy" of Black academic achievement – The persistence of racism and the schooling experiences of Black students in Canada. *Race Ethnicity and Education, 4*(4), 343–75.

Curran, R.E. (2019). "To tear down ... the Corinthian pillars of constitutional liberty": Catholics and Lincoln's Emancipation Proclamation. *US Catholic Historian, 37*(2), 109–36.

Cuyjet, M.J. (Ed.). (2006). *African American Men in College.* San Francisco, CA: Jossey-Bass.

15th Amendment to the US Constitution. United States Library of Congress. www.loc.gov/rr/program/bib//ourdocs/15thamendment.html.

Ericson, B., and Guzdial, M. (2014, March). Measuring demographics and performance in computer science education at a nationwide scale using AP CS data. In *Proceedings of the 45th ACM Technical Symposium on Computer Science Education* (pp. 217–22).

Herring, C.M.P. (2015). *School desegregation: Participant perceptions of a freedom of choice initiative in the South* (Doctoral dissertation).

Joshi, K.D., Kvasny, L., Unnikrishnan, P., and Trauth, E. (2016, January). How do Black men succeed in IT careers? The effects of capital. In *2016 49th Hawaii International Conference on System Sciences* (HICSS) (pp. 4729–38). IEEE.

Kvasny, L., Joshi, K.D., and Trauth, E. (2015). Understanding Black males' IT career choices. *iConference 2015 Proceedings.*

Kvasny, L., Trauth, E.M., and Joshi, K.D. (2016). The role of HBCUs in preparing African American males for careers in information technology. In Prince, C.B., and Ford, R.L. (Eds.), *Setting a New Agenda for Student Engagement and Retention in Historically Black Colleges and Universities* (pp. 234–50). IGI Global. http://doi:10.4018/978-1-5225-0308-8.ch013.

Kvasny, L., Trauth, E.M., and Morgan, A.J. (2009). Power relations in IT education and work: The intersectionality of gender, race, and class. *Journal of Information, Communication and Ethics in Society.*

Lincoln, A. (1863). Emancipation Proclamation, January 1, 1863. *National Archives, 6.*

Martin, G.C. (2014). The effects of cultural diversity in the workplace. *Journal of Diversity Management, 9*(2), 89– 92.

Massey, D.S. (1996). The age of extremes: Concentrated affluence and poverty in the twenty-first century. *Demography, 33*(4), 395–412.

McClelland, S.I., and Holland, K.J. (2015). You, me, or her: Leaders' perceptions of responsibility for increasing gender diversity in STEM departments. *Psychology of Women Quarterly, 39*(2), 210–25.

McGee, K. (2018). The influence of gender, and race/ethnicity on advancement in information technology (IT), *Information and Organization* (28), 1–36.

McGee, E.O., and Martin, D.B. (2011). "You would not believe what I have to go through to prove my intellectual value!" Stereotype management among academically successful Black mathematics and engineering students. *American Educational Research Journal, 48*(6), 1347–89.

Miriti, M.N. (2020). The elephant in the room: Race and STEM diversity. *BioScience, 70*(3), 237–42.

Morris, J.E., and Monroe, C.R. (2009). Why study the US South? The nexus of race and place in investigating Black student achievement. *Educational Researcher, 38*(1), 21–36.

Myers, T.K. (2020). Can you hear me now? An autoethnographic analysis of code-switching. *Cultural Studies ↔ Critical Methodologies, 20*(2), 113–23.

National Science Foundation. (2012). Science and Engineering Indicators. National Science Board. Arlington, VA (NSB 12-01), January 2012. www.nsf.gov/statistics/seind12/start.htm.

Obama, B. (2014). Remarks by the president on "my brother's keeper" initiative. The White House, Washington, DC.

Palmer, R.T., Davis, R.J., and Hilton, A.A. (2009). Exploring challenges that threaten to impede the academic success of academically underprepared Black males at an HBCU. *Journal of College Student Development, 50*(4), 429–45.

Perrin, A. and Duggan, M. (2015). Americans' internet access: 2000–2015. Pew Research Center, June 2015.

Pew Research Center. (2018). Women and men in STEM often at odds over workplace equity.

Prey, J.C., and Weaver, A.C. (2013). Fostering gender diversity in computing. *Computer, 46*(3), 22–3.

Ríos, R. (2007). Embracing a diverse twenty-first century workforce. *Journal – American Water Works Association*, 99(1), 38–41.

Saeedi, S., and Richardson, E. (2019). A Black lives matter and critical race theory – Informed critique of code-switching pedagogy. *Race, Justice, and Activism in Literacy Instruction*, 147–62.

Strayhorn, T.L. (2013). What role does grit play in the academic success of Black male collegians at predominantly white institutions? *Journal of African American Studies*, 1–10.

Tischauser, L.V. (2012). *Jim Crow Laws*. ABC-CLIO.

Trauth, E.M., Cain, C., Joshi, K.D., Kvasny, L., and Booth, K. (2016). The influence of gender-ethnic intersectionality on gender stereotypes about IT skills and knowledge. *The Data Base for Advances in Information Systems*, 47, 3.

Trauth, E.M., and Connolly, R. (2021). Investigating the nature of change in factors affecting gender equity in the IT sector: A longitudinal study of women in Ireland. *MIS Quarterly*, 45(4), 2055–100.

Trauth, E.M., Quesenberry, J.L., and Huang, H. (2009). Retaining women in the US IT workforce: theorizing the influence of organizational factors. *European Journal of Information Systems*, 18(5), 476–97.

Vardi, M.Y. (2015). What can be done about gender diversity in computing? A lot!, *Communications of the ACM*, 58(10), 5–5.

Wen, M., Browning, C.R., and Cagney, K.A. (2003). Poverty, affluence, and income inequality: neighborhood economic structure and its implications for health. *Social Science and Medicine*, 57(5), 843–60.

Whittaker, J.A., and Montgomery, B.L. (2014). Cultivating institutional transformation and sustainable STEM diversity in higher education through integrative faculty development. *Innovative Higher Education*, 39(4), 263–75.

Wilkins, R.D. (2006). Swimming upstream: A study of Black males and the academic pipeline.

Wood, J.L., and Palmer, R.T. (2014). *Black Men in Higher Education: A Guide to Ensuring Student Success*. Routledge.

Zweben, S., and Bizot, B. (2016) 2015 Taulbee Survey: Continued Booming Undergraduate CS Enrollment; Doctoral Degree Production Dips Slightly, *Computing Research News*, 28(5), May 2016.

13. Founding oSTEM: trailblazing for LGBTQA+ communities

Eric Patridge

INTRODUCTION

We are not born with stereotypes, social norms, and biases. These develop over time, partially because our cultures don't adequately prepare us to treat all humans equally. Rather, we learn to selectively embrace a person based on their socioeconomic status, wellbeing, and identity. For example, the United States Declaration of Independence includes the phrase "all men are created equal" (Jefferson, 2009), and most people interpret this to mean all of humanity, but the country's history tells another story: one filled with bias and barriers. Every decade of American history includes elected leaders, policies, rallies, protests, exploitations, massacres, or other events which evidence oppression of other Americans (Alsan et al., 2020; Bloom, 2019; Goodman, 2018; Thomas and Wilcox, 2014; Wallace and Hofstadter, 2012; Wilkins, 2016). Numerous policies enacted before, during, and after the Civil Rights Movement addressed bias and barriers, and these targeted incremental advances for race and ethnicity, national origin and ancestry, citizenship, religious beliefs, sex, age, physical or mental abilities, and veteran status (and later, genetic information) (Hersch and Shinall, 2015). Notably, all of these civil rights policies lacked any mention of sexual orientation and gender identity until perhaps 2010, when the federal government announced that "sex discrimination in education programs" includes these categories. Even this interpretation has already been reversed and subsequently reinstated, demonstrating that any protection of these categories is dependent on the federal administration's interpretation of existing law (Post et al., 2020). Bias and barriers remain salient in America.

The 2010 federal interpretation of "sex discrimination" acknowledges sexual and gender identities beyond heterosexuality and the gender binary (male and female), and these include fluid concepts which we often simplify to linear frameworks like "queer spectrum" and "transgender spectrum." In this chapter, these fluid concepts are embraced by the acronym LGBTQA+, which represents lesbian, gay, bisexual, transgender, queer, and allied individuals, where the plus (+) includes additional sexual and gender identities which resist simple categorization. It is clear that LGBTQA+ people have not yet been universally accepted or supported, but over the last few decades, there have been tangible gains in the fight for equal rights. Even today, while this chapter is being written in the Spring 2021, US congressional leaders are approaching an important decision regarding the Equality Act (H.R. 5, 2021), which has wide-reaching ramifications. Currently, there is no federal law which confers explicit employment protection on LGBTQA+ people, but should the legislation be passed, employment protections would become federal law. As a result of this and several other provisions, this landmark piece of legislation would permanently advance LGBTQA+ communities, especially in the key areas of employment, housing, credit, education, public spaces and services, federally funded programs, and jury service.

Employment protections for LGBTQA+ people have been pursued for more than 40 years (Dreiband et al., 2015), but it has never been easy to collect the data necessary to prompt change. Over the last few decades, members of LGBTQA+ communities appealed for questions on sexual and gender identities to be included in demographic surveys on everything from healthcare to housing to employment (Diamond and Hughes, 2021; Sell, 2017). Still, even health institutions do not routinely collect these metrics (Pinto et al., 2019), and it has been more challenging to find employment-related data on LGBTQA+ people. This paradoxical situation for advancing LGBTQA+ communities, which requires employment metrics that are not actively collected, is one piece of rationale (among many) for establishing an organization dedicated to LGBTQA+ students who demonstrate a penchant for working with data.

In part, data-driven gaps and community needs, like employment protections for LGBTQA+ people, inspired oSTEM (Out in Science, Technology, Engineering and Mathematics), which I founded along with the help of many fine cofounders, colleagues, chapter leaders, volunteers, and corporate sponsors. (Here, I use "cofounders" and "colleagues" out of professional respect, but I consider them to be my closest friends and family, because oSTEM was truly both a collaboration and a labor of love.) Our first executive board was composed entirely of undergraduates, graduates, and postdoctoral associates. Our mission: *to educate and foster leadership among LGBTQA+ students pursuing careers in the fields of Science, Technology, Engineering, and Mathematics (STEM)*. As a chapter-based society, oSTEM was modeled after similar groups like the Society for Women Engineers (SWE), the National Society for Black Engineers (NSBE), the Society for Hispanic Professional Engineers (SHPE), and the American Indian Science and Engineering Society (AISES). Between 2005 and 2009, my colleagues and I independently created five campus groups. After formalizing oSTEM Incorporated as a 501(c)(3) non-profit corporation in 2009, we worked together to cultivate over 60 chapters by 2012, and these have typically been founded by undergraduates or graduates with the help of faculty advisors. The existing framework of higher education helped our chapters to scale across both undergraduate and graduate institutions. Over a short period of time, our total membership grew from 50 to 500 members, and our annual conference event mirrored this growth (and the most recent, in-person conference attracted nearly 1,000 people). The community quickly grew into a skilled and diverse group of LGBTQA+ people, including both students and professionals, from the United States, Canada, the United Kingdom, and beyond. Across all our efforts, we consistently sought ways to advance LGBTQA+ people beyond our organization, but even three years after founding oSTEM, there were still no metrics for LGBTQA+ people in the STEM fields.

In 2012, this prompted a collaboration with my coauthors that produced the first quantitative study to focus on LGBTQA+ faculty in STEM (Patridge et al., 2014), using secondary analyses of the well-established data set: *2010 State of Higher Education for Lesbian, Gay, Bisexual, and Transgender People* (Rankin et al., 2010). We were forced to limit our conclusions to LGBQ faculty, because the subset sample size of 279 faculty, including 47 STEM faculty, did not include enough people who identified as transgender or gender non-conforming (people whose gender expression does not reflect the gender binary), but we made sure to detail all of these findings within the manuscript. Our framework came from the only available academic climate model (Bilimoria and Stewart, 2009), and we focused on themes related to departmental work climate, career consequences of that climate, and the career and mentorship choices of faculty who identified as LGBQ. We also stratified LGBQ faculty by their level of professional *outness*, which we defined as how open individuals were at work with regard to their sexual orientation. The climate model led to odds ratios (ORs) for relationships between identity (*outness*), climate variables (exclusionary behavior; EB), internal experiences (comfort), and retention of LGBQ faculty. (EB included conduct that interfered with the ability to work or learn, such as bullying, harassment, feeling ignored, shunned, etc.).

Our findings indicated that faculty who were mostly or completely *out* to professional colleagues, as well as those who reported EB (whether observed or experienced), were more likely to report being "not comfortable" at the department level (ORs of 14.3, 2.5, and 7.2, respectively; $p < .0001$). In turn, observed EB, experienced EB, and comfort were each directly correlated with more faculty who considered leaving their institution (ORs of 3.4, 4.9, and 2.6, respectively; $p < .0001$). Three other highlights from the manuscript are notable: (1) faculty who reported being "not comfortable" were 14 times more likely to be *out*; (2) as compared to other departments, STEM faculty were significantly more likely to be *out*; and (3) among STEM field respondents, 12 percent of faculty identified within the transgender spectrum. Our findings clearly called for more data and emboldened our efforts to advance oSTEM, in hopes that it would contribute to the advancement of LGBTQA+ populations. Efforts have since made strides in metrics (Cech and Waidzunas, 2021; Hughes, 2018; Jennings et al., 2020; Yoder and Mattheis, 2016), and some have utilized oSTEM as a metric or contributing factor to campus culture for LGBTQA+ STEM students (Abrams and Abes, 2021; Cech et al., 2016; Linley et al., 2018).

When we set out to found oSTEM, there was a mountain of work to do – and this remains true today. At the time, there were no programs dedicated to LGBTQA+ STEM students, and although this was the primary focus for our efforts with oSTEM, so much more is required to bring equity and equality to LGBTQA+ people. Even in 2005, aside from safe spaces and advocacy groups, there was a general dearth of resources for LGBTQA+ people, despite the fact that the first documented gay rights organization was founded in 1924. Online communities were limited, and outside of cities, it was challenging to connect with others (Yarbrough, 2004). For many LGBTQA+ individuals, life was filled with a general haze of fear and anxiety, and it sometimes felt like there was a proverbial witch hunt (i.e., an attempt to find and punish) for anyone aligning with LGBTQA+ communities; we didn't feel safe (O'Connell, 2010). Personally, I first noticed this haze of fear and anxiety in my youth, when it was even more challenging to find spaces that welcomed LGBTQA+ people, which in retrospect is odd, considering how many families are likely to have at least one person who identifies as LGBTQA+ (inferred from LGBTQ+ youth [Conron, 2020]; LGBTQ+ parents [Gates, 2015; Goldberg and Conron, 2018]; LGBTQ+ homeless youth [Durso and Gates, 2012]). Unfortunately, except for perhaps the few spaces that were open to youth (i.e., LGBTQA+ centers, youth groups, and pride parades), essentially no others extended equal treatment to everyone, and even these spaces are often biased. Regardless of whether this was because of bigotry, or perhaps individual lack of awareness, my small world seemed to be crowded with more stereotypes and pejorative terms than there were LGBTQA+ people. Living in such a haze can be debilitating and lead to disappointment, depression, and suicide, and it is hard to fathom how humanity comes to foster such conditions for any group … but I digress. It wasn't until I experienced my first New York City Pride March (2004), Out for Work Conference (2005), and Out and Equal Workplace Summit (2006), that this haze started to lift for me. (My participation in the latter two events was made possible by IBM's LGBT+ Employee Resource Group.) Until this point, fear and anxiety kept me from imagining a future where I could be professionally *out* and networked with an LGBTQA+ community. For the first time, I saw tremendous potential for a STEM community which did not yet exist, and I knew there would be people who felt the same.

But considering our country's history and the history of LGBTQA+ groups, it seemed unlikely that a singular interest in founding oSTEM was sufficient to attract a diverse population of LGBTQA+ STEM students. We weren't simply interested in creating a space to benefit select individuals. We aimed to cultivate a skilled and diverse community of people, and this didn't align with the trend that LGBTQA+ individuals often segregate into heteronormative, homogeneous groups. We knew that the same biases and barriers which exist in the United States also persist within smaller groups (Parmenter et al., 2020), including academia (Rankin et al., 2019; Vaccaro et al., 2015) and the STEM fields (Cech and Waidzunas, 2011). Even when motivated to unlearn stereotypes, social norms, and biases, it can be challenging to do so; they don't simply vanish if we identify as part of a stigmatized group. Instead, individuals often develop coping mechanisms, which include finding ways to benefit by leveraging others, sometimes overtly hurting people within their own communities (McCormick and Barthelemy, 2021). As a result of this behavior and various other traits, including identity alignments, marginalized groups (like LGBTQA+ people) can effectively sort themselves into heteronormative and homogeneous groups, primarily socializing with those who are like-minded or who look like them (Worthen, 2018). (Directly or indirectly, heteronormative social behaviors promote heterosexuality as the preferred sexual orientation, reinforcing the gender binary [van der Toorn et al., 2020]). These segregated, homogeneous groups often reinforce stereotypes, social norms, and biases, especially when people lean into them – whether it is necessary, natural, popular, or advantageous – by building their lives around the physical attractiveness, sexual appeal, and ideology of success within their particular groups. This is a broad generalization, but it is a relevant trend and is still particularly salient.

The narrative given in this chapter is a retrospective lens, capturing our work to cultivate oSTEM and the lessons we learned along the way. The overarching goal of this contribution is to help anyone who is interested in creating diverse communities and to prepare them for some of the questions and challenges they will meet. The next section details how oSTEM was launched and architected, including how the community was built and cultivated. Following this are the lessons we learned, presenting tidbits for would-be trailblazers. Finally, a few concluding thoughts and many of our successes are wrapped into the conclusion.

BACKGROUND

The initial concept for oSTEM emerged from a 2005 focus group of students and stakeholders from various LGBTQA+ communities who attended the first Out for Work Conference. Also present were observers from IBM's LGBT+ Employee Resource Group, who sponsored the event. The guiding question we all sought to answer: "*is there a need for a national technical student society?*" The complete lack of programs geared towards LGBTQA+ STEM students seemed sufficient to answer the question. After some discussion, the idea for oSTEM's name was floated, and we started percolating ideas for the organization. In the months that followed, a logo design was finalized, and we carefully incorporated tangible themes related to gender identity, sexual orientation, STEM, and forward momentum. By 2006, my colleagues and I established oSTEM's first chapter at The Pennsylvania State University. Later that year, I attended the Out and Equal Workplace Summit, which provided great insight into

LGBTQA+ efforts across industry, and this visit connected us with sponsorship opportunities and also served as field research for activities and events we would later bring to oSTEM.

Over the next few years, several LGBTQA+ STEM campus groups independently evolved, and they would serve as the early framework for oSTEM Incorporated. Interested LGBTQA+ alumni and employees motivated each of us, and we were grateful for the sponsorship they secured from companies that are supportive of LGBTQA+ communities, including (alphabetically) Alcoa, Google, IBM, Lockheed Martin, Northrop Grumman, and Raytheon. Fortunately, our initial campus groups were reasonably close in proximity, which made it easy to keep an open dialogue and share ideas. The leaders and members of these campus groups were immersed in academic and professional pursuits at the same time as the groups also contributed to local and regional social justice efforts.

This combination of activities would weave so many important details into the social fabric of oSTEM. Each campus group responded to their own local and regional events for nearly two years, demonstrating at parades, rallies, protests, and sit-ins. We participated in town halls, campus-wide focus groups, as well as institutional reviews that were dedicated to change. At my own school, a series of hateful events further sparked my own involvement and afforded me a steep education as I met with "woke" friends[1] and colleagues to deconstruct my power and privilege as a white, cisgender,[2] gay male. During this time, some of us found close-knit families. We were devoted to supporting and learning about each other's needs, and we worked together to address the pain and suffering which persisted for so many of us. This atmosphere served as a social fabric for oSTEM Incorporated, and without a doubt, these experiences strengthened our vision for the organization as well as our conviction in blazing a trail for people from across LGBTQA+ communities.

Around 2009, leaders from each campus group came together, and we noted there were still no other resources serving the needs of LGBTQA+ STEM students. We also recognized the opportunity for a centralized organization which could both engage new campus groups and sustain those where interest periodically ebbed and flowed. We therefore established oSTEM Incorporated as an all-volunteer, 501(c)(3) non-profit corporation, and our campus groups were immediately rebranded as chapters or affiliates groups. Our overarching goal seemed clear: as a startup non-profit with no seed funding, we would continue bootstrapping an infrastructure dedicated to LGBTQA+ students in the STEM fields. Reflecting on this task, we recognized that there were still years of hard work ahead. Fortunately, our leadership included highly motivated and engaging leaders, each of whom had already created a campus group at their undergraduate or graduate institution. Despite our collective capabilities and imaginations, we knew from the outset that we shouldn't try to succeed alone in our endeavors. One of our goals was to build infrastructure available for *all* LGBTQA+ people, which meant that our resources and services would need to sustain a broadly inclusive atmosphere and attract participants from across LGBTQA+ and STEM communities. To meet this goal, we needed leaders with different perspectives as well as leaders who would represent different spaces within LGBTQA+ communities. This need conveniently aligned with a second goal: to impact as many people as quickly as possible. We spent a great deal of time strategizing how to attract, engage, and retain volunteers in a sustainable way, and we made every effort to build bridges with a variety of communities. Our strategies, combined with the groundwork we had laid over the prior four years, informed and contributed to the infrastructure for oSTEM Incorporated, and our rapid success would quickly make us "experts."

Prior to oSTEM Incorporated, our challenges typically involved raising funds, generating ideas for campus programming, or attracting a critical mass to sustain the groups. I still recall our difficulties in establishing the first oSTEM chapter at The Pennsylvania State University, and there were several events, including invited speakers, where only a handful of students showed up to an otherwise empty auditorium. The ability to engage and retain members continues to be an ongoing concern for individual chapters, but the larger community is now able to support these needs (and many more). Once we filed the formal paperwork for oSTEM Incorporated (i.e., articles of incorporation and tax-exemption requirements), our challenges immediately expanded to include chapter interactions, travel and event logistics, organizational collaborations, and sponsorship alignments.

Rather than detailing the ins and outs of these challenges, it is easier to highlight the ways in which we prioritized students and LGBTQA+ communities. We constantly fielded interactions from our membership, returning immediate and holistic responses. We had professionals recommend that we stop advertising the "Q" (for "Queer"), but of course we didn't listen. We nearly ended a collaboration to ensure our events included all-gender bathrooms (facilities for everyone alike, whether cisgender, transgender, or gender non-conforming). We declined sponsorship from companies which funded anti-LGBTQA+ efforts, and instead, we provided free attendance for them to experience our events and report back to their companies. We advertised as many beneficial opportunities as we could, even if they were competition, because our primary goal was to help people. We gave LGBTQA+ students a voice and represented them at decision-making tables, because there are so many spaces where LGBTQA+ people and/or students are not invited or empowered. Did we make mistakes? Sure, we did. And we lost a few relationships and sponsorships because of them, which I would expect of any trailblazer and transformative effort, but without a doubt, we had far more successes than failures.

As soon as oSTEM Incorporated was established, we created a flexible framework, developing a team of project leaders who could percolate and test ideas as quickly as possible. This allowed us to facilitate many concurrent projects. For the first few years, we worked hard to develop engaging and exciting events, including workshops and casual dinners, with the intention of reaching students at various LGBTQA+ college conferences. These venues, paired with one-on-one appeals, helped us to rapidly build a network through colleges and universities around the country. Interested students would go on to create their own oSTEM chapters or volunteer with oSTEM Incorporated. Our volunteers generated a strong backlog of ideas, more than we could implement, but they also seemed to muster an endless supply of energy when it came to serving students. Examples of these volunteer-led projects include funding mechanisms for both members and non-members, workshops focused on K-12 efforts, a chapter handbook to help jump-start new chapters, mechanisms to survey conference attendees, an automated process to generate chapter logos, and guidelines to help leaders customize giveaway products for members. Most recently, oSTEM awarded scholarships to both undergraduate and graduate students with an entire category dedicated to those attending community colleges.

Many projects resulted in permanent organizational changes while others were integrated into conference programming. One project became a team dedicated to enhancing oSTEM's focus on diversity and inclusion across both the organization and our programming, and this effort was so successful that it eventually evolved an entirely separate, spin-off organization, providing specialized support services for adults who experience personal or professional crisis.[3] Another team was dedicated to managing interactions with chapters and their leadership, and this eventually led to creating a leadership retreat. At one point, volunteers expressed interest in reviewing conference workshop submissions, and this became an annual practice with a focus on enhancing programming to benefit participants while fostering better relationships with presenters. As we rolled out project after project, we honored our volunteers and celebrated their successes whenever possible. (At the same time, the projects and internal transformations did not always come easy.) The impact these projects had on the oSTEM community and the larger LGBTQA+ community was visible, and this renewed our volunteers, encouraging them to reinvest themselves again and again.

Importantly, we quickly learned that while LGBTQA+ communities are comparatively small, they also span every aspect of social, political, and economic influences. We also hoped that through oSTEM, we would be able to evidence an alternative model for diverse LGBTQA+ communities, other than the typical self-constructed, heteronormative or homogenous groups. To meet our goals, we focused on building the motivation and momentum which would incentivize people to professionally align and collaborate across the gamut of sexual and gender identities. Although our efforts were dedicated to molding oSTEM into a sustainable community, another reason for doing this was to build bridges across the divides within LGBTQA+ communities. It was crystal clear to us that reaching across these divides would require meeting a range of needs and celebrating people however they arrive. (It also seemed like most organizations weren't interested in widely adopting this approach.)

As we advanced, our motivation for oSTEM evolved into a desire to facilitate broader transformation. For example, in addition to encouraging individual pursuits, we hoped our success would also popularize STEM across LGBTQA+ communities, thereby introducing more career pathways, inherent connections throughout the STEM fields, and perhaps a greater sense of purpose for some. We also hoped the oSTEM community could benefit from the strengths of each member, including individuals who are both privileged (people with rights and advantages) and underprivileged (people lacking rights and advantages), such that each person would motivate themselves to pursue greater awareness of societal issues. Take a moment to imagine what can be accomplished by a cohesive community, infused with purpose, which aligns with both LGBTQA+ and STEM communities. Once you appreciate how many families are likely to have at least one LGBTQA+ person, then you start to understand the implications. By succeeding with oSTEM, we would not only strengthen LGBTQA+ communities, but there was potential to advance the general population in a variety of ways. These were idealistic thoughts, but directly bridging our divides could arguably have been the shortest route to cultural transformation, and as we rolled out our efforts, the positive outcomes were, indeed, quick to come.

As an all-volunteer organization, no one received compensation for their efforts, and we were always grateful for the time and effort which people gave to oSTEM. Celebrating our volunteers was normalized across our leadership, and this created a culture which our volunteers cherished. We also worked hard to protect our volunteers from sustainability and burnout. We did our best to scope projects and assign tasks appropriately, and we kept momentum by regularly checking in with volunteers. We discouraged everyone from becoming overextended, and we made it clear that the success, health, and happiness of each volunteer came first, before their contributions to oSTEM. We took the time to care for and show up for each other, which was an automatic challenge, since oSTEM has always been an entirely remote organization, with all volunteers contributing from home. Beyond our regular tasks and meetings, we also took advantage of more exciting and motivational opportunities. Whenever possible, volunteers would represent oSTEM at college conferences and they would also join other LGBTQA+ organizations at community-focused events, including a few hosted by the US presidential administration. We invited volunteers to retreats focused on planning oSTEM's future, and we made every effort to address their concerns and incorporate their ideas. The annual conference was always the single most important event, because it was the largest in-person event, and it would energize everyone for the next year. Moreover, this was often the only time for volunteers to meet in person, even after working together for an entire year. The fact that oSTEM remained a volunteer, non-profit organization for so long could raise questions about funding. Over time, our list of corporate, government, academic, and civil sponsors grew, and we used their sponsorship to further cultivate the oSTEM community; our sponsorships funded academic scholarships, annual conference events, chapter awards for programming, and travel awards to help students attend events. Eventually, sponsorship sources came from all sectors and disciplines, and this allowed our members to connect with a variety of employment opportunities. (Moral and ethical sponsorship remains an ongoing conversation, and I believe this is inevitable when sponsors include even just one corporate or government employer.)

As I've already said, we didn't always get things right, but we tried – and we tried hard. We also iteratively improved our efforts because we listened and valued being responsive. Those who volunteered with oSTEM did so for a variety of reasons, and they brought their own motivations and passions. The variety of their contributions echoed their diversity; some volunteers had more time to give, and others cultivated directions which changed the course of the organization. And rather than expecting our volunteers to make great personal sacrifices or meet some arbitrary measure of success, we mindfully valued the efforts of all our volunteers and not just for the results achieved. We understood that long-term outcomes for oSTEM were directly dependent on the ideas and capabilities of diverse volunteers.

In order to maximize momentum for the organization, leadership turnover was deliberately built into oSTEM, thereby retaining opportunities for future leaders. To this end, we carved a path for others to follow, and all of the initial founders have since moved on from the executive board. The trail we blazed continues to grow and to cultivate leaders who are invested in STEM disciplines as well as underrepresented groups within LGBTQA+ communities. Furthermore, those who join the oSTEM leadership are well-positioned to gain organizational experience and learn about a variety of unique dynamics and issues which remain salient within LGBTQA+ communities. At the outset, we never saw ourselves as trailblazers; we simply believed in and built something larger than ourselves – something that could help to advance LGBTQA+ communities. In retrospect, we were, indeed, trailblazers because of our work; we created something innovative which departed from the status quo, and we also blazed a sustainable trail for others to follow. (I would encourage you to read between the lines here and assume that trailblazers also make mistakes and fail frequently.)

It has now been over 15 years since the initial roundtable, and there are still no other organizations which serve LGBTQA+ students so broadly across the STEM fields. However, LGBTQA+ professionals have had the National Organization of Gay and Lesbian Scientists and Technical Professionals (NOGLSTP) since 1983 (Hicks, 2012). Beyond oSTEM and NOGLSTP, there have also been substantial contributions from individuals, programs, and caucuses to advance LGBTQA+ communities within each of the STEM disciplines (Atherton et al., 2016; Bannochie, 2017; Barthelemy, 2020; Belmont, 2014; Britton et al., 2021;

Butterfield et al., 2018; Cooper and Brownell, 2016; Cooper et al., 2020; Farrell et al., 2016; Khadjavi, 2019; Koizumi, 2016; Long, 2016; Ross, 2019; Stallings et al., 2018). In addition, several organizations are programming in STEM with specific focus, and they have done fantastic work and are dominating those spaces (Harris and Daniels, 2018; Gibney, 2019; Jackson et al., 2018; Mackay, 2019; Walker, 2014).

LESSONS LEARNED FROM TRAILBLAZING AND ESTABLISHING OSTEM

Every journey is unique, so before I draw your focus closer, I would first like to paint a broader perspective. Regardless of where your journey goes, you will always meet challenges. And some experiences can be particularly challenging to navigate, such as when you are invested in multiple agendas that conflict. You may have already navigated at least one such experience if you identify as LGBTQ+ and *came out* to your family. This sentiment also applies to contrarian pursuits, meaning those which conflict with current practices or oppose popular opinion, and I would suggest these include any bottom-up effort to achieve equality. In fact, trailblazing often begins as a contrarian pursuit, rather than budding from a trend, because the effort is likely one which few people yet value or one which is particularly challenging from the start.

Overcoming the Naysayers

The first challenges I had to overcome while starting oSTEM were the naysayers. For a variety of reasons, it is typical for startup founders to broadly circulate their ideas, and while circulating the idea for oSTEM, I received a litany of feedback which I assume is commonplace for trailblazers: (1) don't do it; (2) do it differently; (3) if it was worth doing, it would already exist; (4) this is too difficult; (5) someone else is probably doing this; (6) your idea isn't novel; (7) this space is too crowded.

I always welcomed opportunities for feedback, but it did become habitual to tune out messages from naysayers. Moreover, when feedback was exclusionary, I was moved to do the opposite. Eventually, I developed mental notes which were perpetually motivating: (1) an organization like oSTEM is inevitable; (2) no other organization is yet effectively aimed at professionally developing LGBTQA+ students in STEM; (3) from 2005 thru 2012, we could not identify any other individuals working on similar interventions for LGBTQA+ communities; (4) without the groundwork for oSTEM, any similar effort would likely be delayed a decade; (5) ample support for the concept of oSTEM is coming from the right places; and (6) any new effort is likely to leave out the sciences or focus solely on technology or sideline individuals already marginalized within our communities. Sentiments like these were empowering whenever we broke new ground or made key decisions that seemed contrary to mainstream opinion.

While pursuing any endeavor, trailblazers should expect challenging situations, and these will become valuable additions to their stories and histories. Whether or not you are a trailblazer, you will find yourself in professional situations which may be unreasonable, ethically questionable, designed to test you, or simply designed for failure. I wish I could say these situations are few and far between, but colleagues have shared many stories with me. The stories I hear are typically centered on disagreements, harassment, racism, sexism, homophobia, and transphobia, and while these stories are not mine to tell, I do have others to share. For example, orientations where you learn that before addressing inappropriate behavior with a manager or professor, you should consider starting the conversation by praising them. Or those closed-door grant reviews where you realize there is a favored candidate before any reviewer starts talking. Or all kinds of disputes and miscommunications which you can't fully address because "you don't bite the hand that feeds you," malign, or even offend those who have power and influence over your career advancement. Or those "blinded" peer reviews which are inherently unblinded, because the author's name is included and they cite themselves a lot and you know everyone currently publishing in your discipline. Or the complete absence of academic infrastructure for professors who are hospitalized or die, such as formal institutional or funding succession plans.

Most of these stories are from academia, but I've heard similar stories from across all sectors (public, private, and civil). Unfortunately, these aren't even the most egregious stories I have tucked away in some dark corner, but I find them to be so absurd that they act as a spotlight, casting a rather welcoming glow upon the exit sign. And while I would love the opportunity to serve academia, where I might steward knowledge, inspire innovation, and teach people, I have come to realize there are many effective ways to accomplish these things outside academia.

Maintaining the Vision

For trailblazers focused on building community, it is vital to have a vision which can be shared and which creates momentum for others. An effective vision will motivate or build up hope in others, and depending on the goals, it may even create hope in individuals where there was little or none. When these individuals find a community which fills them with hope, whether or not the effect is intentional, it can be a powerful multiplier and become central to a community's growth and purpose. This seemed to happen on a regular basis within oSTEM, which became apparent as more and more students shared with us just how important the organization was and that our vision resonated strongly with them.

Awareness of power and privilege was essential to oSTEM's formation. My own understanding of these themes was largely developed before founding oSTEM, by activism and "woke" colleagues, and this awareness taught me to question myself and my community. Strangely, this aligned well with a basic tenet of science: "question everything." This sentiment stayed with me, and I learned to regularly question everything as an activist, community leader, scientist, and life-long student. When I used this objective lens to find answers beyond the laboratory, I noticed I was increasingly silenced or shut down, and these reactions to my curiosity and persistence begged further scrutiny. I wondered why leaders always had an excuse to avoid collecting data on LGBTQA+ people and why STEM spaces seemed so heteronormative as compared to other disciplines (Cech, 2011; Cooper and Brownell, 2016; Trenshaw, 2013). I wondered why people from STEM disciplines rarely attended campus demonstrations and why these disciplines continued to be dominated by white men, despite a multitude of diversity efforts. I wondered why faculty (especially untenured faculty) were generally absent from campus demonstrations and why student voices seemed to be frequently dismissed and even ridiculed behind closed doors. I wondered why LGBTQA+ communities were so internally divisive and why they sometimes even amplified discrimination which was widespread in the general public. Eventually, I came to understand that objective questions like these are frequently asked by underserved or underrepresented individuals, many of whom are far too familiar with disappointment or with being leveraged for value.[4] This realization became one of many internal struggles which weigh on me, and it's the primary reason oSTEM remained an all-volunteer organization for as long as it did. Moreover, we never stopped asking questions like these, because the way people respond to them can be immensely informative.

These experiences are not unique, but everyone's journey does take a unique path, and it can be quite personal or even isolating. The specific vision we developed for oSTEM grew by aggregating experiences and ideas from many individuals who brought their own stories and identities. Some brought entirely new histories into the fold, although each of us could identify with the pervasive haze of fear and anxiety which resonates with so many LGBTQA+ people. Importantly, each of us wanted to give back to the next generation of LGBTQA+ STEM talent, and we appreciated that our community would span every demographic. Whenever we came together and shared experiences, we always found common ground through our stories, and more often than not, we were transformed in some way. We observed our elders and learned from our peers; we saw what inspires students, how decisions were made, how policies were formed, who is at the decision-making table, and perhaps most importantly, how people treat each other. We knew we would help others if we were successful, and we hoped we could provide a platform that evolves over time and in response to community needs.

Collective Imagination

One of the most extraordinary parts of trailblazing is the opportunity for blue sky thinking, which means to brainstorm without limits, and you should always make room for this kind of thinking. Those planning for a diverse community would gain substantial traction by involving diverse perspectives at the outset. For example, efforts to serve diverse LGBTQA+ communities would benefit from inviting representation from every demographic and from across the rainbow. (There are more than 50 discrete identities under the umbrella of sexual and gender minorities, and some of these were recently incorporated into a technical ontology for LGBTQA+ people, which aims for accessibility and interoperability with existing STEM and medical systems [Kronk et al., 2019].) Not only will their involvement grow your community, but it is also an early opportunity to receive direct input and to demonstrate reciprocity with your members.

If your goals include building out new spaces, like oSTEM did for LGBTQA+ students in STEM, then you are limited only by your collective imagination and the resources you target over time. Trailblazers in this situation are uniquely positioned to set a foundation, direction, and pace for concurrent or subsequent efforts. When oSTEM began, we had the opportunity to create a student contingent within an existing professional group, but our members adamantly sought their own voice; they wanted to move in a different direction and at a different pace. Our community has been extremely vocal about serving the entire spectrum of LGBTQA+ communities, and they yearned to just bring their authentic selves to work. Within oSTEM, there were virtually no restrictions on what we could pursue, and it soon became clear that starting our own group was the right decision; our volunteers flourished, their productivity and creativity were valued across all of our programming, and we were ultimately a success because of their time, energy, and technical expertise.

The important takeaway for trailblazers to know is that efforts which depend on existing relationships or organizations will be limited by external factors; out of necessity, your outcomes will be molded to fit your dependencies. For this and many other reasons, when transformative efforts percolate from within an existing platform, they inevitably meet increasing resistance. If your goal is, for example, to integrate with existing organizations, healthcare, or government bodies, then your efforts will be heavily dependent on relationships, underlying expectations, and domain expertise. In this scenario, your outcomes may specifically be designed to align with or enhance existing infrastructure. Given the nature of academic advancement and the requirements of many funding mechanisms, I suspect this sentiment also widely applies to academic endeavors, which heavily depend on peer reviews for publications and grants as well as peer support for career advancement.

Maintaining Personal and Professional Boundaries

It can be immensely fulfilling to serve as a leader, and certainly, there are easier times ahead, beyond the first few years. However, when problems bubble up to the surface or your plans go completely awry – and they will – it's particularly important for leaders to become adept at triaging or to have someone available who can do this in your stead. In these moments, it would behoove you to stay grounded. When faced with any kind of problem, leaders should know how to calm their team, de-escalate a situation (this is context-dependent), and remain focused on the larger picture. Some of the best leaders will have contingency plans, know how to mitigate more than one problem, be able to harness or pivot the energy, and keep the team appropriately focused and energized. Personally, I am invigorated and moved to action when unexpected roadblocks appear out of nowhere. When I encounter experiences which weigh on me, they eventually help me to understand and define my own personal and professional boundaries for "next time."

Speaking of personal and professional boundaries, as you advance in your career, you may increasingly find cultural rules and expectations imposed on you. One of these may be an unspoken expectation to embrace a stoic detachment from passions and work, and this sentiment seems to be particularly romanticized in academia. Certainly, stoicism teaches us methodologies and tools which help to overcome challenges and to guide certain types of people to success. At the same time, I wonder how many of us assume that this mantle of stoicism requires us to become emotionless and hardened. I believe it is a central mistake to abandon emotions and that we should, instead, teach each other to harness them; emotions can be powerful multipliers for both personal and professional success. When we are able to master our emotions, we are in a better position to manage conflict, to be professionally honest and respectful, to strengthen our personal resolve and conviction, and to ultimately reach success. Unfortunately, I feel this sentiment will not resonate widely in spaces where stoicism is admired – perhaps where academic politics are far more popular than managing conflict. In fact, most spaces seem ill-prepared to help us learn how to master our emotions, and this leaves substantial room for miscommunications and misunderstandings, which I have experienced numerous times on my journey through academia and while leading oSTEM.

Reflecting on my own journey, I know each one of us can be shaken to our core, and there is no one who can adequately prepare you for these experiences. Reasons why this may happen are both situational and personal, and there is no single or correct way to respond or move on. Perhaps your professional efforts are wildly successful, but you realize that you must step away to resolve something personal or realign your own trajectory. Perhaps you are affected by something transformative that you've created, which adds value to some individuals but where others may only perceive losses. Perhaps you've advanced your discipline in important but unconventional ways, and you feel blacklisted from opportunities or that your would-be mentors or colleagues have abandoned you. Perhaps you've had an unexpected experience that would be traumatic for anyone. Should any of these things happen to you, I urge you to take the personal time you need to regain your footing. And I truly hope that your prior efforts and convictions will stand on their own, serving as a beacon or grounding point for you. Regardless of how challenging it may seem, any kind of bad day isn't the end of the world, and you should take as much time as you need to recover or adjust. Taking the space and time you need is important, and if you decide to reinvest yourself, you will be a stronger and more effective trailblazer, whenever you are ready. Many of you will meet challenges like these, and seasoned leaders will become skilled at viewing these as opportunities for problem-solving and innovation.

The Personal and Professional Toll

We had great conviction in our work with oSTEM. Personally, this instilled in me a sense of duty which carried into my professional life; I was dedicated to improving infrastructure as I moved through it, so that those who follow might breathe a little bit easier. Sometimes, our efforts to advance students seemed more focused on generational gaps of LGBTQA+ identities than on community-wide needs. Certainly, focusing on these aspects of academic culture "too early in one's career" (perhaps before tenure) can make an academic trajectory particularly challenging to navigate. Those seeking an academic career should not underestimate the importance of building a supportive network of disciplinary peers; the process of promotion and tenure is like a gauntlet and seems to require broad support within a primary discipline. Those who retain multiple, concurrent "primary" disciplines will encounter specific challenges and likely have an unusual ability to maneuver the academic landscape and garner broad support across two disciplines. In my own case, I consistently had difficulty in building rapport with tenured professionals who did not identify with LGBTQA+ communities. While pursuing relationships with established faculty, I found myself repeatedly distracted by competing agendas, which I had little ability to affect. Many of the stereotypes, social norms, and biases we were trying to resolve within oSTEM also seemed widely present across academia, though, perhaps re-wrapped with stoicism and politics. I remain particularly grateful for substantial dedication and support from collaborative LGBTQA+ faculty.

To date, one of my greatest professional challenges was balancing my work with oSTEM while maneuvering this academic culture, which seemed to prioritize stoicism, resiliency, productivity, and funding over most everything else, including how STEM professionals sometimes treat their students, peers, and colleagues. The further I advanced in academia, the more I became certain of the need for community resources like oSTEM, and this was evidenced by a litany of academic requirements and infrastructure, such as: little time for socialization or mentorship, intradepartmental climate, interdepartmental politics, staff and faculty demographics, inaccessible faculty, campus-wide resources, peer-review processes, promotion and tenure, and access/awareness for funding and career opportunities. On top of all of these things, the stoic and siloed efforts of faculty often seemed to limit their ability to positively influence the academic culture. One thing became clear: no single person had much ability to directly influence change without being explicitly empowered by garnered funding priorities, administrators, and/or tenured faculty. This made me wonder whether there were any circumstances which might be more amenable to moving academia towards change.

Reflecting on this, perhaps it is unsurprising that one of our most challenging (and indirect) goals was to prompt a cultural shift across the STEM disciplines, encouraging professionals to pursue more inclusive practices. Certainly, our manuscript on *Factors Impacting the Academic Climate for LGBQ STEM Faculty* (Patridge, 2014), was a step forward, but I suspect a cultural transformation in academia will only happen when a critical mass of faculty is moved to adopt change, and even then, perhaps just department-by-department. The substantial contributions of colleagues to advance LGBTQA+ STEM communities are certainly examples of ongoing transformation (Atherton, 2016; Bannochie, 2017; Barthelemy, 2020; Belmont, 2014; Britton, 2021; Butterfield, 2018; Cooper and Brownell, 2016; Cooper, 2020; Farrell 2016; Gibney, 2019; Harris and Daniels, 2018; Jackson et al., 2018; Khadjavi, 2019; Koizumi, 2016; Long, 2016; Ross, 2019; Stallings, 2018; Walker, 2014) and several of these specifically focus on developing faculty (Hicks, 2012; Farrell, 2016; Farrell et al., 2018; Linley and Nguyen, 2015; O'Leary et al., 2020; Stallings, 2018).

I wish I could say I've always been filled with the same conviction as we had with oSTEM, but there are many aspects of the STEM fields which give me pause. Along my own journey, I found myself wondering whether some experiences and perspectives were uncommon, out of touch, or perhaps even tied to the same issues discussed in our publication (Patridge, 2014). And when I reflect on how much of my "free time" has been consumed by professional STEM endeavors, I must acknowledge that the obligations I felt to oSTEM and to my success in academia and STEM took a toll on my professional life as well as my personal life. Like so many others, I became all too familiar with giving up my time and neglecting chances to grow closer with people, in both friendships and personal relationships. Whether or not they realize it, my friends and family were all impacted by the time and energy which I redirected away from them, and this also variably affected several long-term relationships.

My story is not unusual; I have heard similar stories over the years, and I'm sure my colleagues and friends have stories of their own. However, I should add that, even though I wanted more quality time with my friends, family, and loved ones, I would take a similar trajectory if I had to do it all over again, because oSTEM is both unique and valuable, and I feel the world is far better off with a community like oSTEM than without. Nonetheless, I find myself wondering how life might be different if I had spent more time cultivating friendships and collegial relationships instead of striving to advance STEM and LGBTQA+ communities.

CONCLUSION

This retrospective narrative captures over ten years of work to cultivate a diverse community (including resources) for LGBTQA+ people, it expands upon lessons we learned, and it underscores the importance and timeliness of our efforts using historical context. Starting from common cultural experiences in the United States, an extensive pattern of discrimination against minority communities is highlighted. Further discussion contextualizes how this pattern similarly impacts LGBTQA+ communities at various levels. Particular focus is drawn to employment protections and related academic experiences, including the findings of our 2014 manuscript, which links identities, cultural climate, and career consequences for

LGBQ faculty in STEM (Patridge, 2014). The goals, strategies, and directions for oSTEM are presented in detail, as are our challenges and lessons learned, which relate to trailblazing for minority communities, programming for diversity, and managing personal and professional experiences. Anyone in pursuit of diverse communities may find useful discussion or even have experiences similar to those shared throughout the narrative.

The oSTEM community is a skilled and diverse group of LGBTQA+ people, including both students and professionals, from the United States, Canada, the United Kingdom, and beyond. At the outset, it may be tempting to focus on the primary differences between oSTEM and other groups like SWE, SHPE, NSBE, and AISES: our sexual and gender identities. While it is true that oSTEM focuses on reaching LGBTQA+ people, our community intrinsically includes diverse identities beyond sexuality and gender; they are every race and ethnicity, they are sexual and gender minorities, and they are women. In reality, oSTEM reaches people from every identity, and the leadership endeavors to provide services and programming as diverse as its community. Given this core aspect of oSTEM, it seems that many of the observations, the challenges, and the lessons presented in this chapter were likely encountered by earlier groups like SWE, SHPE, NSBE, and AISES, and perhaps they still face some of the same issues, despite being explicitly protected by Civil Rights Acts.

Arguably, the most important nuance to working with LGBTQA+ communities is to celebrate people however they arrive, and this also translates into how you value, honor, and include them. This applies not only to theories and future plans but also starting now, wherever you can. Rather than creating assumptions or expectations about how someone stands out or fits into a category that works for you, it can be helpful to let LGBTQA+ people define themselves. When in doubt, simply exchange names and pronouns,[5] but respect that sometimes these are not available. If new resources or novel arrangements are required, then consider making every reasonable effort to implement them, because it is always worth spending a day on something that may be transformational for someone. In situations that depend on collecting or analyzing data on LGBTQA+ communities, even seasoned leaders find it challenging to word questions in a way that both honors identities and facilitates adequate metrics or assessments. Many complex identities resist simple categorization with the gender binary, as well as with the "queer spectrum" or the "transgender spectrum." Instead, some intersectional LGBTQA+ identities may be better described through frameworks like those established by Trauth and Connolly (2021) and Crenshaw (1989), and perhaps through these kinds of lenses, more appropriate questions will lead to more appropriate metrics and analyses.

By any measure, we succeeded in our goals with oSTEM. We cultivated a modest community with a diverse membership, and the organization developed for over a decade, growing more than ten times by several metrics. Along the way, the leadership broke a tremendous amount of new ground, and many of the achievements may be "firsts" for LGBTQA+ communities. For starters, oSTEM was the first STEM organization to focus on LGBTQA+ students, and it was among the first to engage LGBTQA+ people in disciplinary professional development and related leadership positions. By attending both LGBTQA+ and STEM conferences, the organization was among the first to introduce STEM-related material to LGBTQA+ communities and LGBTQA+-related material to STEM communities. Further, our manuscript (Patridge, 2014), was the first quantitative manuscript to report on any LGBTQA+ community in STEM. In addition to these "firsts," the leadership infused a number of differentiators into organizational efforts. For example, although oSTEM was not the first organization to provide all-gender bathrooms, it stood out within the STEM community that these were basic requirements for oSTEM. Eventually, the leadership also created specific training so that our venues, keynotes, and entertainers were adequately prepared to serve LGBTQA+ communities. In addition, oSTEM remains a rare example of a professional community which builds interdisciplinary bridges without having a dedicated silo within the STEM disciplines. Collectively, the organization has coordinated more than ten annual conferences, reaching hundreds of chapters and thousands of people, while disbursing direct chapter or member support in the range of five hundred thousand dollars. These are just a few ways to measure our success.

And yet, there is still so much work to do. In addition to metrics and funding, LGBTQA+ communities also need bold and diverse role models, especially organizational leaders and elected officials, who can work hard and build bridges with so very many people who are willing and waiting to be appropriately engaged. For STEM communities, opportunities for cultural change are abundant in funding mechanisms, in promotion and tenure processes, and at the department level, and many of these also require administrative support. However, before leaning on internal resources to identify problems and solutions, STEM communities should consider approaching external experts who regularly focus on cultural and identity issues. Surely, there are also opportunities for change within oSTEM, and members should get involved or send ideas and feedback to the leadership. It is a certainty that the oSTEM community falls short for some people, just as earlier STEM and LGBTQA+ organizations may have been insufficient; no community is perfectly aligned for every person. When I reflect on our efforts to serve each and every member, even when we tried our hardest, I still recall individuals and budding communities who seemed unfulfilled by oSTEM. This is an inevitability, and I sincerely hope everyone feels invited to join the oSTEM community and leave as they need.

Of course, leaving oSTEM is an eventuality, and I am no exception. In my own case, it was my exit from academic research which prompted me to leave oSTEM after 11 years of service. While finishing up a research position, I realized I might be more effective at influencing both science and culture from outside of academia. Soon after, I planned an exit from oSTEM, because we were already quite successful, and turning over the leadership was extremely important to me. This was the right time for me to leave, because my passions had dimmed, and passion is always a key ingredient for non-profit leadership. Furthermore, it seemed that some of our efforts to advance LGBTQA+ communities had caused friction, and I felt my departure might flush out any remaining issues, allowing oSTEM to emerge stronger. Before leaving, I spent over 18 months strategically and deliberately pulling back from work, encouraging others to step up through conversation, personal appeals, and delegation. By the time I left oSTEM, I had no concerns about the organization's sustainability, but still, I prepared myself for numerous possible outcomes. Needless to say, oSTEM's subsequent growth was impressive, including challenging transformations which took place behind the scenes, and I am so very proud of each and every person.

One last thing to note is that we would not have been able to create oSTEM without making mistakes, without failing, and without causing friction. Seasoned trailblazers know these things are required for success, and it is impossible to please everyone. In spite of this, it is vital to lead with conviction, because you will not build community, garner financial support, or convince others if you can't *lead with conviction*. In the case of oSTEM, I remain grateful to each cofounder, colleague, chapter leader, and volunteer who contributed (and still contributes), and I am proud of everything we accomplished. In all that we did, it was easy to lead with conviction, and I feel privileged to have worked with such fine people.

NOTES

1. Here, "woke" is an adjective describing people who are alert to sensitive social issues in society, especially racial or social discrimination and injustice.
2. Here, cisgender is an adjective describing a person whose gender identity matches their sex assigned at birth.
3. THRIVE Lifeline (founded 2020; thrivelifeline.org).
4. Community leaders often make formal or informal promises to minority groups, and many of these leaders do not deliver on their promises (whether they are disingenuous, disempowered, or incapable of doing so). Therefore, these leaders benefit by receiving political or institutional support from the minority groups they fail to serve.
5. In the last decade, it has become a common practice across LGBTQA+ communities for individuals to share pronouns during introductions or in email signatures, although sometimes people may not feel comfortable to share their pronouns.

REFERENCES

Abrams, E.J., and Abes, E.S. (2021). "It's finding peace in my body": Crip theory to understand authenticity for a queer, disabled college student. *Journal of College Student Development*, 62(3), 261–75. https://doi.org/10.1353/csd.2021.0021.

Alsan, M., Wanamaker, M., and Hardeman, R.R. (2020). The Tuskegee study of untreated syphilis: A case study in peripheral trauma with implications for health professionals. *Journal of General Internal Medicine*, 35(1), 322–25. https://doi.org/10.1007/s11606-019-05309-8.

Atherton, T.J., Barthelemy, R.S., Deconinck, W., Falk, M.L., Garmon, S., Long, E., Plisch, M., Simmons, E.H., and Reeves, K. (2016). LGBT climate in physics: Building an inclusive community. *American Physical Society*, 118–26. www.aps.org/programs/lgbt/upload/LGBTClimateinPhysicsReport.pdf.

Bannochie, C.J. (2017). Alphabet soup and the ACS: The history of LGBT inclusion. In *Diversity in the Scientific Community Volume 2: Perspectives and Exemplary Programs* (pp. 179–87). American Chemical Society. https://doi.org/10.1021/bk-2017-1256.ch016.

Barthelemy, R.S. (2020). LGBT+ physicists qualitative experiences of exclusionary behavior and harassment. *European Journal of Physics*, 41(6), 065703.

Belmont, B. (2014, August). LGBTQ in ACS and the workplace: Status report. In *Abstracts of Papers of the American Chemical Society* (Vol. 248). 1155 16TH ST, NW, Washington, DC 20036 USA: Amer Chemical Soc.

Bilimoria, D., and Stewart, A.J. (2009). "Don't ask, don't tell": The academic climate for lesbian, gay, bisexual, and transgender faculty in science and engineering. *NWSA Journal*, 85–103.

Bloom, J.M. (2019). *Class, Race, and the Civil Rights Movement*. Indiana University Press.

Britton, Ben, Jennifer Carter, Matthew Korey, and Liz Roccoforte. Material goals towards equity along the STEM and LGBTQIA+ spectra. *JOM* (2021). https://doi.org/10.1007/s11837-021-04829-1.

Butterfield, A., McCormick, A., and Farrell, S. (2018). Building LGBTQ-inclusive chemical engineering classrooms and departments. *Chemical Engineering Education*, 52(2), 107–13.

Cech, E.A., and Waidzunas, T.J. (2021). Systemic inequalities for LGBTQ professionals in STEM. *Science Advances*, 7(3), eabe0933. https://doi.org/10.1126/sciadv.abe0933.

Cech, E.A., and Waidzunas, T.J. (2011). Navigating the heteronormativity of engineering: The experiences of lesbian, gay, and bisexual students. *Engineering Studies*, 3(1), 1–24. https://doi.org/10.1080/19378629.2010.545065.

Cech, E., Waidzunas, T., and Farrell, S. (2016). Engineering deans' support for LGBTQ inclusion. https://doi.org/10.18260/p.26633.

Conron, K.J. (2020). LGBT youth population in the United States. https://williamsinstitute.law.ucla.edu/wp-content/uploads/LGBT-Youth-US-Pop-Sep-2020.pdf.

Cooper, Katelyn M., Anna Jo J. Auerbach, Jordan D. Bader, Amy S. Beadles-Bohling, Jacqueline A. Brashears, Erica Cline, Sarah L. Eddy et al. (2020). Fourteen recommendations to create a more inclusive environment for LGBTQ+ individuals in academic biology. *CBE – Life Sciences Education*, 19(3), es6. https://doi.org/10.1187/cbe.20-04-0062.

Cooper, K.M., and Brownell, S.E. (2016). Coming out in class: Challenges and benefits of active learning in a biology classroom for LGBTQIA students. *CBE – Life Sciences Education*, 15(3), ar37. https://doi.org/10.1187/cbe.16-01-0074.

Crenshaw, K. (1989). Demarginalizing the intersection of race and sex: A Black feminist critique of antidiscrimination doctrine, feminist theory and antiracist politics. *U. Chicago Legal Forum*, 139.

Diamond, R., and Hughes, B. (2021, February). Resolving LGBTQ disparities in STEM representation through demographic data. In 2021 Annual Meeting. AAAS.

Dreiband, E.S., Swearingen, B., and Day, J. (2015). The evolution of Title VII – Sexual orientation, gender identity, and the Civil Rights Act of 1964. Jones Day.

Durso, L.E., and Gates, G.J. (2012). Serving our youth: Findings from a national survey of services providers working with lesbian, gay, bisexual and transgender youth who are homeless or at risk of becoming homeless. https://escholarship.org/content/qt80x75033/qt80x75033.pdf.

Farrell, S., Cech, E., Guerra, R., Minerick, A., and Waidzunas, T. (2016). ASEE safe zone workshops and virtual community of practice to promote LGBTQ equality in engineering. www.asee.org/public/conferences/64/papers/14806/download.

Farrell, S., Guerra, R.C., Longo, A., and Tsanov, R. (2018, June). A Virtual Community of Practice to Promote LGBTQ Inclusion in STEM: Member Perceptions and Community Outcomes. In ASEE Annual Conference. www.asee.org/file_server/papers/attachment/file/0009/5073/ASEE_Safe_Zone _Project_2018_V2.pdf.

Gates, G.J. (2015). Marriage and family: LGBT individuals and same-sex couples. *The Future of Children*, 67–87.

Gibney, E. (2019). Discrimination drives LGBT+ scientists to think about quitting. *Nature*, 571(7763), 16–18. https://doi.org/10.1038/d41586-019-02013-9.

Goldberg, S.K., and Conron, K.J. (2018). How many same-sex couples in the US are raising children? https://williamsinstitute.law.ucla.edu/wp-content/uploads/Same-Sex-Parents-Jul-2018.pdf.

Goodman, A. (2017). The long history of self-deportation: Trump's anti-immigrant policies build on more than a century of attempts to create fear and terror within US immigrant communities. NACLA Report on the Americas, 49(2), 152–58. https://doi.org/10.1080/10714839.2017.1331811.

H.R. 5, Equality Act, 116th Congress. (2021).

Harris, A., and Daniels, J. (2018). Lesbians and tech: Analyzing digital media technologies and lesbian experience. https://doi.org/10.1080/10894160.2018.1383799.

Hersch, J., and Shinall, J.B. (2015). Fifty years later: The legacy of the Civil Rights Act of 1964. *Journal of Policy Analysis and Management*, 34(2), 424–56. https://doi.org/10.1002/pam.21824.

Hicks, J. (2012, February). Why awareness of LGBT issues in the physics community makes sense. In *APS March Meeting Abstracts* (Vol. 2012, pp. J20-003). https://meetings.aps.org/link/BAPS.2012 .MAR.J20.3.

Hughes, B.E. (2018). Coming out in STEM: Factors affecting retention of sexual minority STEM students. *Science Advances*, 4(3), eaao6373. https://doi.org/10.1126/sciadv.aao6373.

Jackson, S.J., Bailey, M., and Foucault Welles, B. (2018). #GirlsLikeUs: Trans advocacy and community building online. *New Media and Society*, 20(5), 1868–88. https://doi.org/10.1177/1461444817709276.

Jefferson, T. (2009). *The United States Declaration of Independence (Original and Modernized Capitalization Versions)*. Wildside Press LLC.

Jennings, M., Roscoe, R., Kellam, N., and Jayasuriya, S. (2020, January). A review of the state of LGBTQIA+ student research in STEM and engineering education. In ASEE annual conference. https://par.nsf.gov/biblio/10157897.

Khadjavi, L.S. (2019). The origins of spectra, an organization for LGBT mathematicians. *Notices of the American Mathematical Society*. https://doi.org/10.1090/noti1890.

Koizumi, K. (2016, February). Landscape of LGBT Participation in STEM. In 2016 AAAS Annual Meeting (February 11–15, 2016). AAAS. https://aaas.confex.com/aaas/2016/webprogram/Paper17486 .html.

Kronk, C., Tran, G.Q., and Wu, D.T. (2019). Creating a queer ontology: The gender, sex, and sexual orientation (GSSO) ontology. In *MEDINFO 2019: Health and Wellbeing e-Networks for All* (pp. 208–12). IOS Press. https://doi.org/10.3233/SHTI190213.

Linley, J.L., and Nguyen, D.J. (2015). LGBTQ experiences in curricular contexts. *New Directions for Student Services*, 152(1), 41–53. https://doi.org/10.1002/ss.20144.

Linley, J.L., Renn, K.A., and Woodford, M.R. (2018). Examining the ecological systems of LGBTQ STEM majors. *Journal of Women and Minorities in Science and Engineering*, 24(1). https://doi.org/10.1615/JWomenMinorScienEng.2017018836.

Long, E. (2016, March). Report of the APS Ad-Hoc Committee on LGBT Issues – Presentation of Findings and Recommendations. In *APS April Meeting Abstracts* (Vol. 2016, pp. C13-001). www.aps.org/programs/lgbt/upload/LGBTClimateinPhysicsReport.pdf.

Mackay, A.W., Adger, D., Bond, A.L., Giles, S., and Ochu, E. (2019). Straight-washing ecological legacies. *Nature Ecology and Evolution*, 3(12), 1611. https://doi.org/10.1038/s41559-019-1025-9.

McCormick, M., and Barthelemy, R.S. (2021). Excluded from "inclusive" communities: LGBTQ youths' perception of "their" community. *Journal of Gay and Lesbian Social Services*, 33(1), 103–22. https://doi.org/10.1080/10538720.2020.1850386.

O'Connell, L.M., Atlas, J.G., Saunders, A.L., and Philbrick, R. (2010). Perceptions of rural school staff regarding sexual minority students. *Journal of LGBT Youth*, 7(4), 293–309.

O'Leary, E.S., Shapiro, C., Toma, S., Sayson, H.W., Levis-Fitzgerald, M., Johnson, T., and Sork, V.L. (2020). Creating inclusive classrooms by engaging STEM faculty in culturally responsive teaching workshops. *International Journal of STEM Education*, 7(1), 1–15. https://doi.org/10.1186/s40594-020-00230-7.

Parmenter, J.G., Galliher, R.V., and Maughan, A.D. (2020). LGBTQ+ emerging adults perceptions of discrimination and exclusion within the LGBTQ+ community. *Psychology and Sexuality*, 1–16. https://doi.org/10.1080/19419899.2020.1716056.

Patridge, E.V., Barthelemy, R., and Rankin, S.R. (2014). Factors impacting the academic climate for LGBQ STEM faculty. *Journal of Women and Minorities in Science and Engineering*, 20(1). https://doi.org/10.1615/JWomenMinorScienEng.2014007429.

Pinto, A.D., Aratangy, T., Abramovich, A., Devotta, K., Nisenbaum, R., Wang, R., and Kiran, T. (2019). Routine collection of sexual orientation and gender identity data: a mixed-methods study. *Cmaj*, 191(3), E63-E68. https://doi.org/10.1503/cmaj.180839.

Post, A., Stephens, A., and Blake, V. (2020). Sex discrimination in healthcare: Section 1557 and LGBTQ rights after Bostock. *Calif. L. Rev. Online*, 11, 545.

Randall L. Sell, "Challenges and solutions to collecting sexual orientation and gender identity data." *American Journal of Public Health*, 107(8) (August 1, 2017), 1212–14. https://doi.org/10.2105/AJPH.2017.303917.

Rankin, S., Garvey, J.C., and Duran, A. (2019). A retrospective of LGBT issues on US college campuses: 1990–2020. *International Sociology*, 34(4), 435–54. https://doi.org/10.1177/0268580919851429.

Rankin, S., Weber, G. Blumenfeld, W., and Frazer, S. (2010). *State of Higher Education for Lesbian, Gay, Bisexual, and Transgender People*. Charlotte, NC: Campus Pride.

Ross, A.S. (2019). *"Don't Say Gay. We Say Dumb or Stupid": Queering Prospective Mathematics Teachers' Discussions*. Brigham Young University.

Stallings, D., Iyer, S., and Hernandez, R. (2018). National Diversity Equity Workshops: Advancing Diversity in Academia. In *National Diversity Equity Workshops in Chemical Sciences* (2011–2017) (pp. 1–19). American Chemical Society. https://doi.org/10.1021/bk-2018-1277.ch001.

Thomas, S., and Wilcox, C. (Eds.). (2014). *Women and Elective Office: Past, Present, and Future*. Oxford University Press. https://doi.org/10.1093/acprof:oso/9780199328734.001.0001.

Trauth, E.M., and Connolly, R. (2021). Investigating the nature of change in factors affecting gender equity in the IT sector: A longitudinal study of women in Ireland. *MIS Quarterly*, 45(4), 2055–100.

Trenshaw, K.F., Hetrick, A., Oswald, R.F., Vostral, S.L., and Loui, M.C. (2013, October). Lesbian, gay, bisexual, and transgender students in engineering: Climate and perceptions. In *2013 IEEE Frontiers in Education Conference (FIE)* (pp. 1238–40). IEEE. https://doi.org/10.1109/FIE.2013.6685028.

Vaccaro, A., Russell, E.A., and Koob, R.M. (2015). Students with minoritized identities of sexuality and gender in campus contexts: An emergent model. *New Directions for Student Services*, 2015(152), 25–39. https://doi.org/10.1002/ss.20143.

van der Toorn, J., Pliskin, R., and Morgenroth, T. (2020). Not quite over the rainbow: The unrelenting and insidious nature of heteronormative ideology. *Current Opinion in Behavioral Sciences*, 34, 160–65. https://doi.org/10.1016/j.cobeha.2020.03.001.

Walker, C. (2014). Equality: Standing out. *Nature*, 505(7482), 249–51. https://doi.org/10.1038/nj7482 -249a.

Wallace, M., and Hofstadter, R. (2012). *American Violence*. United States: Knopf Doubleday Publishing Group.

Wilkins, D.E. (2016). A history of federal Indian policy. Talbot, Lobo, and Carlston (Eds.) *Native American Voices* (Vol. III). United Kingdom: Taylor and Francis.

Worthen, M.G. (2018). "Gay equals White"? Racial, ethnic, and sexual identities and attitudes toward LGBT individuals among college students at a Bible Belt university. *The Journal of Sex Research*, 55(8), 995–1011. https://doi.org/10.1080/00224499.2017.1378309.

Yarbrough, D.G. (2004). Gay adolescents in rural areas: Experiences and coping strategies. *Journal of Human Behavior in the Social Environment*, 8(2-3), 129–44. https://doi.org/10.1300/J137v08n02_08.

Yoder, J.B., and Mattheis, A. (2016). Queer in STEM: Workplace experiences reported in a national survey of LGBTQA individuals in science, technology, engineering, and mathematics careers. *Journal of Homosexuality*, 63(1), 1–27. https://doi.org/10.1080/00918369.2015.1078632.

APPENDIX: CHRONOLOGICAL SELECTION OF STEM AND RELATED RESOURCES FOR LGBTQA+ PEOPLE

1. Alias: GLMA
   Name: Health Professionals Advancing LGBTQ Equality
   Website: glma.org
   Incorporated: 1981

2. Alias: NOGLSTP (now renamed Out to Innovate)
   Name: National Org of Gay and Lesbian Scientists and Technical Professionals
   Website: noglstp.org
   Incorporated: 1991

3. Alias: GLADD
   Name: The Association of LGBTQ+ Doctors and Dentists
   Website: gladd.co.uk
   Incorporated: 1995

4. Alias: ALMA
   Name: The Australian Lesbian Medical Association
   Website: www.almas.org.au
   Incorporated: 1999

5. Name: Campus Pride
   Website: campuspride.org
   Incorporated: 2001

6. Name: Out for Work (inactive)
   Website: outforwork.org
   Incorporated: 2004

7. Name: Out for Undergrad
   Website: outforundergrad.org
   Incorporated: 2004

8. Alias: oSTEM
   Name: Out in Science, Technology, Engineering and Mathematics
   Website: ostem.org
   Incorporated: 2009

9. Name: StartOut
   Website: startout.org
   Incorporated: 2009

10. Name: LGBT+ Physicists
    Website: lgbtphysicists.org
    Incorporated: 2009

11. Name: Lesbians Who Tech
     Website: lesbianswhotech.org
     Incorporated: 2012

12. Name: Out in Tech
     Website: outintech.com
     Incorporated: 2012

13. Alias: LGBT Tech
     Name: LGBT Technology Partnership and Institute
     Website: lgbttech.org
     Incorporated: 2012

14. Name: EngiQueers
     Website: engiqueers.ca
     Incorporated: 2013

15. Name: Trans Tech Social
     Website: transtechsocial.org
     Incorporated: 2014

16. Name: InterEngineering
     Website: interengineeringlgbt.com
     Incorporated: 2014

17. Name: Spectra
     Website: lgbtmath.org
     Incorporated: 2015

18. Name: Pride in STEM
     Website: prideinstem.org
     Incorporated: 2016

19. Name: LGBTQ+ STEM
     Website: lgbtstem.wordpress.com
     Incorporated: 2016

20. Name: House of STEM
     Website: houseofstem.org
     Incorporated: 2017

21. Name: 500 Queer Scientists
     Website: 500queerscientists.com
     Incorporated: 2018

22. Name: Queers in Science
     Website: queersinscience.org.au
     Incorporated: 2018

23. Name: LGBTQ+ STEM Berlin
    Website: lgbtqstemberlin.de
    Incorporated: 2019

24. Name: The STEM Village
    Website: thestemvillage.com
    Incorporated: 2019

25. Name: THRIVE Lifeline
    Website: thrivelifeline.org
    Incorporated: 2020

14. The chains that bind: gender, disability, race, and IT accommodations

Eleanor T. Loiacono and Shiya Cao

INTRODUCTION

Around 15 percent of the world's population, or roughly one billion people, live with a disability (International Labour Organization, 2020). About 80 percent are of working age. We want to acknowledge and recognize at the onset of this chapter that there is still a debate in the disability community regarding identity-first (where the disability is recognized before the person (e.g., disabled person) versus person-first language (where the person is recognized before the disability (e.g., person with a disability) (Ferrigon and Tucker, 2019). There are valid arguments on both sides. In this chapter, we use *identity*-first language and will continue to learn from research and seek more guidance from self-advocates to inform our work and the language we use. We hope readers will receive this work as it is intended, to support, celebrate, and empower the disability community.

Diversity includes people with varying abilities as well as different genders and race (Loiacono and Ren, 2018; Nkomo, Bell, Roberts, Joshi, and Thatcher, 2019). A diverse and inclusive workforce can lead to numerous organizational benefits, including economic, legal, and cultural ones (Dong, Oire, MacDonald-Wilson, and Fabian, 2013; Kaye, Jans, and Jones, 2011). For example, diversity adds differing perspectives, which can help organizations create more innovative products that attract a wider consumer base. Additionally, heterogeneity of lived experiences helps organizations see where their business may be inaccessible and susceptible to lawsuits. Finally, there is a moral imperative for organizations to include disabled people into the workplace. Organizations that hire and support disabled workers often increase employee morale. However, the employment rate of disabled people remains low (19.3 percent, compared to 66.3 percent of those without disabilities) (Bureau of Labor Statistics of United States Department of Labor, 2019). In particular, disabled women may face greater hardship than disabled men in employment because they face a triple bind (Moodley and Graham, 2015). Typically, women face a "double bind" – incompatible gender stereotypes in which they are disliked if they are assertive and take charge in the workplace. However, if they are more nurturing and caring, they are seen as less capable. So, women can be seen as competent or likeable, but not both. These often-contradicting messages can cause stress and feelings of failure in women.

Disabled women face even greater hardships (triple bind), because they face stigmas related to their disability as well. Even when employed, disabled women are likely to work part-time at lower skilled jobs (Lindstrom, Harwick, Poppen, and Doren, 2012), earn less than men with and without disabilities, as well as women without disabilities (Doren, Gau, and Lindstrom, 2011; Emmett and Alant, 2006), and receive only 82.5 percent of what disabled men are paid (National Women's Law Center, 2014). They are treated worse in the workforce than women without disabilities (Doren et al., 2011; Emmett and Alant, 2006; Hanna and Rogowsky; 1991;

Lindstrom et al., 2012; Moloney, Brown, Ciciurkaite, and Foley, 2019; National Women's Law Center, 2014). Being a disabled woman of color may create a quadruple disadvantage for some, since some companies feel there are fewer incentives to hiring a disabled woman of color compared to a man or non-disabled female or white individual (Hanna and Rogowsky, 1991; Moloney et al., 2019).

In order to equally include disabled people in the workforce, accommodations are often necessary. Accommodations can be considered an organizational diversity intervention since an accommodation is designed to help disabled employees perform effectively and enjoy work benefits as others (Annabi and Lebovitz, 2018; Annabi and Locke, 2019). Today, many accommodations include an information technology component, such as screen reading software, touch screens, and accessible electronic forms. According to the Equal Employment Opportunity Commission (EEOC) and Title I of the Americans with Disabilities Act (ADA), accommodations need to be requested, negotiated, implemented, and monitored (Job Accommodation Network, 2018; Kofi Charles, 2004). During the process of receiving an Information Technology (IT) accommodation, disabled women may face more challenges because the IT accommodation may be interpreted by some as "special treatment." Besides needing childcare and transportation resources, disabled women, "often seek workplace accommodations and rehabilitation services" (Timmons et al., 2011). But the typical expectation of women is that they perform better than their male counterparts and are rewarded less than they are rewarded (Catalyst, 2007).

These intersectional issues hinder disabled women's employment. However, very little attention has focused on the intersection of gender, disability, and race relative to IT accommodations and employment in the Information Systems (IS) literature (Moodley and Graham, 2015; Nelson and Probst, 2010; Traustadottir and Harris, 1997; Zanoni, 2011). This chapter aims to explore these individual identities and other individual differences relevant to IT accommodations and employment. More specifically, we investigate individuals' experiences and differences in receiving an IT accommodation that helps them integrate into the workplace. The goal of this chapter is to seek a better understanding of individual differences in the accommodation process and how to empower disabled women in the workplace. To do so, by applying the individual differences theory of gender and IT (IDTGIT) (Trauth and Connolly, 2021), we focus on the experiences disabled men and women have with regard to IT accommodations they have asked for in their jobs. We also delve into the interplay of gender, disability, and race in relation to IT accommodations.

By offering insights into the intersectionality of gender, disability, and race relative to IT accommodations and employment, this chapter addresses a significant gap in the IS literature. It focuses on the individual differences in the processes of receiving IT accommodations. This topic is particularly interesting because for disabled women, requesting an IT accommodation can be an even greater concern, since the accommodation may be interpreted by some as "special treatment."

The chapter proceeds as follows. First, we provide the background and literature review in gender, disability, race, intersectionality, individual influences, and IT accommodation as an organizational diversity intervention. Second, we present the research methodology grounded in a critical lens. Third, we discuss our findings of individual differences in the process of receiving IT accommodations. Last, we present the theoretical contributions and the practical implications of this research.

BACKGROUND

This section provides the background and literature review in gender, disability, race, intersectionality, individual influences, and IT accommodation as an organizational diversity intervention. We use Trauth and Connolly's (2021) IDTGIT as a framework to present the relevant literature because it helps us better understand why these bodies of work are relevant and how they address the various dimensions. The IDTGIT framework (Trauth and Connolly, 2021) supports an intersectional approach by highlighting that organizations must account for the diversity and variations among women and move away from a "one size fits all" approach to engaging women in the workforce. Using the three constructs of the IDTGIT framework, the themes of interest in this chapter include: (1) individual identity (gender, disability, race, and intersectionality); (2) individual influences (abilities, past experiences, and supervisors); and (3) environmental influences (IT accommodation as an organizational diversity intervention).

Individual Identity: Gender

Gender has been defined as a "social practice" (Bruni et al., 2004a; 2004b) which is constructed in daily interactions (Acker, 1992). More specifically, it "… is understood as the expected or perceived differences between women and men which are socially constructed, reflecting assumptions about principles, values, and attributes associated with femininity and masculinity" (Alvesson and Billing, 2009; Williams and Patterson, 2019).

Within IT, the role that gender plays in one's IT career, creating opportunities and barriers, has received more attention in recent years. As IT professionals are in high-demand, women have great opportunities to join and advance in the IT industry (Annabi and Lebovitz, 2018). However, the number of women working in IT continues to drop (Annabi and Lebovitz, 2018; Goulden et al., 2011). Ahuja (1995, 2002) and Armstrong et al. (2018) looked at the barriers for women to enter and advance in the IT workforce. Social (e.g., social expectations and work–family conflict) and structural factors (e.g., lack of role models and mentors) were identified which may act as barriers to women entering and advancing within IT. In addition, imposter syndrome[1] remains prevalent among women in IT (Annabi and Lebovitz, 2018; Tapia and Kvasny, 2004). One of the reasons why women doubt their competencies for IT jobs is that they do not see many representations who look like them and have succeed in the IT industry (Tapia and Kvasny, 2004). This creates a negative feedback loop, which causes women to remain underrepresented in the IT industry (Armstrong et al., 2018; Goulden et al., 2011; Ramsey and McCorduck, 2005).

Several recent papers have looked at women in IT academics in particular (Loiacano et al., 2016; Loiacono et al., 2013; Williams et al., 2006). For many women in IT academics (39 percent), their career choice was influenced by their gender and the fact that they faced more challenges than their male counterparts. Women respondents also had a significantly more optimistic outlook for men in the field than they did for women (Loiacano et al., 2016). This is likely related to the double bind women are subject to in the workplace. On the one hand, they are confronted with getting recognition for their work, but on the other hand, they do not want to draw too much attention to themselves lest someone think they require special treatment (Williams et al., 2006). Moreover, a recent study that focused on how gender shapes the careers of women and men in the IS academy in relation to their employing institutions and the Association for Information Systems (AIS) found some interesting insights into what women

IT academics face in their jobs (Gupta et al., 2019). Women had lower levels of job satisfaction and felt less valued and supported by their universities. They perceived greater inequality at both their university and at the association levels. They experienced fewer opportunities for advancement at the association level and greater sexual harassment on the job than their male counterparts as well.

Individual Identity: Disability

Disability is defined by the Americans with Disability Act (ADA)[2] as an individual with "(A) a physical or mental impairment that substantially limits one or more major life activities of such individual; (B) a record of such an impairment; or (C) being regarded as having such an impairment"[3] (Americans with Disabilities Act, 1990). This definition takes a medical model perspective, which highlights a person's limits either physically or mentally. However, there is growing support for the social model perspective, whereby, disability is not perceived as an impairment, but rather a normal variation (Gray, 2009).

In today's workforce, the employment rate of disabled people remains low (19.3 percent, compared to 66.3 percent of those without disabilities) (Bureau of Labor Statistics of United States Department of Labor, 2019). There are multiple major reasons why disabled people are less likely to be hired, including a lack of awareness of disability and relevant responsibilities by organizations, employers' stereotypes and biases towards disabled people, employers' concerns about coworkers' negative reactions, fear of discrimination by disabled individuals, and a lack of work-related training for disabled people (von Schrader, Malzer, and Bruyère, 2014). Even when disabled people enter the workforce, they still face these challenges, which may also impact their decisions to disclose their disabilities and request accommodations.

Much work is needed to reduce these challenges. First, we need to understand and recognize that these barriers and challenges are strongly related to the medical model perspective, which reinforces *ableism*. Recently *ableism*, the dominant attitude in society that devalues disabled people, has entered the lexicon of disability studies. Specifically, it points to the "ideas, practices, institutions, and social relations that presume ablebodiedness and position disability as inability" (Williams and Mavin, 2012, p. 171; first mentioned by Chouinard (1997)). Thus, within the organizational context, ableism puts the burden of seeking work adjustments on the person with a disability and fails to challenge the assumption of the normative (non-disabled) body (Williams and Patterson, 2019). Contrarily, the social model perspective emphasizes the responsibility of the organization or society to enable disabled people to be "bodily present, acknowledged, accommodated and enabled … in organizational life" (Cockburn, 1991, p. 212). Based in the 1975 writings of the Union of Physically Impaired Against Segregation (UPIAS), the social model turns the definition of disability as a factor contained within an individual to one imposed by society. Specifically, the UPIAS claimed that, "… it is society which disables physically impaired people. Disability is something imposed on top of our impairments, by the way we are unnecessarily isolated and excluded from full participation in society. Disabled people are therefore an oppressed group in society" (UPIAS, 1975, p. 4). However, since the establishment of the ADA in 1990, organizations have had to abide by the medical model terminology defined in the Act itself. Unfortunately, this has led to conflict with those who feel the social model is a more accurate definition of disability. More and more researchers, on the other hand, are taking a social interpretation perspective and acknowledging that impairments are simply cognitive and physical variations from normative expectations and associated value judgements (Overboe, 1999; Thomas, 2007).

Second, disabled people need fulfillment and job stratification just like everyone else. People need to feel that they are part of something, contributing to something, and making a difference. Working brings better mental health for disabled individuals and can decrease the need for government support (Dong et al., 2013; Loiacono and Ren, 2018; Mavranezouli et al., 2014). Therefore, it is a moral imperative to provide equal employment opportunity to disabled people. Moreover, developing a diverse workforce benefits organizations (Janssens and Steyaert, 2019; Loiacono and Ren, 2018; Nkomo et al., 2019; Tasheva and Hillman, 2019). Studies have shown that employing disabled people can lead to economic benefits for organizations, such as an increased bottom line and product and service innovations through diverse perspectives and ideas (Accenture, 2018; Loiacono and Ren, 2018). These, in turn, can lead to opening up additional untapped markets (Loiacono and Ren, 2018).

Furthermore, in order to include disabled people in the workplace, accommodations are often necessary, if not required. According to the EEOC and Title I of the ADA, accommodations need to be requested, negotiated, implemented, and monitored (Job Accommodation Network, 2018; Kofi Charles, 2004). Through an effective accommodation process, disabled people can get the support they need, integrate into their work environments, and enjoy the same benefits of employment as others (Lindsay et al., 2018; Solstad et al., 2011; Vornholt et al., 2013). In addition, when disabled employees are provided with necessary accommodations that equitably include them in the workplace, organizations gain tangible and intangible benefits (Austin and Pisano, 2017). Additionally, job satisfaction, commitment, and performance of all employees are improved (Park et al., 2016). Despite the importance of accommodations, companies face many challenges with current processes to provide accommodations. These challenges include mismatches between increasing accommodation demands and limited resources to provide accommodations, little understanding of the accommodation process by supervisors and employees, a lack of a formal corporate accommodation process, and slow information flows among stakeholders in the process (USBLN, 2017).

Individual Identity: Race

Similar to gender and disability, race is the result of the social, historical, and political construction of racial categories (Tapia and Kvasny, 2004). Reducing the diverse voices of people of color ignores the unique challenges they face in the areas of recruitment, retention, and promotion in their IT career (Tapia and Kvasny, 2004). In particular, women of color may face different obstacles from men of color in the male-dominated IT industry.

Although there is still a paucity of research on race for women in IT, some studies have looked at this topic (Kvasny, 2003, 2006; Kvasny et al., 2009; Landivar, 2013; McGee, 2018; Morgan and Trauth, 2013; Tapia and Kvasny, 2004; Trauth et al., 2012, 2016; Windeler and Riemenschneider, 2016). These studies collectively showed that women of color faced barriers at almost every stage of the IT professional pipeline, from secondary education, higher education, to workplace (Windeler et al., 2016). A key reason for these barriers is that gender and racial minorities have limited access to IT resources compared with their majority counterparts (Windeler et al., 2016). Studies showed that women of color are less likely to take IT, science, or engineering-centered courses and they enroll and remain in these majors at much lower rates then men do (Kvasny, 2003, 2006; Landivar, 2013). Other studies found that women of color faced obstacles to advancing to senior IT executive roles influenced by factors such as individual's ability to change, established and gendered beliefs about leadership, as well as less access to mentors, sponsorship, and role models through traditional informal networking and promoting practices (McGee, 2018; Tapia and Kvasny, 2004).

Individual Identity: Intersectionality

Coined by Kimberlé Williams Crenshaw (1989), intersectionality initially referred to the oppression of Black women in society. It has since expanded to include many other social identities, including social class, age, religion, ethnicity, sexual identity, nationality, and ability (Bose, 2012; Trauth, 2013). Taking an intersectional approach allows one to better recognize and understand the reasons for certain instances of oppression in society, which in turn can allow for better policy development to address such inequities (Kitching, 2014; Shields, 2008).

This chapter focuses on the intersection of gender, disability, and race as well as its relevance to one's employment. The extant studies support the importance of moving away from conceptualizing women as a homogeneous group (Díaz García and Welter, 2011; Godwyn and Stoddard, 2011; Morris, 1993; Welter, 2011, 2012), who experience the world in the same way. It is critical to recognize that people experience the world differently based, not only on their gender, but other categories of social difference that they occupy (Holvino, 2010).

There are some studies integrating gender and disability (Annabi, 2018; Annabi and Locke, 2019; Ben-Moshe and Magaña, 2014; Garland Thomson, 2005; Rohrer, 2005; Sheldon, 2004; Thomas, 2006; Traustadóttir, 2006; Williams and Patterson, 2019). A growing number of feminist disability study researchers emphasize the need to include disability in business research as well (Williams and Patterson, 2019). Especially, in the United States, there is a need to study disability in relation to work because "Nowhere is the disabled figure more troubling to American ideology and history than in relation to the concept of work: the system of production and distribution of economic resources in which the abstract principles of self-government, self-determination, autonomy, and progress are manifest most completely" (Thomsom, 1997, 2017, p. 46). Consistent with this line, there is evidence that disabled women fare much poorer than disabled men overall. In terms of education, disabled women are at the lowest level. They are most likely to be unemployed and earn the lowest median income (Erevelles and Minear, 2010; Moodley and Graham, 2015).

To expand this intersectionality to include race, the *Women, Gender, and Families of Color Journal* published a special issue focusing on race, gender, and disability grounded in critical theories in 2014 (Ben-Moshe and Magaña, 2014). It recognized the importance of using a critical framework to study the intersectionality. As indicated in the preceding review, gender, disability, and race are all historically constructed. A critical approach takes the constructed nature of these identities as an entry point, puts them in a social, cultural, and historical context, and challenges the prevalent perspective (Ben-Moshe and Magaña, 2014).

Annabi et al. (2018, 2019) have studied women with autism working in IT. They proposed a framework of organizational interventions mitigating individual barriers framework (OIMIB) to understand the individual barriers and coping methods as well as organizational interventions that are used to mitigate these barriers. Through a critical analysis, these studies empower women with autism in the workplace.

Individual Influences

The IDTGIT framework (Trauth and Connolly, 2021) outlined individual influences in women's participation in the IT field, incorporating *abilities*, *past experiences*, and *supervisors*. These individual influences mainly explain within gender-group variation. Each is discussed in more detail below.

Abilities: Individual interests and abilities for IT topics affect one's engagement with IT. These abilities include technical knowledge and skills as well as "soft" abilities such as seeking help in the workplace, garner resources, and network with people (Annabi and Lebovitz, 2018; Armstrong et al., 2018).

Past experiences: One's past experiences at home, school, and work are very likely to impact decisions about how to interact with IT. These experiences could serve as motivations to overcome negative cultural messages or barriers to preventing people from moving forward (Trauth and Connolly, 2021).

Supervisors: As significant people in one's career, supervisors play an important role in individual choices related to work. Subordinates can reciprocate in the relationship with their supervisors by leveraging leader support to work on their well-being, job stratification, and performance (Annabi and Lebovitz, 2018; Windeler and Riemenschneider, 2016).

Environmental Influences: IT Accommodation as an Organizational Diversity Intervention

In this chapter, we pay particular attention to one environmental influence – IT accommodation as an organizational diversity intervention. Environmental influences refer to contextual factors in society and organizations relative to women's professions (Trauth and Connolly, 2021). Annabi et al. (2018, 2019) did a comprehensive review of existing studies in interventions and proposed a framework for investigating the interplay between organizational interventions and employment barriers. An intervention is defined as "an activity designed to change a state of affairs" (von Hellens, Trauth, and Fisher, 2012, p. 343). Annabi et al. (2018, 2019) identified intervention constructs such as social and structural interventions as well as intervention characteristics such as catalysts and objectives, methods and practices, as well as measurement and evaluation.

As described previously, accommodation is often necessary to equally include disabled people in the workplace. Today, many accommodations include an IT component, such as screen-reading software, touch screens, and accessible electronic forms. Therefore, this chapter focuses on IT accommodation as an organizational diversity intervention. The goal of IT accommodations is to empower disabled employees with the help they need so that they can perform effectively and enjoy work benefits in the same way as others. However, organizations are often concerned about costs of providing IT accommodations (Breen et al., 2019; Coole et al., 2013; Ekberg et al., 2016; Kaye et al., 2011; Khayatzadeh-Mahani et al., 2020; Kuznetsova and Yalcin, 2017; Yosef et al., 2019). When organizational interventions do not work, individual coping methods can mitigate employment barriers and improve individuals' accommodation experiences (Annabi and Lebovitz, 2018). Besides needing childcare and transportation resources, disabled women, "often seek workplace accommodations and rchabilitation services" (Timmons et al., 2011).

Table 14.1   *Description of the participants*

Alias	Gender	Type of disability	Job position
Abby	Female	Cognitive	Librarian
Ben	Male	Vision	Government staff
Beth	Female	Cognitive	Archivist
Bill	Male	Vision	IT office
Chloe	Female	Vision	Customer service
Donald	Male	Vision	IT specialist
Donna	Female	Cognitive	Manager
Ellen	Female	Vision	Government staff
Fran	Female	Mobility	Archivist
Grace	Female	Vision	Consultant
Hope	Female	Hearing	Project manager
Jane	Female	Vision	Teacher
Joe	Male	Vision	IT support technician
Kate	Female	Vision	Attorney
Kevin	Male	Mobility	Adjunct professor
Laura	Female	Vision	IT specialist
Mary	Female	Mobility	IT software analyst
Mike	Male	Hearing	Architect
Monica	Female	Hearing	Archives director
Rachel	Female	Hearing	University staff
Rich	Male	Cognitive	Manager
Sandra	Female	Cognitive	Instructor
Thomas	Male	Vision	Dispatcher

Methodology

To investigate how gender, disability, and race impact IT accommodations and employment, we used a critical lens to guide our study design and method. The critical lens fits the social model perspective of disability and accommodation process because it emphasizes individual experiences of disabled people as well as the responsibility of the organization and society to enable disabled people (Kvasny and Richardson, 2006). We followed the suggestions on how to conduct critical fieldwork made by Trauth and Howcroft (2006). First, we designed a semi-structured interview protocol reviewed by experts in the disability and accommodation field. Overall, the interview questions were about participants' experiences in the workplace accommodation process and IT accommodations that help with their work. The interview protocol was approved by the Institutional Review Board (IRB) prior to the field study.

Second, we sent the recruitment emails to Facebook and LinkedIn disability groups, and other disability organizations in North America.[4] If an individual responded to the email within the recruitment information and indicated interest in participating in an interview, we contacted him/her to verify an interview date and time as well as sent him/her an IRB approved consent form. A total of 23 disabled people (8 males and 15 females), who were employed and had sought an accommodation were interviewed (see Table 14.1. Description of the Participants). The disabilities of the interviewees varied from vision impairments (11), cognitive impairments (5), hearing impairments (4), and mobility impairments (3). Participants who completed the interviews were given a $10 Amazon gift card.

Third, during the interviews,[5] we encouraged participants to critically think about their accommodation experiences (Trauth and Howcroft, 2006). In the meantime, we assured participants that we were not looking for a definite, "correct" answer (Trauth and Howcroft, 2006). Moreover, we adopted an iterative approach to include critical questions that arose from the first few interviews in the interview protocol.

Further, we conducted content analysis to understand individual experiences and differences in the process of receiving IT accommodations, which are appropriate to acquire a rich understanding of this unexplored area (Yin, 2003). The themes of the findings fit with the IDTGIT framework (Trauth and Connolly, 2021), so we use this framework to present our results including individual identity, individual influences, and environment influences. Among these, the quadruple bind disabled women of color faced, a special budget to support IT accommodations, and individual coping methods to mitigate the effect of IT accommodations are critical to understand the intersectional barriers and potential solutions for disabled women in the workplace.

INDIVIDUAL DIFFERENCES IN THE PROCESS OF RECEIVING IT ACCOMMODATIONS

Several inductive content analyses were conducted to better understand the factors that affect people's experience with their IT accommodations and individual differences during such experience. The first was an examination of the words participants used during their interviews in order to understand participants' experiences with their IT accommodations and the processes that provide them with these accommodations. Using the word cloud function in NVivo (version 12.6.0.959), we compared the top ten words used in the interview by men and women. We discovered that for the most part men and women used the similar terminology to discuss their disabilities and the accommodations they received. Men and women used the term *need* with 524 (2.0 percent) and 724 (1.8 percent) of mentions accordingly and used the term *people* with 433 (1.7 percent) and 721 (1.7 percent) of mentions accordingly. The main reason why these two terms are significant may be because the accommodation process is a social process that emphasizes individualized need and interaction with people. *Job*, *process*, and *help* were also used relatively similarly by both groups. However, though most of the frequently stated words were similar for both groups, women were more likely to actually use the words *disability* and *ask* (see Figure 14.1) in their description of the accommodation process. Men were not as likely to use them during their interview (see Figure 14.2). They were more likely to mention technology, such as *software*, and *email* (see Figure 14.2). And, though men and women both mentioned email as a technology they use in the accommodation process, men were clearly focused on the technology terminology (see Figures 14.1 and 14.2). The reason behind this may be partly because three male participants (over one-third of all the male participants) who had IT jobs tended to talk much more about IT solutions than two female participants (two-fifteenths of all the female participants) working in IT and other female participants.

We also compared the top ten words used in the interview by participants who had different disability types. We found that *need, people, job, process, help* were still among the most frequently mentioned words. We observed that participants tended to mention a lot about specific disability terminologies related to their disabilities. For example, for participants

Figure 14.1   Ten most frequent words by female participants

Figure 14.2   Ten most frequent words by male participants

who had hearing disabilities, interpreter, deaf, hearing, sign, and language were ranked 2nd, 4th, 5th, 6th, and 7th most frequently mentioned words. Across disability types, participants who had vision disabilities were more likely to mention technology, such as *email*, *technology*, and *computer*. The reason behind this may be because (1) people who have vision disabilities often use more technology-related accommodations, such as screen reader, electronically formatted material; (2) the three male participants working in IT who talked more about IT solutions all had vision disabilities. It is worth noting that participants who had cognitive disabilities used the term *job* the most (1.3 percent), possibly because people who have cognitive disabilities face more challenges in finding a job. These variations across disability types give us implications that we need to accommodate disabled people to meet their individual needs.

The findings, below, are presented using the IDTGIT framework (Trauth and Connolly, 2021) to understand the issues that influenced participants' experiences in their accommodation process. The goal is to better understand the nuances of how participants experienced the process of receiving an IT accommodation.

Individual Identity

Individual identity factors that came out in the interviews were related to gender, type of disability, race, and, intersectionality. Among these factors, the different types of disability and the intersectionality of gender, disability, and race seem to be impactful in individuals' experiences in the accommodation process.

Although gender differences in individuals' accommodation experiences did not stand out in our data, it is worth noting that three male participants (all had vision disability) who had IT jobs appeared to have much more knowledge in IT accommodations than two female participants (one had vision disability, the other had mobile disability) working in IT and other female and male participants. For instance, Bill was confident that he knew about IT accommodations,

> You have to be able to present and demonstrate why you should get a certain thing and people generally trust that, especially in my role that I know what I'm talking about. (Bill)

Many times, the type of disability can impact how a disabled person is required to act. As Jane[6] points out,

> I know people forget [I'm low vision], especially with my disability. Because they forget I'm disabled at all, because they don't see it. I don't walk with the white cane." This often requires reminding supervisors and others of the need for accommodations and figuring out how to work around one's disability at times. (Jane)

Often times, as Sandra points out, it is necessary to reiterate, especially for those with learning disabilities that,

> [These types of disabilities] are not something you outgrow. A lot of people who have these types of disabilities have grown up potentially if they were diagnosed. At school, they had a lot of resources available to them that when they go into the workspace, they no longer exist. And it's also the first time they have to ask and advocate for themselves, I mean something like we did in college to some degree. But I've often said this is kind of like flying with broken wings and nobody's there to fix them because you don't know which way to fly, and you don't know how to fix your wings and you don't know what to do and there's not a lot of resources or information about how to handle these types of situations in the workplace and it's beyond frustrating to say the least. I'd also say that the ADA is available for any kind of work discrimination scenario, but often people do not disclose their disability at all. And they feel that they will be seen as stupid or less than. The more a company or organization can provide these resources to everyone, the better off it is for the employee. (Sandra)

One participant Abby compared it to pregnancy,

> … when you're in the early stages of pregnancy, like no one really would know, but if they closely observe or like they kind of figure something that is up as you kind of go along and your interactions

that they realized, oh, there's something a little bit different, but they might not know it's a disability really. (Abby)

Also relevant to the individual identity factors mentioned was race, which was combined with gender and disability, suggesting a quadruple bind may actually exist in the workplace. As mentioned in previous literature, experiencing double bind, women face greater hardship than men (Moodley and Graham, 2015). They are expected to perform better than their male colleagues and receive less reward for doing so (Catalyst, 2007). Our empirical data suggests that the quadruple bind may impact the ability of disabled women of color to maneuver the accommodation process. As Monica pointed out that, she has,

> … always been concerned about what they're going to think that I'm [race and gender intersectionality]. (Monica)

And Monica used to be afraid of backlash,

> … because I have always been concerned about what they're going to think that I'm [race and gender intersectionality] … (Monica)

Because of such difficulty and worry, people who face the quadruple bind may experience more mental health issues when they request an accommodation. In order to deal with these issues, Monica has been,

> … working with a therapist. She will help me figure this sort of thing out. A lot of what I've discovered is that because I'm [disabled] I tend to fall back on that model minority kind of thing where I am overly helpful, I am overly nice, I am overly understanding. She has been working with me to say that saying "no" or saying "I will not" is not mean or passive aggressive. It's the way it needs to be. (Monica)

Individual Influences

Participants vocalized the impact that individual influences, such as their abilities, past experiences, and supervisors had on them. A participant's ability to be self-directed is an individual characteristic that influences how they approach the accommodation process. For example, Joe pointed out, that,

> Well, I'm pretty much self-directed. Various webinars, online training, podcast, email list, email groups. There's a few other AT [Assistive Technology] Specialists at work force that is in the same company that I might consult just to kind of get a second opinion. (Joe)

Many acknowledge that the ability to know and articulate your needed accommodations also plays a part in whether and how disabled workers obtain them. Several participants voiced this factor in different ways. For example, Ellen shared that,

> I guess that depends on how well you explain what you need. If you can give a good explanation for why you need it, if you've written out an email in detail for why you need it, or if somebody makes a point and you're able to make a counter point, then you're going to get what you need. But no matter what technology you have, if you can't explain why you need something, then you're probably not going to get it. (Ellen)

Donald stated,

> But I mean generally if somebody needs to make an accommodation request, you make it yourself because you should know what you need … (Donald)

Similarly, the ability to advocate for oneself is a critical individual influence, even though it can be quite intimidating, that can impact one's accommodations and overall experience. For example, Sandra revealed that,

> I've always been a very good advocate for myself. But there's that intimidation factor. It's my first job and I don't know what I can ask him, what I can say. Again, I was open with the fact that I was [disabled]. But every time I requested anything that would help me, I was always told no, partially why I left that job, to be honest. (Sandra)

Monica highlights how she has grappled with self-advocacy over time,

> [R]ight out of the gate, you need to advocate for yourself as opposed to those of us who learn to not advocate for something. It's weird I've been disabled my whole life, but I've only learned to advocate for myself in the last six or seven years. And I was even part of a group that wrote the original disability and accessibility standards for [a disability organization]. And we were so focused on [some types of disabilities] that only forgot to include myself in these accommodations. And when they updated them just a couple years ago, they said why aren't [certain type of disability] included, aren't you [that type]? I go yeah boy, that point we were so focused on just making sure that people [with some types of disabilities] had access to things that we didn't even think about that. And now people get to think about everything at once instead of being so focused on just getting the basics. And we can now think about [other types of disabilities] and the various spectrums of things … (Monica)

Moreover, experiences with past disclosure influences disabled people's choices to disclose in the first place. Jane took a negative view on disclosing upfront at interviews because she had not experienced a "good" disclosure,

> I never used to disclose [during the interview] until it got to a point where they asked me specifically or I got the job in that I had to be like, by the way, I have to tell you about my disability … which every job I've ever had a sort of pull back that information because I know it might be the thing that puts me down on their list … They asked me about my disability and then I tell them and then I get a call back the next few days later, saying like we found someone with more experience. And yet I know that's just sometimes it's true, sometimes I know it's not true. So I haven't had good experiences disclosing … I think I did it like once or twice … if they know, they'll say like well if you did get the job, what accommodations would you need? It seems to be a bigger concern to [them] than it is to me. (Jane)

However, Sandra felt very comfortable opening up about her disability since her first disclosure was understood well by her interview manager,

> I always speak openly about my disabilities from the point that I'm interviewing … the first person I interviewed with, she mentioned she appreciated that I said something because she also had [a disability] and understood those challenges. She said if I'd gotten further into the process, she would do what she could to help me. (Sandra)

Often participants are afraid to disclose their disability to coworkers, given their cumulative past experiences and angst,

Again, sometimes I feel like if I tell my colleagues about my disability, they're going to start treating me different. That they're going to somehow get that idea that person with a disability can't possibly be as good at this job as somebody else. And it does impact like today … I was sort of not willing to tell this new [colleague] who I've never met before like this is going to be harder for me than you think. So it does impact my … like I could have said like you know what, it's going to be difficult for me to [do that] … So yeah, like I could have asked to be accommodated in that way, but I didn't because I just didn't want her to tell people that I was the kind of [professional] who won't do all the work that's asked of me. (Jane)

Experience at university or school is another significant individual influence on disabled people navigating the accommodation process. Many found it difficult to adjust to a world where accommodations were not fully supported. For example, Jane stated,

At the moment … I have to push for [accommodations]. If I want anything, I have to ask for it … I have a lot of accommodations during University. A lot … I had a letter of introduction from the University to tell my professors that I had a disability. So I had a lot of accommodation at University and then the workplace it's sort of I have to really ask for it and I provide a lot of it myself including financially … It's sort of like they're more than willing to help students out to get through University. But once you're in the workplace, you're on your own. (Jane)

In addition, the experience of disabled people asking for accommodations is affected by supervisors' attitudes and behaviors as well. One difficult supervisor situation was explained by Sandra,

I would say in the first job, it was to truly listen and understand that I wasn't bringing this up out of selfishness that it was bringing it up so that I could do my job more efficiently for them and for me and understand that that was also a work balance that was necessary, so listening. And even if he had said to me [participant's name], I hear you. I understand you. I know this can be challenging. I want to help you be a better employee for yourself but also for us and let me see what I can do. I think I would have felt at ease. (Sandra)

In particular, if others in the hierarchy (above the supervisor) do not understand disability, the accommodation process can get derailed. As Fran described,

The [boss's title] didn't think I needed one. She disagreed. She was like, well, I'm not going to order you one. We're done. We're done getting your stuff. This is essentially what happened, even though it caused problems in my [body]. (Fran)

Alternatively, other participants found their supervisors very receptive to their needs and willing to work with them to make necessary accommodations. For example, Thomas stated,

He's very responsive to my requests. In the past we had an issue with something that wasn't working right and he was very proactive in helping me get the situation resolved. (Thomas)

Similarly, Donald noted that there was,

Again, no impact to me because I have a great manager if I explained to her what I need and she thinks it's a good thing, she's going to try to do everything she can to make sure that I get it. (Donald)

Often times it helps to have a supervisor with direct connection to someone with a disability. As Abby illustrated,

> … I was working for a boss whose daughter had some disabilities. So she was a little bit more accommodating and understanding and knows kind of a little bit more what the process was … (Abby)

Environmental Influences

As stated in the literature review, we focus on one environmental influence – IT accommodation as an organizational diversity intervention. A special budget that supports IT accommodations and individual coping methods seems to be impactful in the improvement of individuals' accommodation experiences.

If an organization has a centralized source of funding to support IT accommodations it makes it easier for disabled individuals to get needed accommodations. This is made clear in Kate's comment,

> That program is a centralized source of accommodations … If you need adaptive equipment, my office did not have to pay out of the office budget for any accommodations that I got from [the program]. So that was just a matter of filling out a request form and then it would go through [the program]. And they would purchase the software or the hardware or if I needed training, I could request training to them. So that was pretty easy. (Kate)

Without a special budget, costs are a huge issue that stands often in the way between a disabled person's accommodation request and its fulfillment. This was an issue brought up by numerous participants and brings the most angst to both sides. As Kevin stated,

> The biggest thing I see is talk about the cost of it. But I also think some of these people haven't really looked into the cost. (Kevin)

Sandra added,

> It was a lot of conversations. And basically, I was told that financially it didn't make sense and that if there was a handful of people who wanted something that they could ask for it. (Sandra)

This is what causes many disabled people to support their work with their own personal technology. For example, Beth in an attempt to reduce costs for her employer shared that,

> I told my boss I could use my iPad. I knew because as a student I know what they have set up for disability support. I knew I should be able to get accommodations. That's why I did it. When I worked for [company name], they all were PC-based. The database I had to use were not web-based but PC based. And I always try to make it as least costly to my employer … And I said I could just bring my own Mac because I was there as a volunteer, internship. I always try to do it as least costly as possible. (Beth)

Similarly, Jane pointed out,

> … I have my technology. None of it has been provided by the [organization]. Even when I was under contract, I brought my own technology, but I did have some help setting it up with the [organization] through the IT department. But other than that, basically just what I have. (Jane)

In addition, issues regarding the inaccessibility of technology in general make disabled people's accommodation experiences frustrating and has a profound impact on their ability to be included in work and society. As Rachel indicated,

> It's just related with communication access, access with your family, access in the society and lack of captioning. For example, Facebook live videos, you pull that up and I don't know what they're saying. I've started my own Facebook page ... because I'm frustrated, I'm tired, I'm tired of not having full access to things like other hearing people have access to in the world. They can just sit down and scroll through Facebook and watch many different videos and they can understand everything. But I'm scrolling through and I'm like, where's the captioning? So that's very frustrating. (Rachel)

COVID has provided some additional issues that disabled people have had to deal with. There are still complications that have arisen and companies have missed opportunities to accommodate their disabled employees. Rachel described her accommodation experience during this challenging time,

> Current challenges with everything transitioning to the virtual world because of COVID, when we transitioned in March, from March and then moving forward until now, it's gotten better. But at the beginning of all this, I just noticed so many departments, professors, staff, many of them using or posting or emailing newsletters or videos or trainings with no closed caption. I'm an employee ... I want to know what's going on. I want to be involved in the training in the workshop. Especially because we're transitioning to a virtual world and the training and everything with the cameras, they show the videos, they show how to setup your own virtual desk and how to set up, whatever things are related with this transition. They show how to set up different things online. I can't watch the videos. There's no closed captioning. I don't know how to set up my stuff. I don't know how to do x y z and that has been a big issue ... if you send out something like a mass email with a video, but it's not accessible to everybody, then I just heard like for example, the IT department have IT print out letters, they have things that they show different things, and I'm like, okay, but it's still not the captain for the video. So I can't really follow along with these printouts that you're giving me. And then there was a workshop that was hosted by an outside organization and outside people and when the video came up, I signed it, I was ready to watch, and there was no captioning. There is the chat box and I sent a message saying you know, this workshop/training/webinar, does it have closed captioning? Does it have any transcript? That would be nice. And they were like, oh, no, I'm so sorry. We don't have closed captioning. We'll add captioning later and then we'll send it to you. Now I have to wait until after the workshop's finished instead of previously being involved with the workshop and having the captions right there. Everybody else has access. Why do I have to wait till later? If the college department, if they sent out anything videos or whatever, I would watch, if there is no captioning, if it was the captioning box, I would click it and it wouldn't work. It was there. They would have that out of there. But if you clicked it, it didn't work. (Rachel)

However, Monica has shown how necessity, such as COVID, has opened up companies to reevaluating costs and are willing to take on some of those that they once pushed back on,

> We tend to take on the ownership of responsibility on ourselves. And I know a lot of people who pay for things out of their own pocket. They pay for their own wheelchairs and things like that. And I'm like we shouldn't be doing that. And COVID has shown us that employers are willing to pay for things that they claimed they weren't willing to pay for before. They're willing to pay for us to have a Zoom account so that we can meet in a format that's easier for me than talking on the phone. That means six months ago when I asked for it, you could have done it, then you just didn't agree to. (Monica)

CONCLUSION

In this chapter, we delved into the factors surrounding gender, disability, and race relative to IT accommodations and employment. By applying the IDTGIT framework, we sought a better understanding of individual differences in disabled employees' accommodation experiences. We found that the quadruple bind for disabled women of color existed in the workplace. This quadruple bind may impact the ability of disabled women of color to maneuver the accommodation process. Moreover, as an organizational diversity intervention, IT accommodations need organizational resources support such as a special budget in order to effectively improve disabled employees' accommodation and employment experiences. When the intervention may not work, individual coping methods can help mitigate the barriers to accommodations and thus improve individuals' accommodation and employment experiences. We recommend that a special budget for accommodations is an effective organizational intervention, which mitigates supervisors' concerns over costs of providing accommodations. Moreover, since the experience of disabled people asking for accommodations is affected by supervisors' and coworkers' attitudes and behaviors, we recommend that providing related trainings to supervisors and coworkers is another effective organizational intervention. Sensitivity and disability awareness training on how to support disabled employees can be very helpful (Loiacono and Ren, 2018).

This chapter makes a strong contribution to the IS research and disability studies by extending the application of IDTGIT to the disability community with more intention and offering insights into organizational diversity interventions. This work is multidisciplinary highlighting the relationship between the intersectionality of gender, disability, and race and the IT accommodations intervention to provide a holistic understanding of workplace diversity and relevant interventions. This research also has significant implications for practice. By identifying the quadruple bind disabled women of color face, organizations can minimize these barriers through awareness education for all supervisors and employees. Another practical implication of this research is to inform organizations and individuals about how to use resources and individual coping methods respectively to support interventions.

NOTES

1. Clance and Imes (1978) defined the imposter phenomenon or syndrome as "an internal experience of intellectual phoniness that appears to be particularly prevalent and intense among a select sample of high achieving women" (p. 1).
2. This research focuses on the US law regarding disability. Though some similarities exist with other "western" countries' regulations regarding disabilities do differ across countries and cultures.
3. See www.ada.gov/pubs/adastatute08.htm.
4. The reason for limiting the participants to North America was to have some consistency in terms of legal expectations for accommodation. Both the United States (Americans with Disabilities Act) and Canada (Canadians with Disabilities Act) have similar work accommodation requirements for those with disabilities.
5. We feel it is important to note upfront that interviews were conducted during the COVID-19 pandemic. It was, therefore, impossible to interview participants in person. All interviews were conducted using Zoom online meeting software and when permitted recorded to ensure accuracy of notes taken and to produce a transcript for coding.
6. All participant names have been changed to pseudonyms.

REFERENCES

Accenture. (2018). *Getting to equal the disability inclusion advantage*. Retrieved from www.accenture .com/_acnmedia/pdf-89/accenture-disability-inclusion-research-report.pdf.

Acker, J. (1992). Gendering organizational theory. *Classics of Organizational Theory*, 6, 450–59.

Ahuja, M.K. (2002). Women in the information technology profession: A literature review, synthesis and research agenda. *European Journal of Information Systems*, 11(1), 20–34.

Ahuja, M.K. (1995). Information technology and the gender factor. In *Proceedings of the 1995 ACM SIGCPR Conference on Supporting Teams, Groups, and Learning Inside and Outside the IS Function Reinventing IS*.

Alvesson, M., and Billing, Y.D. (2009). *Understanding Gender and Organizations*. London, UK: Sage Publishing.

Americans with Disabilities Act (1990). Retrieved from www.ada.gov/pubs/adastatute08.htm.

Annabi, H. (2018). The untold story: The masked experiences of women with autism working in IT. *Twenty-fourth Americas Conference on Information Systems*, New Orleans.

Annabi, H., and Lebovitz, S. (2018). Improving the retention of women in the IT workforce: An investigation of gender diversity interventions in the USA. *Information Systems Journal*, 28(6), 1049–81.

Annabi, H., and Locke, J. (2019). A theoretical framework for investigating the context for creating employment success in information technology for individuals with autism. *Journal of Management and Organization*, 25(4), 499–515.

Armstrong, D.J., Riemenschneider, C.K., and Giddens, L.G. (2018). The advancement and persistence of women in the information technology profession: An extension of Ahuja's gendered theory of IT career stages. *Information Systems Journal*, 28(6), 1082–1124.

Austin, R.D., and Pisano, G.P. (2017). Neurodiversity as a competitive advantage. *Harvard Business Review*, 95(3), 96–103.

Ben-Moshe, L., and Magaña, S. (2014). An introduction to race, gender, and disability: Intersectionality, disability studies, and families of color. *Women, Gender, and Families of Color*, 2(2), 105–14.

Bose, C.E. (2012). Intersectionality and global gender inequality. *Gender and Society*, 26(1), 67–72.

Breen, J., Havaei, F., and Pitassi, C. (2019). Employer attitudes toward hiring persons with disabilities in Armenia. *Disability and Rehabilitation*, 41(18), 2135–42.

Bruni, A., Gherardi, S., and Poggio, B. (2004a). Doing gender, doing entrepreneurship: An ethnographic account of intertwined practices. *Gender, Work and Organization*, 11(4), 406–29.

Bruni, A., Gherardi, S., and Poggio, B. (2004b). Entrepreneur-mentality, gender and the study of women entrepreneurs. *Journal of Organisational Change Management*, 17(3), 256–68.

Catalyst. (2007). *The double-bind dilemma for women in leadership: Damned if you do, doomed if you don't.*

Chouinard, V. (1997). Making space for disabling differences: Challenging ableist geographies. *Environment and Planning D: Society and Space*, 15, 379–87.

Clance, P.R., and Imes, S.A. (1978). The imposter phenomenon in high achieving women: Dynamics and therapeutic intervention. *Psychotherapy: Theory, Research and Practice*, 15(3), 241–47.

Cockburn, C. (1991). *In the Way of Women: Men's Resistance to Sex Equality in Organizations*. Ithaca, NY: ILR Press.

Coole, C., Radford, K., Grant, M., and Terry, J. (2013). Returning to work after stroke: Perspectives of employer stakeholders, a qualitative study. *Journal of Occupational Rehabilitation*, 23(3), 406–18.

Crenshaw, K. (1989). Demarginalizing the intersection of race and sex: A Black feminist critique of antidiscrimination doctrine, feminist theory and antiracist politics. *U. Chi. Legal F.*, 139–67.

Diaz Garcia, M.C., and Welter, F. (2011). Female entrepreneurship: Does the conventional approach of study perpetuate certain myths? In *RENT Conference*.

Dong, S., Oire, S.N., MacDonald-Wilson, K.L., and Fabian, E.S. (2013). A comparison of perceptions of factors in the job accommodation process among employees with disabilities, employers, and service providers. *Rehabilitation Counseling Bulletin*, 56(3), 182–89.

Doren, B., Gau, J.M., and Lindstrom, L. (2011). The role of gender in the long-term employment outcomes of young adults with disabilities. *Journal of Vocational Rehabilitation*, 34(1), 35–42.

Ekberg, K., Pransky, G.S., Besen, E., Fassier, J.B., Feuerstein, M., Munir, F., and Blanck, P. (2016). New business structures creating organizational opportunities and challenges for work disability prevention. *Journal of Occupational Rehabilitation*, 26(4), 480–89.

Emmett, T., and Alant, E. (2006). Women and disability: Exploring the interface of multiple disadvantage. *Development Southern Africa*, 23(4), 445–60.

Erevelles, N., and Minear, A. (2010). Unspeakable offenses: Untangling race and disability in discourses of intersectionality. *Journal of Literary and Cultural Disability Studies*, 4(2), 127–46.

Ferrigon, P., and Tucker, K. (2019). Person-first language vs. identity-first language: An examination of the gains and drawbacks of disability language in society. *Journal of Teaching Disability Studies*.

Garland Thomson, R. (2005). Integrating disability. *Gendering Disability*, 73–103.

Godwyn, M., and Stoddard, D. (2011). *Sustainability, Responsibility, Accountability. Minority Women Entrepreneurs: How Outsider Status Can Lead to Better Business Practices*. Sheffield, UK: Greenleaf Publishing.

Goulden, M., Mason, M.A., and Frasch, K. (2011). Keeping women in the science pipeline. *The ANNALS of the American Academy of Political and Social Science*, 638(1), 141–62.

Gray, C. (2009). Narratives of disability and the movement from deficiency to difference. *Cultural Sociology*, 3(2), 317–32.

Gupta, B., Loiacono, E.T., Dutchak, I.G., and Thatcher, J.B. (2019). A field-based view on gender in the information systems discipline: Preliminary evidence and an agenda for change. *Journal of the Association for Information Systems*, 20(12), 1870–1900.

Hanna, William John, and Betsy Rogovsky. (1991). Women with disabilities: Two handicaps plus. *Disability, Handicap and Society*, 6(1), 49–63.

Hirschmann, N.J. (2012). Disability as a new frontier for feminist intersectionality research. *Politics and Gender*, 8(3), 396–405.

Holvino, E. (2010). Intersections: The simultaneity of race, gender and class in organization studies. *Gender, Work and Organization*, 17(3), 248–77.

International Labour Organization. (2020). *Disability and work*. Retrieved from www.ilo.org/global/topics/disability-and-work/WCMS_475650/lang--en/index.htm.

Janssens, M., and Steyaert, C. (2019). A practice-based theory of diversity: Respecifying (in) equality in organizations. *Academy of Management Review*, 44(3), 518–37.

Job Accommodation Network. (2018). *Accommodation and compliance series: Interactive process*.

Kaye, H.S., Jans, L.H., and Jones, E.C. (2011). Why don't employers hire and retain workers with disabilities? *Journal of Occupational Rehabilitation*, 21(4), 526–36.

Khayatzadeh-Mahani, A., Wittevrongel, K., Nicholas, D.B., and Zwicker, J.D. (2020). Prioritizing barriers and solutions to improve employment for persons with developmental disabilities. *Disability and Rehabilitation*, 42(19), 2696–706.

Kitching, J. (2014). *Entrepreneurship and self-employment by people with disabilities*. OECD.

Kofi Charles, K. (2004). The extent and effect of employer compliance with the accommodations mandates of the Americans with Disabilities Act. *Journal of Disability Policy Studies*, 15(2), 86–96.

Kuznetsova, Y., and Yalcin, B. (2017). Inclusion of persons with disabilities in mainstream employment: Is it really all about the money? A case study of four large companies in Norway and Sweden. *Disability and Society*, 32(2), 233–53.

Kvasny, L. (2006) Let the sisters speak: Understanding information technology from the standpoint of the "other." *DATA BASE for Advances in Information Systems*, 37, 13–25.

Kvasny, L. (2003) Triple jeopardy: Race, gender and class politics of women in technology. In *Proceedings of the SIGCPR Conference on Computer Personnel Research*, Philadelphia, PA.

Kvasny, L., and Richardson, H. (2006). Critical research in information systems: Looking forward, looking back. *Information Technology and People*, 19(3), 196–202.

Kvasny, L., Trauth, E.M., and Morgan, A.J. (2009) Power relations in IT education and work: The intersectionality of gender, race, and class. *Journal of Information, Communication and Ethics in Society*, 7, 96–118.

Landivar, L.C. (2013). Disparities in STEM employment by sex, race, and Hispanic origin. *Education Review*, 29(6), 911–22.

Lindsay, S., Cagliostro, E., Albarico, M., Mortaji, N., and Karon, L. (2018). A systematic review of the benefits of hiring people with disabilities. *Journal of Occupational Rehabilitation*, 28(4), 634–55.

Lindstrom, L., Harwick, R.M., Poppen, M., and Doren, B. (2012). Gender gaps: Career development for young women with disabilities. *Career Development and Transition for Exceptional Individuals*, 35(2), 108–17.

Loiacano, E.T., Iyer, L.S., Armstrong, D., Craig, A., and Beekhuyzen, J. (2016). AIS Women's Network: Advancing women in IS academia. *Communications of the Association for Information Systems*, 38(1), 784–802.

Loiacono, E.T., and Ren, H. (2018). Building a neurodiverse high-tech workforce. *MIS Quarterly Executive*, 17(4), 263–79.

Loiacono, E.T., Urquhart C., Beath, C., Craig, A., Thatcher, J., Vogel, D.R., and Zigurs, I. (2013). Thirty years and counting: Do we still need the ICIS women's breakfast? *Communications of the Association for Information Systems*, 33(1), 81–96.

Mavranezouli, I., Megnin-Viggars, O., Cheema, N., Howlin, P., Baron-Cohen, S., and Pilling, S. (2014). The cost-effectiveness of supported employment for adults with autism in the United Kingdom. *Autism*, 18(8), 975–84.

McGee, K. (2018). The influence of gender, and race/ethnicity on advancement in information technology (IT). *Information and Organization*, 28(1), 1–36.

Moloney, M.E., Brown, R.L., Ciciurkaite, G., and Foley, S.M. (2019). "Going the extra mile": disclosure, accommodation, and stigma management among working women with disabilities. *Deviant Behavior*, 40(8), 942–56.

Moodley, J., and Graham, L. (2015). The importance of intersectionality in disability and gender studies. *Agenda*, 29(2), 24–33.

Morgan, A.J., and Trauth, E.M. (2013) Socio-economic influences on health information searching in the USA: The case of diabetes. *Information Technology and People*, 26, 324–46.

Morris, J. (1993). *Independent Lives? Community Care and Disabled People*. Macmillan International Higher Education.

National Women's Law Center. (2014). Retrieved from https://nwlc.org/.

Nelson, N.L., and Probst, T. (2010). *Multiple Minority Individuals: Multiplying the Risk of Workplace Harassment and Discrimination*. The psychology of prejudice and discrimination: A revised and condensed edition, 97–111.

Nkomo, S.M., Bell, M.P., Roberts, L.M., Joshi, A., and Thatcher, S.M. (2019). Diversity at a critical juncture: New theories for a complex phenomenon. *Academy of Management Review*, 44(3), 498–517.

Overboe, J. (1999). "Difference in itself": Validating disabled people's lived experience. *Body and Society*, 5(4), 17–29.

Park, Y., Seo, D.G., Park, J., Bettini, E., and Smith, J. (2016). Predictors of job satisfaction among individuals with disabilities: An analysis of South Korea's National Survey of employment for the disabled. *Research in Developmental Disabilities*, 53, 198–212.

Ramsey, N., and McCorduck, P. (2005). *Where are the women in information technology*. Report of Literature Search and Interviews Prepared for the National Center for Women and Information Technology.

Rohrer, J. (2005). Toward a full-inclusion feminism: A feminist deployment of disability analysis. *Feminist Studies*, 31(1), 34–63.

Sheldon, A. (2004). Women and disability. *Disabling Barriers, Enabling Environments*, 69–74.

Shields, S.A. (2008). Gender: An intersectionality perspective. *Sex Roles*, 59(5), 301–11.

Solstad Vedeler, J., and Schreuer, N. (2011). Policy in action: Stories on the workplace accommodation process. *Journal of Disability Policy Studies*, 22(2), 95–105.

Tapia, A.H., and Kvasny, L. (2004, April). Recruitment is never enough: Retention of women and minorities in the IT workplace. In *Proceedings of the 2004 SIGMIS Conference on Computer Personnel Research: Careers, Culture, and Ethics in a Networked Environment*, 84–91.

Tasheva, S., and Hillman, A.J. (2019). Integrating diversity at different levels: Multilevel human capital, social capital, and demographic diversity and their implications for team effectiveness. *Academy of Management Review*, 44(4), 746–65.

Thomas, C. (2007). *Sociologies of Disability and Illness: Contested Ideas in Disability Studies and Medical Sociology*. Macmillan International Higher Education.

Thomas, C. (2006). Disability and gender: Reflections on theory and research. *Scandinavian Journal of Disability Research*, 8(2–3), 177–85.

Thompson, R.G. (2017). *Extraordinary Bodies: Figuring Physical Disability in American Literature and Culture*. New York: Columbia University Press, 11.

Timmons, J.C., Hall, A.C., Bose, J., Wolfe, A., and Winsor, J. (2011). Choosing employment: Factors that impact employment decisions for individuals with intellectual disability. *Intellectual and Developmental Disabilities*, 49(4), 285–99.

Traustadóttir, R. (2006). Disability and gender: Introduction to the special issue. *Scandinavian Journal of Disability Research*, 8(2–3), 81–4.

Traustadóttir, R., and Harris, P. (1997). *Women with Disabilities: Issues, Resources, Connections*. Revised.

Trauth, E.M. (2013). The role of theory in gender and information systems research. *Information and Organization*, 23(4), 277–93.

Trauth, E.M., Cain, C.C., Joshi, K.D., Kvasny, L., and Booth, K.M. (2016). The influence of gender-ethnic intersectionality on gender stereotypes about IT skills and knowledge. *ACM SIGMIS Database: The DATABASE for Advances in Information Systems*, 47(3), 9–39.

Trauth, E.M., Cain, C., Joshi, K.D., Kvasny, L., and Booth, K. (2012) Embracing intersectionality in gender and IT career choice research. In *Proceedings of the SIGMIS Conference on Computers and People Research*, Milwaukee, WI.

Trauth, E.M., and Connolly, R. (2021). Investigating the nature of change in factors affecting gender equity in the IT sector: A longitudinal study of women in Ireland. *MIS Quarterly*, 45(4), 2055–100.

Trauth, E.M., and Howcroft D. (2006). Critical empirical research in IS: An example of gender and the IT workforce. *Information Technology and People*, 19(3), 272–92.

Union of Physically Impaired Against Segregation. (1975). Retrieved from https://en.wikipedia.org/wiki/Union_of_the_Physically_Impaired_Against_Segregation.

USBLN. (2017). *The 2017 disability equality index*.

von Hellens, L., Trauth, E., and Fisher, J. (2012). Increasing the representation of women in the information technology professions: Research on interventions. *Information Systems Journal*, 22, 343–53.

von Schrader, S., Malzer, V., and Bruyère, S.M. (2014). Perspectives on disability disclosure: The importance of employer practices and workplace climate. *Employee Responsibilities and Rights Journal*, 26 (4), 237–55.

Vornholt, K., Uitdewilligen, S., and Nijhuis, F.J. (2013). Factors affecting the acceptance of people with disabilities at work: A literature review. *Journal of Occupational Rehabilitation*, 23(4), 463–75.

Welter, F. (2012). Entrepreneurship in context. *International Small Business Journal*, 30(6), 728–30.

Welter, F. (2011). Contextualizing entrepreneurship – Conceptual challenges and ways forward. *Entrepreneurship Theory and Practice*, 35(1), 165–84.

Williams, J.G., Higgins, J.P., and Brayne, C.E. (2006). Systematic review of prevalence studies of autism spectrum disorders. *Archives of Disease in Childhood*, 91(1), 8–15.

Williams, J., and Mavin, S. (2012). Disability as constructed difference: A literature review and research agenda for management and organization studies. *International Journal of Management Reviews*, 14(2), 159–79.

Williams, J., and Patterson, N. (2019). New directions for entrepreneurship through a gender and disability lens. *International Journal of Entrepreneurial Behavior and Research*.

Windeler, J.B., and Riemenschneider, C.K. (2016). The influence of ethnicity on organizational commitment and merit pay of IT workers: The role of leader support. *Information Systems Journal*, 26(2), 157–90.

Yin, R. (2003). *Application of Case Study Research* (2nd ed.), Thousand Oaks, CA: Sage Publications.

Yosef, L., Soffer, M., and Malul, M. (2019). From welfare to work and from work to welfare: A comparison of people with and without disabilities. *Journal of Disability Policy Studies*, 29(4), 226–34.

Zanoni, P. (2011). Diversity in the lean automobile factory: Doing class through gender, disability and age. *Organization*, 18(1), 105–27.

15. Gender and work–life balance in the IT field

Manju K. Ahuja

INTRODUCTION

When I was approached to write a chapter about gender in information technology (IT) for this book, I thought for a few days about which study I might write about. In doing so, I thought about the various studies I had done in this area and what occurred to me was that at this point, the journey related to research on this topic was as revealing as any specific study. Given this, my goal in this chapter is to reflect on my career journey regarding gender and IT research themes. I discuss in the following sections how I got started in this research area, describe the various studies done along the way, and where I feel we are today, relative to when I began this journey.

In the second year of my doctoral program in management information systems[1] at a large midwestern US university, I chose the topic of gender in the information technology profession for a class project. The instructor was quite pleased with the choice and she supported me in the writing process. After the class was over, I continued to feel quite excited by the topic and started talking to people about the prospect of using this as my dissertation topic. To my surprise (then, but not now in retrospect), the message I received from those I spoke to was clear – this topic and research are not likely to be perceived as "serious" work, and, if I really want to pursue it, I should do it after I have established myself as a "serious" researcher. For better or for worse, I followed this advice.

Yet, it seemed obvious to me that given the low proportion of women in IT as well as the current labor shortage in the IT industry, it was critical to consider the reasons why women were not choosing, staying in, and advancing in IT careers. It seemed important to identify, discuss, and show evidence of specific factors responsible for loss of valuable human resources, so that they could be used for guiding recruitment and policy decisions concerning women's participation in IT. I believed then, and still do, that women's issues are really business issues (Shwartz, 1992).

The purpose of this chapter is to outline my journey related to gender in IT research through the last 25 years. I note that much of my work fits in the positivist tradition and has been, for the most part (although not entirely), in the context of western, and more specifically, US culture. Hence, unless otherwise specified, any generalization beyond this context should be made with caution. Further, I acknowledge that the viewpoint here is heteronormative, and that not all women fit into this type of work–life. Clearly, the issues are different for same-sex couples (Olbrich et al., 2015). Further, some of this discussion applies to heterosexual couples and may not apply to women who do not have dependents and women who are single parents.

In the paragraphs below, I provide the background outlining the genesis of a conceptual paper on gender in IT and the stream of research it generated. Next, I discuss the research related to the specific issue of work–life balance (WLB) effects on career persistence and advancement of women in IT. Finally, I outline the caregiving crisis as COVID necessitated work from home and its effect on women's employment in general.

BACKGROUND

As mentioned above, this section provides a discussion of the genesis of a conceptual paper on gender in IT and ensuing stream of research that I as well as other colleagues engaged in.

A Framework of Gender in IT

Because this topic continued to occupy my mind after completing my dissertation in 1995, and publishing a couple of elite publications from it, I dusted off the "Gender and IT" folder again. It took me another year or so to craft a publishable manuscript, and a couple more years in the review cycle that included two rejections and two revisions. The paper "Women in the information technology profession: A literature review, synthesis and research agenda" finally found a home in a top journal – the *European Journal of Information Systems* (EJIS) in 2002. In this paper, I highlighted the importance of reducing sources of leakage in the IT career paths of women (Ahuja, 2002) and developed a model of barriers faced by women in the field of IT at three distinct career stages of career choices, persistence, and advancement. The model conceptualized the effects of social (social expectations, work–family conflict [WFC], and informal networks) and structural factors (occupational culture, lack of role models and mentors, demographic composition, and institutional structures), which may act as barriers at each of these stages (Figures 15.1 and 15.2). I suggested that these social and structural factors have a cumulative and incremental effect, resulting in turnover of women in IT, which needed to be addressed urgently. I invited scholars to take on an examination of the factors and issues raised herein, cross-sectionally to investigate the effects at each specific stage as well as longitudinally to examine the cumulative and comparative effects of work–family conflict on performance, innovativeness, and commitment to IT careers.

Interestingly, this became one of my more highly cited papers (over 701 Google scholar citations), and engendered a stream of research related to gender in IT. Many scholars have tested various linkages in this model; most of these are reviewed in detail in Armstrong et al.'s (2018) paper "The advancement and persistence of women in the information technology profession: An extension of Ahuja's gendered theory of IT career stages." Some scholars have examined the career choice and the key factors that lead women to choose a career in IT (e.g., Adya and Kaiser, 2005; Croasdell et al., 2011). Other scholars examining barriers to career advancement examined various factors women frequently encounter in their professional lives that adversely affect their professional advancement. For instance, Allen et al. (2006) addresses work stress, job qualities, and other "perceived barriers to promotion." Another

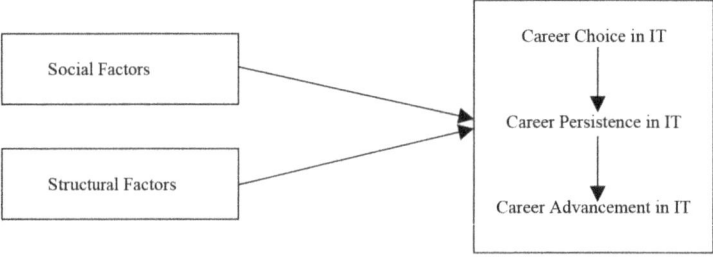

Figure 15.1   *Model of the effects of social and structural factors*

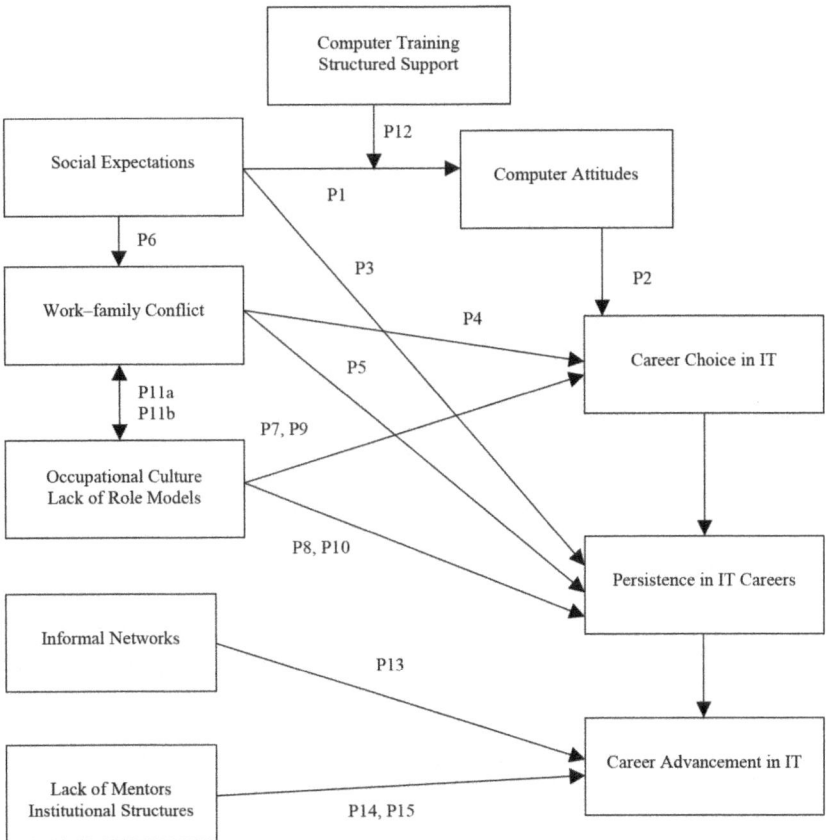

Figure 15.2   Detailed model of the effects of social and structural factors

stream has examined the issues of retention and turnover (e.g., Ahuja et al., 2005; Armstrong et al., 2007). Other related research based on the ideas in the EJIS paper spans a broad array of gender-related issues prevalent in the IT industry. Examples of these include studies of how gender influences the speed of technological learning in firms (e.g., Buche and Scillitoe, 2007), and the perceived masculine and feminine characteristics that are associated with top performers in IS operations (e.g., Joshi and Kuhn, 2007).

Rodhain and I (2000) wrote a conceptual article that proposed an interaction between two of the key factors in Ahuja (2002) – mentoring and WLB. Mentoring is key for women's persistence and advancement. It is also considered important for a number of organizational and individual-level outcomes. In this paper, we examined certain factors that may increase the likelihood that the mentoring relationship will be lasting and beneficial for the individual and organization. Specifically, we examined the role of similarities based on gender and work–family conflict in determining the effectiveness of mentoring relationships.

Although there have been over 701 papers citing Ahuja et al. (2002), many of which directly tested one or more of the linages in the model, no single empirical validation of this model existed until Armstrong, Riemenschneider, and Giddens (2018). These authors explored the factors and linkages in Ahuja (2002) related to the advancement and persistence of women

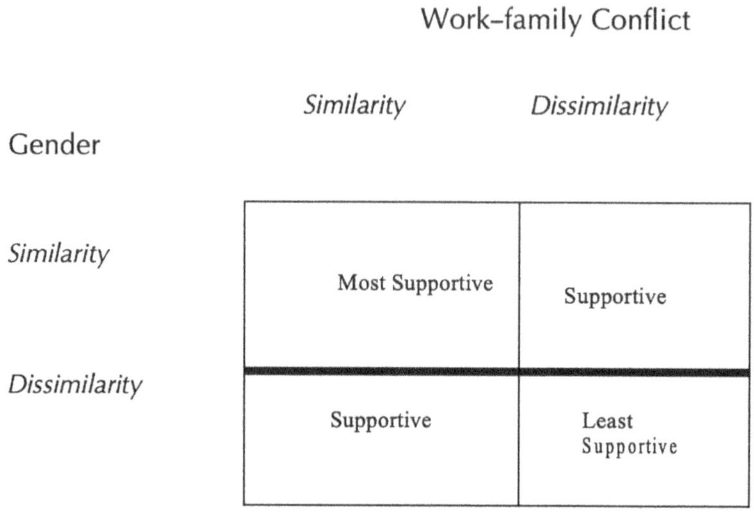

Work–family Conflict

	Similarity	Dissimilarity
Gender		
Similarity	Most Supportive	Supportive
Dissimilarity	Supportive	Least Supportive

Figure 15.3   Work–family conflict and gender

working in the IT field. Given that the women in their study had already chosen an IT career, their model focused on the persistence and advancement career stages. Conducting interviews from women working in IT departments at three organizations, they proposed an updated model to Ahuja (2002). Mapping their results onto the model, Armstrong et al. (2018) proposed several implications for research in light of the changes in society and the IT work environment since Ahuja (2002) was published. Armstrong et al. (2018) confirmed many relationships consistent with my framework as well as finding some new relationships evoked from our data and discussed how the new relationships fit within the original framework. Overall, their interviews (conducted in the United States) indicated that the social expectations, work–family conflict, lack of informal networks and institutional structures constructs influenced advancement in IT careers. They also found that the social expectations, occupational culture, and institutional structures constructs influenced persistence in IT careers. It is important to note here that these results should be interpreted in the context of the United States because as I discuss later, cultural issues greatly affect WFC.

Armstrong et al. (2018) concluded that there seems to be a shift in focus such that "work–family conflict" appears to not to affect a woman's "persistence in IT career" as strongly it might affect her "advancement opportunities in an IT career." This could be due to the recent organizational imperative on implementing WLB policies and practices (Brough et al., 2005; Butts et al., 2013). These authors (2018) also conjectured that the mobile and telecommuting technologies might also account for this finding because they allow women to "manage their work and non-work responsibilities more seamlessly so as to be able to remain in the IT field" (p. 14).

Despite this shift in focus, unfortunately, not much has changed in terms of the status of women in IT in practice. In fact, in 2020–21 when this chapter was written, the situation worsened and most of the job losses were incurred by women. One recent analysis of over 6,000 companies, 180,000 employees and 15,000 founders, found that women not only make less money than men, but also own less equity in Silicon Valley.[2] I find it stunning that men

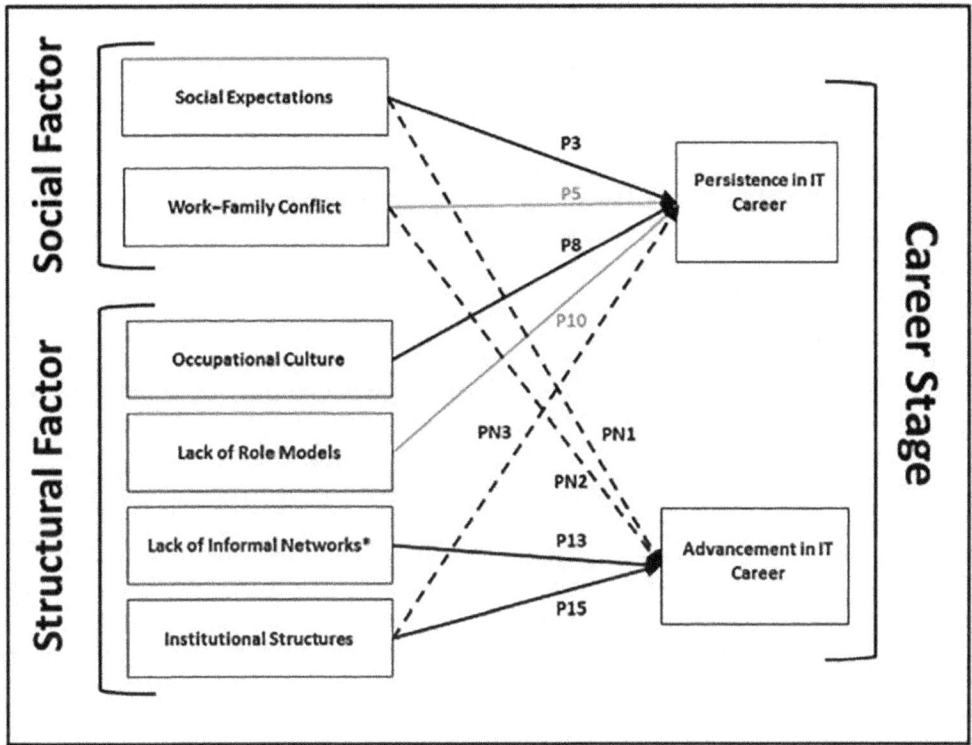

Figure 15.4   Social and structural factors by career stage

own 91 percent of employee and founder equity in Silicon Valley, while women own a mere 9 percent. Even more depressing are the reports showing that the approaches taken by tech companies are having the opposite effect (Wynn, 2019). For example, unconscious bias[3] training and mentorship programs – two popular approaches to culture change – are proving not only to be ineffective, but may be exacerbating bias. Wynn (2019) conducted a year-long, in-depth case study of one large Silicon Valley technology company implementing a gender-equality initiative, finding that these programs place blame for inequality only on individual people (as opposed to culture or structures) and also placed the responsibility for addressing it on individuals.

Wynn (2019) points out the limitation of this approached and the underlying assumption that men can be taught to limit unconscious bias and women can be taught to behave more assertively and demonstrate valued skill. This assumption has the effect of reinforcing the gender stereotype that men are more suited to technology than women, rather than motivating actions that could initiate change. Recognizing that individuals are limited in how much they can do to change their own biases, and the research findings that providing training without also emphasizing a shared commitment to broader organizational change can actually exacerbate bias, she recommends structural changes over these measures.

The Career Choice Stage

The career choice stage was examined in a large-scale, NSF-funded study that explored the "shrinking pipeline" of women in the IT-related programs. We (Jean Robinson, Christine Ogan, Susan Herring, and I) noted that the decision to pursue an IT career typically takes place years before an individual enters the workforce, and therefore examined the factors that lead to career choice of IT (Ahuja, 2002), including educational choices and experiences (Ahuja et al., 2006). Our team investigated tertiary education programs in information science, information systems, instructional systems technology and informatics, with computer science as a baseline comparison, in major IT degree-granting institutions across the United States in order to determine which are most successful at recruiting and retaining female students, and what factors favor success over time. The goal was to inform programmatic recommendations aimed at moving more women into the IT pipeline through a diverse range of educational programs.

This project was funded by the Information Technology Workforce (ITWF) directorate of the National Science Foundation in 2003. We collected and analyzed data on the experiences, attitudes, and outcomes of women and men in IT programs at five large public universities in the United States at undergraduate, master's, and doctoral levels, majoring in applied IT disciplines, using web-based surveys, telephone interviews, and face-to-face interviews with students, faculty and staff in the IT programs. We found several gender-based variations in our study. Our results showed that (in the United States) applied IT fields tend to attract more and different kinds of women than the more technical, computer science, programs in units that combine technological skills with applications to real-world problems. We also found that women (self)reported less skill, comfort and engagement with computing than do men, and that female applied IT majors are no more confident than female computer science majors. These are discouraging findings and indicate that it will take strategic efforts to solve the deeply ingrained problems women face in deciding to enter the traditionally masculine world of computing. It is also possible that the findings reflect the cultural effects and apply not just to computing fields but also to other traditionally masculine professions in the United States.

We also examined WLB issues for women enrolled in IT-related disciplines. We asked students to assess the relative importance they place on their career and their personal life. While most people answered that the two are equally important, females were more likely to choose the balanced response, while males were more likely to say that their careers are more important. This corroborated previous findings about social expectations in that men are more focused on their work, while women seek a more balanced lifestyle (Frenkel, 1990; Margolis and Fisher, 2002) and that IT disciplines were no exception in this regard. When asked about work–life balance as a potential cause of stress, women were more likely to say that school strain affects their personal life all the time, and men were more likely to respond that it seldom does. This is a reflection of women having greater domestic responsibilities (cf. Netemeyer et al., 1996); more of the female than male survey respondents said they were married and/or had children at home.

Finally, we noted gender differences in early computer experience and in computer self-efficacy (Durndell and Haag, 2002; Fromme, 2003). We also looked at the masculine culture of many computer science programs and IT workplaces, which can be unwelcoming to many women (Bentson, 2000; Spertus, 1991; Turkle, 1988). Our social norms and expectations conventionally associate computers with masculinity and create educational environments in which girls feel marginalized (Turkle, 1988). Margolis and Fisher (2002) suggest that "women are further alienated by a stifling 'geek culture' that celebrates obsessive computing at the expense of broad interests." Expectations of long hours and late nights, and manifesting "highly focused, almost obsessive behavior" associated with the geek and hacker cultures we saw are still prevalent today (Frenkel, 1990). Life stages contribute to the WLB issues as well, since this work ethic conflicts with women's desire to start a family or their actual family responsibilities. Rasmussen and Hapnes (1991) suggest that this type of culture is important in producing and reproducing male domination in higher education in computer-related fields, and that it influences the integration of women and their position within the field of computing. Despite the interventions institutions are using to recruit more women, the attitudes within the institutions and in society at large prevail, discouraging women from entering IT disciplines.

WLB EFFECTS ON CAREER PERSISTENCE AND ADVANCEMENT

My personal scholarly journey along this path took me to research WLB and WFC issues as they relate to gender. In addition, I started to wonder about the role of technology in WLB and WFC outcomes for women. It seems appropriate to discuss here the relationship between WLB and WFC. WLB is the balance between the domains of work and life in which the individual performs two different roles (Greenhaus, 1988). This is determined by the ratio of time spent between work and family. WFC, is a "form of inter-role conflict in which the role pressures from the work and family domains are mutually incompatible in some respect" (Greenhaus and Beutell, 1985, p. 77). Adams et al. (1996) found that family–work conflict might be associated with level of involvement with work. When WFC is high, the work spills over into the family domain. Work and life pull the individual in opposite directions and therefore, when WFC is present beyond a reasonable level, a satisfactory level of WLB is difficult to achieve.

While I was at Florida State University, we heard from a company during our board of advisors' meeting that they implemented telework and immediately faced turnover to the tune of 30 percent. Kathy Chodoba, Harrison McKnight, Joey George, and I set out to explore what might be contributing to this turnover and conducted a set of interviews, followed by a company-sponsored survey (another co-author, Charles Kacmar, joined us later). We found that WFC was a major contributing factor, because balancing family and job responsibilities is more difficult for road warriors (IT professionals who spend most of their workweek away from home at client sites), since they work at distant client sites.

We adopted and empirically tested Moore's (2000) turnover intention model in the context of IT road warriors. Our interviews showed that for road warriors the key source of stress was work–family conflict (WFC), and therefore we substituted this for Moore's role stressors and explored its effect on organizational commitment, which present as unique challenges when workers are not at the company site. Like Moore, we examined the effects of autonomy, perceived work overload, work exhaustion, and fairness of rewards, all of which are still relevant in the road warriors' context. As one would expect, our interviews suggested that gender played a significant role in explaining WFC. The broader survey results confirmed that WFC explained turnover (Ahuja et al., 2005).

Although we did not measure a moderating role of gender here, Jason Thatcher and I did so in another study (Ahuja and Thatcher, 2005), in which we examined the influence of the work environment and gender on trying to innovate with information technology. We utilized the theory of trying, and argued that work environment impediments render intentions inappropriate for examining emergent uses of IT. We proposed the goal-based construct of trying to innovate with IT as an appropriate dependent variable for examining emergent IT use – a substitute for examining intentions. More germane to the discussion here, we explored whether gender influences the relationship between perceptions of the environment and trying to innovate with IT. We found that autonomy interacts with overload (which can be a strong source of WFC) to influence trying to innovate with IT and that these relationships vary by gender.

In our paper, we theorized WFC as a source of occupational stress. In this commentary on the "electronic briefcase" era, and the contexts in which workers can work from anywhere, anytime, and can communicate with colleagues electronically, the family domain in our study was expanded to include family, significant others, and close friends. WFC arises when demands of one domain of life are incompatible with demands of the other domain, and that this conflict can affect the quality of both the domains (Greenhaus, 1988; Greenhaus and Beutell, 1985; Netemeyer et al., 1996). Salaff (2002) was likely the first scholar to suggest that in virtual settings, a blurring of home and work boundaries leads to stress and exhaustion (Salaff, 2002).

We found that work overload was associated with higher levels of WFC and that road warriors were especially susceptible to perceptions of higher work overload because of tensions inherent in their boundary-spanning role (Singh et al., 1994), as they navigate between needs of both employer and client. The road warriors were connected with their employers via information communication technologies (ICTs) and therefore share many of the same characteristics of today's Zoom-based workers. We found that work overload will have a positive influence on work–family conflict (WFC) of the road warriors, which is similar to the effects of Zoom fatigue on the workforce today as they work from home (WFH). All of the effects are much more pronounced for women.

In today's WFH, as back then, there seems to be a tension between autonomy and control. In 2005, we (Ahuja and Thatcher, 2005) proposed and found that job autonomy (defined as the degree to which the job provides substantial freedom, independence and discretion in scheduling and setting procedures (Hackman and Oldham, 1975, p. 162)) interacts with overload to determine individuals' efforts to innovate with IT and that these relationships vary by gender. One would think that autonomy allows workers to manage WFC in a way that makes sense for them personally. Early research suggested that higher autonomy (Thomas and Ganster, 1995) and autonomy over when work is conducted (Goldstein 2003) was associated with lower levels of WFC. In a meta-analysis, Lee and Ashforth (1996) demonstrate a relationship between perceived work overload and lack of autonomy.

Wynn and Rao (2020) show that in consulting firms, where employees often have a sense of autonomy over their own selves, careers, and lives face a "flexibility stigma," and shapes employees' preference to rely on themselves to manage any work–life challenges rather than to avail themselves of flexibility policies. Using interview data with 50 management consultants from five prestigious firms based in the United States, these authors found that in the context of flexible work programs, these workers glorified their work passion and suitability for a profession that they view as requiring overwork. They also emphasize their self-dependence and skill and use a "choice" rhetoric. Wynn and Rao go on to suggest that in order to gain their sense of autonomy, consultants manage their work–life conflict individually and avoid flexibility policies. They found that in this context, perceived control operates as a powerful force supporting the individualization of work–life conflict. Paradoxically, assuming individual responsibility appears to offer a sense of control, but, in fact, it undermines employees' ability to demand the change they really need.

Wynn and Rao (2020), as well as other literature on gender and IT, assume that organizational factors affect all women in the same ways. Trauth et al. (2009) suggest that within-gender differences can offer additional insights into the gender imbalance in the IT profession. Based on 92 in-depth interviews conducted with women employed in the IT sector in the United States, these authors show that work–life balance, organizational climate, and mentoring affected the women's career development in various ways, rather than monolithically. They suggest that individual identity as well as individual and environmental influences affect women's career choices, participation, and progression. These findings are important because they offer guidelines for flexible interventions that take into account within-gender differences within the IT workforce.

WLB of Globally Distributed Software Developers and Knowledge Workers

Around 2008, Saonee Sarker, Suprateek Sarker (who had conducted a WFC study of their own at a major software company) and I sought and received NSF funding for studying WFC of software developers. As Quesenberry et al. (2006) have argued, women in IT experience higher WFC since they find it extremely difficult to balance "domestic responsibilities while trying to keep pace with a rapidly changing field" (p. 136). The findings of this work are summarized in a paper that was published in *Information Systems Research* (Sarker et al., 2018) and more detail is provided in a book (Sarker et al., 2021) that came out a decade later. Here, I would like to summarize the gender-related findings from this study as they pertain to geographically distributed work[4] and the need for coordination across multiple time zones and locations, which can potentially increase WLB.

In this study of WFC of IT professionals involved in globally distributed work, we drew on border theory, which relies on the notion of regions divided by boundaries within the employee's lifespace (such as work and life) (Lambert et al., 2006). This theory relies on the concept of border keepers' roles (e.g., the organization), the flexibility and permeability of the borders (as in, characteristics related to the distributed environment), and the role of border crosser (as in IT professionals). Gender fits the last category of border crosser. The distributed nature of this knowledge work and the context of IT work, which includes challenges such as long working hours and a constantly changing landscape (Armstrong, 2018; Sarker et al., 2018), provides insights which can apply to many other types of work environments.

Interestingly, we found that gender had no significant effect on WFC. This is not entirely inconsistent with prior research, which often argues that "women have been socialized over the generations to the nurturing role of the family. No matter how achievement orientated the women is," she is able to balance these two domains more easily (Gambles et al., 2006, p. 77). Korabik et al. (2003) also report from their focus group studies that women tend to be better at multitasking than men, as a result they are able to "fulfill many work and family duties without much support" (p. 294) and not be burdened by the conflict between the two domains. Our findings can also be explained by a self-selection effect such that women who survive in this profession are better able to handle two different domains of responsibilities than men.

While some researchers have suggested that gender does not have a significant effect on WFC, others have emphatically argued (consistent with what our interviewees described) that gender does play an important role, with female employees experiencing greater WFC than male employees (e.g., Lyness and Kropf, 2005). Eddleston and Mulki (2017) as well as Kwok (2016) state that gender differences inevitably emerge when discussing the ability to manage work and family demands, and Wood and Eagly (2010) assert that gender role theory maintains that time allocated to work has a clear pattern that is incongruent with female gender role expectations (Wood and Eagly, 2010). Rothbard (2001) identified female roles as "integrators" and male roles as "segmentors." Greenhaus and Powell (2006) argue that "individuals who participate in multiple roles (such as work and family) inevitably experience conflict and stress that detract from their quality of life" (p. 72). Given that women (especially in traditional families and as a part of dual-earner couples) tend to more often play both roles simultaneously (Webster, 2002), they experience higher WFC. Kossek et al. (1999) state that "women consistently work more hours (combined paid and unpaid), spend more time on family responsibilities, and experience greater role overload and work–family conflict" (p. 110).

Dual-earner couples face even more challenges.[5] Kossek et al. (1999) notes that in "dual earner families with multiple children, women worked ninety hours a week (combined paid and unpaid chores) compared with sixty hours for men" (p. 110). Although couples have become more egalitarian in recent years, it is clear that gender roles and expectations seem to persist. Allen et al. (2019) suggest that gender egalitarianism plays a role in work demands as well as micro-role transitions (Ashforth et al., 2000), which in turn affects WFC. Once again, interviews suggested that different expectations existed for men and women, the strains related to being a working mom were significant, and many women choose to step away from the workforce for motherhood. For our context of global software development, the companies involved in such distributed work tend to have professional cultures where 16-hour workdays and working through weekends is the norm, and taking calls during what is traditionally considered personal time is expected. Cohabitation appeared to result in higher WFC for such workers because partner/spouse expectations tend to create competing demands for time and attention.

The Caregiving Crisis During COVID and its Effect on Women's Employment

Previous literature has provided evidence related to caregiving roles contributing to WFC (Allen et al., 2019). Research suggests that caregiving roles can have strong implications for the WFC of women (Lyness and Kropf, 2005; Greenhaus and Beutell, 1985; Netermeyer et al., 2004), because the time and energy required for and pressure arising from caregiving can be particularly harmful to remote work productivity and leads to more family–work conflict (Eddleston and Mulki, 2017). Employees with families face higher demands in their caregiving roles (Kossek et al., 1999). The demands of the family domain coupled with the intense demands at the workplace that general IT workers specifically engaged in distributed work face cause serious WFC (Messersmith, 2007).

In our study (Sarkar et al., 2018), we examined the effects of caregiving roles on WFC of remote knowledge workers. In the specific context of the IT industry, Armstrong et al. (2018) have discussed certain distinctive characteristics that are unique to this profession. These include intensity of the work environment, extent of change these workers must deal with, and the need for continuous learning, all of which make caregiving much more challenging. The IT-related work environment is focused around a complex, multifaceted knowledge base and continuous adaptation of that knowledge base. In addition, the globally distributed IT workers must deal with time zone differences. Such a domain becomes challenging for individuals who have parenting and other caregiving responsibilities.

Consistent with the effect of WFH phenomenon in the COVID era, many more female than male participants reported having stepped back from intense work after parenthood, especially in situations where both parents are working (Sarker et al., 2018). Others reported feelings of intense guilt about working from home and for feeling distracted by work situations during family time, resulting in high WFC.

Zhang (2020) examined the extent to which telework was frequency associated with life stages, and their relationships with work–life conflict perspective.[6] Accounting for individual, household, job-related and environmental characteristics, their results showed three patterns. First, parents, regardless of gender and marital status, are less likely to telework compared to those without children. Single individuals without children were found to be more likely to telework than married ones, and males more likely than females. They found that the presence of children plays a more significant role, than other factors, increasing both work-to-family conflict and family-to-work conflict.

These and other scholars suggest that family demands can interfere with job performance, which increases WFC. Caregiving tasks range from not only "Activities of Daily Living (ADL) such as eating, dressing, bathing," etc., but also Instrumental Activities of Daily Living (IADLs) such as companionship, and transportation (Kossek et al., 1999, p. 116). Increases in job demands or ADLs and IADLs can cause serious WFC. Such a conflict appears to be common in the offshoring context, wherein individuals are often required to work at times that are traditionally allocated for personal activities and set aside for families.

Consistent with the research outlined above, our study showed a significant effect of caregiving role, on WFC of knowledge workers. We also found that IT workers in our study who valued their leisure time and privileged the family and time spent with family experienced higher WFC.

National Culture and Women's WLB

While it is tempting to draw universal conclusions about the antecedents and effects of WLB on women, many of these effects can be attributable to national culture. For example, Trauth et al. (2008) demonstrated the effect of cultural attitudes about maternity, childcare, parental care and working outside the home on a woman's choice of an IT career. In addition, several additional socio-cultural factors served to add further variation to gendered cultural influences: gendered career norms, social class; economic opportunity, and gender stereotypes about aptitude.

Research has largely examined individual-level differences in this domain (such as gender-related expectations, relationship status, caregiving role, priority of time with family, etc.). Much of this research shows mixed results, which can also be attributed to a failure to account for cultural effects (Rajadhyaksha et al., 2020; Trauth and Connolly, 2021; Trauth et al., 2008). Expectations vary within cultures in terms of gender, the strains of being a working mother, and mothers facing the decision of whether to continue working outside the home intersect with culture. These variations in norms of work and family roles can, directly or indirectly, lead to different levels of WFC.

Also, in line with the metaphor of border crossing, Ashforth et al. (2000) offered a conjecture that the culture in which individuals are embedded may affect boundary (borders) management processes, because culture reflects norms and expectations that guide how individuals separate, define, and draw boundaries between interpersonal relationships and different roles. Indeed, a meta-analysis by Allen et al. (2019) showed that national culture meaningfully moderates the relationship between demands of the domain and WFC. For instance, they found the relationship between WFC and satisfaction outcomes was significant across regions, except South Asia, a region that has been described as highly group oriented, humane, male dominated, and hierarchical (Gupta et al., 2002). The findings of the meta-analysis suggest that relationships between work–family conflict and satisfaction outcomes are weaker in societies that are more collectivistic and exhibit higher power distance than those that are more individualistic and exhibit lower power distance. In a similar vein, Rajadhyaksha et al. (2020) showed that the dynamics of social expectations and availability of social support can be different from those in the eastern countries. Most Anglo/European countries score much lower on the long-term orientation dimension, and high on value of leisure time (Hofstede, 2001; Aryee et al., 2005). In the countries where leisure time is considered important, the need to work during outside of working hours is likely to result in experiences of higher WFC because they are likely to consider such spillover as interference that prevents them from enjoying leisure (Aryee et al., 2005),

Yet, some eastern countries such as India, score high on the long-term orientation dimension of Hofstede (2001). In such cultures, organizations and governments tend to value leisure time much less than other cultures and do not make a concerted effort to protect leisure time for their employees. This is reflected in long working hours (Gambles et al., 2006), and high work devotion (Hofstede, 2001) of IT outsourcing workers in India, particularly compared to that of Anglo/European countries. In countries where gender norms are more traditional, even when fathers engage in WFH or hybrid work for care purposes, there is still an expectation that they will maintain their work devotion and protect their work spheres and prioritize it over family time and childcare roles (Chung and van der Lippe, 2018).

In terms of outcomes of WFC, both Ahuja et al. (2005) and Sarker et al. (2018) found that regardless of gender, high WFC was associated with turnover intention. We found that higher WFC was associated with higher performance. We do not have data on causality on this link. It could be tempting to conclude that women with nonwork responsibilities/conflict also learn how to manage work more efficiently in order to get things done, and that may also help in performance. It is, however, more likely to be the case that high performers are more likely to experience work–family conflict, and that this is particularly true for women (Bolino and Turnley, 2005). It is also possible that those with higher level of conflict at home might take solace in work and overwork to avoid conflict at home.

COVID, Remote Work, and Gender

In their empirical test of Ahuja's (2002) model, Armstrong et al. (2018) explained findings that were different from those I proposed back in 2002 in terms of the IT tools (telecommuting technologies, virtual private networks, mobile ubiquity, and consumerization) that had become available in recent years and allows for remote work. They suggested that IT might be providing opportunities for women to manage their work and non-work responsibilities more seamlessly so as to be able to remain in the IT field. While this seems to make intuitive sense, the WFH phenomenon in the COVID era tells a different, cautionary tale in this regard. My current research on multitasking and interruptions shows that these factors can have a detrimental, rather than beneficial, effect on women's WLB, wellbeing, productivity, and turnover.

Let's take a deeper look at the effect of WFH during the COVID era. COVID has made apparent the WLB issues as it has forced changes in family and work roles. Singles, couples, and parents worked from home, which became not only where they lived and worked, but also schooled and entertained their children.

These developments lead to a disproportionate increase in women's workload. We have long known that despite women's increasing participation in the paid labor force, they continue to bear a disproportionate burden of the household physical and mental labor (Arendell, 2001; Bianchi et al., 2006; Craig and Bittman, 2008). For example, women tend to be more likely than men to combine both housework and childcare with other activities (Bianchi et al., 2006; Craig, 2006; Ironmonger, 2004). One cultural artifact that has led to such multitasking is the "modern-day virtue" of intensive parenting (i.e., the contemporary pressures to interact with the children in highly active and engaging ways that promote development) (Arendell, 2001; Nelson, 2010). Women have shouldered a disproportionate share of the load and the impact of intensive parenting.

Even as women's participation in the workforce has increased and gender roles have evolved, the domains related to family, friend, and romantic relationships continue to belong to women and, compared to men, they place greater importance on these roles (Ahuja, 2002; Eagly and Wood, 2012; Greenhill and Wilson, 2006; Wood and Eagly, 2012). Eddleston et al. (2006) showed that masculine self-schemas weigh the career role more highly whereas the feminine self-schemas emphasize the family role (Powell and Greenhaus, 2010), even when they are the higher earning partner (Bittman et al., 2003). This pattern appears consistent across professions. For example, Starmer et al. (2019) showed that female pediatricians in their study spent more time on household responsibilities than male pediatricians, and had lower levels of WLB satisfaction.

Plenty of reports have shown that the pandemic and the ensuing WFH phenomenon, has affected women's work productivity and employment status more adversely (e.g., Fazackerley, 2020; Kitchener, 2020). One telling statistic shows that most job losses due to COVID were incurred by women. Aspen (2021a) reported a National Women's Law Center statistic that women accounted for almost 80 percent of US adults who stopped working or looking for work in January 2021. Aspen (2021b) further reports that during COVID, women have lost 5.4 million jobs (as of February, 2021), while almost 2.1 million women have left the labor force entirely (i.e., they are not even looking for work).

A significant factor accounting for these job losses is the caregiving crisis. A survey of more than 2,900 adults in 2019 found that while more than 80 percent of men, and 91 percent of all survey respondents, reported believing that men and women should equally share in care work, only 46 percent of all respondents reporting evenly sharing such responsibilities (Aspen, 2021). In fact, 45 percent reported that these responsibilities fell primarily on women, and only 4 percent reported that they fell primarily on men. These economic losses are incurred by mothers across race and occupation, but Black and Latina women were affected most adversely.

Interestingly, though, men who have high-intensity caregiving experiences (such as parents or caregivers of adults or children with medical or behavioral conditions or other special needs), were found to be more likely to take on a larger share of caregiving tasks. These men faced many of the same work–life conflicts that broadly undermine working women and reported leaving the workforce at the same rate as women. Among men, 41 percent of "high-intensity" caregivers put off work because of demands on their time at home, compared with 33 percent of other fathers, and 10 percent of men who are not caregivers. Thus, all women and men with high-intensity caregiving experiences left the workforce or retired early (50 percent of men and 51 percent of women), missed work to provide assistance (64 percent men vs. 71 percent women), or reduced their work hours (42 percent men vs. 45 percent women), as a result of their caregiving obligations. Men without children or "high-intensity" caregiving responsibilities report much less WFC. These results underscore the role the caregiving is playing in the employment crisis facing women (and some men). Much more work, by scholars, and managers is urgently needed to address how to change these social norms, expectations, and roles.

Social expectations and gender

In a recent report by New America's Better Life Lab[7] Brigid Schulte, director of the lab states, "If everybody expects women to drop out or expects women to hold back at work, they expect men to keep working. They're supposed to be the breadwinners," and that "we need public policies, and we need workplace practices that support all people with caregiving responsibilities." Another study shows that gender and children at home explained workers' control over time, technology usefulness, and WFH conflict at a higher level for women. Women also found technology to be less useful than did men (Ajjan et al., 2020). Although ICT was found to be critical in performing work, it also disrupted the balance between personal and work lives as it blurs prior work–life boundaries. Gender-related work styles or patterns, or conflicting demands placed on parents or caregivers are less than compatible with work demands relying on ICT.

Therefore, the increase in caregiving load implies increased role requirements for women (Anderson and Kelliher, 2020). Various studies (Alon et al., 2020; Carlson et al., 2020) and media reports (e.g., Ascher, 2020; Manzo, 2020) have demonstrated the challenges to and difficulties of working women's WLB since this pandemic seems to have exacerbated traditional gender stereotypes and inequalities in families and societies.

The role of multitasking and attentional capacities

Our attentional capacities and reservoirs offer a cogent explanation of these findings. Human beings have a limited capacity for attentional and processing resources that may be allocated to task performance (Wickens, 1991). The notion here is that that human beings possess a finite general reservoir of resources (Allport, 1980; Neisser, 1976) available for information processing and resource allocation to one task can result in diminished resources available for other task(s) in which an individual may be engaged.

As women have taken on a greater proportion of household and childcare responsibilities during COVID, they are essentially continually multitasking. Multitasking has been shown to deplete people's self-regulatory resources over time – including their self-control regarding multitasking itself (Zijlstra et al., 1999), thus reducing the task effectiveness. Robertson et al. (2019) showed that the mental work needed to identify and divide up the tasks at home also lay on mothers' shoulders, exhibiting that the cognitive burden of this was also a part of their gendered realities, which can contribute to the exhaustion and stress. Bolino and Turnley (2005) also found that personal cost of citizenship behavior[8] is higher for women. Similarly, Offer and Schneider (2011) found that mothers were more likely than fathers to multitask at home as well as in public, creating more WFC and consequently, adversely affecting their well-being.

Some research shows that as workers try to become more efficient in the use of their time as a means of coping with their growing responsibilities, this will increase productivity (O'Leary et al., 2011). However, this is only the case if the tasks are fairly similar to each other – cooperation among tasks improves when they use separate versus shared resources (Wickens, 1991). As differences across tasks increase, more time and cognitive effort are needed to cope with the effort required to switch from one task context to another (Fruchter et al., 2010; Matthews et al., 2012), increasing stress and WFC. Constantly switching between different types of tasks increases perceptions of higher work overload because the differences may increase the amount of cognitive effort.

Moreover, research suggests that once an optimal number of tasks is exceeded, individual productivity may start to decrease as employees spend time managing and tracking activities across teams rather than finding creative solutions for the software development tasks (O'Leary et al., 2011). As the task list grows longer, a degrading effect in performance of the primary task should occur as a result of a reduction in the resources activated. This would suggest that a higher number of competing tasks can result in a higher demand competition for available resources.

Thus, human attentional and cognitive capacity limitations explain why the work productivity of women has been decreasing drastically since COVID (Feng and Savani, 2020). In one of our studies (Ahuja et al., working paper), we found that women were more affected and frustrated by the work and family permeability enabled by IT in this period. This challenges the previous research showing that flexible work settings help women balance work and family more effectively as they are able to perform their work on their own schedule and with more autonomy (Hardwick and Salaff, 2003).

Andrew et al. (2020) empirically measured changes in time-use patterns on men and women based on a field survey of parents with small children during pandemic lockdown measures in the United Kingdom. They report a substantial increase in the time squeeze of parents but this time squeeze is higher for mothers, resulting from an increase in unpaid work hours as well as from work intensification (due to multitasking and interrupted work patterns). They further found that there is a widening of the gender difference in paid work due to higher rates of employment disruption for mothers. Another recent study by Ajjan et al. (2020) similarly suggests that women typically experience higher WFC than men. As Kossek (2020) has suggested, work–life inclusion should, therefore, be viewed as a potentially valuable intervention and an essential aspect of an inclusive climate.

CONCLUSION

This chapter has shown that WLB factors are a significant part of the explanation for the gender gap in IT. A deeper study of WLB factors is essential in the upcoming era of hybrid work, where at least part of the work week is spent working from home. A final note is that this "work–family narrative" itself can be more inclusive (Padavic et al., 2019). Individuals as well as couples (both heterosexual and homosexual) experience it and more attention is needed on how couples navigate these dynamics. There is much work to be done to better understand the diversity of experiences relating to work–life configurations.

It is also clear that gender and role expectations play a role in WFH and hybrid work outcomes. It is critically important for firms to create conditions for long-term persistence and advancement of women, including conditions for a balanced work and life (Armstrong et al., 2018; Ahuja, 2002). If left unaddressed, we will lose more talent due to WFC issues and, mostly, it will be women who will pay the price (Chung and van der Lippe, 2018). As I wrote earlier, it would be mistake to consider this as "their choice," and instead should be viewed as a constraint under which women are negotiating their work spheres (Lott, 2018).

Many employers are now attempting to address these issues and are making efforts to stop women's exodus from their own workforce through internal policies as well as advocating better public policies such as federal paid family leave and medical leave. It is important that organizations continue to promote policies that make this a less burdensome practice for women. Thus far, however, organizations' efforts to mitigate WFC through flexible work policies have not improved women's advancement prospects. Taking advantage of these policies can, in fact, feed into the social narrative that women are less "serious" about work, and can hurt their advancement. To address this, Padavic (2019) developed a multilevel theory that explains this conundrum faced by organizations. They suggest that the work–family explanation has become a "hegemonic narrative," that is, a pervasive, status-quo-preserving story that prevails despite countervailing evidence. Their evidence shows that organizations use this narrative and attendant policies and practices as an unconscious "social defense," sustaining workplace inequality. The phenomenon of "social defense" relies on the prevailing and unchallenged assumption of the necessity of long work hours and the inescapability of women's stalled advancement. As scholars, we must examine and challenge the utility of the 24/7 culture.

Another explanation of this phenomenon is offered by Wynn and Rao (2020) (a description of this study was offered earlier). As organizations constrain consultants' decisions about how to manage their work–life conflict, and workers themselves express little faith in flexibility programs, the workers reclaim a sense of control by emphasizing their professional skills and natural ability in managing WLB. The implication here is that if contradicting structures remain in place, the addition of formal policies will not necessarily result in a fundamental change, unless work structures are fundamentally redesigned to support WLB. The cultural issue I discussed earlier implies that in some cultures, these effects will not work the same way. WLB issues operate in a complicated system of individual, organization, society and culture, which also means that change in one part of the system would reverberate in the entire system.

Along similar lines, Kossek et al. (2001) have pointed out a need to consider work–life inclusion as "a form of diversity inclusion that intersects with social identities involving gender, caregiving ambition, and multi-culturalism" (p. 1). She defines a work–life inclusive climate to be one in which a member would not feel that success in one's job role could mean sacrificing their family and non-work identities in order to succeed in their job role (Kossek et al., 2001). Kossek et al. (2011) suggest that workers with higher identification with the family role and greater involvement with nonwork demands should not face implicit biases. I call on scholars to examine ways in which WFH practices, time use, cognitive burden, and multitasking can contribute to the WLB research stream. Such efforts can deepen our knowledge in this area and promote organizational policies that make WLB a virtue for both men and women alike.

Governmental interventions and public policy at all levels will be key here. A recognition of women's job losses during COVID not only must be recognized, but serious measures must be taken to address it. Misty Heggeness, a principal economist with the US Census Bureau suggests, "For women, there has been some scarring here that we may never recover from. We need to do policy intervention sooner rather than later – and accelerate it as much as we can" (Aspan and Hinchliffe, 2021, para 4). Nicole Mason, of the Institute for Women's Policy Research, argues that federal and state policies should focus specifically on rehiring and, if necessary, retraining women affected by the pandemic-era recession (Aspan and Hinchliffe, 2021). And Michael Madowitz, a labor economist at the Center for American Progress, cites previous efforts to reskill manufacturing workers and those in other male-dominated occupations. I ask the reader to consider Madowitz's assertion that "We've had very gendered job losses before, and I've seen a lot of proposals about what you do about the 60,000 coal miners in the world. But we've never done that for women" (Aspan and Hinchliffe, 2021, para 23). The question is, why not? Perhaps COVID provides an opportune moment to start doing exactly that!

In summary, this chapter has provided a reflection on my career journey regarding gender and IT research themes. In the backdrop of various studies related to this issue, I discussed my views on where we are today, and what more needs to be done. I provided the context and circumstances surrounding the genesis of my conceptual paper on gender in IT (Ahuja, 2002) and the research it generated. I dove deeply into the research related to the specific issue of work–life balance (WLB) and its relationship to career persistence and advancement of women in IT. I concluded with a discussion of the caregiving crisis as the pandemic gave rise to the work from home phenomenon, transforming the nature of work and how it affected women's employment status.

NOTES

1. By, MIS, I refer to the IS profession discipline within the overall IT field that includes computer science, information science, etc.
2. See www.cnbc.com/2018/09/19/in-silicon-valley-women-face-an-equity-gap-that-is-far-larger-than-the-pay-gap.html.
3. Unconscious bias is prejudice in favor of or against a group of people compared with another usually in a way that is considered to be unfair.
4. Defined as work done by multiple individuals across different locations and time.
5. I acknowledge that not all women are married, have children, are dual-earner couples, or are not in heterosexual couples.
6. The study is based on pre-pandemic data.
7. See www.newamerica.org/better-life-lab/reports/providing-care-changes-men/.
8. Citizenship behavior is defined as "activities that employees engage in to benefit their organization even though the activities generally fall outside the formal job description of the employee" (Bolino and Turnley, 2005, p. 748).

REFERENCES

Adams, G.A., King, L.A., and King, D.W. (1996). Relationships of job and family involvement, family social support, and work–family conflict with job and life satisfaction. *Journal of Applied Psychology*, 81, 411–20.

Adya, M., and Kaiser, K.M. (2005). Early determinants of women in the IT workforce: A model of girls' career choices. *Information Technology and People*, 18(3), 230–59.

Ahuja, M. (2002). Women in the information technology profession: A literature review, synthesis and research agenda. *European Journal of Information Systems*, 11(1), 20–34.

Ahuja, M.K., Ogan, C., Herring, S.C., and Robinson, J.C. (2006). Gender and career choice determinants in information systems professionals. In Niederman, F., and Ferratt, T. (Eds.), *IT Workers Human Capital Issues in a Knowledge Based Environment*. Information Age Publishing, Greenwich, CT.

Ahuja, M., and Rodhain, F. (2000), The effects of mentoring and work–family conflict on IT workers' careers, Proceedings of the Association of Information Conference (AIS) (Long Beach, CA).

Ahuja, M., and Thatcher, J. (2005). Moving beyond intentions and toward the theory of trying: Effects of work environment and gender on post-adoption information technology use. *MIS Quarterly*, 29(3), 427–59.

Ajjan, H., Abujarour, S., Fedorowicz, J., and Owens, D. (2020). Working from home during the COVID-19 crisis: A closer look at gender differences. International Research Workshop on Women, IS, and Grand Challenges. International Research Workshop on Women, IS, and Grand Challenges.

Allen, M.W., Armstrong, D., Riemenschneider, C.K., and Reid, M.F. (2006) The effects of mentoring to reduce stress in a state IT department during times of transformational change. *International Journal of Learning and Change*, 1(4).

Allen, T.D., French, K.A., Dumani, S., and Shockley, K.M. (2019). A cross-national meta-analytic examination of predictors and outcomes associated with work–family conflict. *Journal of Applied Psychology*, 105(6), 539–76.

Allport, D.A. (1980). Attention and performance. In G. Claxton (Ed.), *Cognitive Psychology: New Directions*. London: Routledge and Kegan Paul. 112–53.

Alon, T., Doepke, M., Olmstead-Rumsey, J., and Tertilt, M. (2020) The impact of Covid-19 on gender equality. NBER Working Paper No. w26947, Available at SSRN: https://ssrn.com/abstract= 3569411Anderson and Kelliher, 2020.

Andrew, A., Cattan, S., Costa Dias, M., Farquharson, C., Kraftman, L., and Krutikova, S. (2020). How are mothers and fathers balancing work and family under lockdown? The Institute for Fiscal Studies. Retrieved from www.ifs.org.uk/uploads/BN290-Mothers-and-fathers-balancing-work-and-life-under -lockdown.pdf.

Arendell, T. (2001). Gender, emotion, and the family by Leslie Brody. *Contemporary Sociology*, 30(1), 4–50.

Armstrong, D.J., Riemenschneider, C.K., and Giddens, L.G. (2018). The advancement and persistence of women in the information technology profession: An extension of Ahuja's gendered theory of IT career stages. *Information Systems Journal*, 28(6), 1082–124.

Armstrong, D.J., Riemenschneider, C., Reid, M., and Allen, M. (2007) Voluntary turnover and women in IT: A cognitive study of work–family conflict. *Information Management*, 44(2), 142–53.

Aryee, S., Srinivas, E.S., and Tan, H.H. (2005). Rhythms of life: Antecedents and outcomes of work– family balance in employed parents. *Journal of Applied Psychology*, 90(1), 132–46.

Ascher, D. (2020). Coronavirus: "Mums do most childcare and chores in lockdown." BBC News. Retrieved from www.bbc.com/news/business-52808930?fbclid=IwAR1pbW4amJI45txjsL2rjIhiiVn hSfx0nWGvHG6qoBgR2tu0P6SuCk9Zg7Q. May 27.

Ashforth, B.E., Kreiner, G.E., and Fugate, M. (2000). All in a day's work: Boundaries and micro role transitions. *Academy of Management Review*, 25, 472–91.

Aspan, M. (2021). Nearly 80% of the 346,000 workers who vanished from the US labor force in January are women. Fortune. https://fortune.com/2021/02/05/covid-unemployment-rate-january-jobs-report -2021-jobless-job-loss-us-economy-working-women/.

Aspan, M., and Hinchliffe, E. (2021). 5 steps the US government could take to tackle the crisis facing working women. *Fortune*. https://fortune.com/2021/02/01/women-unemployment-covid-gov ernment-help-labor-policy/?fbclid=IwAR3fAIQzFawT20eH_Ef3DfWz5YdJcpD6xavEv3mym7ITh1 mJjaLaEWszEb4.

Bentson, C. (2000). Why women hate IT. *CIO Magazine*, September 1. www.cio.com/archive/090100/ women.html.

Carlson, D.L., Petts, R., and Pepin, J. (2020). US couples' divisions of housework and childcare during COVID-19 pandemic (Working Paper). Retrieved from https://osf.io/preprints/socarxiv/jy8fn.

Bianchi, S.M., Robinson, J.P., and Milke, M.A. (2006). The changing rhythms of American family life. *The American Psychological Association*.

Bittman, M., England, P., Sayer, L., Folbre, N., and Matheson, G. (2003). When does gender trump money? Bargaining and time in household work. *American Journal of Sociology*, 109(1), 186–214.

Bolino, M.C., and Turnley, W.H. (2005). The personal costs of citizenship behavior: The relationship between individual initiative and role overload, job stress, and work–family conflict. *Journal of Applied Psychology*, 90(4), 740–48.

Brough, P., O'Driscoll, M.P., and Kalliath, T.J. (2005). The ability of "family friendly" organizational resources to predict work–family conflict and job and family satisfaction. *Stress and Health*, 21(4), 223–34.

Buche, M.W., and Scillitoe, J.L. (2007). Influence of gender and social networks on organizational learning within technology incubators. *American Journal of Business*, 22(1), 59–68.

Butts, M.M., Casper, W.J., and Yang, T.S. (2013). How important are work–family support policies? A meta-analytic investigation of their effects on employee outcomes. *Journal of Applied Psychology*, 98(1), 1–25.

Carlson, D.L., Petts, R., and Pepin, J.R. (2020). Changes in parents' domestic labor during the COVID-19 pandemic. https://doi.org/10.31235/osf.io/jy8fn.

Chung, H., and van der Lippe, T. (2018). Flexible working, work–life balance, and gender equality: Introduction. *Social Indicators Research*, 151, 365–81.

Craig, L. (2006). Does father care mean fathers share? A comparison of how mothers and fathers in intact families spend time with children. *Gender and Society*, 20(2), 259–81.

Craig, L., and Bittman, M. (2008). The incremental time costs of children: An analysis of children's impact on adult time use in Australia. *Feminist Economics*, 14(2), 59–88.

Croasdell, D., McLeod, A., and Simkin, M.G. (2011). Why don't more women major in information systems? *Information Technology and People*, 24(2), 158–83.

Durndell, A., and Haag, Z. (2002). Computer self efficacy, computer anxiety, attitudes towards the internet and reported experience with the internet, by gender, in an East European sample. *Computers in Human Behavior*, 18(5).

Eagly, A.H., and Wood, W. (2012). Social role theory: A biosocial analysis of sex differences and similarities. In Van Lange, P.A.M., Kruglanski, A.W., and Higgins, E.T. (Eds.), *Handbook of Theories of Social Psychology*. Sage Publications Ltd.

Eddleston, K.A., and Mulki, J. (2017). Toward understanding remote workers' management of work–family boundaries: The complexity of workplace embeddedness. *Group and Organization Management*, 42(3), 346–87.

Eddleston, K.A., Veiga, J.F., and Powell, G.N. (2006). Explaining sex differences in managerial career satisfier preferences: The role of gender self-schema. *Journal of Applied Psychology*, 91(2), 437–45.

Fazackerley, A. (2020). Women's research plummets during lockdown – but articles from men increase. *The Guardian*. www.theguardian.com/education/2020/may/12/womens-research-plummets-during -lockdown-but-articles-from-men-increase.

Feng, Z., and Savani, K. (2020). Covid-19 created a gender gap in perceived work productivity and job satisfaction: Implications for dual-career parents working from home. *Gender in Management: An International Journal*, 35(7/8), 719–36.

Frenkel, KA. (1990). Women and Computing. *Communications of the ACM*, 33(11), 34–46.

Fromme, J. (2003). Computer games as a part of children's culture. *Game Studies – The International Journal of Computer Game Research*, 3(1).

Frone, M.R., Russell, M., and Cooper M.L. (1992). Antecedents and outcomes of work–family conflict: Testing a model of the work–family interface. *Journal of Applied Psychology*, 77(1), 65–78.

Fruchter, R., Bosch-Sijtsema, P., and Ruohomäki, V. (2010). Tension between perceived collocation and actual geographic distribution in project teams. *AI and Society*, 25, 183–92.

Gambles, R., Lewis, S., and Rapoport, R. (2006). *The Myth of Work–Life Balance: The Challenge of Our Time for Men, Women, and Societies*. John Wiley and Sons, Ltd, Sussex.

Gibboner, C. (2020). Women academics seem to be submitting fewer papers during coronavirus. "Never seen anything like it," says one editor. *The Lily*. www.thelily.com/women-academics-seem-to-be -submitting-fewer-papers-during-coronavirus-never-seen-anything-like-it-says-one-editor/.

Goldstein, J.S. (2001). *War and Gender: How Gender Shapes the War System and Vice Versa*. Cambridge University Press.

Greenhaus, J.H. (1988). The intersection of work and family roles: Individual, interpersonal, and organizational issues. *Journal of Social Behavior and Personality*, 3(4), 23–44.

Greenhaus, J.H., and Beutell, N.J. (1985). Sources of conflict between work and family roles. *Academy of Management Review*, 10(1), 76–88.

Greenhaus, J.H., and Powell, G.N. (2006). When work and family are allies: A theory of work–family enrichment. *Academy of Management Review*, 31(1), 72–92.

Greenhill, A., and Wilson, M. (2006). Haven or hell? Telework, flexibility and family in the e-society: A Marxist analysis. *European Journal of Information Systems*, 15(4), 379–88.

Gupta, V., Surie, G., Javidan, M., and Chhokar, J. (2002). Southern Asia cluster: Where the old meets the new? *Journal of World Business*, 37(1), 16–27.

Gutek, B.A., Searle, S., and Klepa, L. (1991). Rational versus gender role explanations for work–family conflict. *Journal of Applied Psychology*, 76(4), 560–68.

Hackman, J.R., and Oldham, G.R. (1975). Development of the job diagnostic survey. *Journal of Applied Psychology*, 60(2), 159–70.

Hardwick, D., and Salaff, J. (2003). Managing work–family boundaries: How do teleworkers make it work? [Paper presentation]. Academic Conference Series on Work and Family Business and Professional Women's Foundation and the Community, Families and Work Program, Brandeis University, Waltham, MA, USA.

Hofstede, G. (2001). *Culture's Consequences: Comparing Values, Behaviors, Institutions, and Organizations across Nations* (2nd ed.). Sage Publications, London.

Ironmonger, D. (2004). Bringing up Bobby and Betty: The inputs and outputs of childcare time. In N. Folbre and M. Bittman (Eds.), *Family Time: The Social Organization of Care* (pp. 93–109). Routledge.

Joshi, K.D., and Kuhn, K.M. (2007). What it takes to succeed in information technology consulting: Exploring the gender typing of critical attributes. *Information Technology and People*, 20(4), 400–424.

Korabik, K., Lero, D.S., and Ayman, R. (2003). A multi-level approach to cross cultural work–family research: A micro and macro perspective. *International Journal of Cross Cultural Management*, 3(3), 289–303.

Kossek, E.E. (2020). Exploring an organizational science view on faculty gender and work–life inclusion: conceptualization, perspectives, and interventions. In E. Kossek and K.-H. Lee (Eds.), Fostering gender and work–life inclusion for faculty in understudied contexts: An organizational science lens (pp. 28–32). West Lafayette, IN: Purdue e-Pubs. DOI: 10.5703/1288284317255. Retrieved from https://docs.lib.purdue.edu/worklifeinclusion/2018/gwlibsufc/3/.

Kossek, E.E., Colquitt, J.A., and Noe, R.A. (2001). Caregiving decisions, well-being, and performance: The effects of place and provider as a function of dependent type and work–family climates. *Academy of Management Journal*, 44(1), 29–44.

Kossek, E.E., Noe, R.A., and DeMarr, B.J. (1999). Work–family role synthesis: Individual and organizational determinants. *The International Journal of Conflict Management*, 10(2), 102–29.

Kwok, J. (2016). The relationship between working from home and the time parents spend on child care. Available at SSRN 2867773.

Lambert, C.H., Kass, S.J., Piotrowski, C., and Vodanovich, S. J. (2006). Impact factors on work–family balance: Initial support for border theory. *Organization Development Journal*, 24(3), 2006.

Lee, R.T., and Ashforth, B.E. (1996). A meta-analytic examination of the correlates of the three dimensions of burnout. *Journal of Applied Psychology*, 81(2), 123–33.

Lott, Y. (2018). Does flexibility help employees switch off from work? Flexible working-time arrangements and cognitive work-to-home spillover for women and men in Germany. *Social Indicators Research*, 151, 471–94.

Lyness, K.S., and Kropf, M.B. (2005). Work–family balance: A study of European managers. *Human Relations*, 58(1), 33–60.

Manzo, L.K.C., and Minello, A. (2020). Mothers, childcare duties, and remote working under COVID-19 lockdown in Italy: Cultivating communities of care. *Dialogues in Human Geography*, 10(2), 120–23. https://doi.org/10.1177/2043820620934268.

Margolis, J., and Fisher, A. (2002). *Unlocking the Clubhouse: Women in Computing*. The MIT Press.

Matthews, R.A., Swody, C.A., and Barnes-Farrell, J.L. (2012) Work hours and work–family conflict: The double-edged sword of involvement in work and family. *Stress and Health*, 28(3), 234–47.

Messersmith, J. (2007). Managing WFC among information technology workers. *Human Resource Management*, 46(3), 429–51.

Moore, J.E. (2000b). One road to turnover: An examination of work exhaustion in technology professionals. *MIS Quarterly*, (1), 141–68.

Neisser, U. (1976). *Cognition and Reality*. San Francisco: Freeman.

Nelson, M.K. (2010). *Negotiated Care: The Experience of Family Day Care Providers*. Temple University Press.

Netemeyer, R.G., Boles, J.S., and McMurrian, R. (1996). Development and validation of work–family conflict and family–work conflict scales. *Journal of Applied Psychology*, 81(4), p. 400.

Netemeyer, R.G., Brashear-Alejandro, T., and Boles, J.S. (2004) A cross-national model of job-related outcomes of work role and family role variables: A retail sales context. *Academy of Marketing Science*, 32(1), 49–60.

Offer, S., and Schneider, B. (2011). Revisiting the gender gap in time-use patterns: Multitasking and well-being among mothers and fathers in dual-earner families. *American Sociological Review*, 76(6), 809–33.

Olbrich, S., Trauth, E., Niederman, F., and Gregor, S. (2015). Inclusive design in IS: Why diversity matters. *Communications of the Association for Information Systems*, 37, 767–82. 10.17705/1CAIS.03737.

O'Leary, M.B., Mortensen, M., and Woolley, A.W. (2011). Multiple team membership: A theoretical model of its effects on productivity and learning for individuals and teams. *The Academy of Management Review*, 36(3), 461–78.

Padavic, I., Ely, R.J., and Reid, E.M. (2020), Explaining the persistence of gender inequality: The work–family narrative as a social defense against the 24/7 work culture, 2019, *Administrative Science Quarterly*, 65 (1), https://doi.org/10.1177/0001839219832310.

Powell, G.N., and Greenhaus, J.H. (2010). Sex, gender, and decisions at the family–work interface. *Journal of Management*, 36(4), 1011–39.

Quesenberry, J.L., Trauth, E.M., and Morgan, A.J. (2006). Understanding the "mommy track": A framework for analyzing work–family balance in the IT workforce. *Information Resources Management Journal*, 19(2), 37–53.

Rajadhyaksha, U. (2017). Examining the interaction of societal culture and contextual variables on work–family conflict and work–family positive spillover. In K. Korabik, Z. Aycan, Roya Ayman (Eds.), *The Work–Family Interface in Global Context*. New York. NY: Routledge.

Rasmussen, B., and Hapnes, T. (1991). Excluding women from the technologies of the future: A case study of the culture of computer science. *Futures*, 23, 1107–1119.

Robertson, L.G., Anderson, T.L., Hall, M.E.L., and Kim, C.L. (2019). Mothers and mental labor: A phenomenological focus group study of family-related thinking work. *Psychology of Women Quarterly*, 43(2), 184–200.

Rothbard, N.P. (2001). Enriching or depleting? The dynamics of engagement in work and family roles. *Administrative Science Quarterly*, 46(4), 655–84.

Salaff, J.W. (2002). Where home is the office: The new form of flexible work. In Wellman, B., and Haythornthwaite, C. (Eds.), *The Internet in Everyday Life* (pp. 464–95). Blackwell Publishers Ltd.

Sarker, S., Ahuja, M.K., and Sarker, S. (2018). Work–life conflict of distributed software development personnel: An empirical investigation using border theory. *Information Systems Research*, 29(1), 103–26.

Sarker, S., Ahuja, M., Sarker, S., and Bullock, K.M. (2021). *Navigating Work–Life Balance in an Evolving and Expanding Workforce: Challenges and Opportunities for Organizations*, Palgrave Macmillan, New York, USA.

Schwartz, F.N. (1992) Women as a business imperative. *Harvard Business Review*, 70, 65–76.

Singh, J., Goolsby, J.R., and Rhoads, G.K. (1994). Behavioral and psychological consequences of boundary spanning burnout for customer service representatives. *Journal of Marketing Research*, 31(4), 558–69.

Spertus, E. (2004). Why are there so few female computer scientists? MIT AI Lab.

Starmer, A.J., Frintner, M.P., Matos, K., Somberg, C., Freed, G., and Byrne, B.J. (2019). Gender discrepancies related to pediatrician work–life balance and household responsibilities. *Pediatrics*, 144(4).

Taylor, P., and Bain, P. (2005). "India calling to the far away towns": The call centre labour process and globalization. *Work, Employment and Society*, 19(2), 261–82.

Thomas, L.T., and Ganster, D.C. (1995). Impact of family-supportive work variables on work–family conflict and strain: A control perspective. *Journal of Applied Psychology*, 80(1), 6–15.

Trauth, E.M., and Connolly, R. (2021). Investigating the nature of change in factors affecting gender equity in the IT sector: A longitudinal study of women in Ireland. *MIS Quarterly*, 45(4), 2055–100.

Trauth, E.M., Quesenberry, J.L., and Huang, H. (2009). Retaining women in the US IT workforce: Theorizing the influence of organizational factors. *European Journal of Information Systems*, Special Issue on Meeting the Renewed Demand for IT Workers 18, 476–97.

Trauth, E.M., Quesenberry, J.L., and Huang, H. (2008). A multicultural analysis of factors influencing career choice for women in the information technology workforce. *Journal of Global Information Management*, 16(4). 1–23. 10.4018/jgim.2008100101.

Turkle, S. (1988). Computational reticence: Why women fear the intimate machine. In C. Kramarae (Ed.), *Technology and Women's Voices*. New York: Routledge and Kegan Paul.

Webster, J. (2002). Widening women's work in information and communication technology: Integrative model of explicative variables. EU Programme for Information Society Technologies.

Wickens, C.D. (1991). Processing resources and attention. In D. Damos (Ed.), *Multiple-Task Performance*. Taylor and Francis Inc.

Wood, W., and Eagly, A.H. (2012). Biosocial construction of sex differences and similarities in behavior. *Advances in Experimental Psychology*, 46, 55–123.

Wood, W., and Eagly, A.H. (2010). Gender. In S.T. Fiske, D.T. Gilbert, and G. Lindzey (Eds.), *Handbook of Social Psychology* (5th ed., Vol. 1, pp. 629–67). New York, NY: Oxford University Press.

Wynn, A. (2019). Pathways toward change: Ideologies and gender equality in a Silicon Valley technology company. *Gender and Society*, 34(1), 106–30.

Wynn, A., and Rao, H. (2020) Failures of flexibility: How perceived control motivates the individualization of work–life conflict, *ILR Reviews*, 73, 1.

Zhang, S. 2020. A work–life conflict perspective on telework. *Transportation Research Part A: Policy and Practice*, 141, 51–68.

Zijlstra, F.R.H., Roe, R.A., Leonora, A.B., and Krediet, I. (1999). Temporal factors in mental work: Effects of interrupted activities. *Journal of Occupational and Organizational Psychology*, 72(2), 163–85.

PART III

INDIVIDUAL INFLUENCES

16. Empowering Techgirls: role modeling and mentoring the next generation in STEM

Tricia Massey, Jenine Beekhuyzen and Sue Nielsen

INTRODUCTION

This chapter tells the story of the Tech Girls Movement Foundation – a not-for-profit educational outreach program aimed at shifting the Australian science, technology, engineering and mathematics (STEM) landscape – and its efforts to create cultural change. Specifically, the chapter investigates how role models can have a profound personal influence on individuals, and examines how mentoring can facilitate strong and lasting relationships that contribute to identity construction and long-term change. The chapter offers research evidence from a longitudinal evaluation of the Techgirls program over seven years to support the notion that counter-stereotypical role modeling is instrumental in encouraging and fostering girls' interest in and pursuit of careers in technology and STEM. The application of a theatrical lens to the empirical data suggests that role modeling and mentoring enables girls to re-imagine and then re-design their individual personas to include elements of technology into their identities.

For many years, research from the United Kingdom, the United States, and Australia has reported that women make up only about a quarter of all technology positions in their countries, with even fewer women choosing to study STEM subjects at university (often less than 20 percent). With predictions for an ever-increasing reliance on a STEM-trained workforce beyond 2020, the Tech Girls Movement Foundation was born to bring about social change and equity. This chapter offers a narrative of the movement's evolution and provides a theoretical analysis of its achievements, measured in real-life stories of creating actual change and providing pathways for young women to indeed "*be* it, once they can *see* it."

Techgirls was established to give girls choices in life. It does this through imaginative, fun, free adventure story books featuring real-life women as STEM role models titled *Tech Girls Are Superheroes*, in addition to hands-on design thinking and entrepreneurship programs in which teams of school girls work with a coach to solve a local community problem they care about. The organization gives girls awareness of STEM through access to skills development, which enables confidence with technology, leading to girls making informed choices about careers in STEM. While many young people's professional career choices will be STEM-related by default due to the proliferation of STEM across workplaces and sectors, the ultimate goal of Techgirls is to prepare young people to not only actively participate in the myriad of career opportunities before them in the next decade, but to stand as leaders among their peers in this space. This is especially important at a time when new technologies such as artificial intelligence are laden with complex biases that need to be dismantled before they do significant damage to minority and vulnerable populations in particular (Criado-Perez, 2019; Strengers and Kennedy, 2020).

This chapter details the second author's, Jenine Beekhuyzen's, ground-breaking, grassroots work in establishing and growing Techgirls in Australia and New Zealand, as well as the experiences contributing to her commitment to be a trailblazer for inspiring young women to pursue careers in STEM. Beekhuyzen has been running intervention programs for school girls, and researching them, for 20 years. The other contributing authors, Tricia Massey and Sue Nielsen, have a long-standing relationship with both Beekhuyzen and Techgirls in various capacities. Massey, a mother to two young girls, has many years of experience as a lecturer in intercultural communication and drama. Her research explores theatre and constructions of individual identity. She edited the *Tech Girls Are Superheroes* fiction books which are a key part of the campaign and brand of Techgirls. She contributes an understanding of the profound impact of role-playing and role modeling with regard to identify formation and challenging gender stereotypes. Nielsen's pioneering work in the participation of women in the world of information technology (IT) in Australia starting back in 1997 helped bring attention to gender inequity in the Australian information technology industry. Further, Nielsen's and Liisa von Hellens's leadership in the WinIT (Women in IT) research project and authorship of many of the *Tech Girls Are Superheroes* stories have variously supported this work. Nielsen was also Beekhuyzen's undergraduate lecturer and doctoral supervisor. The team began working together as university teaching colleagues and their paths have continued to overlap in numerous projects and they have united for this chapter to align their research and theoretical perspectives about the impact of Beekhuyzen's vastly successful brainchild: The Tech Girls Movement Foundation.

In our combined experience, we have found that role modeling can have a profound effect on personal and career choices, with timeless stories like those of Ada Lovelace, Hedy Lamarr, and Grace Hopper showing women's potential and opportunity in STEM. Contemporary local accessible role models who are also involved in mentoring can have a similarly influential impact to those we revere in history: we have watched those young people who practiced at playing out the role models eventually, in turn, become role models themselves. Together we have both been part of and played witness to a ripple effect, in which Beekhuyzen herself was positively influenced by her female lecturers, like Nielsen, in her undergraduate studies at university. During this time, she was experiencing vulnerability as a minority (there were only a handful of women in the degree) and she seriously considered quitting her studies. The support and positive ethos offered by her female lecturers enabled her to graduate and led her to become a role model and mentor: first to secondary school students (as an undergraduate student); then to undergraduate students (as a doctoral student and chair of the newly established alumni association in her university faculty); and later by community volunteering in STEM for 20 years. This contribution to the community was recognized with an honorable Order of Australia Medal endorsed by Queen Elizabeth II in June 2020 for Beekhuyzen's contribution to information technology, and to women.

Techgirls has continued to evolve over the last nine years. As evidenced in the literature, over the last decade, it has become apparent how important it is to introduce both typical and counter-stereotypical STEM role models to young people to help shape how they see women and their roles in work, family, and society. Techgirls' overarching goal is to continue to inspire girls and women to nurture their curiosity and ingenuity in areas of IT. As we say at Techgirls: *if you can see it, you can be it.*

This chapter firstly reviews the research on female participation in STEM, focusing on data around mentoring and role modeling, and girls' commitment to study and career paths in areas of information technology. The discussion then moves on to unpack the concepts of identity construction, introducing Judith Butler's (1990) powerful notion of gender as performative.

The chapter continues with an overview of the origins and impact of Techgirls, and highlights some of the integral work being done to expand girls' opportunities to engage with STEM. We then conclude by considering the powerful implications that role models and mentors may have on young girls and women and their career paths.

BACKGROUND

Mentoring and Role Modeling

The topic of mentoring has received far more attention in the academic literature than role modeling, so it is necessary to distinguish between them.

Mentoring: providing "advice and support to a protege through an interactive relationship" (Gibson, 2004, p. 137).

Role models: "(a) show us how to perform a skill and achieve a goal – they are *behavioral models*; (b) show us that a goal is attainable – they are *representations of the possible*, and (c) make a goal desirable – they are *inspirations*" (Morgenroth, Ryan, and Peters, 2015, p. 467).

In short, role models may perform mentoring, but mentoring is not a necessary function of a role model. Mentors are, however, often role models. Merton (1957) coined the term "role model" to refer to individuals in specific roles (e.g., surgeons) serving as examples of the behavior associated with this role. Shapiro et al. (1978, p. 52) saw role models as "individuals whose behaviors, personal styles and specific attributes are emulated by others" and they argued that modeling contributes to identity construction. Sealy and Singh (2006, p. 1) expanded this definition of a role model as a "symbolic entity"; a person who is "inspirational and/or motivational" and "someone from whom one can learn and model desired behaviours". More broadly, role modeling is "a cognitive process in which individuals actively observe, adapt and reject attributes of multiple role models" (Gibson, 2004, p. 136).

Forming in early childhood and continuing to influence behavior through adolescence and adulthood, gender roles may impact the choice of academic majors and careers (Olsson and Martiny, 2018). It is argued that "the current unequal distribution of women in various occupational roles acts as a psychological barrier to women's entry into certain academic and high-status professional fields" (p. 10) and Olsson and Martiny (2018, p. 1) find it reasonable "that if observing men and women in gender-congruent roles fosters gender-congruent aspirations and behaviour, then frequently observing gender-incongruent role models (e.g., male kindergarten teachers or female scientists and leaders) should reduce gender stereotyping and promote gender-counter stereotypical aspirations and behaviour".

The notion of counter-stereotypical identity modeling can be understood through Sealy and Singh's (2010) work on identity formation, which suggests that social theories of identity formation can better help one to understand the importance of role models: they explain that role models are cognitive constructions based on attributes of people in social roles. They argue that for the role modeling to work, the individual needs to perceive the role model to be similar enough to themselves, and then increase their perceived similarity to them by emulating those attributes. Their research found that this similitude can result in a behavioral change. Thus, role-model-based interventions can positively impact young girls' perceptions of gender stereotypes. Further, in their study, González-Pérez et al. (2020) found that counter-stereotypical role models can increase self-expectations of success in maths and the choice of STEM for study.

The underlying reasons for the success of some role model interventions are not always clear (Olsson and Martiny, 2018) and there is little academic research into the gendered aspects of role modeling. However, we may glean some insights from Kneeskern and Reeder (2020) who examined the impact of fiction literature on children's gender stereotypes. They found that fiction literature can influence the beliefs and attitudes of readers of all ages; therefore, fiction may well be a useful tool for helping individuals overcome potentially damaging beliefs, such as rigid gender stereotypes.

Influencing Girls in Technology

Research over the past two decades has pointed to two main reasons why girls choose not to engage in STEM: a lack of visible female role models and a lack of understanding of what people working in technology do. It is widely discussed that girls opt out of STEM around the start of primary school because they do not feel they belong. This can be accounted for by the brilliance bias that girls experience in relation to maths. A US study found that at six years old, girls suddenly see themselves as less competent at maths compared to boys, and this gender-brilliance bias is difficult to correct (Criado-Perez, 2019). A New York poll of 200 kindergartens confirms this phenomenon, finding that a gender-brilliance stereotype already exists among students at this young age, with the characteristic of being "really, really smart" associated more with white men than white women (Jaxon et al., 2019).

Erikson (1968) suggested eight stages of psychosocial development in young people, each characterized by a unique developmental issue or crisis. In adolescence (stage 5), identity formation is a critical issue and the results of personal exploration during adolescence may be the development of a strong sense of self, independence, and control, or a weaker sense of self, insecurity, and role confusion. Girls (and boys) are faced with choices about what they will bring from their earlier life into maturity and what they will discard. Bowing to parental and peer pressure, many girls turn away from activities often considered to be stereotypically affiliated with masculinity – such as active participation in sports, video games, or pursuits that are often presented as incompatible with feminine heterosexual development (Carr, 2007). It is these choices during adolescence that may wind up discouraging young women from exploring careers in technology.

While the reasons these choices occur have been investigated for decades (Trauth, 2002; von Hellens and Nielsen, 2001), there has been less exploration of the reasons as to why a small number of women resist the pressure to conform and instead develop and maintain a strong sense of self during the crucial adolescent period, going on to thrive in successful careers in these areas. The experiences of these women are accounted for by theories of individual differences (Trauth and Connnolly, 2021).

Identity and the Power of Role Modeling

Let's consider identity formation through a theatrical lens. Moving forward from the rich history of Western theatre as providing a mimetic, or mirroring, effect, to investigate the power of role-playing and role-modeling in everyday life provides a unique lens through which to consider the disruption and re-imagining of gendered identity construction to include women as key players in technology fields.

Role-playing and role-modeling can be instrumental in shaping one's identity: indeed, role-models act as anchors in our lives. As Giddens (1991) suggests, the formation of our sense of "being in the world" (p. 37) depends on emotional as well as cognitive anchoring; we require confidence in both our cultural setting (the norms, values, and institutions associated with the areas we are interested in) and trust in the reliability of persons we depend on (family, friends, peers and authority figures). Erikson (1968) calls this basic trust, which forms our emotional and cognitive connection to others and to ourselves, self-identity. This trust is achieved partly through "the symbolic interpretations of existential questions" (p. 38); these may take the form of myths and legends, media portrayals, films, computer games, magazines, and books.

Regarding career choices, we may interpret this to mean that our cognitive confidence is supported by information and explanations about potential careers provided by industry organ-izations, educational and counseling services, and news items. Such information purports to be objective, providing generalized, accurate information. In contrast, our emotional confidence is supported by more symbolic means, and these are sadly lacking and frequently derogatory in information technology sectors in comparison with other career areas such as medicine and law (Clayton et al., 2012). There is a lack of stories that allow women to emotionally identify with reliable others in the STEM world. In other words, the anchoring role models are not often women in technology; thus, there are few possibilities for *role-playing* women in technology. Techgirls was designed as an intervention to provide role models for girls who are facing decisions about tertiary education and careers, connecting students with mentors to personally guide them through the journey. In altering the anchors, Techgirls is working to provide a range of identities that do show women in technology, thereby expanding the space for girls to role-play such identities.

Girls and boys are equally competent in STEM areas; despite this, girls still tend to be unin-terested and avoid them as study and career choices, which Morgenroth et al. (2015, p. 468) surmise can be interpreted "as a motivational rather than a performance issue." The impor-tance of motivating forces and, in particular, role models, should not be underestimated. Sealy and Singh (2008) assert that role models are necessary for the procurement of "professional identity, personal growth and career success, as they provide a source of learning, motivation, self-definition and career guidance." Understanding the significance of role modeling through a lens often employed in the realm of theatre scholarship can be insightful. Judith Butler (1990, p. 179) proposes that "gender is in no way a stable identity or locus of agency from which various acts proceed; rather, it is an identity tenuously constituted in time – an identity instituted through a *stylized repetition of acts*." This notion – that identity is formed through the rehearsal and performance of behaviors modeled to individuals – offers an explanation as to how and why role modeling works so effectively to foster and inspire girls to pursue careers in STEM. This, in turn, highlights why Techgirls has worked so tirelessly to ensure powerful role models are continuously provided for its participants.

Performance and Habituation

Identities are formed through performing and re-performing acts (Butler, 1990) in a process that Wehrle (2020) refers to as habituation. So, if identity is formed through habituation and the performing and re-performing of what is socially construed as normal, it follows that few girls would be forming an identity that embraces a tech-savvy, computing-keen woman because this is not a very common identity portrayed by women. The initial process of becoming a self involves modeling what is most consistently shown in daily life and passively accepting or adhering to it, which Butler (2015) refers to as the first stage of performativity. Many girls, for example, might want to wear make-up or high-heeled shoes because they see this modeled by grown-up women around them, particularly in traditional and social media. While times are certainly changing, in 2022, the images that are presented most consistently are still generally heteronormative and stereotypically feminine. Fortunately, Butler's second stage of identity formation suggests something more optimistic: identity construction as voluntaristic, so girls may opt to deny or alter the previously performed aspects of identity which are absorbed via continual performance through childhood.

Within the world of computing, technologists tend to be portrayed or imaged as predominantly heterosexual male and "geeky" (Cheryan et al., 2013). In other words, because role modeling fails to portray women as technology workers, this possible aspect of identity is excluded from both the passive and active components of identity formation. Further, regardless of whether one falls in or outside of the normative, hetero, and feminine qualifiers which are seen as typically female, the technology worker's identity choices remain largely rooted in traditional male-focused stereotypes. With most role modeling rooted in traditional stereotypes, children are likely to perform what they have been presented with – certainly in the passive phase of identity formation. As Cheryan et al.'s (2013) research highlights, even a brief interaction with a negative stereotype could have lasting effects on subject choices and career possibilities. However, if counter-stereotypical role models are presented and they elicit and encourage a sense of belonging to the technology world *and* there are active interventions in the stereotypical portrayal of technology workers, more successful attraction and retention of girls in STEM fields will follow. This would allow for both levels of identity formation to encapsulate tech careers, and may proffer a perpetuation of new, more inclusive technology identities through positive role modeling and habituation.

THE STORY OF TECHGIRLS

Techgirls was launched at a public event on International Women's Day –March 8, 2014 – at the Queensland State Library in Australia with a host of supporters. Based on more than a decade of research, the program featured an entrepreneurship competition where female students build an app to solve a local community problem they care about, and a free adventure book for schoolgirls called *Tech Girls Are Superheroes*. The book featured 16 women in STEM as superhero characters along with a fun adventure story and an originally designed avatar of each role model. Subsequent books featuring more role models have been created and distributed, with 80,000 of them now in circulation. To date, more than 6,000 female students have participated in the competition, and approximately 11,000 in total have participated across the hands-on design thinking Techgirls programs.

Back in 1997, American academic Tracy Camp declared that "the ratio of women in computer science from high school to graduate school has been dwindling at a startling pace over the past decade" (Camp, 1997, p. 104). In Australia, Nielsen and von Hellens responded by establishing the WinIT project (Women in IT), which was partially motivated by the lack of women graduating in the Information Technology (IT) degree program at their university. It prompted the question: why are Australian women not studying in technology higher degrees in similar numbers to their male counterparts? In line with Camp they wondered, "Can this problem be solved?"

Beekhuyzen joined the WinIT team in 2000 while completing the final year of her undergraduate degree. By this time, the concept of the "shrinking pipeline" of computing graduates was evident with female students a minority in the IT degree. It became difficult to ignore the potential long-term consequences of the dearth of female students, and the impact this would have on the development of future technologies.

The WinIT team interviewed undergraduate and postgraduate students about their motivations to study in a technology-focused field, and the challenges they faced in their studies. As part of this, the team travelled to Hannover, Germany by invitation to deliver a course based on this research, and while there, researched university and high school students (von Hellens et al., 2009). These findings formed the basis for several Australian outreach programs: a secondary school mentoring program; a university alumni mentoring program for undergraduate students; and an annual outreach event *Technology Takes You Anywhere* for high school students to meet and hear from women working in technology. The findings of these programs were published in local and international outlets and informed the design of the longitudinal "Go Girl, Go for IT" program, an annual two-day outreach program to educate female students about career possibilities in technology, hosted on a university campus (Gorbacheva et al., 2014). From 2000 until 2022, Beekhuyzen has been directly involved in all of these programs and a variety of other initiatives and research projects to understand the continuing lack of female participation in computing, and now, more widely, in STEM.

The Impact of Techgirls

Like the project carried out by Gonzales-Perez et al. (2020), more than 1,000 professional women across various technology roles have participated as mentors in the Techgirls competition and, by default, as role models for the participating girls. Through this, they are fostering interest and providing real-life inspiration. One mentor described her involvement as "the optimal way to encourage young girls to pursue emerging high-growth roles, particularly those requiring STEM skills, is to expose them to the professional and personal experiences of actual female role models with a successful professional trajectory in STEM fields."

We can view role models as superstars: "individuals of outstanding achievement who inspire and motivate others to do their best" (Lockwood and Kunda, 1997, p. 91). Role models need to be seen as relevant, with their success seemingly attainable (Olsson and Martiny, 2018). The ideal role model is "a person who is somewhat older and at a more advanced career stage than the target individuals and who has achieved what these individuals hope for – outstanding but not impossible success at an enterprise in which they too wish to excel" (Lockwood and Kunda, 1997, p. 102). In Techgirls, these role models are presented with avatars in the *Tech Girls Are Superheroes* series of books, and through mentors matched with teams of girls in the competition, in the coaches of the teams, and in the Techgirls teams themselves.

Identity formation as a quest

At the highest level, the 12-week Techgirls STEM entrepreneurship competition represents a quest. The quest structure is found in seminal works such as Homer's *Odyssey*, Tolkien's *Lord of the Rings*, as well as many popular action movies. Quest stories are popular with adolescents, as exploration is fundamental to the transition through this stage of life. According to Erikson (1968), the remaking of personal identity from child to adult is dependent on exploration.

In the quest, the hero(ine) receives a call – an internal or external message – to undertake a journey. This invitation to participate in the competition is extended annually by Techgirls. Girls join as a team with up to four of their peers, register their team via a coach (a parent or teacher), and embark on the 12-week quest. As the program is often not a mandatory part of their school curriculum, this is an unconventional path for those participating.

The team of hero(ine)s set out on their journey to first identify a problem in their local community that they would like to solve (after brainstorming many problems). Then they research how others have tried to solve that problem, design a solution in the form of an app, build a business plan, construct a working prototype of the app, and pitch it in a 30-second and a two-minute video. Teams are encouraged to align the problem they are solving with the United Nations Sustainable Development Goals (SDGs). It is truly inspiring to witness, and a privilege to play a part in, 11-year old children writing 50-page business plans and getting their evidence-based apps onto the Google Play store in a mere 12 weeks.

This quest story mirrors the search for identity which Erikson (1968) suggests all adolescents face. The hero(ine)s in the Techgirls program forget their personal issues in their fascination with the opportunities that the technology presents, and the opportunity to solve problems they care about. This results in young people solving young people's problems. The appeal of Techgirls derives from the fact that a large proportion of its audience is confronting the need to establish their own identity in the face of pressure from peers and society, and they are inspired by the stories of other women's struggles and success. As Giddens's research conveyed, the everyday anchors that one is seeing need to be repeatedly shown to become a mainstay in the construction of identity. The role models that Techgirls provide both as real-life mentors and coaches, and through the *Techgirls Are Superheroes* stories, offer a myriad of examples for the students to observe and aspire to. In this complex process of witnessing, rehearsing, and performing for themselves, as Butler's (1990, 2015) work suggests, the participants are able to procure a more malleable identity than that which is typically presented in the more mainstream, hetero-normative media and even daily lives of participants.

From students to role models

Beekhuyzen has been role modeled to in our journey, and we have been fortunate to pass that role modeling on to not only the young people in our hands-on program, but, in some instances, even to those who only experience a casual engagement with the program. For example, one mother wrote to us to explain that her "daughter attended one of your events and even though she didn't do your program, she now has direct entry into medical science after being inspired by your program" (email from a parent, 2020).

We have also witnessed a role modeling rippling effect that will influence generations of young women to come. One sign that role models are effective is when those being modeled to become the role models themselves. To give a few examples, we have compiled some first-hand accounts of our participants.

SUNFUN: The team of five 10- and 11-year-old girls were our first public school winners in 2018. Previously all winning teams were from independent/private schools. SunFun is a sun safety app educating 10–14-year-old children to avoid sunburn and stay safe in the sun. The team reside in the skin cancer capital of the world (the Gold Coast, Australia) and have been personally affected by skin cancer through family members. This empathy prompted their solution to a local community problem they cared about. On commencing the Techgirls program, the inventive team had no coding or entrepreneurial experience; they taught themselves to code via YouTube with the help of their industry mentor and the Techgirls program. The girls opted to spend their spare time working on their project during lunch times and after school with their coach and mentor.

SunFun workshopped their "user experience vignettes" as they called them, discussing why it was important to fail (bad) ideas fast. Their professional partnership with the Queensland Cancer Council gave them added confidence. The primary school team went on to confidently pitch to ten tech companies, executives, and inventors in Silicon Valley in 2019 and positioned themselves as Ambassadors for Australia. The team also developed strong branding: identically dressed in hair braids, clothes, shoes, and even nail polish that reflected imagery of their app. The iconic team wore sports-like tracksuits with their app name and the Australian flag when traveling to Silicon Valley to pitch their apps as part of their prize for winning the competition. When strangers asked, "what sporting team are you part of?" they proudly declared they were the "Techgirls from Australia!" The girls are now in secondary school and are considered the "cool girls" in school by their peers.

VOCABULARY VOYAGERS: A team of three 14-year-old girls from Perth, Australia created an app to help their peers to prepare for their school's standardized test, NAPLAN. The Vocabulary Voyagers secondary school team were state and then national winners over two years. One team member, Kira, is now interning for a global technology company while studying supported by a scholarship at university. Kira's participation in Techgirls was mandatory as part of her school studies in Year 9. Claiming their prize, the team pitched their app in Silicon Valley. In her final year of secondary studies, Kira impressively coached a winning team in the Techgirls competition, and she hosted a week-long coding camp for her school peers.

Kira credits her participation in the Techgirls program with her choice of tertiary studies in medicine and technology and her new passion of combining these fields in her new career. Kira always dreamed of being a doctor, and after her positive and confidence-building experience in Techgirls, she realized she could have more impact by utilizing technology in the medical field. In her first year at university, her work in breast cancer screening was awarded the best undergraduate computer science project in Australia, and she presented her work overseas in Asia at an International Awards ceremony. Kira in particular credits her mentor as having a tremendous impact on her interest to pursue a career in STEM.

P-CUBED: The Plastic Pollution Preventers or the P-Cubed team created an app to help users track their single-use plastic use. The team of 9- and 10-year old students worked tirelessly during lunch times and on weekends to build their business plan, their app and their pitch. At just 10 and 11 years old they pitched their work for a week at Google, Facebook, and Ebay, etc. in Silicon Valley, and they have left a legacy in their school by inspiring an environmental school-wide initiative titled "wrapper- free Wednesday," encouraging all students to leave plastic out of their lunch boxes one day a week. At a very young age, they have presented to politicians and to rooms full of industry professionals with the confidence and spirit of true leaders.

They have since empowered more than 20 other young girls in their school to join Techgirls in subsequent years, who say they are 100 percent involved because of P-Cubed's inspiration. However, all of this may never have happened if it wasn't for one persistent parent. Chloe, one of the team members, declined to enter the competition, after an invitation from her mother, as Chloe said she "would not be good at it." Fortunately, her mother persisted, convincing her daughter to participate. Chloe found two friends, and worked on an issue they really cared about, with a great coach and a supportive mentor. The team were then more successful than 1,000+ other students and were awarded as national primary school winners. They are now strong, supportive role models for their peers.

The Entrepreneur as Role Model

The students' personal stories and experiences, and those of the lead Techgirl herself, Beekhuyzen, show the compounding influence of strong role models and mentors, and a supportive learning environment. Let us consider Jenine's personal story and experiences in light of these theoretical perspectives to understand how she, despite many odds working against her, managed to forge a successful career in technology and plant the important seeds for growing the next generation of female technologists. To begin with, her mother offered an initial spark of interest in technology for Jenine: as Sealy and Singh (2010) suggested, her mother presented as a role model who was similar to herself, thus achievable. While they were a working-class family from an underprivileged area, her mother bucked trends and ensured that her very modest salary found ways to afford new technological gadgets. This seemingly trivial exposure to a significant woman in her life valuing technology may have acted as that initial intervention in the more conventional, stereotypical women she was being exposed to, and thus planted a seed of interest in tinkering with technology. Having a key figure from her life early on, who upset some stereotypical elements of gendered identity, may have vastly impacted her self-performance by destabilizing the more accepted social performance seen outside the home. Her childhood rehearsals of identity performance thus encapsulated elements outside the norm. As the practice of "doing" and "speaking" are key processes in performing gender (Butler, 1990, p. 25), such ruptures quite possibly altered her worldview from childhood. Wehrle (2020, p. 366) argues that "through habituation [norms] become part of what one is or becomes." Therefore, the more Jenine "rehearsed" as a girl who tinkered with technology, the more embedded, or normalized, within herself, this notion of identity became.

Other positive role models include the triumvirate of successful, stereotype-defying women in university leadership positions Jenine studied under in her faculty. Continued exposure to such counter-stereotypical identities would also have impacted and then worked to nourish her burgeoning interest in technology. According to Ollson and Martiny (2018, p. 11) in their review of the literature on gender role modeling, this is the exact type of "longitudinal exposure" that can reap the greatest benefits as it "facilitate[s] active engagement ... [and] enhance[s] role model effects among young adults, particularly among highly motivated students." The profound effect of continual exposure to such powerful role models and mentors should not be overlooked in its efficacy to encourage gender equality and stereotype defiance in computing.

Outcomes of Techgirls

Indeed, it is this longitudinal and engaged role-modeling that Techgirls has endeavoured to ensure that girls throughout Australia and New Zealand are being exposed to over the past nine years. The broad range of role models and mentors that Techgirls offers provides a counter-narrative to traditional, outdated stereotypes and widens the notion of technology-worker identity. As Cheryan et al. (2013) suggest, this is necessary to disrupt contemporary pervasive stereotypes. Techgirls does this in a similar way to Trauth, et al.'s social activist play, *iDream*, which encourages girls to pursue technology careers based on empirical research that indicates young girls typically "lacked exposure to the range of IT educational options that were available" (Trauth et al., 2016, p. 5). *iDream* offers several role models who worked across different aspects of technology. Like *iDream*, Techgirls has worked to ensure that mentors come from a broad range of fields in computing and this has the power to increase awareness and understanding about the many possibilities for work in the world of computing and can further broaden the potential identity options for young girls to mirror (Trauth et al., 2016). In seeing these identities played out before them, girls can then "try them on" for themselves. In the repeated performance of such acts, these behaviors become embodied and work to actively constitute the identity. As Butler (1990, p. 278) surmises, "the act is not contrasted with the real but constitutes a reality." The girls who perform in and see *iDream*, and the girls who read about, work with, and engage with the Techgirls program, its ambassadors, mentors, and coaches, are being invited to try on and then perform, repeatedly, these new identities; through this, their realities, and their *identities*, are shifted.

The process of voluntarily adopting these nuanced and diverse identities is more likely with continual exposure. Recent research by Gonzales-Perez et al. (2020) highlights significant correlation between regular mentoring sessions with successful female STEM role models for schoolgirls and their pursuit of STEM subjects. Interestingly, these results indicate that when a role model displayed stronger counter-stereotypical qualities, the likelihood for expectations of success and pursuit in STEM increased. This was witnessed with Finn, who participated in the Techgirls program three times starting at ten years old. At the time, she was the only girl interested in technology in her class at school. During those three years, Finn role modeled her interest in technology to her friends, and then she was no longer the only tech girl in the class. Now when asked what she wants to do as a career, Finn replies, "I want to do what Jenine does."

In the Techgirls competition, teams of girls meet with their industry mentors for at least one hour per week for 12 weeks. This is a strong example of the type of longitudinal exposure that Olsson and Martiny (2018) refer to. Significantly, post-competition survey results suggest that the process of weekly meetings has not just role-modeled technology career options for the girls, but the girls have, in turn, acted as inspiration for the role models, with some mentors and coaches commenting that participating in the program "was the highlight of my year." Mentor accounts of the students' endless creativity and the pride mentors felt in seeing their students through to the end of the competition were common.

Some mentors have established long-term relationships with the teams of girls they worked with in the Techgirls program. One took her team to the movies to bond early in the relationship, as she felt initially somewhat ill-equipped to guide those she was mentoring, commenting, "I haven't spent time with eleven-year-olds since I was eleven years old, myself!" Another invited her team to her house every Saturday for the 12 weeks to work on their project, with her husband making them wood-fired pizzas. Another team of regional girls memorably visited their mentor on the fortieth floor of one of the tallest buildings in downtown Sydney – an experience made more remarkable because they had never visited the city and had certainly not been in such a tall building before. They most enjoyed the revolving door to the office and an in-house "waiter" who brought them drinks and snacks. By providing mentors and role models, the Techgirls program is helping to foster positive experiences and creating life-long memories with much-needed opportunities for girls to see counter-stereotypical identities role modeled to them, which successively enables the girls to potentially elect for voluntaristic gendered identity construction. Through these role models, the girls are being influenced to follow steps towards careers in technology and more widely in STEM.

Empirical evidence to support such claims of success regarding Techgirls role modeling and mentoring exists in an array of feedback collected by the organization throughout its existence. The data collected from participants once they completed the Techgirls competition showed that 88 percent of students felt they were well supported by their mentors and coaches (teachers) throughout the competition. There are hundreds of significant, compelling, specific stories that exemplify this. For instance, Nirmala (a young woman who works as a security consulting manager) and Daya (a senior analyst in application development) are mentors who both grew up without any STEM role models, and in their experience, girls were not encouraged to pursue careers in technology. They joined Techgirls to inspire the next generation: "I hope to be able to help girls understand that they can have a job in tech, that they are smart and capable enough, and to help them deal with some of the problems they might face."

The mentors reportedly loved watching the students they worked with grow new skills, nourish their curiosity, and develop their confidence while they worked to build their apps, business plans, and pitches. The girls being mentored reported feeling a great sense of pride and accomplishment on account of the supportive relationship offered by the mentor. We at Techgirls have seen firsthand the profound impact that having an adult that they do not know take them and their ideas seriously can have on these young people. Through the perspectives of Nirmala and Daya, two real-life, accessible examples of women in technology careers, we can see how they actively invited their mentored girls to perform what is being role modeled for them. They are constructing identities that portray girls as technically competent. These are not isolated cases: the Techgirls post-competition survey from 2020 saw many comments indicating that the mentors were genuinely motivated by the opportunity to make a difference in the lives of young people, such as: "[I joined] to bolster the confidence of our future women in STEM and encourage them to be curious, tenacious, driven and assertive – qualities that men have been encouraged to adopt, but women have typically not." This suggests that the mentors, who are role modeling successful women in technology across a broad range of careers, play a pivotal role in allowing girls to re-imagine their identities to include a future in technology.

One last example of how the Techgirls program is re-writing the notion of identity to include girls as technically capable is through the book series *Tech Girls Are Superheroes*. Featuring fictional adventure stories about technology, each story has a protagonist based on a real woman who works in a technology-based role; in the books, she has a "super-power" to help her solve problems. One mother commented on picking up the book for the first time at an expo that her daughter would not leave the stationary car once home until she had finished every last word of the book. Another mother was overjoyed that her daughter was sleeping with the book to be close to her favorite superhero character at three years old. Teachers have dressed up as their favourite character in the book for "book week" at school, with one teacher regularly wearing a skirt to primary school featuring the Techgirl superhero book avatars. In seeing real women capable of working in various fields of technology around the world, the stereotype of technologists as male is being eroded, re-imagined, and re-written.

CONCLUSION

This chapter contributes to the important discussion about our globally ubiquitous gender inequity problem in three ways. Firstly, and most importantly, the chapter details the significant work that the Tech Girls Movement Foundation is doing at a grassroots level to invoke systematic cultural change through its program, mentors, and role models, and connection to industry and schools. Secondly, our theoretical analysis highlights how and why the program holds so much potential for real world, ongoing, long-standing sustainable change, offering a distinctive interstice in the discussion around empowering young women and encouraging them to pursue careers in technology and STEM more broadly. Thirdly, the chapter's uniting of the seemingly disparate elements of the sciences and the arts to unpack the Techgirls' success shows readers how sciences and arts are, of course, inextricably linked and can work in tandem with each other – supporting and pushing each to go further. Indeed, as the girls in our programs work in ever-creative, artistic ways to solve problems, scientifically, through their technology interruptions, they showcase the way the arts and the sciences intersect. The chapter's cross-disciplinary approach may provoke further fissures in much-needed discussions about enduring change for the representation of women in technology and STEM.

The chapter contributes a nuanced understanding of role modeling and mentoring, through a novel perspective: a theatrical lens. For the girls who complete the Techgirls competition, and particularly for those who become finalists and winners, what is then reflected to them is their ability to conquer previously untried areas quickly and effectively. The competition offers showcase events for participants to publicly share their creations at the end of the competition, which are held in various states and attended by government officials, industry leaders, as well as family and friends. This powerfully denotes just how valued and important their work is. Moreover, the annual trip to Silicon Valley for the regional, state, and national winning teams enables them not only to meet inspiring and competent women and like-minded girls from all over the world, but also to "pitch" their apps to influencers at some of the most significant global technology companies (such as Amazon, Facebook, Google, and eBay). In this way, too, the girls are being shown powerful images of successful women in technology-based roles and are thus able to actively try on, perform, and re-perform identities that counter the stereotypes they might have previously been exposed to.

Allowing girls to see a range of examples of women across various STEM fields, in such a myriad of contexts, is another important aspect of the work Techgirls is doing to empower girls and encourage them to pursue technology careers. The fact that Techgirls has a strong and competent counter-stereotypical role model at its helm doing regular public speaking engagements, distributing the Techgirls book, working with strategic partners, and commandeering the mentors and coaches, ensures that girls are being exposed to these "superstar" role models visually, orally, aurally, virtually, and physically. In other words, all of these girls' senses are being engaged with counter-stereotypical versions of technology workers across various media. Like the goal for Eileen and Suzanne Trauth's play *iDream*, the Techgirls programs invite "awareness and attitude change [which may] set in motion behavior change specific to each individual's life and circumstances" (Trauth et al., 2016, p. 10). This type of wide-reaching exposure may have the power to bring about transformative learning and, ideally, may invite and encourage societal change.

This discussion aims to highlight the powerful implications that role models and mentors may have on young girls and women and their career paths. Further, if we consider identity as performative – both passively and actively created – these role models have allowed girls' identities to be transformed by encouraging and fostering their interest in computing careers. Ostensibly, this could allow the next generation of girls to already be passively performing counter-stereotypical identities to change the trajectory of individual identity development for young women: women will become commonplace and normalized within the fields of computing and STEM to include a broader range of gendered identity options. The move toward Butler's second phase of actively being able to choose to perform any identity may become more accessible with continued programs like Techgirls in place by shaking the foundations of individual identity in STEM, thus drawing new, more hopeful, more inclusive identities.

For now, the Techgirls program will continue to trailblaze, with coaches and mentors contributing "to the community to make the world a better place." Next, we will embark on a "where are they now?" campaign to further examine the longitudinal impact the program has had. We are aware that some of our student alumni are now studying in engineering, computer science, and gender studies at university. The campaign will bring their stories and journeys to life. Lastly, Techgirls is forever grateful for the more than ten thousand hours that all our coaches and mentors have contributed; and for this truly priceless contribution in shaping the future of technology, we thank them. We invite you, the reader, to contact us if you have any questions or if you would like to be involved in Techgirls.

REFERENCES

Butler, J. (2015). *Notes Toward a Performative Theory of Assembly*. Cambridge, UK: Harvard University Press.

Butler, J. (1990). *Gender Trouble: Feminism and the Subversion of Identity*. New York/London: Routledge.

Camp, T. (1997). The incredibly shrinking pipeline. *Communications of the ACM, 40*(10).

Carr, L. (2007). Where have all the tomboys gone? Women's accounts of gender in adolescence. *Sex Roles, 56*, 439–48.

Cheryan, S., Drury, B.J., and Vichayapai, M. (2013). Enduring influence of stereotypical computer science role models on women's academic aspirations. *Psychology of Women Quarterly, 37*(1), 72–79. https://doi.org/10.1177/0361684312459328.

Clayton, K., Beekhuyzen, J., and Nielsen, S. (2012). Now I know what ICT can do for me! *Information Systems Journal, 22*(5), 375–90. https://doi.org/10.1111/j.1365-2575.2012.00414.x.

Criado-Perez, C. (2019). Invisible women : Data bias in a world designed for men. Retrieved from www.overdrive.com/search?q=CC516C4E-B1B8-45A2-B5D6-DE8C0608FC58.

Erikson, E.H. (1968). *Identity, Youth and Crisis*. New York, NY: Norton.

Gibson, D.E. (2004). Role models in career development: New directions for theory and research. *Journal of Vocational Behaviour, 65*, 134–56.

Giddens, A. (1991). *Modernity and Self-Identity: Self and Society in the Late Modern Age*. Cambridge, UK: Polity Press.

González-Pérez, S., Mateos de Cabo, R., and Sáinz, M. (2020). Girls in STEM: Is it a female role-model thing? *Frontiers in Psychology, 11*(September). https://doi.org/10.3389/fpsyg.2020.02204.

Gorbacheva, E., Craig, A., Coldwell-Neilson, J., and Beekhuyzen, J. (2014). ICT interventions for girls: Factors influencing ICT career intentions. *Australasian Journal of Information Systems, 18*(3), 53–74.

Jaxon, J., Lei, R.F., Shachnai, R., Chestnut, E.K., and Cimpian, A. (2019). The acquisition of gender stereotypes about intellectual ability: Intersections with race. *Journal of Social Issues, 75*(4), 1192–215. https://doi.org/https://doi.org/10.1111/josi.12352.

Kneeskern, E.E., and Reeder, P.A. (2020). Examining the impact of fiction literature on children's gender stereotypes. *Current Psychology*. https://doi.org/10.1007/s12144-020-00686-4.

Lockwood, P., and Kunda, Z. (1997). Superstars and me: Predicting the impact of role models on the self. *Journal of Personality and Social Psychology*, *73*(1), 91–103. https://doi.org/10.1037/0022-3514.73.1.91.

Merton, R.K. (1957). *Social Theory and Social Structure*. New York, NY: Free Press.

Morgenroth, T., Ryan, M.K., and Peters, K. (2015). The motivational theory of role modeling: How role models influence role aspirants' goals. *Review of General Psychology*, *19*(4), 465–83. https://doi.org/10.1037/gpr0000059.

Olsson, M., and Martiny, S.E. (2018). Does exposure to counterstereotypical role models influence girls' and women's gender stereotypes and career choices? A review of social psychological research. *Frontiers in Psychology*, *9*(DEC), 1–15. https://doi.org/10.3389/fpsyg.2018.02264.

Sealy, R., and Singh, V. (2008). The importance of role models in the development of leaders' professional identities. In James, K., and Collins, J. (Eds). *Leadership Learning: Knowledge into Action*. London: Palgrave Macmillan, pp. 208–22.

Sealy, R.H.V., and Singh, V. (2010). The importance of role models and demographic context for senior women's work identity development. *International Journal of Management Reviews*, *12*(3), 284–300. https://onlinelibrary.wiley.com/doi/10.1111/j.1468-2370.2009.00262.x

Shapiro, E.C., Haseltine, F.P., and Rowe, M.P. (1978). Moving up: Role models, mentors, and the "patron system." *Sloan Management Review*, *19*(3), 51–58.

Strengers, Y., and Kennedy, J. (2020). *The Smart Wife: Why Siri, Alexa, and Other Smart Home Devices Need a Feminist Reboot*. London, UK: MIT Press.

Trauth, Eileen M. (2002). Odd girl out: An individual differences perspective on women in the IT profession. *Information Technology and People*, *15*(2), 98–118. https://doi.org/10.1108/09593840210430552.

Trauth, E.M., and Connolly, R. (2021). Investigating the nature of change in factors affecting gender equity in the IT sector: A longitudinal study of women in Ireland. *MIS Quarterly*, *45*(4), 2055–100.

Trauth, E., Keifer-Boyd, K., and Trauth, S. (2016). *iDream*: Addressing the gender imbalance in STEM through research-informed theatre for social change. *Journal of American Drama and Theatre*, *28*(2), 1–18.

von Hellens, Liisa, Nielsen, Sue, and Beekhuyzen, J. (2004). An exploration of dualisms in female perceptions of IT work. *Journal of Information Technology Education*, *3*, 103–16.

von Hellens, L., Clayton, K., Beekhuyzen, J., and Nielsen, S. (2009). Perceptions of ICT careers in German schools: An exploratory study. *Proceedings of the 2009 InSITE Conference*, 8. https://doi.org/10.28945/3348.

von Hellens, L., and Nielsen, S. (2001). Australian women in IT. *Communications of the ACM*, *44*(7). Retrieved from http://dl.acm.org/citation.cfm?id=379310.

von Hellens, L., Nielsen, S., Clayton, K., and Beekhuyzen, J. (2007). Conceptualising gender and IT: Australians taking action in Germany. Wellington, New Zealand: 4th QUALIT Conference Qualitative Research in IT and IT in Qualitative Research.

Wehrle, M. (2020). "Bodies (that) matter": The role of habit formation for identity. *Phenomenology and the Cognitive Sciences*. https://doi.org/10.1007/s11097-020-09703-0.

17. Intervention organizations to increase women's engagement with IT: a case study of NCWIT[1]

Roli Varma

INTRODUCTION

There has been almost exponential growth in information technology (IT) employment opportunities, yet the IT sector continues to have a major problem with diversity in the United States. For instance, in 2017 women constituted 47 percent of the overall workforce, but only 25 percent of the computer and information sciences workforce. The racial and ethnic distribution of employed women in computer and information sciences was 12.6 percent for White, 7.5 percent for Asian, 3.4 percent for Black, and 1.4 percent for Latinx (National Science Foundation 2019). White people make up approximately 60 percent of the US population; the numbers for Latinx, Black, Asian, Native Peoples, and others are 18.5 percent, 12.5 percent, 5.8 percent, 0.9 percent, and 2.2 percent respectively. In other words, women in all racial and ethnic groups except for Asian are underrepresented as their representation in computer and information sciences is less than their representation in the population as a whole.

There are various reasons why underrepresentation of women in IT is important. For one thing, it is a sector with great careers, especially for women. Economically, graduates in IT fields such as computer science, computer engineering and computer information systems receive high starting salaries. There is a gender pay gap in most science, technology, engineering and math (STEM) occupations, but it seems to be shrinking in IT. For instance, new female graduates in computer science average $79,223 in pay, whereas their male counterparts average $82,159 (Pratt, 2021). Location-wise, IT skill and knowledge are used in all sectors of the US economy and they are not limited to just the IT industry itself. Similarly, IT work occurs throughout the country and not limited to the Silicon Valley in California or Route 128 in Massachusetts – popularly known as the leading centers of electronics innovation. So, women with a degree in an IT field can find a job where they would like to live. In terms of work, IT jobs are increasingly office based and have flexible hours. This goes well with a desire to have work–life balance. Most importantly, it is a social equity issue that women are missing from these best well-paying jobs.

It should be noted that lack of diversity in IT has serious implications for the US economy. According to the US Bureau of Labor Statistics (2021), employment in IT occupations is projected to grow 11 percent from 2019 to 2029, much faster than the average for all occupations. These occupations are projected to add about 531,200 new jobs. Without an adequate supply of IT workers, the United States is likely to face challenges to its global competitiveness. Thus, a larger participation of women is likely to meet the growing demand for IT workers. This will also enhance innovation since a wider variety of people will bring diverse perspectives.

Scholarly literature on the reasons for underrepresentation of women in IT has grown in the last three decades. In contrast, there are a few studies on forces that have shaped efforts to broaden participation in IT (an exception is Aspray 2016a, 2016b). This chapter focuses on organizational interventions to increase the proportion and experiences of women in IT education and thus in employment. In particular, it presents a case study of the National Center for Women in Information Technology (NCWIT), the largest organization focused on advancing innovation by correcting underrepresentation in computing.

> NCWIT convenes, equips, and unites nearly 1,500 change leader organizations nationwide to increase the influential and meaningful participation of girls and women – at the intersections of race/ethnicity, class, age, gender identity, sexual orientation, disability status, and other historically marginalized identities – in the field of computing, particularly in terms of innovation and development.[2]

In many ways, NCWIT is unique as it has sought to build the organization's work on a strong foundation of social science research. Historically, social sciences and IT fields have been viewed in terms of "two cultures" that are mutually exclusive. Social sciences are concerned with value, culture, belief, subjectivism, emotion, personality, ethics, and other human aspects, whereas, IT fields are concerned with fact, reason, logic, objectivism, rationality, and other aspects of STEM (Perlman and Varma 2005). NCWIT, on the other hand, recognized that to address diversity in IT, one needs to know factors that influence how IT fields and sectors operate, which has been social scientists' areas of expertise. To this end, NCWIT has its own social science researchers, and it formed the Social Science Advisory Board (SAAB). According to William Aspray (2016a, p. 189), reliance on social science research "has led to a better understanding of such issues as unconscious bias, stereotype threat, collective intelligence, and changing mindsets – all of which have helped illuminate the causes of under-representation in computing." Unconscious bias occurs when people consciously reject stereotypes but still unconsciously make evaluations based on stereotypes (American Association of University Women 2016). Stereotype threat is understood as a situational predicament in which individuals are at risk of confirming negative stereotypes about their group (Inzlicht and Schmader 2012). Collective intelligence holds that when groups of people work together, they create intelligence that cannot exist on an individual level (Malone 2004). Changing mindsets promote the idea that one's talents and abilities can be developed over time through effort and persistence (Dweck 2007). Most importantly, social sciences focus on the nature of economic and political systems and how they impact underrepresentation in computing.

The author of this chapter has been a member of SAAB since 2008 and was a co-chair from 2010 to 2012. This chapter is the author's perspective of NCWIT as a gender scholar and a SAAB member with a focus on areas of particular importance for women of color – a term used mostly for American Indian, Alaskan Native, Asian, Black, Latinx, Native Hawaiian, and Pacific Islander. It is mostly because there is a need to question an underlying assumption that what applies to white women also applies to women of color. While white women have been discriminated against based on gender, they still enjoy privileges based on race which women of color do not.

The goal of this chapter is to present the case study of NCWIT as an intervention organization. First, the chapter gives a brief overview of women's movements in the United States as they laid the foundation of women's collective actions to improve their conditions. It shows that historically American women's movements have had a problem with class, ethnicity and race. Second, it briefly discusses the status of women in STEM, and a number of professional organizations that formed after World War II to broaden participation in STEM. Without these organizations laying the groundwork, we would not have organizations principally broadening participation of women in IT. It is followed by the section on why there is such a separate push solely for the status of women in IT. The next section presents how NCWIT evolved, its organizational structure, and some of its projects that are contributing to broadening the participation of girls and women in IT. Finally, the chapter discusses NCWIT's approach to broaden ethnic representation within its overall goal to encourage all women to pursue IT. To this end, it looks at women's movements and the role they played to improve the status of women of color. The chapter concludes with some of the lessons learned towards the future work for women in IT especially for women of color.

BACKGROUND

Women's Movements in the United States

With the rise of industrialization, urbanization, and liberalism in the United States in the nineteenth century, the first wave of major women's movements took place from the 1840s to the 1920s. During this time period, American men dominated commercial and political spheres, while women were consigned to the domestic domain. The women's suffrage movement demanded the improvement of women's status in the family and society through women's right to vote. The Seneca Falls Convention – the first women's rights convention held in 1848 in Seneca Falls, New York – protested women's lower economic, political and social positions in the country, and launched the women's suffrage movement (O'Connor 1996). In the temperance movement, white women primarily belonging to the upper and middle classes sought to limit the consumption of alcoholic beverages which were seen to cause domestic violence against women (Edman 2015). The abolitionist movement sought to end the practice of slavery everywhere. American abolitionists saw slavery as an abomination and an affliction on the United States and had many women such as Harriet Beecher Stowe, Susan Anthony, Harriet Tubman and Sojourner Truth (Bessler 2015). The anti-lynching movement composed of mainly Blacks and women (e.g., Ida Well, Mary Burnett Talbert and Angelina Weld Grimke) aimed to eradicate the practice of public killing of Blacks (Smith, 1996).

The depression of the 1930s, World War II in the 1940s, the civil rights and the farmworkers unionizations of the 1960s, and the anti-war and anti-nuclear movements of the 1970s had a major impact on women's movements. From the 1960s to the 1980s, women's movements focused on women's liberation. Their major effort was on passing the Equal Rights Amendment to the Constitution to secure social and economic equality regardless of sex. Women's subjugation was seen to center on the prevalence of patriarchy which viewed women as limited to being wives and mothers. Sex as a biological construct (e.g., chromosomes, sex organs, hormones and other physical attributes categorizing people as female and male) was differentiated from gender which was seen as a social construct (e.g., social role, position, behavior, or identity). The main feminist motivation for making this distinction was to counter biological determinism or the view that biology is destiny (Mikkola, 2019).

American women's movements contributed to reducing discrimination, harassment and prejudice against women at home and in the workplace. The 13th Amendment to the US Constitution ratified in 1865 abolished slavery in the United States. The 18th Amendment to the US Constitution ratified in 1919 prohibited consumption of alcohol in the United States; however, it was repealed by the 21st Amendment in 1933. The 19th Amendment to the US

Constitution, ratified in 1920, prohibited the government from denying the right to vote to its citizens based on sex. Since 1918, over 200 anti-lynching bills were introduced, all of which were voted down. Finally, in 1920 Congress moved to make lynching a Federal crime. The US Equal Employment Opportunity Commission (EEOC) established in 1965 made it illegal to discriminate against people because of their race, color, sex, national origin, religion, age, ability status, and genetic information. In 1973, the Supreme Court in *Roe v. Wade* legalized abortion. In the 1970s, women's studies emerged in academia to change the basis from which women's position in society is understood.

Though a lot of progress has been made to deal with injustice and discrimination which women have faced throughout history, not all women were benefiting. For instance, women earned the right to vote in 1920. However, minority women (and men) were blocked from voting due to Jim Crow laws – those laws that enforced racial segregation from the 1870s to 1950s – and other barriers such as poll taxes and literacy tests (Mack, 1999). It was not until the Voting Rights Act of 1965, which banned several discriminatory voting practices at the state and local levels, that it became easier for minority women to exercise their right to vote. Martha Jones (2020) has argued that Black women were not included in the women's suffrage movement to hold onto the support of many white southern women. The whole process of the Seneca Falls Convention was premised on voting rights for white women. Betty Friedan's (1963) *The Feminine Mystique*, known to spark the second wave of feminism, focused on the problem endured by white upper- and middle-class women. bell hooks (2000) has argued that Friedan was mostly concerned with the needs of white privileged women and excluded women of other races and classes as though they simply did not exist. Though largely excluded from the major white-led women organizations, minority women were engaged in race and gender-related issues. For instance, Sarah Remond filed one of the earliest lawsuits protesting race segregation; Ida B. Wells advocated for federal laws against lynching; Mary Church Terrell established the National Association of Colored Women; Frances Ellen Watkins Harper challenged Susan B. Anthony and Elizabeth Cady Stanton at an 1866 women's rights meeting; and Mary McLeod Bethune led voter registration drives after women gained the right to vote (Schuessler, 2019).

In the 1980s, with post-colonialism – how people in formerly colonized countries continue to be marginalized by western power – (see, Loomba, 2005) and post-modernism – questioning of the ideas and values associated with modern progress and innovation – (see, Lyotard, 1984), many previous constructs especially the notion of "universal womanhood" were deconstructed. Feminism was seen as multicultural and global where differences due to class, ethnicity, nationality, genetic information, and so forth were to be celebrated. Gender was no longer seen as a social construct in terms of rigid power relations; instead, it became a fluid concept based on social norms, societies' determination of men's and women's roles and activities, and how individuals understand their identities. Women of color challenged the growing inequalities with their white sisters inside the United States, and between western and non-western women outside the United States. For instance, Alice Walker (1982) depicted what it was like for a Southern Black woman to grow up and come to self-realization. Patricia Hill Collins (1990) presented a rich intellectual tradition of Black feminists despite confronting race, gender and class oppression. Chandra Talpade Mohanty (1988) critiqued western feminists for reducing third-world women into a single collective "other."

In short, if it were not for women's movements, we would not have reforms on issues such as domestic violence, equal pay, maternity leave, reproductive rights, right to vote, sexual harassment, and so forth. Nonetheless, women's movements in the United States have had a problem with class, ethnicity, and race. They have assumed that white women are the norm, and have not been inclusive of the needs, perspectives and concerns of women of color. In short, historically the women's movements have lacked an inter-sectional analysis.

Women in STEM

In 1969 when Margaret Rossiter, an American historian of science, asked, "Were there ever women scientists?" She received an authoritative answer: "no, never, none" that reflected the oppression and devaluing of women of the times (cited in Dominus, 2019). Before World War II, very few women had received bachelor's degrees in science or engineering which were mostly from all women's colleges (Rossiter, 1982). Further, opportunities for higher education in STEM were mostly available to white women from wealthy families. After World War II, rapid growth took place in STEM areas such as computers, microwaves, missiles, penicillin, radar, satellite, and a host of technologies, which gave STEM a new image with the American public. On October 4, 1957, the Soviet Union launched Sputnik – the first artificial satellite put into orbit – which created a perception of the US weakness in STEM education in general and in space science in particular. The US government began providing money and facilities for growth in STEM on a larger scale than before. The National Defence Education Act of 1958 focused on mathematics, science and foreign language education to meet the demands posed by national security needs. It encouraged students to continue their education in STEM beyond high school. By making federal student aid available to women for the first time, the Act removed financial need as a barrier to higher education in STEM for women (Rose, 2016). Prior to the 1958 Act, on the one hand, families mostly invested in higher education for their sons who were seen as breadwinners; women, on the other hand, were seen as not remaining in the labor market after getting married and having children.

Various women's organizations were formed to broaden women's participation in STEM. For instance, the Society of Women Engineers (SWE)[3] was founded in 1950. Its mission is to give women engineers a unique place and voice within the engineering industry. The Association for Women in Science (AWIS)[4] was created in 1971. It is the leading organization that advocates on behalf of women in STEM to achieve business growth, social change, and innovation. It is dedicated to driving excellence in STEM by achieving equity and full participation of women in all disciplines and across all employment sectors. The Women in Engineering ProActive Network (WEPAN)[5] was established in 1990. It leverages research and best practices to propel the inclusion of women in the field of engineering. WEPAN's network connects advocates who actively pursue strategies and implement solutions to increase participation, retention, and success of women and other underrepresented groups in engineering from college to executive leadership. Many professional STEM organizations also established women's committees within them.

National Science Board (2018) data shows that since the 1980s, women have out-numbered men in undergraduate education in the United States. Since the late 1990s, they have earned about 57 percent of all bachelor's degrees and half of all STEM bachelor's degrees. In 2015, women earned half or more of the bachelor's degrees in psychology, biological sciences, agricultural sciences, and all the social science fields within social sciences except for economics. In the same year, women earned 45 percent of all science and engineering (STEM) master's and 51 percent of doctoral degrees. Field selection at masters and doctoral levels followed the same pattern as bachelors level. In 2015, women in all racial and ethnic groups earned the majority of bachelor's degrees and more than half of the master's and doctoral degrees in the social and behavioral sciences and non-STEM fields. Overall men have been earning the vast majority of degrees awarded in engineering, computer science, physics, mathematics, and statistics; whereas, women are earning lower proportions of bachelor's, master's, and doctorate degrees in computer science, engineering, and physics. So, the question remains, why is there a gender gap in some STEM education and thus in occupations?

Scholarly literature on women in STEM is enormous. Earlier, Sue Berryman (1983) coined the phrase the STEM pipeline – young women are engaged with STEM in elementary, middle, and high school education, select a STEM major in higher education, and then go into STEM careers. The leaky pipeline metaphor has been used for the likelihood of women leaving STEM fields because of missteps. The STEM pipeline metaphor has been criticized as being too simplistic, without much empirical support (Cannady et al., 2014). The "STEM pathway," which suggests more fluid steps women take to careers in STEM, has been proposed as an alternative to the pipeline metaphor (Wang and Degol, 2013). Most studies on the gender gap in STEM have focused on various barriers that women face such as gender stereotypes (Carli et al 2016), masculine educational and work environments (Simon, Wagner and Killion 2017), unconscious gender bias (Corbett and Hill 2015), the dearth of female role models (Drury et al., 2011), questioning of mathematical abilities (Sax et al., 2015), lack of self-efficacy (MacPhee, Farro and Canetto 2013), and a sense of not belonging (Stets et al., 2017).

Often generalizations are made about all women even though the sample may be limited to white and privileged women. The classic report, *The Double Bind* (Malcom, Hall and Brown 1975) asserted that women of color are victims of two inter-related problems in science: racism and sexism. Some studies have emerged which document the lived experiences of women of color in both predominately white universities and minority-serving institutions (Ong et al., 2011; Borum and Walker, 2011; Martinez and Sriraman, 2015; Alexander and Hermann, 2016). There remains a need to do more research on women of color in STEM. A recent book *Hidden Figures* by Margot Lee Shetterly (2016) reveals the untold stories of Black women who were hired during World War II to work as "human computers" to run calculations that would launch rockets and astronauts into space.

In response to the US Civil Rights Movement – a struggle for social justice that took place from 1954 to 1968 for Blacks to gain equal rights – multiple organizations were founded in the 1970s to help minorities to engage in STEM education and careers (Aspray, 2016a). For instance, the National Action Council for Minorities in Engineering, the National Society of Black Engineers, and the National Consortium for Graduate Degrees for Minorities in Engineering and Sciences were set up to encourage Blacks to study STEM. Similarly, the Society for Advancement of Hispanics, Chicanos and Native Americans in Science, Latinos in Science and Engineering, and the Society of Hispanic Professional Engineers seek to advance the educational and career opportunities for Hispanics. The American Indian Higher Education Consortium and the American Indian Science and Engineering Society serve Native Americans in STEM education. These organizations assist both minority men and women in STEM education, though it is not clear if they have established subgroups focused specifically on women within their organizations. Nonetheless, the presence of these organizations and the affirmative action programs to make up for past racial and gender discrimination have led to claims of favouritism to minorities and women of color and the existence of reverse discrimination against whites.

Push for Women in IT

After World War II, the United States was characterized as entering what Daniel Bell (1973) called the post-industrial society. In this stage, the service sector was seen as contributing more to the US wealth than the manufacturing sector and employing more white-collar than blue-collar workers. The use of IT was seen to contribute to the country's structural shift to a service economy (Castells, 1996). Since the 1970s, IT has had a profound impact on the productivity, globalization, and growth of the US economy (Alberts and Papp, 1997). Many revolutionary technologies in the past (e.g., the printing press, the steam engine, the transistor) have brought profound changes to society. But the key qualitative difference with IT is not only that it has made fundamental changes, but more importantly, these changes have taken place at a very fast and unprecedented pace. IT has contributed to growth in demand for labor as well as an overall skill upgrade in the workplace (National Science Board, 2000).

With the rapid growth of IT, concerns were raised about the supply of IT workers. The IT industry itself reported a large shortage of IT workers (Information Technology Association of America, 1997, 1998). The US Department of Commerce (1997) issued a similar warning. The perception of the shortage of IT workers resulted in Congress adopting legislation to increase the number of temporary H-1B visas that can be awarded annually to foreign skilled workers from 65,000 to 115,000 for 1999 and 2000 fiscal years. In 2000, H-1B visas expanded to 195,000 for 2001, 2002, and 2003 fiscal years (Varma, 2020). Without strengthening its supply of IT workers, the United States was seen to risk the loss of its worldwide leadership in IT and other high-technology industries (Freeman and Aspray, 1999). It was argued that without an adequate IT workforce, these industries would experience slow innovation and product development and may become less competitive globally. Also, companies dependent on the IT workforce may move offshore to meet the competitive challenge, which may result in a negative impact on employment. A lack of IT workers may slowdown the productivity improvement, resulting in companies hiring fewer qualified people and reducing the range of products manufactured. The bottom line was that an IT worker shortage was seen to have a significant adverse impact on the US economy.

One important way to produce a continuous supply of IT workers is to expand the representation of women in IT education and thus in the workforce. However, in IT fields the gender gap has remained large. For instance, in the United States, women make up about 18 percent of the bachelor's, 30 percent of master's, and 20 percent of the doctorate degrees in computer science. These statistics have gone down since the mid-1980s. Women's share of bachelor's degrees in computer science peaked at 15,126 in 1986 and came down to 7,063 by 1995 (National Science Board, 2000). It started to increase and went up to 10,522 in 2000. Since then, the share of bachelor's degrees awarded to women in computer science has declined by 10 percent in 2015 (National Science Board, 2018). Black women's share of bachelor's degrees in computer science has declined from 4.9 percent in 1996 to 2.2 percent in 2016; Latina women's share of bachelor's degrees in computer science has not gone up from 1.87 percent over the past two decades (National Science Foundation, 2019). The proportion of master's degrees earned by women declined in computer science from 33 percent in 2000 to 31 percent in 2015. Although women earned less than one-third of the doctorates awarded in computer science, they increased their numbers from 131 in 2000 to 439 in 2015 (National Science Board, 2018). A recent study by Frieda McAlear and Allison Scott (2019) shows that among all women employed in computer and information science occupations, only 12 percent are Black or Latina women.

Given the importance of IT to the US economy and society, those interested in the US global competitiveness and those interested in social justice for women found a common platform to increase women's representation in IT. The National Science Foundation (NSF) – an independent federal agency created by US Congress in 1950 to promote the progress of science – has supported studies to improve underrepresentation in the STEM disciplines since the 1970s. In the 1990s, NSF began to invest funds specifically in computing education and human resources through various programs namely IT Workforce (2000–2005), Broadening Participation in Computing (2006–2012), and Computing Education for the 21st Century (2010–present). Since the beginning of the twenty-first century, women's underrepresentation in IT was being scrutinized from many angles (Ahuja, 2002; Margolis and Fisher, 2002; Katz et al., 2003; Larsen and Stubbs, 2005; Cohoon and Aspray, 2006; Singh et al., 2007; Papastergiou, 2008; Lang, 2010; Cheryan, 2011; Good et al., 2012; Quesenberry and Trauth, 2012; Beyer, 2014; Aspray, 2016a). Some studies have addressed class, ethnicity and race factors on women in IT (Varma, 2007a, 2007b, 2010).

A CASE STUDY OF NATIONAL CENTER FOR WOMEN IN INFORMATION TECHNOLOGY (NCWIT)

After World War II, several organizations cropped up to increase women's representation in STEM. When computer science emerged as a distinct academic discipline, these organizations provided support for women in computing. Since the late 1980s, new organizations began to surface to help women to build computing careers. For instance, the Anita Borg Institute[6] started a digital community for women in computing in 1987. The Computing Research Association Committee on the Status of Women in Computing Research (CRA-W)[7] was established in 1991 to increase the success and participation of women in computing research. In 1993, the Association for Computing Machinery started its Committee on Women in Computing (ACM-W).[8] It supports, celebrates, and advocates internationally for the full engagement of women in all aspects of the computing field.

In 2004, NCWIT was founded by Lucy Sanders, Robert Schnabel, and Telle Whitney as an American non-profit 501(c)(3) organization to focus on women's participation in IT across the entire ecosystem with the $3.25 million award from NSF (2004). Since then, NCWIT has grown to help over 1,400 organizations recruit, retain, and advance women from kindergarten to 12th grade (K-12) and higher education through industry and entrepreneurial careers by providing support, evidence, and action. At present, NCWIT is the largest organization that assists women to build a career in IT both in scope and participation.

NCWIT's success has been in its vision to move beyond an NSF-funded project at a university to a full-fledged non-profit organization. This way, NCWIT could both raise and distribute funds as well as control intellectual property generated in the process (Aspray, 2016a). NCWIT has been extremely successful in raising funds from IT and traditional technology companies. Currently, it has (i) Apple as lifetime partner, (ii) NSF, Microsoft, Bank of America, Google, Intel, Merck, AT&T, and Cognizant US Foundation as strategic partners, (iii) Avaya, Pfizer, Bloomberg, Hewlett Packard Enterprise, Qualcomm, Facebook, Morgan Stanley, Amazon, and Johnson and Johnson as investment partners, and (iv) individuals.

Though their main office is on the University of Colorado campus, NCWIT staff is distributed throughout the United States. This way, NCWIT keeps a fluid organizational structure, that is, increasing importance of flexibility and dynamism and decreasing importance of organizational boundaries, structures and processes (Jarvi et al., 2018). In addition, NCWIT is able to hire a workforce that is likely to be diverse as they come from different places.

Instead of building infrastructure based on physical plant, NCWIT has been mostly functioning through "Alliances" which are groups of like-minded organizations. The Academic Alliance,[9] NCWIT's first alliance, focuses on changing the local conditions that create barriers to attracting and graduating women in postsecondary computing by adopting and contributing research-based practices. It brings together more than 1,700 representatives from more than 500 colleges and universities nationwide such as the Boston University, California Institute of Technology, Harvey Mudd College, and United States Naval Academy to focus on local barriers to attracting and graduating women in postsecondary computing. Its main goal is to address all issues of gender and computing in higher education (Aspray, 2016a).

In 2005, NCWIT started its K-12 Alliance[10] which is made up of youth-serving organizations, professional educator associations, academic institutions, and businesses. It leverages the reach and diversity of its members to create national outreach programs that increase the participation of girls in computing. For organizations such as Girl Scouts, Campfire and YWCA who are focused on girls, the K-12 Alliance brings them together to learn how to run a successful computing program for girls by showing gender and culturally responsive STEM/computing hands-on activities. Organizations that serve educators such as the Computer Science Teachers Association and National Girls Collaborative Project, the K-12 Alliance offers research-based practices and resources to do a better job in engaging girls in computing. For organizations such as Code-org, Bootstrap and Black Girls Code who teach coding to students and teachers, the K-12 Alliance ensures that coding is taught in an inclusive way (Aspray, 2016a). NCWIT believes that the K-12 Alliance has the potential to reach 100 percent of the girls in the US.

The Workforce Alliance[11] started in 2006. It focuses on internal corporate culture change to promote more inclusive environments, build stronger technical teams, and enhance technical innovation. NCWIT provides the Workforce Alliance members such as Amazon, Pfizer, Samsung, Sony, and Wells Fargo guidance in applying research-based strategies to mitigate bias in systems such as hiring, task assignment, performance evaluation, and promotion, making company cultures more inclusive, so every voice is heard. Unlike the Academic Alliance and K-12 Alliance, however, the Workforce Alliance has been slower to grow. It is partially due to the fact that many companies are reluctant to collect/provide internal data and reveal their best practices (Aspray, 2016a). NCWIT believes that companies have gradually become interested in improving their internal operations to broaden participation. At the end of 2020, NCWIT introduced the magazine *e:think*, which advocates for a more diverse, equal and inclusive technology industry.

NCWIT has numerous successful programs. For instance, its Aspirations in Computing Program[12] started in 2007 uses awards, scholarships, internships, and professional opportunities to get girls involved in computing. So far 90 percent of its award winners have reported a college major or minor in an STEM field. NCWIT's Extension Services Program[13] uses research-based tools, processes, and best practices to increase women's participation in undergraduate and graduate computing programs. NCWIT's Pacesetter Program[14] started in 2010 is a fast-track program in which industry and university leaders committed to increasing their numbers of technical women. These organizations set for themselves a numeric goal of "Net New Women" to recruit and retain women. NCWIT's Sit with Me[15] is a national advocacy program in which people are invited to sit on a red chair to create spaces where everyone can reflect on the value of diversity and inclusion. The red chair gives all individuals an engaging way to sit in solidarity with technical women and invite others to participate. Every year at the beginning of summer, NCWIT holds an annual summit on women and IT that is free and open to the public. The summit brings educators, scholars, people working in the industry, university administrators, and the community interested in improving diversity and equity in computing. It should be noted that NCWIT has many other programs than what has been mentioned in this section.

Organizationally, NCWIT has over 30 experienced boards of directors representing academia, industry, policy, and the community. They approve the NCWIT annual budget, strategies and operation plans, audit finance, ensure legal and ethical integrity, and enhance NCWIT's public standing. NCWIT research teams consists of almost 15 scholars with doctorates in a variety of fields including education, sociology, communication, and psychology. They create research-based resources to increase awareness for systemic change, translate existing research into actionable practices, and conduct original research and evaluation. NCWIT has a number of consultants who provide expert analysis and recommendations. In addition, NCWIT has created the Social Sciences Advisory Board (SSAB) which consists of approximately 20 non-NCWIT scholars who bring expertise from the areas of anthropology, education, evaluation, gender studies, history, policy, psychology, sociology, technology, and workforce studies. They provide a support for NCWIT initiatives and goals through their knowledge of research and theory at the intersection of women and computing.

Over the years, the SSAB members have reviewed NCWIT resources and suggested improvements; contributed to setting NCWIT's research agenda; collaborated on reports about relevant theoretical and empirical issues which are published as white papers; and reviewed computing courses to remove gender bias. Often, annual summits' agendas have reflected the discussions that have previously taken place in SSAB's meetings. Some of the titles of SAAB members presentations at NCWIT are (Aspray, 2016a):

- The performance vs persistence paradox: Myths about women in IT
- Changing the image of computing to increase female participation
- Programs for undergraduate women in STEM: Issues, problems and solutions
- Crocheting the way to math equality: The effects of teaching style on math performance
- Preventing stereotype threat in standardized testing
- Beyond the double bind: Women of color in STEM
- Persistence research in STEM
- Female recruits explore engineering (free project) and free pathways
- The STEM agency initiative for STEM learning among marginalized youth
- Gender and computing: A case study of women in India.

Historically, the value of social sciences has not always been appreciated by those in physical or natural sciences. John Platt (1964), for instance, considered some scientific fields which rely on systematic methods of scientific thinking to be more productive than others. Norman Storer (1967) differentiated some science fields as "hard" and others as "soft" based on the degree of rigor directly related to the use of mathematics. Those in astronomy, biology, chemistry, geology, mathematics, and physics tend to consider social science research less legitimate as it focuses on society and people (Editorials, 2012). When NSF was established in 1950, social sciences were included under the rubric of "other sciences." Often, questions were raised about lack of rigorous methods, mathematical rigor, and verification of results in social science research. Many at NSF viewed the field of social sciences as not a "real" science (Aspray, 2016b). It was not until 1991 when a separate Directorate of Social, Behavioral and Economic Sciences was established, that social sciences began enjoying recognition and visibility like other fields at NSF. By the 1990s, NSF had also implemented various programs to increase participation of underrepresented groups in STEM. However, the outcome was not what NSF had hoped for. Some of the NSF supported programs had local success, but very little progress was made to broaden participation at the national level. There was a concern that the funding which NSF was putting into diversity was not leading to increased participation of women and underrepresented minorities in STEM (Aspray, 2016b). Most of these NSF-supported programs were limited to research conducted by social scientists on the underrepresentation of women and minorities in STEM disciplines. When the Presidential Information Technology Advisory Committee (PITAC) issued a report on September 8, 1999, that called for broadening participation in IT, NSF started its IT Workforce Program to support multi-disciplinary collaborative projects among investigators in IT and the social sciences and/or education. NSF sought to harness social science research to generate findings that would offer guidance about how to address pressing social issues such as representation of women and minorities in IT (Rosenbloom and Ginther, 2016). It is, therefore, no surprise that NCWIT sought a support from NSF with multi-disciplinary and collaborative research efforts using a variety of methods.

NCWIT has been very successful in growing from a small to a very large organization in a very short time period. NCWIT's success has been rooted in its dynamic leader Lucy Sanders, who has done most of the operational work – how to handle day-to-day activities and how to address the problem of underrepresentation of women in an effective and efficient way. According to Sanders, her goal is to make enough changes that NCWIT can put itself out of business in 20 years (cited in Aspray, 2016a). It means there should be high involvement of women in IT. However, as previously described, the problem of gender diversity in IT education and workforce in the United States has yet to be solved

Women of Color and SAAB

NCWIT's agenda has been to address "all" women – at the intersections of race, ethnicity, class, age, sexual orientation, and disability status – in the field of IT. It is similar to recent debate that instead of "Black Lives Matter" – a global organization in the United States, the United Kingdom, and Canada whose mission is to eradicate white supremacy and build local power to intervene in violence inflicted on Black communities by the state and vigilantes – focus should be on all lives matter. Many SSAB members especially those who have been chairs and co-chairs in the past feel that NCWIT ought to have a special focus on women of color. It is because women of color have a different history of oppression and injustice, and thus are not in positions in computing similar to their white sisters. According to SAAB leadership, NCWIT's focus on "all" women with intersectionality is not the best way to express its purpose and priorities because of racial history of the United States and the persistence of racial/ethnic bias. For instance, NCWIT's theory of change argues that women's absence from computing is rooted in societal systems. It discusses bias, sexism, and stereotypes and, as pointed out by Jane Margolis (2019), misses out a key variable – "racism." In August 2018, the Kapor Center and the Center for Gender Equity in Science and Technology at the Arizona State University have formed the Women of Color in Computing Research Practitioner Collaborative.[16] It is working across a broad range of stakeholder groups to build a detailed and robust foundational body of research on the barriers, trends, and effective interventions specific to underrepresented women of color in technology to inform and scale new initiatives, policies, and funding priorities.

In its 2017 summit in Tucson, Arizona, NCWIT attempted to engage the Black community by holding a workshop "NCWIT Resources Review: Ensuring Relevance for Black Communities." According to Enobang Branch (2017), one of the speakers of the workshop, it was mostly symbolic and lacked deep discussion on underlying issues which Blacks face. A similar workshop "Ensuring Engagement and Relevance in Black Communities: From Kindergarten to Workplace" was repeated at NCWIT 2018 summit in Grapevine, Texas, with similar outcome. As an example, women of color were trying to make an argument about why Black women in technology companies needed more support. However, they felt undercut when NCWIT declared that it is for "all" women. This led SAAB to discuss what "all" women means in historical times. At NCWIT 2019 summit in Nashville, Tennessee, Jane Margolis (2019) made a "Presentation to SAAB on All Women." She argued that why talking about "all" women can appear to be undercutting to the most marginalized women in the American society. SAAB members believe that an "all" women approach is limited to promoting the meaningful inclusion of girls and women from diverse backgrounds. According to them, holding such workshops, recruiting SSAB members belonging to different race/ethnicity, and having some Black speakers at its summits, NCWIT merely gives an impression that it is doing something for women of color, when in reality it is not doing it in a meaningful way.

Intersectionality, the term coined by Kimberlé Crenshaw (1989), stresses the importance of paying attention to complex interactions between multiple identities and dynamics of power and oppression. It should be noted that there is a growing concern within women's movements to re-evaluate how women face bias along with multiple identities such as: ability, age, class, ethnicity, gender identity, genetic information, nationality, race, and sexual orientation. In fact, Robin DiAngelo,[17] one of the plenary speakers at the NCWIT summit in 2019, discussed how hard it can be for white people to "see the racial water," that is, to become conscious of the constructions of whiteness and white racial identity that stand-in for "normal" in a society that is deeply divided by race. In other words, to speak for "all" women ignores the needs of women of color; instead, it ends up addressing issues of upper- and middle-class white women. Emphasis on "all" women recognizes the voices of women of color as long as it appears inclusive of the aims of white women. Even Michelle Bernard (2013), a Black woman, has argued that "Friedan's unintentional exclusion of African American women in *The Feminine Mystique* was in fact a gift because it might be what led to the development of a black feminist movement and scholarship in the area of black feminist thought" (paragraph 16). In other words, Black women owe their gender consciousness to white women. This shows how deep the problem is for women of color.

Nonetheless, SAAB's push for a specific focus on women of color appears to have resulted in two outcomes. First, NCWIT is figuring out how to address the long-standing issue in women's movements which have centred on issues relevant to white and privileged women. NCWIT is listening to the critique that various categories like race and ethnicity need to be looked into in relational terms rather than as separate categories in isolation from each other, and there is a need for grassroots activism. For instance, summits held 2019 onwards had a number of diverse speakers discussing issues pertaining to women of color. NCWIT Color of Our Future[18] is a new program which anchors NCWIT programs, initiatives, and research-based resources focused on broadening the meaningful participation of underrepresented women and girls of color (Black, Latinx, and Native American) to positively impact the future of computing. Its Modern Figures Podcast[19] guest stars Black women in computing who share their stories and perspectives on technical, societal, and personal topics. Similarly, TECHNOLOchicas[20] presents powerful stories of Latina technologists from diverse background backgrounds and environments. The Color of Our Future Conversation Series[21] explores the advancement and inclusion of Black women and girls from K-12 through career. In response to high-profile incidents of Black Americans being killed at the hands of police in 2020 and protest against racial/ethnic injustice, Sanders has sent out two newsletters: "An NCWIT Community Call to Action for Racial Justice" in January 2021, and "Beyond Thoughts and Prayers: Dismantling the Racist, Sexist Root Causes of Violence" in March 2021.

Nonetheless, NCWIT has begun to question the usefulness of SSAB. This is despite the fact that a lot of discussions about race/ethnicity have been happening in the SAAB. To this end, NCWIT is planning to significantly restructure the SSAB as an advisory body, alter individual SSAB members' activities and membership lengths, and limit support for SSAB members to the annual summit. NCWIT has its own social science research team which has grown from two to almost 15 researchers with doctorates. They work on the projects that are generated by them within the current and future goals of NCWIT. In this sense NCWIT research team does not have the absolute freedom which SSAB members enjoy. While SSAB members can engage in intellectual debate without any constraint and can do critical evaluation of NCWIT programs, the NCWIT research team remains influenced by the NCWIT context. It seems NCWIT prefers to work with its own social scientists, rather than SSAB who provide access to advice from experts. SSAB has been providing multiple perspectives and is able to critically evaluate every part of NCWIT activities.

CONCLUSION

This chapter has presented a brief case study of NCWIT as an example of the interventions that organizations can conduct in promoting the participation of advancement of women in computing. The perspective on NCWIT is by the author who has been a member of NCWIT and SAAB for the last 12 years, and was a co-chair of SAAB for two years. This chapter is the author's perspective of NCWIT as a gender scholar and a SAAB member with a focus on areas of particular importance for women of color. The chapter has framed the case study in the context of three underlying themes: women's movements in the last 200 years; the status of women in STEM and organizational efforts to change that for the better; and the importance of IT for the US economy and the need to have a focused organizational effort to increase the representation of women in computing. The chapter has described the sharp expansion of NCWIT in a very short time period. The credit for this goes to the leadership's vision, flexible organizational structure, innovative programs, and continuous flow of resources from the corporate sector. Most importantly, NCWIT has laid the groundwork for women's organizations to base their activities and best practices on social science research. The chapter has highlighted recent tension between SAAB members and NCWIT on "all" versus specific focus on women of color. NCWIT has began to address diversity among women in IT.

NOTES

1. I would like to thank Jane Margolis for sharing slides on her presentation to SAAB in 2019, and Sarah Kuhn and the reviewers for constructive comments. This chapter was partially supported by a grant from the National Science Foundation (1937849).
2. See ncwit.org/about-ncwit/.
3. See https://swe.org/about-swe/.
4. See https://awis.org/about-awis/awis-history/.
5. See https://wepan.org/page/aboutwepan.
6. See https://anitab.org/about/.
7. See https://cra.org/cra-wp/history.
8. See https://women.acm.org/.
9. See ncwit.org/higher-ed/#alliance.
10. See ncwit.org/k-12/#alliance.
11. See ncwit.org/workforce/#alliance.
12. See ncwit.org/program/aspirations-in-computing/.
13. See ncwit.org/program/highered-programs/extension-services/.
14. See ncwit.org/ncwit-program/pacesetters-program/.
15. See ncwit.org/program/sit-with-me/.
16. See www.wocincomputing.org.
17. See ncwit.org/video/2019-ncwit-summit-robin-diangelo-white-fragility.
18. See ncwit.org/program/color-of-our-future/.
19. See ncwit.org/program/color-of-our-future/modern-figures-podcast/.
20. See ncwit.org/program/aspirations-in-computing/technolochicas/.
21. See ncwit.org/program/aspirations-in-computing/the-color-of-our-future-conversation-series/.

REFERENCES

Ahuja, M.K. (2002). Women in the information technology profession: A literature review synthesis and research agenda. *European Journal of Information Systems*, 11, 20–34.

Alberts, D.S. and Papp, D.S. Eds. (1997). *The Information Age: An Anthology on its Impact and Consequences*. Washington, DC: National Defense University.

Alexander, Q.R. and Hermann, M.A. (2016). African-American women's experiences in graduate science, technology, engineering, and mathematics education at a predominantly white university: A qualitative investigation. *Journal of Diversity in Higher Education*, 9, 307–22.

American Association of University Women. (2016). *Barriers and Bias: The Status of Women in Leadership*. Washington, DC: AAUW.

Aspray, W. (2016a). *Women and Under-Represented Minorities in Computing*. Switzerland: Springer.

Aspray, W. (2016b). *Participation in Computing: The National Science Foundation's Expansionary Programs*. Switzerland: Springer.

Bernard, M. (2013, February 21). Betty Friedan and Black women: Is it time for a second look? *Washington Post*.

Berryman, S.E. (1983). *Who Will Do Science? Minority and Female Attainment of Science and Mathematics Degrees: Trends and Causes*. New York: Rockefeller Foundation.

Bessler, J.D. (2015). The abolitionist movement comes of age from capital punishment as a lawful sanction to a peremptory, international law norm. *Montana Law Review*, 79, 8–48.

Beyer, S. (2014). Why are women underrepresented in computer science? Gender differences in ste-reotypes, self-efficacy, values, and interests and predictors of future CS course taking and grades. *Computer Science Education*, 24, 153–92.

Borum, V.O. and Walker, E. (2011). Why didn't I know? Black women mathematicians and their avenues of exposure to the doctorate. *Journal of Women and Minorities in Science and Engineering*, 17, 357–69.

Branch, E. (2017). Personal communication.

Cannady, M.A., Greenwald, E. and Harris, K.N. (2014). Problematizing the STEM pipeline metaphor: Is the STEM pipeline metaphor serving our students and the STEM workforce? *Science Education*, 98, 443–60.

Carli, L., Alawa, L., Lee, Y., Zhao, B. and Kim, E. (2016). Stereotypes about gender and science: Women ≠ scientists. *Psychology of Women Quarterly*, 40, 244–60.

Castells, M. (1996). *The Rise of the Network Society*. Cambridge: Blackwell.

Collins, P.H. (1990). *Black Feminist Thought*. London: Harper Collins Academic Publishers.

Collins, P.H. (2015). Intersectionality's definitional dilemmas. *Annual Review of Sociology*, 41, 1–20.

Cohoon, J.M. and Aspray, W. Eds. (2006). *Women and Information Technology: Research on Underrepresentation*. Cambridge: MIT Press.

Corbett, C. and Hill, C. (2015). *Solving the Equation: The Variables for Women's Success in Engineering and Computing*. Washington, DC: American Association of University Women.

Crenshaw, K. (1989). Demarginalizing the intersection of race and sex: A Black feminist critique of anti-discrimination doctrine, feminist theory and antiracist politics. *University of Chicago Legal Forum*, 1989, 139–67.

Dominus, S. (2019). Women scientists were written out of history: It's Margaret Rossiter's lifelong mission to fix that. *Smithsonian Magazine*. www.smithsonianmag.com/science-nature/unheralded -women-scientists-finally-getting-their-due-180973082/.

Drury, B.J., Siy, J.O. and Cheryan, S. (2011). When do female role models benefit women? The impor-tance of differentiating recruitment from retention in STEM. Psychological Inquiry, 22, 265–69.

Dweck, C.S. (2007). *Mindset: The New Psychology of Success*. New York: Ballantine Books.

Editorials. (2012). A Different Agenda. *Nature*, 487, 271.

Edman, J. (2015). Temperance and modernity: Alcohol consumption as a collective problem, 1885–1913. *Journal of Social History*, 49, 20–52.

Freeman, P. and Aspray, W. (1999). *The Supply of Information Technology Workers in the United States*. Washington, DC: Computing Research Association.

Friedan, B. (1963). *The Feminine Mystique*. New York: W.W. Norton.

hooks, b (2000). *Feminist Theory: From Margin to Center*. London: Pluto Press.

Information Technology Association of America. (1997). *Help Wanted: The IT Workforce Gap at the Dawn of a New Century*. Arlington: ITAA.

Information Technology Association of America. (1998). *Help Wanted: A Call for Collaborative Action for the New Millennium*. Arlington: ITAA.

Inzlicht, M. and Schmader, T. (Eds.). (2012). Stereotype Threat: Theory, Process, and Application. Oxford University Press. DOI:10.1093/acprof:oso/9780199732449.001.0001.

Jarvi, K., Almpanopoulou, A., and Ritala, P. (2018). Organization of knowledge ecosystems: Prefigurative and partial forms. *Research Policy*, 47, 1523–37.

Jones, M.S. (2020). *Vanguard: How Black Women Broke Barriers, Won the Vote, and Insisted on Equality for All*. New York: Basic Books.

Katz, S., Aronis, J., Allbritton, D., Wilson, C. and Soffa, M.L. (2003). Gender and race in predicting achievement in computer science. *IEEE Technology and Society Magazine*, 22(3), 20–27.

Lang, C. (2010). Happenstance and compromise: A gendered analysis of students' computing degree course selection. *Computer Science Education*, 20, 317–45.

Larsen, E.A. and Stubbs, M.I. (2005). Increasing diversity in computer science: Acknowledging yet moving beyond gender. *Journal of Women and Minorities in Science and Engineering*, 11, 139–69.

Loomba, A. (2005). *Colonialism/Post-Colonialism*. London: Routledge.

Lyotard, J.F. (1984). *The Post-Modern Condition: A Report on Knowledge*. Minneapolis: University of Minnesota Press (trans. G. Bennington and B. Massumi).

Mack, K.P. (1999). Law, society, identity, and the making of the Jim Crow South: Travel and segregation on Tennessee railroads, 1875–1905. *Law and Social Inquiry*, 24, 377–409.

MacPhee, D., Farro, S. and Canetto, S.S. (2013). Academic self-efficacy and performance of underrepresented STEM majors: Gender, ethnic, and social class patterns. *Analyses of Social Issues and Public Policy*, 13, 347–69.

Malcom, S.M., Hall, P.Q. and Brown, J.W. (1976). *The Double Bind: The Price of Being a Minority Woman in Science*. Washington, DC: American Association for the Advancement of Science.

Malone, T. (2004). *The Future of Work: How the New Order of Business Will Shape Your Organization, Your Management Style and Your Life*. Boston: Harvard Business School Press.

Margolis, J. (2019). Personal communication "Presentation to SAAB on All Women."

Margolis, J. and Fisher, A. (2002). *Unlocking the Clubhouse: Women in Computing*. Cambridge: MIT Press.

Martinez O.A. and Sriraman, V. (2015). Exploring faculty insights into why undergraduate college students leave STEM fields of study – A three-part organizational self-study. *American Journal of Engineering Education*, 6, 43–60.

McAlear, F. and Scott, A. (2019). Women of color in computing: A researcher-practitioner collaborative. *Proceedings of American Society for Engineering Education*, https://peer.asee.org/by_year/2019, Paper ID 25024.

Mikkola, M. (2019). Feminist perspectives on sex and gender. *The Stanford Encyclopedia of Philosophy*, https://plato.stanford.edu/archives/fall2019/entries/feminism-gender/.

Mohanty, C.T. (1988). Under western eyes: Feminist scholarship and colonial discourses. *Feminist Review*, 30, 61–88.

National Science Board. (2000). *Science and Engineering Indicators*. Arlington: National Science Foundation.

National Science Board. (2018). *Science and Engineering Indicators*. Alexandria: National Science Foundation.

National Science Foundation. (2019). *Women, Minorities, and Persons with Disabilities in Science and Engineering*. Alexandria: National Science Foundation.

NSF (2004, October 19). NSF Announces $3.25 Million Award to Increase Participation by Women in Information Technology Careers. *News Release 04-144*. www.nsf.gov/news/news_summ.jsp?cntn_id=100461.

O'Connor, S.D. (1996). The history of the women's suffrage movement. *Vanderbilt Law Review*, 49, 657–75.

Ong, M., Wright, C., Espinosa, L. and Orfield, G. (2011). Inside the double bind: A synthesis of empirical research on undergraduate and graduate women of color in science, technology, engineering, and mathematics. *Harvard Educational Review*, 81, 172–209.

Papastergiou, M. (2008). Are computer science and information technology still masculine fields? *Computers and Education*, 51, 594–608.

Perlman, B. and Varma, R. (2005), Barely managing: Attitudes of information technology professionals on management technique. *Social Science Journal*, 42, 583–94.

Platt, J.R. (1964). Strong inference. *Science*, 146(3642), 347–53.

Quesenberry, J.L. and Trauth, E.M. (2012). The (dis)placement of women in the IT workforce: An investigation of individual career values and organizational interventions. *Information Systems Journal*, 22, 457–73.

Rose, D. (2016). The public policy roots of women's increasing college degree attainment: The National Defense Education Act of 1958 and the Higher Education Act of 1965. *Studies in American Political Development*, 30, 62–93.

Rosenbloom, J.L. and Ginther, D.K. (2016). The effectiveness of social science research in addressing societal problems: Broadening participation in computing. *Economics Working Papers*, Iowa State University. https://lib.dr.iastate.edu/econ_workingpapers/1.

Rossiter, M. (1982). *Women Scientists in America: Struggles and Strategies to 1940*. Baltimore: Johns Hopkins Press.

Sax, L.J., Kanny, A.M., Riggers-Piehl, T.A., Whang, H. and Paulson, L.N. (2015.) But I'm not good at math: The changing salience of mathematical self-concept in shaping women's and men's STEM aspirations. *Research in Higher Education*, 56, 813–42.

Schuessler, J. (2019, August 15). The complex history of the women's suffrage movement. www .nytimes.com/2019/08/15/arts/design/the-complex-history-of-the-womens-suffrage-movement.html.

Shetterly, M.L. (2016). *Hidden Figures: The American Dream and the Untold Story of the Black Women Mathematicians Who Helped Win the Space Race*. New York: Harper Collins Publishers.

Simon, R.M., Wagner, A. and Killion, B. (2017). Gender and choosing a STEM major in college: femininity, masculinity, chilly climate, and occupational values. *Journal of Research in Science Teaching*, 54, 299–323.

Singh, K., Allen, K.R., Scheckler, R. and Darlington, L. (2007). Women in computer-related majors: A critical synthesis of research and theory from 1994 to 2005. *Review of Educational Research*, 77, 500–533.

Smith, B.H. (1996). Lynching, federalism and the intersection of race and gender in the progressive era. *Yale Journal of Law and Feminism*, 8, 31–78.

Stets, J.E., Brenner, P.S., Burke, P.J. and Serpe, R.T. (2017). The science identity and entering a science occupation. *Social Science Research*, 64, 1–14.

Storer, N.W. (1967). The hard sciences and the soft: Some sociological observations. *Bulletin of the Medical Library Association*, 55, 75–84.

US Bureau of Labor Statistics. (2021). www.bls.gov/ooh/computer-and-information-technology/home .htm.

US Department of Commerce. (1997). *America's New Deficit: The Shortage of Information Technology Workers*. US Government, Office of Technology Policy.

Varma, R. (2020). To be or not to be on HJ-1B visas. *Perspectives on Global Development and Technology*, 19, 281–302.

Varma, R. (2010). Why so few women enroll in computing? Gender and ethnic differences in students' perception. *Computer Science Education*, 20, 301–16.

Varma, R. (2007a). Women in computing: The role of geek culture. *Science as Culture*, 16, 359–76.

Varma, R. (2007b). Decoding the female exodus from computing education. *Information, Communication and Society*, 10, 181–93.

Walker, A. (1982). *The Color Purple*. New York: Harcourt Publishers.

Wang, M. and Dejol, J. (2013). Motivational pathways to STEM career choices: Using expectancy-value perspective to understand individual and gender differences in STEM fields. *Developmental Review*, 33, 304–40.

18. Lessons from women coping with IT workplace barriers

Hala Annabi

INTRODUCTION

Gender equity in the workplace continues to dominate the world's business, social, and humanitarian agenda. Across the globe, women lag in access to employment, equal pay, rate of advancement, and overall sense of belonging and empowerment at work (Catalyst, 2021). Challenges related to access and empowerment reflect women's status in society and their ability to exercise their human rights (Joshi et al., 2015; Trauth et al., 2006). Public and private organizations that recruit, retain, and advance women and men at equal rates increase social access and receive the benefits of corporate social responsibility (Bear et al., 2010). Furthermore, these organizations reap the organizational benefits of gender diversity in the workplace, including higher productivity, innovation, more patents, broader market penetration, higher market shares, and overall higher financial performance (see Annabi and Lebovitz, 2018 for complete review).

At the core of the disparity women experience in the workplace are organizational and societal structures that form barriers for them to secure and maintain employment, especially in STEM workplaces (Annabi and Lebovitz, 2018; Ahuja, 2002). Organizations across the private and public sector invest greatly in interventions to improve the participation of women, especially in information technology (IT) (Annabi and Lebovitz, 2018). Over the past four decades, diversity, equity, and inclusion (DEI) initiatives have made little impact on the status of women in IT. In the 2020 Scorecard report, The National Center for Women and Information Technology (NCWIT) revealed that little has changed in terms of the status or representation of women in IT in the US, the focus of this chapter.

In the United States, the focus of this chapter, the status of women in the IT workplace remains far below expectations. Women make up only 26 percent of computing professions compared to 57 percent of all occupations in 2019 (DuBow and Gonzalez, 2020). Women CIOs hold only 19 percent of CIO positions in the top 100 firms in the United States (NCWIT, 2021). Closer investigation of the data reveals further disparities among women. Reports show that women of color, lesbian, gay, bisexual, transgender, queer or questioning, and other (LGBTQ+) women, and disabled women have weaker representation in education, workforce, and leadership highlighting the complexity of intersectionality of race, gender, and disability affecting access and advancement.

NCWIT 2020 Scorecard report on representation does show promise in the growth of white women earning computing degrees in post-secondary education (DuBow and Gonzalez, 2020). The promise of an increase in the supply (pipeline), however, does not address the more vexing retention problem plaguing the IT industry (Annabi and Lebovitz, 2018). Various reports suggest that IT women leave the workplace at a 45 percent higher rate than men (Bowman Williams, 2020). In fact, *56 percent of women in IT leave their work within the first five years*, which is twice the turnover rate of men in IT and women in other professional fields (Ashcraft et al., 2016; Glass et al., 2013). Simply put, while it is important to address the leaky pipeline, it is just as prudent to plug the gaping holes in the tank.

While we do see some improvements in the pipeline, we still cannot retain women and advance them through the ranks. Over the last ten years I have focused my research and consulting practice on assessing and improving the efficacy of interventions aimed at improving retention of women in IT in the United States (e.g., Annabi and Lebovitz, 2018; Annabi and Tari, 2018; Bhatia and Annabi, 2020). My research revealed a gap between the interventions and the barriers women experience. Most interestingly, my research showed the perseverance and resourcefulness women exhibit in developing their own coping methods to persist and beat the odds in the IT workplace. In this chapter, I surface barriers women experience and how they use their own methods to address them. To do so I draw on my research and share my own experiences as an Arabic academic woman leader and immigrant in the United States. The aim of this chapter is twofold. First, I aim to review the academic and popular literature to synthesize what women continue to develop as coping methods to persist in IT. Second, I offer critical lessons to inform how the research and practitioner communities may further investigate coping methods, and rethink how interventions are designed and deployed in order to serve a broader range of women in the IT industry.

BACKGROUND

The alarming attrition rate of 56 percent, explained in the Introduction, is especially troubling considering that 74 percent of women in technology indicate "loving their work" (Ashcraft et al., 2016). If women "love their work," why are 56 percent of them leaving IT work for non-IT work? Scholars attribute attrition to organizational and societal structures that present barriers for many women to stay in the IT workplace (e.g., Wajcman, 1991; Ahuja, 2002; Quesenberry and Trauth, 2012; Annabi and Lebovitz, 2018). A workplace barrier is defined as an "obstacle that impedes progress regarding an individual's job achievement" that often leads to attrition (Riemenschneider et al., 2006, p. 60). Such barriers lead to unequal pay, lack of growth opportunities, and overall dissatisfaction with the work.

Social Structures

As early as the 1970s, scholars recognized the different nature of IT work and culture from other types of professional work (Armstrong et al., 2007; Panteli et al., 1999) and its effects on individual women (Trauth, 2002). Ahuja (2002), and the subsequent research suggested that societal expectations about women's roles at home and work influence stereotypes about women, which commonly lead to mistaken assumptions about their IT interests, capabilities, and skills (Foust-Cummings et al., 2008; Accenture, 2010; Tapia, 2006; Ahuja, 2002). For example, Gallivan (2004) found that managers perceived women as having lower technical skills. Recent studies further confirm that women in IT are subject to stereotyping on a more regular basis than their men counterparts (Annabi and Lebovitz, 2018; Mannix and Neale, 2005; Accenture, 2010), limiting career development and advancement opportunities. Furthermore, research findings revealed women being pushed toward stereotypically "feminine" and less technical work roles (Yoder, 1997; Boldry et al., 2001; Scott-Dixon, 2004) and passed over for promotions due to assumptions about being "family-focused" and "unwilling to travel" (Blum and Smith, 1988). Ahuja and others recognized the "double burden" women face when balancing work and family (Glass and Estes, 1997; Trauth et al., 2009; Armstrong et al., 2007; Blair-Loy, 2007; Desvaux et al., 2007; Ashcraft and Blithe, 2009). While this barrier is not unique to IT women, specific qualities in the IT field make work–life balance particularly challenging and tightly correlated to the promotion and high turnover rates (Ahuja, 2002; Armstrong et al., 2007; Riemenschneider et al., 2006; Simard et al., 2008). IT demands long work hours, unpredictable travel schedules, constant availability, and keeping current with rapidly changing technology (Ahuja, 2002; Scott-Dixon, 2004; Orser et al., 2007; Simard et al., 2008; Hewlett et al., 2008). Several studies demonstrated that the IT work environment increases stress and conflict for women, leading to dissatisfaction and turnover (Ahuja 2002, Armstrong et al., 2007), and negatively impacted women's persistence and advancement in the field.

Organizational Structures

IT organizations reflect the broader societal views (social structures) discussed above and contain organizational structures that exclude and marginalize women (Ahuja, 2002; Tapia, 2006; Trauth et al., 2009; Annabi and Lebovitz, 2018). IT occupational culture – predominantly based on masculine values and power – creates discrimination, legitimacy, and access barriers for women (Ahuja, 2002). In this culture, women are viewed as intrinsically less capable than men (legitimacy) (Wilson, 2004; Tapia, 2006) and their performance is evaluated and rewarded differently than that of their men peers (discrimination) (Galivan, 2004; Simard et al., 2008; Hewlett et al., 2008; Riemenschneider et al., 2007). Research finds men's achievements are credited to personal effort and skill, while women's are attributed to luck and ease of assignments (Galivan, 2004; Simard et al., 2008). The impact of such views is most evident in the persistent pay gap women experience in IT. Despite recent improvements, on average IT women continue to make 3 percent less than their men counterparts upon entry to their workforce and the pay gap is even higher for senior women in the field (Gonzalez, 2021).

Furthermore, studies reported that women's views and perspectives were undervalued and not heard (Foust-Cummings et al., 2008), which is problematic due to the relationship between perceptions of voice and fairness, job satisfaction, level of commitment, access to development and advancement opportunities, and performance appraisals (Konovsky, 2000; Avery and Quinones, 2002; Riemenschneider et al., 2006). Legitimacy and access barriers are compounded by the flattening of the structure of IT organizations and marketplace globalization (Ahuja, 2002). IT's institutional culture and structures lead to few women in supervisory roles and poor supervisor–supervisee relationships in general (e.g., Foust-Cummings et al., 2008; Deloitte, 2010). Inadequate supervisor–supervisee relations have been widely discussed in the literature and result in a lack of training, regular feedback, and long-term relationships due to higher turnover (Foust-Cummings et al., 2008; ILM, 2011; Trauth et al., 2009; Reid et al., 2010; Annabi and Lebovitz, 2018). Women often feel unable to communicate openly and honestly with direct supervisors, limiting discussions of career goals and personal development opportunities (Foust-Cummings et al., 2008; ILM, 2011). Men managers often lack both the awareness to recognize the challenges women face and the training to constructively address these challenges (Reid et al., 2010). Poor supervisor–supervisee relationships, however, are not limited to men supervisors. Annabi and Lebovitz (2018) suggest IT women did not feel supported by their women supervisors, referring to inherent inequities within the socio-cultural construction of IT which may exacerbate biases in women managers' minds relative to other women. Moreover, supervisors often sponsor employees, advocating for their advancement and highlighting their accomplishments (Foust-Cummings et al., 2008; Deloitte, 2010). Lack of sponsorship makes it difficult for women to successfully navigate career advancement in many organizations (Hewlett et al., 2008; Hewlett et al., 2008; Deloitte, 2010; ILM, 2011). For the underserved employee (Kvasny, 2006), the supervisor plays an invaluable role by offsetting isolation, suggesting mentors, acting as a sponsor, and enabling necessary flexibility (Baron, 1984; Ashcraft and Blithe, 2009; Tapia, 2006). Furthermore, women, being a minority in IT, lack mentors (Burt, 1998; Ahuja, 2002; Trauth et al., 2006; Orser et al., 2007; Foust-Cummings et al., 2008; Hewlett et al., 2008), have limited successful role models (Burt, 1998; Ahuja, 2002; Hewlett et al., 2008; Simard et al., 2008), and thus have a limited professional network (Granovetter, 1995; Igbaria and Chidambarma, 1995; Ibarra, 1997; Morgan and Trauth, 2006; Orser et al., 2007; Hewlett et al., 2008) which leaves them marginalized (Kvasny, 2006).

Organizational and Societal Responses

Per the review above, our understanding of barriers women experience due to organizational and social structures has been the focus and study of industry and academia for at least two decades. Such research posited various interventions to address the barriers highlighted above. To alter the nature of IT and combat the effects of discrimination, access barriers, and legitimacy, research recommends interventions that build a culture of acceptance within organizations (ILM, 2011; Forbes Insights, 2011). Such interventions include sensitivity and bias training for all employees within the organization, especially management training (Tapia, 2006; Ahuja, 2002; Panteli et al. 1999; Thomas and Ely, 1996; Cox and Blake, 1991; Simard et al., 2008; Simard and Gilmartin, 2010). This work emphasized the importance of executive sponsorship and for all employees to participate in sponsored activities to ensure cultural principles extend across the entire organization (Ashcraft and Blithe, 2009; Desvaux et al., 2010). Research also emphasizes flexible work, including part-time and flexible schedules, parental leave, and telecommuting arrangements, to alleviate some of the stress women experience in balancing work and family, resulting in higher employee satisfaction and reduced absenteeism (Dalton and Mesch, 1990; Desvaux et al., 2007; Simard et al., 2010). This work emphasized that people of all genders must participate in such offerings (Foust-Cummings et al., 2008; Deloitte, 2010) and highlight employees in management roles who also utilize these arrangements (Foust-Cummings et al., 2008; Desvaux et al., 2007; Simard et. al., 2008; ILM, 2011). Researchers also stressed the importance of ensuring that women with flexible schedules are not penalized when considered for promotions or other opportunities (Simard et al., 2008; Desvaux et al., 2010).

To alleviate isolation and exclusion, research suggests professional development opportunities including mentoring and networking on company time and adjusting workloads to encourage participation (Tapia, 2006, Desvaux et al., 2007; Hewlett et al., 2008; Simard et al., 2008) as a means to increase confidence and ambition, enhance promotion opportunities, and create sponsorship (Laudicina, 1992; Mannix and Neale, 2005; Orser et al., 2007; Desvaux et al., 2007; Hewlett et al., 2008; Ashcraft and Blithe, 2009; Desvaux et al., 2010; Simard and Gilmartin, 2010). For these efforts to be successful, they must be integrated into the organization's rewards structure, providing incentives for mentoring (Mannix and Neale, 2005). Furthermore, the literature stressed the importance of emphasizing and fostering healthy supervisor–supervisee relationships to improve the retention of women in IT (Desvaux et al. 2007; Simard and Gilmartin, 2010; ILM, 2011). Studies found that supervisors who regularly discuss career development, planning, and strategy with supervisees increase advancement rates and overall employee satisfaction (Lee, 2001; Desvaux et al., 2007). Other studies found that regular coaching sessions empower employees by building self-confidence, developing clear career goals, and encouraging calculated risks (Simard and Gilmartin, 2010; ILM, 2011). Supportive supervisory relationships could be achieved through supervisor training and reviews of supervisors (Armstrong et al., 2007).

Lastly, research suggests that setting an inclusive cultural tone requires senior leadership to demonstrate support, visible commitment, and accountability (Laudicina, 1992; Cox and Blake, 1991; Kalev et al., 2006; Forbes Insights, 2011) by establishing metrics to monitor success and identify opportunities (Laudicina, 1992; Desvaux et al., 2010; ILM, 2011). Generating gender diversity indicators raises awareness and focuses on action (Desvaux et al., 2010). Research on gender diversity interventions concludes that evaluation of current practices across levels and functions to eliminate biases related to task assignments, promotion criteria, performance evaluations, and gender distribution is critical to the success of such interventions (Dalton and Mesch, 1990; Blum et al., 1994; Simard et al., 2008; Desvaux et al., 2010; Simard and Gilmartin, 2010).

Limitations of Organizational Interventions

The last decade resurfaced the important role women played in pioneering computing. Academic and popular publications boasted stories and images of women pioneers in early computing such as Loveless, Lamarr, Bartik, Hopper, and Boyd Granville. More reports, books, and efforts emphasized the important role women can and should play to "lean in" or "lift" one another in computing and more broadly to achieve the representation and wage equity they deserve. Furthermore, IT organizations invested heavily in interventions (reviewed above) in an attempt to support women in the workplace. Most notable of such interventions were women affinity groups. In fact, a Mercer report indicated that 93 percent of companies deployed women affinity groups (Mercer 2011) and rely heavily on them to address the representation and advancement of women (Bhatia et al., 2020; Ward 2012). Despite the well-intentioned nature of such efforts, interventions fail to address the attrition problem of women in IT for several reasons discussed below.

First, initiatives are too often not empirically or theoretically grounded (Annabi and Lebovitz, 2018; Trauth, 2013; von Hellens et al., 2012; Craig, 2016). In my work (e.g., Annabi and Lebovitz, 2018; Bhatia et al., 2020), we found a gap between what women identify as the most significant barriers (legitimacy, access, poor supervisory relationships, and bias) and the most common interventions deployed (e.g., affinity groups, networking, and professional development). For instance, while most organizations rely heavily on women affinity groups to advance equity and inclusion (Mercer. 2011; Goode et al., 2016; Bhatia et al., 2020), affinity groups do little to address the most significant barriers such as bias and discrimination from peers and the poor supervisory relationships that inhibit their ability to grow and contribute to their teams.

Second, global and organizational efforts often seemed to center on "fixing the women" rather than fixing the societal and organizational barriers inherent in the environment that leave women isolated and marginalized (Annabi and Lebovitz, 2018). Few organizations require bias training related to the inclusion of women in the workplace beyond one-time ineffective sexual-harassment training. Furthermore, efforts like "lean in circles"[1] and women's affinity groups are often grounded in narratives that continue to place the burden of equality on women expecting them to lead the struggle. Equity for women is an organizational problem and all genders are responsible for it and will stand to benefit when it is achieved.

Third, in my early research (Annabi and Lebovitz, 2018), it was clear that intervention efforts were often fragmented at best in their approach to addressing systemic bias. Few organizations required bias training related to the inclusion of women in the workplace. Very few, if any organizations utilized a systematic approach to identifying barriers, address barriers, and measure the efficacy of their interventions (Annabi and Lebovitz, 2018, Craig, 2016).

Fourth, the narrative around equality in the workplace emphasizes a more homogeneous representation of women and their stories, overrepresenting white, heterosexual mothers without disabilities. The emphasis on motherhood is critical as it is a dominant identity affecting persistence in IT (Trauth and Connolly, 2021). Motherhood identity and related barriers, however, are often used as a crutch that inhibits more holistic interventions that address deep barriers about who should work in and lead IT (legitimacy and access). Furthermore, barriers affect individual women differently (Quesenberry and Trauth, 2012; Trauth, 2002; Kvasny, 2006; Trauth and Connolly, 2021) and women may need different types of interventions to mitigate the barriers (Annabi and Lebovitz, 2018). Women of color, for instance, experience harassment and marginalization at higher rates than their white counterparts (Bowman Williams, 2020). One quote from one of our focus groups (detailed below) says this best:

> Intersectionality is an important topic and I feel like the problem with women's groups in tech is it tends to be a lot about white women … they don't necessarily focus on the specific needs of women of color … just sort of the default is to white women.

The overemphasis on white mothers in intervention methods and narrative, while important, ignores the diversity of women's career aspirations, family structures, and goals, and the discrimination they experience due to biases stemming from intersectional identities such as race, sexuality, or disability. Interventions fall short largely because they are embedded in essentialist views of gender which consider inherent and consistent differences across men and women with little consideration for individual differences based on experience and other identity characteristics.

In the absence of holistic approaches to interventions that address the barriers inherent in society and organizations, women developed their own "coping" methods to "deal" with barriers. In the remainder of this chapter, I address the limitations of current interventions highlighted above by drawing on how women cope with barriers. I reflect on the consequences of women expending energy and resources to take on barriers on their own. I conclude each discussion with implications for research and practice to inform how we can learn from women's coping methods to study how to incorporate them in interventions and further research efforts.

Research-Informed Observations

In writing this chapter I provide a synthesis of the literature relevant to "individual coping methods" and draw on my research which investigates interventions aimed at the retention and advancement of IT women conducted over the last 12 years. I draw on data from interviews, online discussions, and focus groups collected in four different studies. The first study investigated the effectiveness of gender diversity interventions and included 23 semi-structured interviews across nine case studies including women in IT, human resources professionals for IT roles, and IT diversity intervention leaders (Annabi and Lebovitz, 2018). The second pilot study investigated the effectiveness of women affinity groups in IT and included one focus group and eight interviews with women in IT (Bhatia et al., 2020). The last set of interviews is research in progress including six autistic women in IT (Annabi, 2018). I also draw on an analysis of the topic "career and employment" (including 623 posts across 40 discussion threads) in the online affinity group for autistic women "AsperGirls"[2] on the Reddit website. The chapter also draws on recent themes from the popular press related to how women use individual coping methods to persist and even change the IT culture. Interviews include women working in IT consulting, banking, manufacturing, healthcare, insurance services, retail, IT start-up, and software. Lastly, I include reflections from my own career as a woman of color working in IT and holding several leadership positions over the years including primary investigator of significant grants, department chair, program chair, and director of honors and leadership programs.

WOMEN'S INDIVIDUAL COPING METHODS

In the absence of interventions that mitigate the barriers inherent in the social and organizational structures within the IT workplace, women exercise agency and deploy individual-level strategies to respond to barriers. "Individual coping methods," is an individual-level construct first identified in Trauth's (2002; 2006) work and elaborated on by others. Trauth's (2002; 2006) individual differences theory of gender and IT (IDTGIT) emphasizes that an individual woman's perceptions of barriers and potential coping strategies to mitigate them are influenced by individual sociocultural factors (Trauth, 2002; Trauth, 2006; Kvasny, 2006). Trauth and Connolly (2021) also highlighted that national and regional environmental factors (e.g., economy and culture) also influence whether and to what extent these gendered notions of IT manifest. This line of study also illustrated the lengths to which women go "to fit" into IT organizations by developing individual coping methods such as self-regulating behavior to either adopt (Joshi and Kuhn, 2007) or overemphasize what are often considered masculine behaviors or traits (Trauth, 2002). Table 18.1 summarizes the various coping methods women leverage, that were presented in Annabi and Lebovitz's (2018) organizational interventions mitigating individual barriers (OIMIB) framework. The OIMIB framework provides a holistic approach to investigate and guide the assessment of gender-diversity interventions in IT and their effects on the diversity of IT women and barriers that exist in the workplace. OIMIB includes individual-level constructs, intervention-level constructs, and organizational-level constructs and how these constructs are interrelated. My observations throughout this chapter are informed by this framework.

Table 18.1   *Individual coping methods from Organizational Interventions Mitigating Individual Barriers framework (OIMIB) (Annabi and Lebovitz, 2018)*

Create informal networking and find mentors	Identifies key decision makers and important individuals in the organization and creates personal opportunities to build lasting relationships. Locates and establishes relationships with peers or superiors whom women can approach for coaching.	Morgan and Trauth (2006); Annabi and Lebovitz (2018)
Make individual-level changes and mask	Women make personal adjustments in their actions, behaviors, work ethic, and/or some aspect of their personality. This is sometimes referred to as social masking and camouflaging: learning/mimicking socially acceptable behaviors.	Morgan and Trauth (2006); Adam (2006); Annabi and Lebovitz (2018); Annabi and Locke (2019); Lai, Lombardo et. al. (2011); Bargiela, Steward, and Mandy (2016); Hull et. al. (2017); Hickey, Crabtree and Stott (2018); Lorenz, Frischling, Cuadros, and Heintz (2016)
Ignore barriers	Women can ignore the issues without becoming negatively affected.	Orser et al. (2007); Adam (2006); Annabi and Lebovitz (2018)
Leave the IT field	Leave the organization after trying to find a compromise or after seeing situations worsen rather than improve, or seek better opportunities elsewhere.	Orser et al. (2007); Joshi and Kuhn (2007); Annabi and Lebovitz (2018)
Choices regarding motherhood	Women make different choices regarding motherhood (some chose motherhood over career, some chose career over motherhood, and others chose to balance both)	Quesenberry et al. (2006); Trauth (2002)
Make themselves heard and visible	Find ways to ensure their voice is heard and their work is recognized.	Annabi and Lebovitz (2018)
Deliver consistent high performance	Women work harder than men and develop a reputation for performing at a high level by regularly producing high-quality work.	Trauth et al. (2009); Annabi and Lebovitz (2018)
Confront the problem individual	Identify the individual with the personal problem and approach them one-on-one to resolve the issue.	Annabi and Lebovitz (2018)

Trauth (2002, 2006) suggested that women employ individual informal strategies to overcome barriers: women reported building relationships by creating and leveraging personal and professional social channels independent of organizational interventions (Quesenberry et al., 2006; Morgan and Trauth, 2006). Women also reported making changes in individual actions, behaviors, work ethic, education, and/or some aspect of personality to better assimilate into the majority (Orser et al., 2007; Adam, 2006). Furthermore, Quesenberry et al. (2006) reported on the different choices women made regarding motherhood: some chose motherhood over career, some chose career over motherhood, and others chose to balance both. Orser et al. (2007) observed that some women can ignore barriers without becoming bitter or resentful. Annabi and Lebovitz (2018) identified two new individual coping methods that revealed women's agency in addressing problematic behavior from colleagues, "confronting the problem individual" and "making themselves heard" in order to persist. When organizational interventions and informal methods fail to minimize barriers, studies report a tendency for women to leave the organization or the IT field entirely (Foust-Cummings et al., 2008; Simard and Gilmartin, 2010; Deloitte, 2010).

Over the years, in formal and informal interviews and conversations with IT women, I have learned so much about and been so impressed by women's ingenuity and ability to adapt and persist. Research interviews almost always revealed at least one informal coping method that women use to overcome barriers. In the next section, I elaborate on the coping methods I have learned about, how they are manifested, and how they have evolved over the years.

The Cost of Building Relationships

The literature, echoed by many of my interviewees, emphasizes the importance of building relationships with colleagues, supervisors, and mentors to address issues of isolation and exclusion women often feel in IT (Annabi and Lebovitz, 2018; Ahuja, 2002). These networks build the social capital of employees, their confidence and ambition, enhance their promotion opportunities, and create sponsorship (Laudicina, 1992; Mannix and Neale, 2005; Orser et al., 2007; Desvaux et al., 2007; Hewlett et al., 2008; Ashcraft and Blithe, 2009; Desvaux et al., 2010; Simard and Gilmartin, 2010). For organization-sponsored interventions aimed at building relationships to be successful, these must occur on company time with adjustments to workload and be integrated into the organization's rewards structure to provide incentives for participation (Mannix and Neale, 2005; Tapia, 2006, Desvaux et al., 2007; Hewlett et al., 2008; Simard et al., 2008). In the absence of these elements in networking and mentoring interventions, women adopt their own networking channels. Some women even begin affinity groups through grassroots efforts to address the challenges in informal networking. In one of my interviews, Jenna from the affinity group study explained:

> We kind of realized that while men had this informal opportunity to get to know each other while they just go out for a smoke or something like that, right, we needed something similar. But there wasn't really a very feasible informal forum to bring people [women] together. That's when we thought, okay, let's come up with a women-focused group.

A recent informal conversation with Rula, a vice president (VP) in an IT firm, founded a women affinity group elaborated on the drive and need to create a space for women stating:

I started Women in [omitted for anonymity] because I kept hearing similar stories across companies and in my organization. I thought it was important to bring people together. I thought, if I don't [do it] who will?

Women reported building relationships by creating and leveraging personal and professional social channels independent of organizational interventions (Quesenberry et al., 2006; Morgan and Trauth, 2006). In my interviews, building the right relationships was the most frequently cited individual coping strategy followed by informally finding mentors. Building such relationships on one's own time with limited support comes at a cost to women. Already taxed with balancing multiple roles (family and work) and having to invest more time proving their legitimacy at work, women must build relationships and seek informal networks inside and outside of work. This adds another burden on women's time and energy and relates to the hidden work women do. For women who mentor in informal ways outside of organization-sponsored and rewarded interventions, this adds to the invisible labor they carry. The mentee and mentor are dedicating time that does not contribute to productivity and may affect advancement in the long run. These efforts are even more time-consuming and emotionally taxing for women of color who have even fewer women or men with similar ethnic and racial identities to identify with. One interviewee who belonged to an IT women affinity group explained the cost of participating in formal organizational interventions or informal networking opportunities:

As a woman, I am already facing prejudice in different ways at work and that already is hindering my progress … by spending time participating in women affinity groups, they will be more disadvantaged while their male counterpart is just focusing on work.

The cost of building relationships goes beyond time investment for some women who may be more introverted, have social anxiety, or who may be neurodivergent.[3] For these women, expending energy on building informal relationships taxes their mental and physical energy and affects their overall wellbeing (Baldwin and Costley, 2016). For example, one Chinese-American autistic program manager at a technology firm explained her challenge to me and how she approached building relationships in a methodical way.

I am using this kind of format when people have birthdays I give gifts, when … hm … when … they need help I am happy to offer help. But sometimes to start a conversation I feel like … what kind of talking should I choose, those are the kind of [a] challenge. Then I try to have a good list for people to start conversation from. Checking on weather … like the weekend, what they like to do, their interests their hobbies so … I can use those kind of guidelines to need to start the conversation.

She later explains the impact of utilizing those tactics and the pressure she feels to build relationships in a way that does not come as naturally to her.

[I]t is a lot of energy-consuming. 'Cause I know my nature is just … I just want to do the things right. So, when people get very emotional, sometimes I have to go back and rely on my previous psychological study to think – maybe […] some of the demands and true deep needs. Follow the Maslow Hierarchy of needs and try to think, "hmm what [does] that fit?" So I know I am using a lot of skills and learning from the book to try to accommodate. But still, a lot of people will think I am a good manager, I am a good leader.

Building informal mentoring relationships across genders, which is necessary for women to advance in IT, is often fraught with sensitivities and misconceptions of intent due to the sexually charged masculine culture of the IT workplace (Lin and Besten, 2019). When a man colleague asks a superior to grab a beer after work, there is no subtext. When a female subordinate asks a male superior to grab a coffee to discuss new projects, these conversations and perceptions are fraught with suspicion. The man-dominated culture which is grounded in notions of the woman being less capable, and the often-unconscious bias around successful women being less legitimate, imposes a different layer and burden to build such relationships.

In recent years, building networks and informal relationships at work became even more fraught for women with the raised attention given to sexual harassment by the MeToo movement.[4] Men supervisors and superiors expressed hesitation to develop one-on-one relationships with female employees (Bower, 2019).

In a recent *Harvard Business Review* article, the authors reported on the results of a survey that explored the impact of the MeToo movement on harassment in the workplace. Consistent with my observations, the study revealed that "22% percent of men and 44% of women predicted that men would be more apt to exclude women from social interactions, such as after-work drinks; and nearly one in three men thought they would be reluctant to have a one-on-one meeting with a woman" (Bower, 2019, p. 4). These findings are even more troubling when we explore how they may affect different types of women in different proportions. The same study revealed that both men and women making hiring decisions might be less likely to hire attractive women to avoid harassment altogether. Findings from the study, consistent with my interviews, suggest that the top two key strategies (building relationships and finding mentors) women utilize to address the most significant barriers they experience (isolation, legitimacy, and access to opportunities) may now be harder than ever before as we continue to work towards equity and accountability for harassment.

The prevalence and complexity of building informal relationships for women in IT reveal the need for interventions that facilitate relationship building for women with more care and intentionality. Such interventions must take a holistic approach that accounts for the intersectionality of identities and needs. These interventions must also be weaved into the fabric of organizational structures, reward systems, and development. Relationship building for the underserved may be included in onboarding, coaching, and overall developmental efforts. Furthermore, in recognizing the critical nature of informal relationships and the burden they pose on underrepresented identities in the organization, the burden of creating such relationships should be shifted to those in leadership and be a critical aspect in their overall performance expectations and rewards. Leaders at all levels should be required to mentor and champion underserved employees and be trained to do so properly. Measures of accountability must also be built into the new structures.

Women Outperforming to Perform

In the background section, I explained that women in IT experience barriers related to their legitimacy as IT professionals. These challenges to legitimacy are rooted in the masculine culture of IT (Trauth and Connolly, 2021) that undervalues women's contributions and skills, and attribute technical skills to men over women, drives women to work harder than necessary to prove themselves (Annabi and Lebovitz, 2018). Studies report that women's views and perspectives were undervalued and not heard (Foust-Cummings et al., 2008), which is problematic due to the relationship between perceptions of voice and fairness to job satisfaction, level of commitment, and performance appraisals (Konovsky, 2000; Avery and Quinones, 2002). This is explained in Julie's sentiments:

> I'm not sure if it's just the technical aspects, kind of the engineering mentality that men are more logical and women are more emotional. Men seem to just be looked upon [more] as subject matter experts than women are. That can be frustrating.

The characterizing of IT as a "masculine occupation" results in women having access to limited opportunities and being overlooked relative to technical roles. One of my greatest achievements as a chair of an MIS department was the fact that we had over 45 percent women in our student population. My colleagues (mostly men) and I were intentional about breaking down the barriers for women to enter the program. We worked on the education problem to challenge the perceptions about IS careers and developed a culture that celebrates differences and encourages women. The more challenging part of our efforts to attract and retain women in the major and the field was when our female students would return from internships and express how it was difficult to fit in. For example, one of my high-achieving students, Sara, returned after a summer internship feeling disparaged. She explained how her supervisor and other men interns would often restrict her to documentation roles and prohibit her contributions to writing code. She loved to code and felt she was not given the opportunity to learn and develop in that capacity. She kept saying over and over to me:

> I like the technical! I want to learn but they won't let me! But I get better grades on technical assignments than the guys! I just don't get it!

She ended up working as a consultant in one of the big four accounting firms. She stayed in IT, but in a less technical role. There are many similar stories and anecdotes I heard over the years that reveal how deeply engrained these legitimacy and access barriers are for women in our field.

Across various interviews and in the online discussion boards we analyzed, women shared that they cope with legitimacy issues by delivering consistent high-quality work. Like Sara, they get into "I will show you" mode or focus on perfectionism to prove their expertise and value. These findings are consistent across other STEM fields as well (Howe-Walsh and Turnbull, 2016).

The exact rates of women advancing throughout the IT pipeline are not always available, though global and national data suggests that, in general, the burden of working harder is amplified for women of color and mothers (McKinsey, 2021). Reports suggest that motherhood and being a woman of color present a double burden. For mothers, there is a consistent questioning of priorities related to the challenges women face balancing home and work responsibilities. The challenging nature of the IT field was a key factor identified in almost half of the interviews and messages we analyzed. One respondent expressed her struggle with managing work and the perception of not being as committed, while still having to manage motherhood:

> That's during the time I'm starting a family – you're having to juggle your children … and, in IT, it's a very demanding career, right? You're on call. Your project goes amuck, you're working late. You know your schedule is crazy and very demanding.

It is important to note that the perceived lack of commitment attributed to mothers and legitimacy barriers experienced by women (especially women of color) is also imposed on women by supervisors and colleagues of all genders equally. In fact, some respondents emphasized that working for women bosses did not make a difference and most women supervisors had "a pay your dues" mentality. Maggie elaborates on the nature and impact of such dynamics:

> I've had times where I've worked for other women, and I felt like that was at times tougher because they've done it, they didn't complain, you know … they had two kids and they seem to do it all, so they don't want to hear my excuses of why I can't get it done. So that in itself is another challenge, working with other women. I think that could be a bigger challenge.

On the other hand, women who were supervisors emphasized the pressure they faced to treat everyone equally without seeming partial to other women. A few explained that they sometimes felt that they were being watched more or that some male and female subordinates expected them to fail. One IT vice president (VP) explained:

> I have to set the example and be on guard all the time.

For women of color, legitimacy barriers may be compounded in biased stereotypes at the intersection of race and gender. Our legitimacy is questioned relative to our technical skills and the merit of our ideas. The legitimacy of our ideas and authority also gets challenged and attributed to stereotypes relative to our cultural identity and ways in which we communicate and express ourselves. For example, assertive women of color are described as aggressive. As an Arabic woman, who grew up in the Middle East and spent much of her formative educational experience in the northeast of the United States, I tend to be direct in my communication and feedback. But I have often been described by peers of all genders as aggressive or curt. My white husband, who has an even more direct communication and work style, in contrast, is described as a confident and to-the-point kind of person in a complementary way.

The fact that women feel the need to work harder to minimize biased legitimacy and commitment perceptions, highlights the impact of societal and occupational structures such as gendered family roles, stereotypes of race, and the masculine IT profession. The only way to address these deeply rooted barriers and eliminate the need for women to work harder, which eventually leads to disengagement and burnout, is through long-term educational programs that change the prevalent conception of the nature of IT work and the role of women in society. This calls for organizational as well as policy-level interventions that are holistic, long-term, and repetitive (Annabi and Lebovitz, 2018; Trauth and Connolly, 2021). We must continue to change the ecosystem of IT and embed accountability measures to reinforce our goals for equity.

I'll Do What It Takes; Even If It Means Changing Who I Am

As I indicated above and supported by other researchers and practitioners over the years (Ahuja, 2002; Trauth and Connolly, 2021), the *IT occupational culture* – predominantly based on masculine values and power – forms discrimination, legitimacy, and access barriers for women. What is often referred to as the "brogrammer" culture in IT, emphasizes socially constructed masculine traits centering on competition and fraternity-like behaviors in the workplace. This environment is also often sexually charged, thereby making it less welcoming to women. Most women we interviewed referenced making individual-level changes to cope with the IT occupation culture. Women expressed paying attention to what they wear to blend in or not stand out. I heard so many women, especially younger women, express being conscious of what they wore to work to look older and be understated to blend in and avoid attention. Some would emphasize needing to wear modest or conservative attire. A recent *Wall Street Journal* article did a brilliant job highlighting the stereotype associated with tech geeks being men who dress down in hoodies and sneakers. Any person who does not play to that image of the brilliant tech geek is surely not capable. The article critiques the impact on women's self-expression and the rise of a culture war underway using fashion to break down the stereotype that tech geeks are understated men in hoodies (Satran, 2021).

From my research and interactions with students, I learned that women of different intersecting identities used different strategies to blend in with the IT environment. I heard Black women speak of having to wear their hair differently than what they prefer to blend in and not "show up too ethnic," as my student Faye stated recently. Kathy, an autistic respondent shared that she found the IT space more of a refuge because she could just wear her comfortable unisex clothing and fit in better with the IT engineering culture. In all the women's narratives, the common theme is that women in IT use mental energy to carefully blend in a masculine-dominated culture. On the other hand, two autistic women leaned into the feminine representation of their identity to mask their autism. Fashion style changes are certainly not ideal as they reveal that women do not feel free to express their styles, cultural identities, and true selves.

Perhaps more troubling is that some women felt they must make sometimes harmful, individual changes to overcome barriers and succeed in IT. They would refer to changing their personality and the way they communicate or interact. One IT manager in a global manufacturing organization described how she felt the need to change her personality. She observed other women who were successful in IT and she attributed their success to the fact that they exhibited what is socially perceived as masculine traits:

> The women that can kind of suppress their emotional side and just rely on the logical side do get ahead. There are a handful of women who are respected and looked upon as leaders within the IT organization, but when you look at those women, it's the "nonstereotypical woman" – they're the ones who are, I hate to say "robotic," but the ones … [who] suppress that emotional side; they suppress the feminine side.

Although these perceptions prescribe to socially constructed feminine versus masculine traits men and women should exhibit, the emphasis is on the respondent feeling the need to change her personality characteristics within gendered dichotomy in order to succeed – an extremely problematic belief.

For women of color, how we present our emotions is often met with attributions to negative and harmful stereotypes. One colleague once referred to my expressive nature and the way I challenged authority when I felt that our department was being wronged as my "hot-head Arab nature coming out." He quickly apologized when he realized what he had said. It was the type of reaction that makes me second guess my responses and often filter my authentic self at work. Such reactions take an emotional toll on one's emotional health and waste cognitive energy.

What is especially peculiar in the oppressive nature of IT culture is the double bind it places on women concerning when and how women can express socially constructed feminine roles. On the one hand, the workplace overemphasizes socially constructed masculine traits and expects women to be less assertive, more logical, and less expressive of emotion. On the other hand, it *wants* women to carry the burden of creating a nurturing supportive environment. These roles are emotionally taxing, time-consuming, and undervalued in IT until one achieves a managerial role. Yet if women do not express such traits they are often stereotyped negatively. In some cases, women who tend to be less social and nurturing are penalized for it in performance reviews. This prohibits further advancement and leads to dissatisfaction. For some women, taking on the social task management and nurturing roles can be especially challenging. For many autistic women, small talk, showing interest in colleagues' personal lives, or softening feedback does not come naturally. Wei, a product manager in a tech firm, illustrates this best:

> I worked at a Fortune 50 company for 25 years, first as a developer, then as a manager. I'm also officially diagnosed [as] autistic. I can't count the times I've been told I'm "talking down" to someone when I'm showing them how to do something. I think I'm being nice and patient, but my autistic communication style isn't perceived as I intended. Another example is that I tend to focus on the work, and am not sufficiently sensitive to the feelings of people I work with. These are two broad examples, there are dozens more. I was always good at the technical aspects of my job, but I wasn't good at maintaining a happy workplace, no matter how hard I tried. If I had it all to do over again, I would not have pursued the manager's job. I was happier (and less stressed) as a developer.

This double bind we place on IT women to take on nurturing social roles while remaining assertive and logical is based on socially constructed feminine versus masculine traits. It is simply oppressive and slowly erodes women's sense of self. Once again, organizational interventions rarely reveal these deeply rooted problematic expectations we place on women or address the stereotypes around how women express themselves. Organizational interventions must include long-term educational programs to surface the double bind. Organizational reward structures should recognize the nurturing team maintenance behavior exhibited by all genders that contribute to engagement at work. Furthermore, evaluation and advancement instruments should be assessed to identify biased criteria based on stereotypes. Lastly, manager training should incorporate socially constructed feminine behaviors expected of women to nurture teams to be carried out by managers of all genders.

Finding their Voice, Place, and Style

As organizational interventions fail to address the barriers women experience, more women are taking it upon themselves to address the challenges inherent in the IT workplace (Annabi and Lebovitz, 2018). Along with the coping methods above, women are increasingly finding their voice, place, and style in the IT workplace.

Finding their voice

In my research, some women explained that it takes them a long time to recognize bad behaviors and microaggression. They often spend a long time questioning themselves and what they are doing wrong. When they gather the courage and find their voice and call out problematic behavior, sometimes they take that task on at significant risk to their employment and advancement. Halima, a young woman of color in a tech firm explained that she had recently called out her manager for asking her, more often than he does her male colleagues, to take on administrative tasks such as setting up meetings. After one too many of these instances, she approached the manager politely asking: "Are you asking me to schedule this meeting because you expect me to lead it?" His response was defensive and that he just needed her help. She later followed up and suggested that he may need administrative help as she does not see this being her role. In response, the manager accused her of being unreasonable and not a team player. Halima felt that she was left with no choice but to elicit the support of a senior woman manager to help her navigate her options and rights. As a response to the event, Halima took a leave from the team while she explored other employment options within and outside the firm. The manager's behavior was addressed by HR, with little consequence to him or the culture of the team. Such events place significant stress on women and lead to their elimination from the team.

It is important to note that not all women find their voices easily. Disability, ethnicity, and race may affect agency significantly. In the case of some autistic women, a lack of understanding of sexual innuendoes or degrading humor often limits their ability to recognize how to exercise voice. In one of my interviews with Marcy, a talented autistic IT program manager, she stated that it took her a long time to understand that advances from male colleagues were not appropriate.

Like the senior female manager in Halima's story, more women feel compelled to set the tone for inclusion in their teams and organizations. Recently, while I was providing consultation related to inclusion in the IT workplace, a female senior VP expressed the need to be visible to the executive team to set the tone for the organization. Paraphrased from my notes, she said:

> I am the highest-ranked woman in IT on the leadership team. It is my responsibility to set the tone for the inclusion of women … that is one more thing I must do … I want to do more, but there is never enough time.

Around the same time, in a conversation with another female VP, she explained to me that she created an inter-organizational women's affinity group focused on women in [specific domain eliminated for anonymity] to bring women together and find support for one another. Most women's affinity groups emerge due to a grassroots effort with little support from the organization. As organizational structures fail to address barriers to women, more women are expending energy to create affinity groups, mentor one another, call out bad behavior, and take on the task of inclusion. Yet another set of unrecognized and undervalued things for women to do.

Finding their place

Like finding one's voice is finding the right role to feel included and empowered. This was a pronounced theme in my interviews with autistic women in IT. In most of these interviews, the women said they found their place in IT by leveraging their strengths. In one of my interviews with Katrina, a data scientist in a technology firm, she expressed that she had always felt isolated as a child and was often bullied. Katrina grew up in an eastern European country that values mathematical skills. The first time she felt included was later in high school when she excelled in math. Being recognized by her teacher, and later her peers, who needed her help with homework, gave her self-confidence and the mechanism to create social relationships based on respect with her peers. She referred to her talent in math and data science as her superpower. She was better than most people in that area, and that enabled her to belong. Katrina's story is similar to the post below from AsperGirls' on Reddit:

> My time is spent researching solutions to business problems, asking questions, analyzing data, documenting requirements and processes, establishing rules for the system, and configuring or developing business applications. I'm hyper-specialized in one specific software and the work is very open-ended and autonomous. I've found this to be an amazing fit for me. All the "aspie traits" that caused problems in work before are now a huge asset.

For disabled women, finding a role and team that empowers them to leverage their superpower rather than focus on the social or communication differences is key to inclusion and empowerment. Organizations that understand and adopt a social model for inclusion rather than continue to dwell in a deficit view of disability and masculine culture benefit from the power of diversity (Annabi and Locke, 2019; Nordfors et al., 2019).

Another aspect of finding support to cope with barriers is finding one's place with one's tribe. While women affinity groups are a powerful place to find support and network for opportunities in IT, many of these groups cater to dominant women identities, such as white mothers. My analysis showed that participants' intersecting identities led them to seek and engage with multiple affinity groups. Five participants mentioned this concept and compared their experiences across affinity groups for different identities. Tammara explained her dilemma:

> Intersectionality is an important topic and I feel like the problem with women's groups in tech is it tends to be a lot about white women … they don't necessarily focus on the specific needs of women of color … it just sort of the default is to white women.

Some respondents with intersecting identities stressed the importance of allyship in women affinity groups. Some expressed that most women affinity groups that cater to white mothers do not do well in supporting women of color, disabled women, or LGBTQ+ women who participate in them. Fran explained that was necessary for lesbian women to feel that others in their chosen women affinity group were allies for their intersecting identity for them to feel supported and for the affinity group to meet its goal.

The academic and practitioner communities seem to have little understanding of the dynamics of affinity groups. This is disturbing considering how prominent and powerful these groups tend to be in setting the agenda for diversity and inclusion strategy in organizations.

Finding their style

A recent article in the *Wall Street Journal* highlighted the culture war some women in Silicon Valley are waging on the brogrammer culture plaguing our IT occupation (Satran, 2021). In this article, the author highlights a trend in the tech industry where the younger generation of engineers and IT women leaders challenge stereotypes of IT professionals as hoodie-wearing geeks through bold feminine fashion statements. Over the last ten years, we have seen similar pushback in other countries like the rise of "Tech Girls are Chic" in Australia (Beekhuyzen and Dorries, 2008). In many IT programs, we display women of various ages and look to break down the deeply rooted, often idealized notion of the disheveled IT genius. These images not only keep women out of IT, but they also convey notions about legitimacy of who has technical aptitude. Furthermore, these images set the stage for a masculine culture that would condone a sexually charged, competitive, and aggressive environment not conducive to women, or sensible people of any gender. Thanks to the talents of tech women and the prevalence of social media, they are dispelling this insidious notion of who belongs in tech to open doors for all women and all genders with all sorts of self-expression. Tech companies should take note of these efforts and bolster them in recruitment and training efforts.

No Way Out But Out of IT

The prevalent feelings of having to work harder, pressure to change their natural authentic self, feeling isolated and undermined, or bearing the double burden of home and a demanding career leave many women with one option – to leave IT. As indicated in the background section of this chapter, 57 percent of women leave IT for other fields after five years of work. This attrition rate is twice that of men in IT, and women in other professions. Amina a woman in IT, suggests the burden of having to work harder leads to attrition:

> Some women pull out because they feel it's not, for many reasons, it's not worth having to be perceived as always having to work harder, do the extra things because they're the lone female or one of the minorities.

For other women like Saheefa, the double burden of motherhood and a demanding profession like IT leaves them feeling out of options:

> It's not even just a matter of sacrifice for you, right? It's a matter of "this will not work. I can't breastfeed my child for a year and travel 70 percent of the time." So, I think that is something I still struggle with and would probably be my main reason for leaving technology if I did.

To reduce attrition, interventions must address the root causes of women feeling helpless against societal expectations, the masculine nature of the IT culture, and the lack of true support for motherhood.

CONCLUSION

Women have deployed coping methods to deal with barriers for decades. Some of these coping methods, such as networking and the creation of affinity groups, are replicated in organizational interventions, other methods are not. In fact, some of the individual coping methods are alarming and reveal the extent of harm the IT workplace inflicts on women, in some cases leading to a deliberate change in personal characteristics. To minimize the attrition of women in IT, it may be time to pay close attention to barriers and how women themselves avoid them. The coping methods identified above are critical techniques that are often ignored in organizational interventions (Annabi and Lebovitz, 2018). Organizational interventions stand to gain from understanding and incorporating the strategies women use systematically and devise methods to surface such methods more intentionally. A better understanding of such methods will not only provide us with better means to improve the empowerment, and therefore retention of women, they will also provide a better understanding of the nuanced needs of the diversity of women based on their complex intersecting identities.

Throughout this chapter, I have identified specific interventions or characteristics of interventions needed to address equity for women. Most noteworthy to highlight are the following few. First and foremost, interventions must be aimed at addressing the barriers women experience rather than trying to fix the women to assimilate to the IT culture. Making women assimilate is not only immoral, but also defeats the purpose of bringing a diversity of perspectives to the workplace. Second, what is needed in the way of interventions aimed at women is training to build self-advocacy, negotiation, and self-care skills. Third, interventions must be holistic. Organizations must identify and change barriers inherent in structures (e.g., culture, rewards, processes and procedures, evaluations systems, criteria used in hiring and advancement), provide inclusion training for all employees of all genders and at all levels of the organization, measure outcomes of interventions, and create systems of accountability. Fourth, interventions aimed to support women such as affinity groups, networking events, and others must be weaved into the fabric of the organization, done on company time, required of all genders to support, and rewarded appropriately. Lastly, the role of supervisors as key mechanisms to set the tone, reinforce expectations, and champion women, cannot be emphasized enough. Supervisors must be trained to lead their teams with an inclusive lens which requires significant and continuous training. These are a few of the key interventions and intervention characteristics that are foundational to creating lasting change.

This chapter provides a starting place to think more critically about efforts to identify the interplay between women's complex identities and organizational interventions. In addition to changes in the workplace, comprehensive multilevel studies are also necessary to advance this area of research. Theoretical frameworks such as Trauth's IDTGIT and Annabi and Lebovitz OIMIB provide the foundation for critical theoretically grounded work that will advance our understanding of interventions and coping methods. This, however, may only be achieved through partnerships between industry and academia. Developing empirical efforts in partnership with industry to conduct systemic, multilevel empirical investigations of interventions in which we elicit the needs of the diversity of IT women will increase women's participation in interventions and their recruitment, retention, and advancement in IT.

NOTES

1. Lean in Circles are women affinity groups that grew from the book *Lean In* by Sheryl Sandberg and became a global movement to empower women to lead for equity https://leanin.org/circles.
2. AsperGirls is an online affinity group for women and girls who identify as autistic hosted on Reddit www.reddit.com/r/aspergirls/.
3. Neurodiversity is a term coined by Judy Singer in the late-1990s and refers to the neurological/cognitive differences amongst humans such as autism, ADHD, and dyslexia. Neurodiversity as a term also refers to a movement that promotes the view that neuro-differences are a natural and important aspect of our human diversity and evolution.
4. #MeToo Movement is a social movement that draws attention and raises awareness about sexual abuse and sexual harassment allegations https://metoomvmt.org/.

REFERENCES

Accenture. (2010). *Millennial women in the workplace success index: Striving for balance.* Accenture. (www.accenture.com/us-en/company/people/women/Pages/womens-research-2010-millenial-wom en-summary.aspx). Retrieved October 1, 2011.

Ahuja, M. (2002). Women in the information technology profession: A literature review, synthesis, and research agenda. *European Journal of Information Systems*, 11, 20–34.

Annabi, H., and Lebovitz, S. (2018). Improving the retention of women in the IT workforce: An investigation of gender diversity interventions in the USA. *Information Systems Journal*, 28(6), 1049–81.

Annabi, H., and Locke, J. (2019). A theoretical framework for investigating the context for creating employment success in information technology for individuals with autism. *Journal of Management and Organization*, 25(4), 499–515. doi:10.1017/jmo.2018.79 IF: 1.167.

Annabi, H., and Tari, M. (2018). Are women affinity groups enough to solve the retention problem of women in the IT workforce? *Proceedings of the 51st Hawaii International Conference on System Sciences 2018*, 5146–54.

Armstrong, D., Riemenschneider, C., Allen, M., and Reid, M. (2007). Advancement, voluntary turnover and women in IT: A cognitive study of work–family conflict. *Information and Management*, 44, 142–53.

Ashcraft, C. ,and Blithe, S. (2009). *Women in IT: The Facts*. National Center for Women and Information Technology, Boulder, CO.

Ashcraft, C., McLain, B., and Eger, E. (2016). *Women in Tech: The Facts*. National Center for Women and Information Technology, Boulder, CO.

Avery, D., and Quinones, M. (2002). Disentangling the effects of voice: The incremental roles of opportunity, behavior, and instrumentality in predicting procedural fairness. *Journal of Applied Psychology*, 87, 81–86.

Baldwin, S., and Costley, D. (2016). The experiences and needs of female adults with high-functioning autism spectrum disorder. *Autism*, 20, 483–95.

Bargiela, S., Steward, R., and Mandy, W. (2016). The experiences of late-diagnosed women with autism spectrum conditions: An investigation of the female autism phenotype. *Journal of Autism and Developmental Disorders* (46), 3281–94 (doi: 10.1007/s10803-016-2872-8).

Baron, J. (1984). Organization perspectives on stratification. *Annual Review of Sociology*, 10, 37–69.

Bear, S., Rahman, N., and Post, C. (2010). The impact of board diversity and gender composition on corporate social responsibility and firm reputation. *Journal of Business Ethics*, 97(2), 207–21.

Beekhuyzen, J., and Dorries, R. (2008). Tech girls are chic! (not just geek!). [Brisbane, Qld.]: tech2morrow.

Bhatia, S., Tari, M., and Annabi, H. (2020). What women really think of women affinity groups in tech. AMCIS 2020 Proceedings.https://aisel.aisnet.org/amcis2020/social_inclusion/social_inclusion/10.

Blair-Loy, M. (2007). *Competing Devotions: Career and Family among Women Executives*. University of California Press, Oakland, CA.

Blum, L., and Smith, V. (1988). Women's mobility in the corporation: A critique of the politics of optimism. *Signs*, 13, 528–45.

Boldry, J., Wood, W., and Kashy, D. (2001). Gender stereotypes and the evaluation of men and women in military training. *Journal of Social Issues*, 57, 689–705.

Bower, T. (2019). The #MeToo Backlash, *Harvard Business Review*, Sept–October 2019.

Bowman Williams, J. (2020). Maximizing #MeToo: Intersectionality and the movement, *Boston College Law Review*, June 5, 2020) Available at SSRN: https://ssrn.com/abstract=3620439.

Burt, R. (1998). The gender of social capital. *Rationality and Society*, 10, 5–46.

Catalyst (2021). Women in the workforce – global: Quick Take (2021, February 11). www.catalyst.org/research/women-in-the-workforce-global/.

Cox, T., and Blake, S. (1991). Managing cultural diversity: Implications for organizational competitiveness. *Academy of Management Executive*, 5, 45–56.

Craig, A., Fisher, J., and Dawson, L. (2011). Women in ICT: Guidelines for evaluating intervention programmes. *ECIS 2011: European Conference on Information Systems*, 1–13.

Dalton, D., and Mesch, D. (1990). The impact of flexible scheduling on employee attendance and turnover. *Administrative Science Quarterly*, 35, 370–87.

Deloitte. (2010). *Unleashing Potential: Women's Initiative Annual Report*. Deloitte Development, LLC. (www.deloitte.com/assets/dcom-unitedstates/local%20assets/documents/war_sm%20final.pdf). Retrieved October 20, 2011.

Desvaux, G., Devillard, S., and Sancier-Sultan, S. (2010). Women at the top of corporations: Making it happen. McKinsey and Company (www.mckinsey.com/~/media/mckinsey/). Retrieved October 1, 2011.

Desvaux, G., Devillard-Hoellinger, S., and Baumgarten, P. (2007). Women matter: Gender diversity, a corporate performance driver. McKinsey and Company. (www.mckinsey.com/~/media/mckinsey /dotcom/client_service/Organization/PDFs/Women_matter_oct2007_english.ashx). Retrieved October 1, 2011.

DuBow, W., and Gonzalez, J.J. (2020). NCWIT Scorecard: The Status of Women in Technology. Boulder, CO: NCWIT.

Forbes Insight. (2011) *Global Diversity and Inclusion: Fostering Innovation through a Diverse Workforce*. Forbes Insight, New York, NY.

Foust-Cummings, H., Carter, N., and Sabattini, L. (2008). *Women in Technology: Maximizing Talent, Minimizing Barriers*. Catalyst, New York, NY.

Gallivan, M. (2004). Examining IT professionals' adaptation to technological change: The influence of gender and personal attributes. *ACM SIGMIS Database*, 35, 28–49.

Glass, J., and Estes, S. (1997). The family responsive workplace. *Annual Review of Sociology*, 23, 289–333.

Glass, J., Sassler, S., Levitte, Y., and Michelmore, K. (2013). What's so special about stem? A comparison of women's retention in stem and professional occupations. *Social Forces*, (92), 723–56.

Gonzalez, C. (2021). Men got higher pay than women 59% of the time for same tech jobs. *Bloomberg*, May 19, 2021. (www.bloomberg.com/news/articles/2021-05-19/gender-pay-gap-in-tech -male-job-candidates-paid-3-higher-than-women#:~:text=Men%20Got%20Higher%20Pay%20Than ,counterparts%2C%20a%20new%20survey%20finds).

Goode, S., and Dixon, I. (2016). Are employee resource groups good for business? *HR Magazine*, 61, 24–25.

Granovetter, M. (1995). *Getting a Job: A Study of Contacts and Careers*. University of Chicago Press, Chicago, IL.

Hewlett, S., Buck Luce, C., Servon, L., Sherbin, L., Shiller, P., Sosnovich, E., and Sumberg, N. (2008). The Athena factor: Reversing the brain drain in science, engineering, and technology. *Harvard Business Review Research Report*, 10094.

Hewlett, S., Jackson, M., Sherbin, L., Sosnovich, E., and Sumberg, K. (2008). *The Under-Leveraged Talent Pool: Women Technologists on Wall Street*. Center for Work-Life Policy, New York, NY.

Hickey, A., Crabtree, J., and Stott, J. (2018). "Suddenly the first fifty years of my life made sense": Experiences of older people with autism, *Autism*, 22(3), 357–67 (doi:10.1177/1362361316680914).

Howe-Walsh, L., and Turnbull, S. (2016). Barriers to women leaders in academia: Tales from science and technology, *Studies in Higher Education*, 41:3, 415–28, DOI: 10.1080/03075079.2014.929102.

Hull, L., Petrides, K.V., Allison, C., Smith, P., Baron-Cohen, S., Lai, M.-C., and Mandy, W. (2017). "Putting on My Best Normal": Social camouflaging in adults with autism spectrum conditions. *Journal of Autism and Developmental Disorders*, (47), 2519–34 (doi:10.1007/s10803-017-3166-5).

Ibarra, H. (1997). Paving an alternative route: Gender differences in managerial networks. *Social Psychology Quarterly*, 60, 91–102.

Igbaria, M., and Chidambaram, L. (1997). The impact of gender on career success of information systems professionals: A human-capital perspective. *Information Technology and People*, 10, 63–86.

Institute of Leadership and Management. (2011). *Ambition and Gender at Work*. Institute of Leadership and Management, London, UK.

Joshi, K.D., and Kuhn, K.M. (2007). What it takes to succeed in information technology consulting: Exploring the gender typing of critical attributes. *Information Technology and People*, 20(4), 400–424.

Joshi A., Neely B., Emrich C., Griffiths D., and George G. (2015). Gender research in AMJ: An overview of five decades of research and call to action. *Academy of Management Journal*, 58: 1459–75.

Kalev, A., Dobbin, F., and Kelly, E. (2006). Best practices or best guesses? Assessing the effectiveness of corporate affirmative action and diversity policies. *American Sociological Review*, 71, 589–617.

Konovsky, M. (2000). Understanding procedural justice and its impact on business organizations. *Journal of Management*, 26, 489–511.

Kvasny, L. (2006). Let the sisters speak: Understanding information technology from the standpoint of the "other." *ACM SIGMIS Database*, 37(4), 13–25.

Lai M.C., Lombardo M.V., Pasco G., Ruigrok A.N.V., Wheelwright S.J., et al. (2011). A behavioral comparison of male and female adults with high functioning autism spectrum conditions, *PLoS ONE* 6(6), e20835 (doi:10.1371/journal.pone.0020835).

Laudicina, E. (1992). Diversity and productivity: Lessons from the corporate sector. *Fifth National Public Sector Productivity Improvement Conference*. Newark.

Lee, P. (2001). Technopreneurial inclinations and career management strategy among information technology professionals. *Proceedings for the 34th Annual Hawaii International Conference on System Sciences*.

Lin, Y.W., and den Besten, M. (2019). Gendered work culture in free/libre open source software development. *Gender, Work and Organization*, 26(7), 1017–31.

Lorenz, T., Frischling, C., Cuadros, R., and Heinitz, K. (2016). Autism and overcoming job barriers: comparing job-related barriers and possible solutions in and outside of autism-specific employment. *PLoS ONE*, 11(1), e0147040 (doi:10.1371/journal.pone.0147040).

Mannix, E., and Neale, M. (2005). What differences make a difference? The promise and reality of diverse teams in organizations. *Psychological Science in the Public Interest*, 6, 31–55.

Mercer. (2011). "ERGs come of age: The evolution of employee resource groups." January 26, 2011, www.afscmeinfocenter.org/blog/2011/01/ergs-come-of-age-the-evolution-of-employee-resource-groups.htm#.YONTFehKj-g.

Morgan, A., and Trauth, E. (2006). Women and social capital networks in the IT workforce. In Trauth, E.M. (Ed.), *Encyclopedia of Gender and Information Technology, Vol. 2*. Idea Group Publishing, Hershey, PA, pp. 1245–51.

NCWIT (2021). Women in information technology: By the numbers. https://ncwit.org/resource/bythenumbers/.

Nordfors, D., Grundwag, C., and Ferose, V.R. (2019). A new labor market for people with "coolabilities." *Communications of the ACM*, 62, 7 (June 2019), 26–28. DOI: https://doi.org/10.1145/3332809.

Orser, B., Riding, A., Dathan, M., and Stanley, J. (2007). Gender challenges of women in the Canadian advanced technology sector. In *CATAWIT Forum Women in Technology*.

Panteli, N. (2012). A community of practice view of intervention programmes: The case of women returning to IT. *Information Systems Journal*, 22, 391–405.

Quesenberry, J., and Trauth, E. (2012). The (dis) placement of women in the IT workforce: An investigation of individual career values and organizational interventions. *Information Systems Journal*, 22, 457–73.

Reid, M.F., Allen, M.W., Armstrong, D.J., and Riemenschneider, C.K. (2010). Perspectives on challenges facing women in IS: The cognitive gender gap. *European Journal of Information Systems*, 19(5), 526–39.

Riemenschneider, C., Armstrong, D., Allen, M., and Reid, M. (2006). Barriers facing women in the IT work force. *ACM SIGMIS Database*, 37, 58–78.

Satran, R. (2021). No hoodies and allbirds for these women in tech. *The Wall Street Journal*, February 12, 2021.

Scott-Dixon, K. (2004). *Doing IT: Women Working in Information Technology*. Sumach Press, Toronto, Ontario.

Simard, C., and Gilmartin, S. (2010). *Senior Technical Women: A Profile of Success*. Anita Borg Institute for Women and Technology, Palo Alto, CA.

Simard, C., Henderson, A., Gilmartin, S., Schiebinger, L., and Whitney, T. (2008). *Climbing the Technical Ladder: Obstacles and Solutions for Mid-Level Women in Technology*. Stanford University, Anita Borg Institute for Women and Technology, Palo Alto, CA.

Tapia, A., and Kvasny, L. (2004). Recruitment is never enough: Retention of women and minorities in the IT workplace. In *Proceedings of the 2004 SIGMIS Conference on Computer Personnel Research: Careers, Culture, and Ethics in a Networked Environment*, pp. 84–91.

Thomas, D., and Ely, R. (1996). Making differences matter: A new paradigm for managing diversity. *Harvard Business Review*, 74, 79–90.

Thomas, R., Cooper, M., Mcshane Urban, K., Cardazone, G., Bohrer, A., Mahajan, S., Yee, L., Krivkovich, A., Huang, J., Rambachan, I., Burns, T., and Trkulja, T. (2021). Women in the workplace 2021. McKinsey and Company. Available: www.mckinsey.com/~/media/mckinsey/featured% 20insights/diversity%20and%20inclusion/women%20in%20the%20workplace%202021/women-in-the-workplace-2021.pdf?shouldIndex=false.

Trauth, E. (2013). The role of theory in gender and information systems research. *Information and Organization*, 23, 277–93.

Trauth, E. (2011). Rethinking gender and MIS for the twenty-first century. In Galliers, R. and Currie, W. (Eds.), *The Oxford Handbook on MIS*. Oxford University Press, Oxford.

Trauth, E. (2006). Theorizing gender and information technology research. In Trauth, E. (Ed.), *Encyclopedia of Gender and Information Technology*. Idea Group Publishing, Hershey, PA, pp. 1154–9.

Trauth, E. (2002). Odd girl out: An individual differences perspective on women in the IT profession. *Information Technology and People*, 15, 98–118.

Trauth, E., and Connolly, R. (2021). Investigating the nature of change in factors affecting gender equity in the IT sector: A longitudinal study of women in Ireland, *MIS Quarterly*, 45(4), 2055–100.

Trauth, E.M., Huang, H., Morgan, A.J., Quesenberry, J.L., and Yeo, B. (2006). Investigating the existence and value of diversity in the global IT workforce: An analytical framework. In Niederman, F., and Ferratt, T. (Eds.), *Managing Information Technology Human Resources*. Information Age Publishing: Hershey, Pennsylvania.

Trauth, E., Quesenberry, J., and Huang, H. (2009). Retaining women in the US IT workforce: Theorizing the influence of organizational factors. *European Journal of Information Systems*, 18, 476–97.

Wajcman, J. (1991). *Feminism Confronts Technology*. Penn State University Press, University Park, PA.

Ward, G. An investigation of performance and participation in employee resource groups at a global technology company. (2012). Wayne State University Dissertations. Paper 714.

Wilson, M. (2004). A conceptual framework for studying gender in information systems research. *Journal of Information Technology*, 19, 81–92.

Yoder, J. (1997). Looking beyond numbers: The effects of gender status, job prestige, and occupational gender-typing on tokenism processes. *Social Psychology Quarterly*, 57, 150–59.

19. Job crafting to recruit and retain women in the IT workforce: what would it take to keep you here?

Mari W. Buche

INTRODUCTION

"What would it take to keep you here?" Wouldn't it be thrilling to hear that question directed at you from a supervisor with authority and organizational power? Generally, it means that you have proven yourself to be valuable (or invaluable), and they can't imagine losing your skills, knowledge, connections, and/or expertise. What it also implies is that you have signaled in some tangible way that you are not content with the status quo. Either you have announced your imminent departure, visibly pulled back from organizational commitments, demonstrated significant mood changes in the work environment, or increased your flow of complaints to coworkers.

Following decades of developments of numerous psychological theories focused on curing mental health challenges and debilitations, positive psychology has recently emerged as an approach that nurtures and reinforces desirable outcomes (Achor, 2011; Seligman and Csikszentmihalyi, 2000). One example from this research domain is job crafting. Job crafting is a method of countering or mitigating the negative effects experienced by mildly unhappy, disgruntled employees. It encourages participants to take an active role in establishing a more acceptable work environment that will enhance individual work engagement.

"How can we solve the perplexing and persistent challenge of attracting talented, underrepresented employees in information technology (IT)?" Even when the representation of women in computer science was relatively healthy in the mid-1980s within the United States (e.g., Cohoon and Aspray, 2006), the long-term academic projections were sparse beyond the bachelor's degree level. Women often choose IT careers initially, but many professionals either drop out of the IT workforce or transition to different career disciplines (Stone, 2008). The science, technology, engineering, and math (STEM) fields have struggled with both recruitment and retention of underrepresented groups for generations. Females opting out of the IT field was a phenomenon that quickly produced a seismic quake, creating talent shortages across the industry (e.g., Stone, 2007; Trauth, 2002). It doesn't appear that industry or higher education have achieved measurable success in increasing diversity of the IT workforce.

"Where have we gone wrong?" And, more importantly, *"What can be done to reverse the negative trend?"* Both societal and individual factors influence career choices very early in the decision-making process (Joshi et al., 2013). Underrepresented groups might lack the self-efficacy needed to pursue IT jobs, contributing to perceived differences between themselves and successful IT professionals (Trauth et al., 2004). Recruitment is a single aspect responsible for constricting the potential meteoric advancements IT is capable of producing. Organizations must actively strive to retain talent, especially of underrepresented IT profes-

sionals (Quesenberry and Trauth, 2012). A substantial body of research has been devoted to this challenge over several decades.

The IT workforce needs to build momentum based on the results of tangible successes. A simple, yet effective, solution might be found in job crafting research. In a nutshell, how might we guide underrepresented IT professionals through the process of modifying their individualized work experience, thereby strengthening their work engagement (Wrzesniewski and Dutton, 2001)? One technique that could prove beneficial is job crafting. Thus, the goal of this chapter is to introduce job crafting as a possible solution for attracting, mentoring, and retaining women in the IT workforce.

This chapter is structured as follows. The background section explains the origins of job crafting within the positive psychology field (see Lopez, 2011; Peterson, 2006), including foundational theories and applications intended for both employers and employees, regardless of industry or level of responsibility. This includes a brief review of the theoretical foundations related to this topic, definitions, and an abbreviated literature review. The chapter then shifts to a more reflexive analysis regarding tools and techniques to apply job crafting to improve diversity within the IT workforce. The focus is on the usefulness of job crafting in recruitment, retention, and management of women within the IT workforce. Some of the approaches have been investigated empirically, while others are offered as possibilities for future research. Implications are presented to highlight management techniques formulated to improve diversity, equity, inclusion, and social belonging of the IT workforce.

This chapter is intended to benefit both workers and managers, providing actionable information and techniques that contribute to improvement of work engagement at all levels and in all industries. Moreover, in order to contain the scope of this chapter to a specific context, I consider job crafting in organizations seeking to broaden the diversity in their IT workforce. Specifically, I focus on techniques designed to attract and retain women IT professionals across a broad spectrum of corporate institutions. The information reported in this chapter is collected primarily from surveys conducted in the United States, and might not generalize to global populations.

BACKGROUND

Job Crafting in the Workforce

The three components that are key to job crafting are (1) motives, (2) strengths, and (3) passions (Wrzesniewski and Dutton, 2001). In general, it is presumed that people self-select career paths that satisfy these three personal drivers. However, many people are so eager for employment, they fail to critically assess their own motives. As time passes, these workers often become disillusioned and miserable. Based on the Conference Board survey of 5,000 households in the United States, only 45 percent of respondents reported happiness with their jobs (Achor, 2011).

So why do people persist in jobs that make them unhappy? One explanation can be found in determining the work orientation of the individual: job, career, or calling (Wrzesniewski, 2003). Someone with a job orientation would focus primarily on the salary. They essentially "work to live." The other two work orientations are more accurately represented by the adage "live to work." People with career orientations are motivated by recognition, achievements, and increasing power. They seek opportunities that lead to promotions and accolades, or other extrinsic rewards. In contrast, workers exhibiting the calling orientation are satisfied by elements associated with intrinsic motivation (e.g., Duffy et al., 2014). Employees with this orientation seek challenging opportunities in order to satisfy a desire to fulfill a purpose, that is, something bigger than themselves. Historically, the purpose was often reflective of a spiritual nature (e.g., Duffy and Dik, 2013; Warren, 2002) – a calling originated by a supreme deity for a higher purpose (Wrzesniewski, 2003). In more recent years, a calling is often related to facilitating and improving the world in some meaningful way (Wrzesniewski, 2003). Although workers in all types of orientations may experience work disillusionment, it is more commonplace for workers with a job orientation (Maher, Igou, and Van Tilburg, 2020).

In order to position job crafting within the broader discipline, positive psychology originated in the 1980s as a split from the focus on abnormal or dysfunctional human behavior that found prominence following World War II (Achor, 2011; Seligman and Csikszentmihalyi, 2000). Fixing what is broken is only one approach to establishing favorable mental and behavioral health. What if people are encouraged and nurtured to increase what is not broken in order to perpetuate a healthy lifestyle? It would decrease the costs of medical care, mental rehabilitation, and stunted emotional development (Achor, 2011). Coaching and mentoring provide positive solutions for adjusting or reframing work experiences (e.g., Rose and Kram, 2007; Cohoon and Aspray, 2006; Ensher and Murphy, 2005; Kram, 1988). Alternatives such as job turnover and reskilling are more drastic responses, and might be necessary as a final resort accelerated by job dissatisfaction (Buche, 2006). Interacting with the employee before problems escalate is much less dramatic and often leads to unexpected improvements that benefit the organization as well as the individual (Wrzesniewski et al., 2010). One way to model this positive reaction is to consider a contagion that infects the entire organization. Unlike a virus that causes disruptions and loss, this contagion is positive and produces higher yields for greater outcomes (Cavanaugh and Miller, 2016). In other words, an employee who experiences happiness and enthusiasm will "infect" co-workers, supervisors, clients, and suppliers with their positive energy (Achor, 2011).

There are three core aspects of the work environment that are relevant for job crafting: (1) tasks; (2) relationships; and (3) perceptions (Wrzesniewski and Dutton, 2001). Job crafting can lead to modifications in any, or all, of these areas. First, in order to break out of the routine, mundane or "daily grind," some workers simply choose to take on more job tasks. The opposite also occurs in which the employees choose to perform fewer activities due to feelings of inadequate compensation, invisibility, physical or mental fatigue. Adapting an employee's activities to fill idle time or, conversely, to reduce feelings of anxiety, one might simply adjust the number of tasks required of the worker. Second, interactions at work can either be enjoyable or annoying. Workers can often modify their reactions to their jobs by managing the frequency of communications – or methods of communication – in the office setting. For example, emotionally draining conversations might be reduced to a minimum with deliberate effort and establishing personal boundaries. Switching from in-person meetings to virtual conference calls will often help to curtail unproductive small talk. Third, overall attitude and perspective fundamentally impact work engagement. Discovering a sense of purpose, or meaningfulness, can be an amazing catalyst to improve a disappointing work experience. Many employees change jobs frequently searching for deeply rooted purpose. Finding value and meaning in work is a strong motivator for fulfillment and significance (Wrzesniewski, 2003). Job crafting is a process of cultivating work engagement by planting seeds of satisfying activities, pruning destructive relationships, and fertilizing perceptions of positive development.

A positive technique for managers and workers is to hold a job crafting session to identify proactive steps and actions to redesign work tasks and responsibilities. When a manager initiates a job crafting session, the worker is asked to diagram current job tasks, using larger boxes to represent activities that require the most time or attention (see Appendix 19A.1, Figure 19A.1: "Before" sketch). This visual representation becomes the starting point for analyzing and identifying opportunities for improvement (Wrzesniewski and Dutton, 2001). The individual is encouraged to map out improvements to the diagram, envisioning aspects that could be more satisfying if they were more accurately aligned with the person's ideal image (Wrzesniewski, 2003; Wrzesniewski and Dutton, 2001). Visualization of a more rewarding and enjoyable work experience provides a road map for change. It might also benefit employees who honestly don't know where to begin the process. After creating the diagram, potential areas of improvement must then be implemented following a deliberate set of actions. Reevaluation at a later date would be useful for assessing the impact on the employee's work engagement.

Another positive technique is for managers to encourage their workers to adjust overall thinking about their jobs in order to broaden their perspective and consider the meaning as it relates to providing a valuable service. For IT workers, this might mean focusing on how a new system will save time and money for clients, rather than focusing on the tedious tasks of testing and debugging a new system module. Service to colleagues and clients, providing personal expertise to help others avoid calamitous and frustrating setbacks, or providing societal impact that is genuinely life-saving, would all be examples of changing cognitive perceptions of the job. Instead of searching for program errors and defects – a potentially negative cognitive orientation – managers could encourage IT professionals to revel in the efficient optimization and artful system improvements benefiting everyone involved. This change in perspective could potentially lift spirits and contribute to happiness in the workplace (Achor, 2011). In case that sounds too unattainable, significant improvements don't always require radical change. Minor adjustments can also deliver noticeable payoffs if implemented consistently and habitually over time. For example, dedicating the first 15 to 30 minutes of the workday to reviewing previous successes and achievements will often generate personal satisfaction that is empowering for the employee. Like other behavioral modification techniques, positive outcomes reinforce the desire to continue the practice (Hantula, 2015). This cycle of improvement can become inherently rewarding for the participants (Wei and Yazdanifard, 2014).

Theoretical Foundations of Job Crafting

There are several relevant theories that have contributed to (or could contribute to) job crafting research. This section highlights some of the salient aspects of each theoretical underpinning and how it can be utilized within a job crafting context. With respect to space constraints, the theories, presented in alphabetical order, will not be covered in depth. Interested readers are encouraged to visit references cited for more information.

Fit theory

Fit theory (Edwards and Cooper, 1990) has been studied in the IT discipline in various forms (e.g., person-environment, PE fit; person-organization, PO fit; person-job, PJ fit). Regarding fit and gender, Venkatesh and his colleagues found that: "(1) the effect of extrinsic outcomes on PO fit was moderated by gender, such that it was more important to men in determining their PO fit perceptions; (2) the effects of social outcomes on both PO fit and PJ fit was moderated by gender, such that it was more important to women in determining their fit perceptions; and (3) intrinsic outcomes influenced perceptions of PJ fit for both men and women" (Venkatesh et al., 2017:525).

In terms of demographic differences, gender has also been found to moderate the effects of job crafting on work engagement outcomes (Berg et al., 2008); but older workers were reported to benefit from a job crafting intervention related to strengths crafting, while younger employees did not benefit from the intervention (Kooij et al., 2017). The researchers interpreted the results as follows: "As employees age, they gain more insights in their strengths and interests and develop a tendency to create environments that fit their identity" (Kooij et al., 2017:977). Future research should focus on adaptations that might prove effective for younger workers such as focusing on purposes related to societal impact (specifically, diversity, equity, and inclusion). Social responsibility is a strong motivator, especially in the early stages of one's career. In order to retain employees, managers could assist the workers in developing cognitive shifts to appreciate their contributions to purposes related to societal impact. Although some suggestions might seem idealistic at first, building on employees' passions can produce sustainable organizational commitment.

Job characteristics model

The job characteristics model (Hackman and Oldham, 1980) states that jobs are not strictly the elements that make up the work environment – such as task variety, task characteristics, and autonomy. All of these components are important and play an active part in the employee's job satisfaction within the work setting. However, jobs and job elements are most often designed by supervisors (not the employee) who control the dimensions of job expectations and assessment. While the idea is to design a job that increases motivation, satisfaction, and performance, job crafting goes beyond the basic structure of the job, to include the relationship the employee has with the job – the work engagement component.

Organizational citizenship behavior theory

The organizational citizenship behavior (OCB) theory (e.g., Suwanti et al., 2018; Podsakoff and MacKenzie, 1997) explains positive/constructive actions that are not part of an individual's job description. Engaging in these behaviors has positive outcomes for the individual and the organization (Podsakoff et al., 2009). OCB encompasses all three elements of job crafting: tasks, relationships, and purpose. Workers, especially people from underrepresented groups, may be pressured by their peers to fit in by trying to impress others. Job crafting would assist employees to identify areas that contribute to work engagement and job satisfaction. Special care needs to be taken to restrain overextending the individual, potentially leading to burnout. More research should be conducted with this unique population to investigate the impact of job crafting on job satisfaction as related to organizational citizenship behavior.

Role identity theory

Role identity theory (Ashforth, 2001; Ashforth and Mael, 1989, 1996) provides a context for the internalization of experiences and perceptions for the individual. Workers develop a number of aspects of identity based on life experiences, social interactions, individual roles, positional authority, etc. A person's work identity is sometimes emotionally stabilizing and may assist workers in gaining a sense of understanding and comfort in the work environment. In other cases, the individual's personal identity may conflict with the work environment. For example, a working mom (personal identity) may be un-stabilizing when facing pressures at work. Each role an individual enacts may have an associated identity (e.g., database administrator, mother), and will often represent a composite of individual differences (e.g., self-perceptions) that represent both technical and non-technical (e.g., business) dimensions (Buche, 2008, 2006; Trauth, 2002). Job crafting encourages workers to adapt their jobs to encompass skills and knowledge in a way that manifests as an ideal role for that individual. Presumably, the crafted job will result in a role (or roles) that leads to greater job satisfaction.

Self-efficacy theory

The self-efficacy theory (Lai and Chen, 2012; Wood and Bandura, 1989) evaluates the influence of perceived confidence when undertaking activities. Self-efficacy can be increased through four practices (Wood and Bandura, 1989: 364–65): (1) mastery through practice; (2) modeling through observation (comparing one's self to others); (3) social persuasion (encouragement); and (4) assessing psychological states (tension and calmness). Job crafting is influenced by this concept such that employees will more naturally incorporate new tasks as long as the work is compatible with their perceptions of current skills. This is often a subconscious decision, tied to a feeling of self-assurance and certainty.

Social cognitive theory

Social cognitive theory (SCT) (Bussey and Bandura, 1999; Wood and Bandura, 1989; Bandura, 1986, 1997) provides a general focus on human development at the intersection of biological, cognitive, and social factors. With SCT, social influence is the key to acquiring and maintaining behavior. SCT has been applied to studies of gender, indicating that gender roles are learned by observing (and imitating) others within a social context. With regard to gender roles, Bussey and Bandura (1999) state that changes occur over the life of the individual. Therefore, job crafting could be a valuable technique for employees of all ages, in all stages of career development, and effective in a variety of work roles.

Social learning theory

Social learning theory (Bandura, 1977) describes how individuals learn and become part of an organization by observing and imitating others. This practice could be blended into the job crafting exercise by having the worker reflect on an ideal coworker, one that seems to be embracing aspects of the job that are appealing and fulfilling. Incorporating some of those desired elements could improve the worker's perceptions of their job.

JOB CRAFTING FOR WOMEN IN THE IT WORKFORCE

Overview

Job crafting research has stretched over two decades (Wrzesniewski and Dutton, 2001), and has been examined within many industry settings (Wrzesniewski, 2003). Most of the work outside of this stream of research has focused on the implementations and actions managers could make to help to motivate employees (e.g., Hackman and Oldham, 1976, 1980). For example, job design focuses on changes that are made to the structure of the job to improve the experience of the employees. Task- and role-based characteristics of the job don't necessarily become internalized by the worker (Hackman and Oldham, 1976, 1980). The focus is on the static structure of the work, including key elements such as autonomy, task variance, and task identity. The emphasis is placed on the influence of managers in creating jobs that optimizes employee satisfaction. Job crafting is not at odds with job design, but can be employed as a supplemental practice. Managers and employees, in partnership, can improve work engagement through dynamic modifications to required tasks, human interactions, and perceptions of purpose. When individuals are unable to relate to the overall mission of the corporation, a tension or disconnect may develop. Deliberate action can stabilize the overall situation and reestablish a harmonious climate.

Job crafting has been studied in a variety of disciplines and industry settings. From the seminal work of Wrzesniewski and Dutton (2001), researchers in psychology, sociology, and human resources have applied the concepts within various work settings. In many instances, this activity occurs without cognitive awareness of any parties involved. It seems to be "common sense" for some employees. When an individual voluntarily seeks increased responsibilities beyond the formal job description, pursues new opportunities to build professional skills, or enhances their role in a given work environment, we are witnessing examples of job crafting in practice. However, relying on employees to voluntarily adopt these activities assumes they have the agency to do so, and leads to both unpredictable and non-repeatable outcomes.

Although the original model of job crafting has evolved over time, the basic principles have remained constant. When workers feel trapped or bored in their current positions, job crafting is one technique that can break the bonds of drudgery or tedium. To initiate the process, workers might be asked to describe their ideal job, or a perfect day at work. Depending on the specific individual, the response might focus on professional challenges, pleasant interactions with colleagues and clients, or the demonstration of newly acquired skills that result in appreciation or recognition. The response for each employee will likely take a different direction. Intrinsic motivation combined with extrinsic incentives will sufficiently guide the person into new areas of interest and restore a sense of wellbeing, energy, and vitality. This type of outcome clearly produces a positive impact on the organization (Wrzesniewski, 2003). Unfortunately, that is not always the case. As will be discussed in more detail below, job crafting can also produce negative effects for work groups and the entire organization (e.g., Dierdorff and Jensen, 2018). This often occurs when employees choose to reduce tasks or involvement, refrain from collaborative interactions with coworkers, or develop negative attitudes toward their work. Proactive and deliberate interventions initiated by supervisors or managers will help guide IT professionals in following acceptable growth paths.

Work identity is another component that could be improved by job crafting. As individuals become more engaged in their work roles, they may generate new or modified personal identities (Ashforth and Mael, 1996, 1989). Job satisfaction improves when the identity fits with the individual's desires and values. Contentment can develop a greater sense of happiness (Achor, 2011). In the IT profession, work identity includes both technical and non-technical (e.g., business) dimensions (Buche, 2008, 2006). Job crafting could be used to encourage employees to optimize and balance the two factions. Understandably, some IT professionals gravitate to the technical skills, enjoying programming and modeling computer applications. Other IT professionals prefer problem solving and designing solutions that meet the business users' needs. Both sets of workers might be permitted to craft their jobs (i.e., tasks, collaborations/roles, and overall purpose), so they find their niche within the organization.

Recruitment

Job crafting is most often used by employees to enhance and modify their work situations. Some employees seem intrinsically motivated to seek greater challenges and opportunities. Additionally, there are some occupations that naturally lend themselves to job crafting, with opportunities to pursue professional development and to bring those skills to the work environment. With unlimited resources for skills building, motivated learners can (and do) register for training in just about any topic imaginable, both technical and nontechnical. Adventurous and ambitious employees are never content to simply perform a job as described in the announcement.

There is a talent in writing a job description that implies job crafting which can be used to entice IT workers. Key words and phrases could be embedded in IT job postings to signal opportunities for personal and professional development. Interviewers should consider including information and questions related to personal initiative. For example, applicants who include completed online courses and certifications on their resumes imply an innate interest in continuous learning. Those potential employees should be encouraged to maintain that inquisitive nature by incentivizing developmental efforts should they join the organization. Experimental studies focused on analyzing responses to variations in job descriptions may assist in developing templates for more effective recruiting.

Retention

The IT profession, by its nature, generally attracts a large proportion of intrinsically motivated self-starters who eagerly accept the challenge of maintaining technical skills throughout their career. It is obvious to employees engaging in this discipline that technologies, applications, and skills become obsolete over time. In order to remain in high demand, employees must pursue professional development activities. "What can you do for our organization?" quickly becomes "What have you done for us lately?" Dedication and professional commitment must, rightfully, include implementing advanced techniques and complex approaches. Opportunities to practice new skills help workers to integrate their knowledge in novel ways. Knowing what motivates the employee (i.e., recognition, bonuses, and awards) will incentivize the growth and perpetual learning. Supervisors should ensure that the professional development training is relevant to the needs of the organization so the effort does not detract from worker expectations.

Workshops and tutorials should be available to encourage employees to consider job crafting in order to enhance their work experience. By actively promoting job crafting, employers are more likely to motivate workers to adopt practices that lead to positive outcomes for both the employee and the organization. Not all job crafting activities are positive, however. When employees reduce their number of tasks, and those activities are critical for organizational productivity, the results can be unacceptable. Guidance and mentoring can help direct employees to modify their work in ways that are beneficial to everyone. For example, observations of detrimental task avoidance and/or procrastination should be addressed immediately by supervisors before these negative behaviors become problematic. Using bonuses and incentives can help to encourage acceptable work changes, as well. Future research could explore the influence of the worker (motives, strengths, and passions) and work environment (tasks, relationships, and perceptions) in the job crafting process.

Management and Professional Development

Job design has been used for decades to encourage managers to engage in modifications that researchers say will increase job satisfaction (Hackman and Oldham, 1976, 1980). Presumably, these management efforts will help to produce happy employees who are more productive. Job crafting differs from job design and the job characteristics model in that the worker is responsible for the job changes, not the manager. Annual performance appraisals would be ideal points in times to discuss opportunities for job crafting. Managers can make suggestions and/or provide guidelines/boundaries, but they should refrain from directing the worker's introspective activity. In other words, employees must believe they are in control of the process in order to benefit from changes evoked through the job crafting activity.

Job Crafting for Women in IT and their Organizations

Successfully completing academic milestones is not the problem; females have graduated with computer science and information systems degrees in fairly high numbers (Joshi et al., 2013). Sometime after graduation, retention in computing careers seems to become a problem. Studies have identified time commitments, identity profiles (e.g., nerd or geek), and work–life balance as drivers that push women out of the IT field (Quesenberry and Trauth, 2012). Job crafting might provide a path for revising and replenishing interest in IT professions. Building on the foundations that attracted IT professionals to the field in the first place, job crafting initiatives could help focus on identifying the positive elements of the job. Managing routine tasks and obligations could mitigate the negative effects that drained their enjoyment from the work experience.

Comparing college graduates' exit questionnaires with their impressions at the three-, five-, and ten-year points could help to reveal what is changing that makes women and underrepresented groups leave the IT field. These gaps can form the basis for meaningful job crafting. Identifying which aspects (tasks, relationships, or purpose) create the negative impressions would help managers formulate initiatives to broaden perspectives well before turnover intent takes root.

Imposter syndrome (Clance and Imes, 1978; Crawford, 2021; Sherman, 2013; Topping and Kimmel, 1985) is characterized as feelings of being an intellectual phony and is a common complaint and affliction of underrepresented groups. They often internalize pressures to outperform the comparison groups, and typically find faults with themselves – both real and exaggerated (Collins et al., 2020; Ling et al., 2020). Job crafting can help to reduce the debilitating effects of imposter syndrome by making the job more uniquely satisfying for the worker. By actively engaging in job crafting, tasks, relationships, and cognitive impressions become better aligned with the individual's strengths, skills, and interests. Job crafting can help to overcome imposter syndrome by providing space to be your unique and authentic self (Rosenstein et al., 2020).

One way to keep individuals healthy and vital is through a sense of belonging (Baumeister and Leary, 1995; Lee and Robbins, 1998; Rubin et al., 2019). Workers often feel unfulfilled if they are lonely or isolated at work. Another related research area is social connectedness (Townsend and McWhirter, 2005). They search for colleagues who accept and welcome them into their peer groups (Brands et al., 2017). Job crafting is especially relevant in the area of building rewarding relationships. Negative interactions with colleagues are cited as a common reason for voluntary termination (Lee et al., 2010) or opting out (Good et al., 2012; Lewis et al., 2016). This might also impact work identity since associations with other workers validate perceptions of personal characteristics. Similarities might be recognized as professionally fortifying, while not fitting in with the work group can lead to a sense of failure and depression. Job crafting empowers workers to directly change the situation, strengthening the sense of belonging.

Implications and Limitations

Job crafting has been researched in many settings, with a variety of participants. Anecdotal examples continue to demonstrate improvement in employee perceptions and outlook. Many of the examples emerged as a personal experiment of trial and error. Developing a semi-prescriptive formula for proactive modification could increase the likelihood of repeated successes in the workplace. A limitation of this work is generalizability when the studies are based on small sample sizes. The studies referenced throughout this chapter tend to be snapshots, lacking the ability to assess causality of longitudinal studies with some exceptions (e.g., Tims et al., 2015). This phenomenon should be studied further to determine the causality and potential interaction effects within the IT field.

Mentorship practices have become valuable approaches to employees' professional development both in industry and academia. Encouraging workers, whether they are peers or subordinates, to attain personal fulfillment is often an effective means to improve workplace interactions. Naturally, there are mixed results, and may require skills that are not found in abundance in all industries. Once a mentor has developed a bond with the mentee, job crafting could be introduced as a method of developing latent skills and abilities. As a personal example, the author had a former student who was grappling with frustration due to workplace boredom. She had achieved all of the goals established for her position, and there was little opportunity for advancement or professional development. As is common in the IT field, challenging projects that test skills are often motivating in addition to monetary bonuses. In this instance, the employee was encouraged to speak with her supervisor and explain her lack of enjoyment with the position. The supervisor asked her to remain in that role for nine more months, and they'd see what new position was available at that time. It wasn't surprising to hear that before the nine months elapsed, she had left that company for another position that encouraged greater skills development and corporate mobility. When supervisors are alerted to mild discontent by a stellar performer, it should be considered a "wake-up call." Outstanding employees often have lower tolerance for boredom than mediocre workers. This phenomenon should be studied further to determine the antecedents and outcomes of boredom's influence on turnover intentions for women within the IT field.

It is often difficult for applicants to understand being rejected when they are deemed "over-qualified". From a naïve perspective, it may seem advantageous to hire exceptionally talented workers who can easily learn the necessary skills to complete the work assigned.

However, what industry has discovered is that over-qualified workers tend to become discontent employees soon after they master their tasks (e.g., Arvan, Pindek, Andel, and Spector, 2019). Job crafting might not resolve this situation, but it could be used to help talented workers broaden and develop their positions into something more rewarding. If the position has the elasticity to expand in new directions, along with acceptable compensation to reward the worker, both the employee and the organization benefit from the outcome. Too often, however, jobs are not flexible, so the over-qualified worker usually leaves quickly and often with negative memories of the experience.

The modification of jobs to retain workers who are technically or fundamentally over-qualified has not been adequately investigated. IT professionals might not be a good fit for the job as it is currently defined and described. However, with a little ingenuity, managers could engage in job crafting in collaboration with the applicant in order to cultivate an acceptable compromise. The critical issue then becomes negotiating a reasonable compensation. Further research is warranted to develop this recommendation in depth.

It is good advice for job seekers to ask about opportunities to grow the position once the necessary skills are mastered. Not only should job seekers ask, but be clear about expectations and desires for professional growth and explain how the enhancements will benefit not only the employee, but also the employer. If the growth requires financial investment, employees should be willing to show loyalty to the employer and demonstrate that the investment was worthwhile. Employers, likewise, should recognize the needs of employees to learn and grow, and not become stagnant in a "dead-end job." What gets you excited to come to work in the morning? Is it the people – colleagues, clients, or employer? Is it the challenge of completing tasks to the very best of your ability – pride in one's work? Is it the purpose or meaning of the work you perform – benefiting the greater good? If you are a happy worker, it is probably a combination of all three. If you are dreading your work routine, can you pinpoint which element is dragging you down? Is it all three?

Awareness of the problem is the place to start. Grab a sheet of paper and list the tasks that make up your work environment. Size the boxes by how much time each one consumes. On another sheet of paper, diagram the ideal work setting with boxes sized to how you would choose to perform given the chance. From that vantage point, begin to lay out a plan to change aspects of your job. Be proactive and dream big. You can always scale back the changes or lengthen the timeline. (See Appendix 19A.1 for details.) The very act of considering changes can bring about joy and optimism. Thinking one's way out of a gloomy box can begin the process of growth and achievement. Keep a journal to maintain motivation when you reach plateaus. Some goals will require more time and effort than others. Some will create pain points that tempt you to give up. Some might benefit from partnerships to hold you personally accountable in meeting your goals. Don't try to fulfill anyone else's expectations on this journey. It is your story and your experience.

Realtors have a special language they use to describe properties listed for sale – "lived in look," "requires some TLC" (i.e., tender loving care, or significant renovations), "fixer-upper" which communicate properties that require work to make them ready for habitation. Job descriptions can also make use of phrases that indicate a suitable environment for job crafting. Words like "opportunity," "development," "broadening," and "challenge" indicate an inherent growth potential. Job crafting is all of those things, and much more. Interviewers must also be on the alert for applicants who signal their desire for "more" – more challenges, more opportunities, or more responsibility. If the position includes flexible boundaries, that candidate would be an ideal fit. Intentionally provide incentives to encourage new hires to make the job their own. Nearly every task has multiple ways to achieve successful results; avoid clipping the wings of self-motivated workers. Encourage workers to impress and surprise you with their talent. Innovation and creativity are born of opportunity to experiment, and is the general antithesis of micro-managing. As Hackman and Oldham (1976) concluded more than four decades ago, autonomy is an important characteristic of job satisfaction. This approach should be studied in a variety of industries, and in the IT field in particular. Todd and his colleagues (1995) studied job descriptions in ads over a 20-year period, identifying an increase in technical requirements listed for systems analysts and a slight decrease in business and systems knowledge. Their study might be replicated to determine the suggested potential for enhancing opportunities within IT-related job postings. The results could be used to develop a theory on how job descriptions are related to the diversity of applicant pools.

Corporations need to determine why women reject their job offers or leave the company. Don't ignore the signs of dissatisfaction – increased absences, distracted, irritability, avoidance, lack of preparedness in meetings, and a general poor attitude. Treat every employee like a valuable asset, help them recognize their strengths, and motivate them to achieve greater heights of performance. Create innovative (technical and non-technical) tracks for employees who demonstrate promise. When an employee exceeds the required tasks of the position, be sure to recognize that performance. Positive reinforcement is a powerful tool when combined with job crafting. It is like adding flammable accelerant to the natural inclination to succeed.

CONCLUSION

The purpose of this chapter was to introduce the research area of job crafting, and to identify ways that it can be used to improve recruitment and retention of underrepresented IT professionals. Specifically, this chapter introduced job crafting as a proactive technique that might provide benefits in the ongoing challenge of recruiting and retaining women in the IT workforce. Job crafting encompasses all "*active changes* employees make to their own job designs" (Wrzesniewski and Dutton, 2001:1). Berg, Dutton and Wrzesniewski (2008) emphasize that job crafting is a process, an ongoing process – not an isolated exercise or singular event. Job crafting is a bottom-up approach that focuses on changes instituted by the employee, rather than the managers (Berg et al., 2008). Adaptable across all industries, IT professionals might increase their personal engagement and career commitment, sparking a renewal of professional enthusiasm.

As part of an annual review, managers could invite IT employees to perform a voluntary self-assessment of their perceived work engagement. A draft worksheet is provided (see Appendix 19A.2). The conversation could be part of a mentoring program, supporting the worker's positive growth that might lead to organizational loyalty and retention. Feeling appreciated and recognized are effective methods that encourage underrepresented employees to achieve their greatest potential. A corollary to this message is that job crafting cannot be imposed or approached as a one-size-fits-all solution. Mandatory participation is unlikely to produce the intended outcome. More research is needed to evaluate the boundaries and outcomes of various types of job crafting interventions within the IT workforce.

Although there are mixed empirical findings on the importance of developing happy employees, Achor (2011) states that happiness precedes success. He builds the argument on the foundations introduced by the field of positive psychology (see Lopez, 2011; Peterson, 2006). With that assumption as a starting point, job crafting presents a potentially powerful approach to attracting, retaining, and coaching women IT professionals. Optimizing work engagement for all employees, and encouraging them to reach their highest potential, is an important objective for successful organizations. Job crafting is one method that can fortify and sustain a diverse talent pool within the IT workforce.

REFERENCES

Achor, S. (2011). *The Happiness Advantage: The Seven Principles the Fuel Success and Performance at Work*. Virgin Books, NY.

Ahuja, M.K. (2002). Women in the information technology profession: A literature review, synthesis and research agenda. *European Journal of Information Systems*, *11*(1), 20–34.

Armstrong, D., Riemenschneider, C., Buche, M.W., and Armstrong, K.R. (2018). Mitigating turnover intentions: Are all IT workers warriors? *AIS Transactions on Replication Research*, *4*(1), 10–31.

Arvan, M.L., Pindek, S., Andel, S.A., and Spector, P.E. (2019). Too good for your job? Disentangling the relationships between objective overqualification, perceived overqualification, and job dissatisfaction. *Journal of Vocational Behavior*, 115–29, 103323.

Ashforth, B.E., and Mael, F.A. (1996). Organizational identity and strategy as a context for the individual. *Advances in Strategic Management*, *13*, 19–64.

Ashforth, B.E., and Mael, F. (1989). Social identity theory and the organization. *Academy of Management Review*, *14*(1), 20–39.

Bandura, A. (1986). *Social Foundations of Thought and Action: A Social Cognitive Theory*. Prentice-Hall, Englewood Cliffs, NJ.

Bandura, A. (1977). *Social Learning Theory*. Prentice-Hall, Englewood Cliffs, NJ.

Baumeister, R.F., and Leary, M.R. (1995). The need to belong: desire for interpersonal attachments as a fundamental human motivation. *Psychological Bulletin*, *117*(3), 497–529.

Berg, J. M., Dutton, J. E., and Wrzesniewski, A. (2013). Job crafting and meaningful work. In B.J. Dik, Z.S. Byrne and M.F. Steger (Eds.), Purpose and meaning in the workplace (pp. 81–104). Washington, DC: American Psychological Association.

Berg, J.M., Dutton, J.E., and Wrzesniewski, A. (2008). What is job crafting and why does it matter. *Retrieved from the website of Positive Organizational Scholarship on April*, *15*, 2011.

Brands, R.A., Rattan, A., and Ibarra, H. (2017). Underrepresentation, social networks and sense of belonging to organizational leadership domains. In *Academy of Management Proceedings* (Vol. 2017, No. 1, p. 12798). Briarcliff Manor, NY 10510: Academy of Management.

Buche, M.W. (2008). Influence of gender on IT professional work identity: Outcomes from a PLS study. In *Proceedings of the 2008 ACM SIGMIS CPR Conference on Computer Personnel Doctoral Consortium and Research*, 134–40.

Buche, M.W. (2006). Gender and IT professional work identity. In *Encyclopedia of Gender and Information Technology* (pp. 434–39). IGI Global.

Bussey, K., and Bandura, A. (1999). Social cognitive theory of gender development and differentiation. *Psychological Review*, 106(4), 676–713.

Cameron, K.S., Dutton, J.E., and Quinn, R.E. (2003). *Positive Organizational Scholarship: Foundations of a New Discipline*. Berrett-Koehler Publishers, CA.

Cavanaugh, A., and Miller, C. (2016). *Contagious Culture: Show Up, Set the Tone, and Intentionally Create an Organization that Thrives*. McGraw-Hill.

Clance, P.R., and Imes, S.A. (1978). The imposter phenomenon in high achieving women: Dynamics and therapeutic intervention. *Psychotherapy: Theory, Research and Practice*, *15*(3), 241–47.

Cohoon, J.M., and Aspray, W. (2006). *Women and Information Technology: Research on Underrepresentation*. The MIT Press.

Collins, K.H., Price, E.F., Hanson, L., and Neaves, D. (2020). Consequences of stereotype threat and imposter syndrome: The personal journey from stem-practitioner to stem-educator for four women of color. *Taboo: The Journal of Culture and Education, 19*(4), 161–80.

Crawford, J.T. (2021). Imposter syndrome for women in male dominated careers. *Hastings Women's Law Journal, 32*(2), 26–75.

Duffy, R.D., Allan, B.A., Autin, K.L., and Douglass, R.P. (2014). Living a calling and work well-being: A longitudinal study. *Journal of Counseling Psychology, 61*(4), 605–15.

Duffy, R.D., and Dik, B.J. (2013). Research on calling: What have we learned and where are we going? *Journal of Vocational Behavior, 83*(3), 428–36.

Duffy, R.D., Dik, B.J., Douglass, R.P., England, J.W., and Velez, B.L. (2018). Work as a calling: A theoretical model. *Journal of Counseling Psychology, 65*(4), 423–39.

Dutton, J.E., Debebe, G., and Wrzesniewski, A. (2000). A social valuing perspective on relationship sensemaking. Working paper. University of Michigan, Ann Arbor. (In Wrzesniewski and Dutton, 2001.)

Edwards, J.R., and Cooper, C.L. (1990). The person–environment fit approach to stress: Recurring problems and some suggested solutions. *Journal of Organizational Behavior, 11*(4), 293–307.

Ensher, E.A., and Murphy, S.E. (2007). E-mentoring. *The Handbook of Mentoring at Work*, 299–322.

Dierdorff, E.C., and Jensen, J.M. (2018). Crafting in context: Exploring when job crafting is dysfunctional for performance effectiveness. *Journal of Applied Psychology, 103*(5), 463–77.

Good, C., Rattan, A., and Dweck, C.S. (2012). Why do women opt out? Sense of belonging and women's representation in mathematics. *Journal of Personality and Social Psychology, 102*(4), 700–716.

Hackman, J.R., and Oldham, G.R. (1980). *Work Redesign*. Reading, MA: Addison-Wesley.

Hackman, J.R., and Oldham, G.R. (1976). Motivation through the design of work: Test of a theory. *Organizational Behavior and Human Performance, 16*, 250–79.

Hantula, D.A. (2015). Job satisfaction: The management tool and leadership responsibility. *Journal of Organizational Behavior Management, 35*(1–2), 81–94.

Joshi, K.D., Trauth, E., Kvasny, L., and McPherson, S. (2013). Exploring the differences among IT majors and non-majors: Modeling the effects of gender role congruity, individual identity, and IT self-efficacy on IT career choices. *Thirty-Fourth International Conference on Information Systems, Milan 2013*.

Kooij, D.T., van Woerkom, M., Wilkenloh, J., Dorenbosch, L., and Denissen, J.J. (2017). Job crafting towards strengths and interests: The effects of a job crafting intervention on person–job fit and the role of age. *Journal of Applied Psychology, 102*(6), 971–81.

Kram, K.E. (1988). *Mentoring at Work: Developmental Relationships in Organizational Life*. University Press of America.

Lai, M.C., and Chen, Y.C. (2012). Self-efficacy, effort, job performance, job satisfaction, and turnover intention: The effect of personal characteristics on organization performance. *International Journal of Innovation, Management and Technology, 3*(4), 387–91.

Lee, K., Joshi, K., and Bae, M. (2010). An investigation of the influence of the IS context on the determinants of turnover intentions in Korea. *Journal of Organizational Computing and Electronic Commerce, 20*(1), 45–67.

Lee, R.M., and Robbins, S.B. (1998). The relationship between social connectedness and anxiety, self-esteem, and social identity. *Journal of Counseling Psychology, 45*(3), 338–45.

Lewis, K.L., Stout, J.G., Pollock, S.J., Finkelstein, N.D., and Ito, T.A. (2016). Fitting in or opting out: A review of key social-psychological factors influencing a sense of belonging for women in physics. *Physical Review Physics Education Research, 12*(2).

Ling, F.Y., Zhang, Z., and Tay, S.Y. (2020). Imposter syndrome and gender stereotypes: Female facility managers' work outcomes and job situations. *Journal of Management in Engineering, 36*(5).

Llorente-Alonso, M., and Topa, G. (2019). Individual crafting, collaborative crafting, and job satisfaction: The mediator role of engagement. *Journal of Work and Organizational Psychology, 35*(3), 217–26.

Lopez, S.J. (Ed.). (2011). *The Encyclopedia of Positive Psychology*. John Wiley and Sons.

Maher, P.J., Igou, E.R., and Van Tilburg, W.A. (2020). Disillusionment: A prototype analysis. *Cognition and Emotion, 34*(5), 947–59.

Oliner, S.D., and Sichel, D.E. (2000). The resurgence of growth in the late 1990s: Is information technology the story? *Journal of Economic Perspectives, 14*(4), 3–22.

Peterson, C. (2006). *A Primer in Positive Psychology*. Oxford University Press.

Podsakoff, N.P., Whiting, S.W., Podsakoff, P.M., and Blume, B.D. (2009). Individual- and organizational-level consequences of organizational citizenship behaviors: A meta-analysis. *Journal of Applied Psychology*, 94(1), 122–41.

Quesenberry, J.L., and Trauth, E.M. (2012). The (dis) placement of women in the IT workforce: An investigation of individual career values and organizational interventions. *Information Systems Journal*, 22(6), 457–73.

Ragins, B.R., and Kram, K.E. (2007). *The Handbook of Mentoring at Work: Theory, Research, and Practice*. Sage Publications.

Rosenstein, A., Raghu, A., and Porter, L. (2020, February). Identifying the prevalence of the impostor phenomenon among computer science students. In *Proceedings of the 51st ACM Technical Symposium on Computer Science Education*, 30–36.

Rubin, M., Paolini, S., Subašić, E., and Giacomini, A. (2019). A confirmatory study of the relations between workplace sexism, sense of belonging, mental health, and job satisfaction among women in male-dominated industries. *Journal of Applied Social Psychology*, 49(5), 267–82.

Seligman, M.E.P., and Csikszentmihalyi, M. (2000). Positive psychology: An introduction. *American Psychologist*, 5–14.

Sherman, R.O. (2013). Imposter syndrome: When you feel like you're faking it. *American Nurse Today*, 8(5), 57–58.

Stone, P. (2008). *Opting Out? Why Women Really Quit Careers and Head Home*. University of California Press.

Suwanti, S., Udin, U., and Widodo, W. (2018). Person-organization fit, person-job fit, and innovative work behavior: The role of organizational citizenship behavior. *European Research Studies Journal*, 21(3), 389–402.

Tims, M., Bakker, A.B., and Derks, D. (2015). Job crafting and job performance: A longitudinal study. *European Journal of Work and Organizational Psychology*, 24(6), 914–28.

Todd, P.A., McKeen, J.D., and Gallupe, R.B. (1995). The evolution of IS job skills: A content analysis of IS job advertisements from 1970 to 1990. *MIS Quarterly*, 1–27.

Topping, M.E., and E.B. Kimmel. (1985). The imposter phenomenon: Feeling phony. *Acad. Psychol. Bull.*, 7(2), 213–26.

Trauth, E.M. (2002). Odd girl out: An individual differences perspective on women in the IT profession. *Information Technology and People*, 15(2), 98–118.

Trauth, E.M., Quesenberry, J.L., and Morgan, A.J. (2004). Understanding the under representation of women in IT: Toward a theory of individual differences. In *Proceedings of the 2004 SIGMIS Conference on Computer Personnel Research: Careers, Culture, and Ethics in a Networked Environment*, 114–19.

Townsend, K.C., and McWhirter, B.T. (2005). Connectedness: A review of the literature with implications for counseling, assessment, and research. *Journal of Counseling and Development*, 83(2), 191–201.

Venkatesh, V., Windeler, J.B., Bartol, K.M., and Williamson, I.O. (2017). Person-organization and person-job fit perceptions of new IT employees: Work outcomes and gender differences. *MIS Quarterly*, 41(2), 525–58.

Warren, R. (2002). *The Purpose Driven Life: What on Earth am I Here For?* Zondervan, MI.

Wei, L.T., and Yazdanifard, R. (2014). The impact of positive reinforcement on employees' performance in organizations. *American Journal of Industrial and Business Management*, 4, 9–12.

Wood, R., and Bandura, A. (1989). Social cognitive theory of organizational management. *Academy of Management Review*, 14(3), 361–84.

Wrzesniewski, A. (2003). Finding positive meaning in work. In Cameron, K.S., Dutton, J.E. and Quinn, R.E., Eds., *Positive Organizational Scholarship: Foundations of a New Discipline*. Berrett-Koehler Publishers, San Francisco, 296–308.

Wrzesniewski, A., Berg, J.M., and Dutton, J.E. (2010). Turn the job you have into the job you want. *Harvard Business Review*, 88, 114–17.

Wrzesniewski, A., and Dutton, J.E. (2001). Crafting a job: Revisioning employees as active crafters of their work. *Academy of Management Review*, 26(2), 179–201.

APPENDIX 19A.1: JOB CRAFTING EXAMPLE (LOOSELY BASED ON BERG, DUTTON, AND WRZESNIEWSKI, 2013)

I am using myself as the participant in this example. The context is from a previous work experience, when I was the accounting clerk for a do it yourself lumber and supplies retailer. Although I liked most aspects of the position, it wasn't challenging enough for me. I was continually taking on more tasks and broadening my relationships with coworkers and customers. Both of these actions represent characteristics of job crafting. Specifically, my original job description was approximately 200 words in length describing the following responsibilities: Create folders for vendors; create folders for corporate clients; update sales invoices; calculate LIFO and FIFO values for lumber products; perform weekly ordering for three distribution centers; reconcile orders with delivered products; perform other tasks as requested by my supervisor. By the time I left the company less than one year later, my list of tasks and descriptions filled more than four typed pages, single-spaced.

I was definitely overqualified for the position from the start. This was evident when I started experiencing boredom after one month on the job. Instead of allowing my attitude to become negative, though, I started looking for more challenges and learning opportunities. Some of the additional tasks I took ownership of were printing shelf labels for floor sales staff; developing predictive analytics to forecast lumber sales; filling in for coworkers who were on vacation; stepping in for my supervisor during departmental leadership meetings; assisting store security with tracking loss and asset management; training cashiers; and assisting customers during the busiest shopping season. The one thing I wanted most of all, I wasn't able to add to my portfolio – a management role.

When I put in my two-week notice, the CEO asked me the question that appears in the title for this chapter: "What will it take to keep you here?!" I succinctly explained that I wanted to move into management. While pounding a fist on the desk, the CEO exclaimed that it wasn't possible, so I left that otherwise enjoyable job. If I would've completed the job crafting exercise during that work experience, it would look something like this (adapted from Berg, Dutton and Wrzesniewski, 2013):

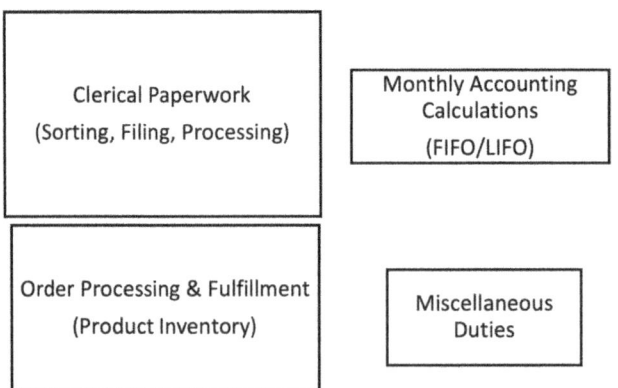

Note: Box size indicates amount of time occupied by that activity relative to other responsibilities.

Figure 19A.1   Tasks – "before" sketch

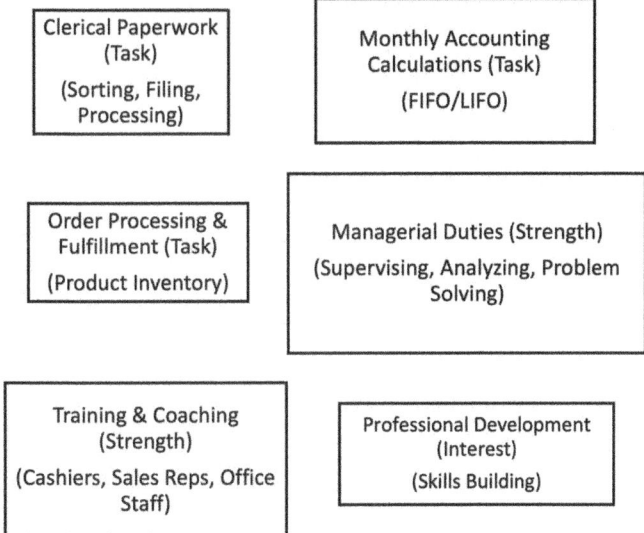

Note: Box sizes indicate amount of time the participant would prefer to spend on activities relative to other responsibilities.

Figure 19A.2   Tasks – "after" sketch

Analysis: Skills and interests are not being used effectively in the "before" figure. Strengths and Interests (passion) clearly include working with people and adopting a management role. Without an opportunity to craft the job to encompass these changes, there is likely to be an irreconcilable impasse. I often refer to these critical dimensions as "non-negotiables." We don't always recognize them at the outset, however. Once the initial orientation and training experiences have been satisfied, job enhancing desires often surface.

APPENDIX 19A.2: SELF-ASSESSMENT WORKSHEET

Worksheet: Completion of this questionnaire is optional. The purpose is to encourage employees to perform a self-assessment of their work engagement.

Instructions: Respond to the following questions naturally, without engaging in personal censorship. Candid, honest responses will be more helpful in crafting and redesigning your current job to maximize your positive experience.

1. What I like best about my current job is: _____.
2. What I like least about my current job is: _____.
3. I feel most satisfied when: _____.
4. The most important change I would like to make in my job is:_____.
5. I feel appreciated by my supervisor when: _____.
6. My strength that isn't being used currently is: _____.
7. One skill I would like to develop is: _____.

8. The biggest waste of time, the one thing I would eliminate, is: _____.
9. My short-term (1-year) goal is: _____.
10. My longer-term (3–5 years) goal is: _____.

On the back of this worksheet, list words that come to mind when you think of your job today. When we meet in person, feel free to discuss only those items that you choose to share. This worksheet is strictly voluntary, and its contents will not be used in the performance review.

20. Applying a feminist ethics of care in conducting internet-based archival gender research: the case of studying Gamergate Reactions

Florence M. Chee, Todd Suomela, Bettina Berendt and Geoffrey Martin Rockwell

INTRODUCTION

How can a feminist ethics of care, when applied at the early stages of research, serve to mitigate potential harms for both researchers and the researched? In this chapter, we tell the story of how our research group employed an Ethics of Care as a means of intervention in a research project examining the data gathered from events where gendered harassment featured prominently in driving the contours of popular understanding of an online movement, now known as Gamergate. We document this as a sociohistorical series of events that involve Gamergate Reactions (a 2015 study of a 2014 event), and the responses of the researchers by 2016, which then became an article published in 2019. In the light of 2022 and onwards in a continuing pandemic, the gender framework we discuss in this chapter is even more essential for garnering insight into events, research frameworks, and the epistemological standpoints of the researchers themselves.

How should humanities researchers approach the intersectional challenges of putting research ethics into practice? How can gender affect the dynamics of analysis when engaged in the creation of contemporary cultural artifacts developed on the Internet? As a global professional community of internet researchers, we must continue to analyze our ethical relationships in concert and in dialogue with our subjects, especially as these contexts of study continue to shift. In this chapter, we use the collection of data objects (such as tweets from Twitter and archived materials from websites) to show how the examination, collection, and curation of objects pertaining to a controversial cultural phenomenon (like Gamergate) requires a deeper examination of the ethical guidelines that shape and drive our research.

One purpose of this chapter is to describe a research project as it evolved in terms of an ethical reflection throughout the project life cycle, to provide an example of how feminist ethics of care praxis can be considered in gendered harassment. The initial research project upon which we base our analysis, Gamergate Reactions (Rockwell and Suomela, 2015), examined the Gamergate online conflict from its emergence in 2014. Since this time the Gamergate controversy, which erupted during the summer of 2014, became known as a flashpoint in how niche online game communities thought of themselves, quickly engulfing the online gaming community in a legitimacy crisis and intense debate about what and who belonged in gaming culture (Romano, 2021). The conflict quickly escalated into another battlefront in the culture wars involving gender, identity, and political beliefs (Hathaway, 2014; Wagner 2014). Overt harassment of women and others who were critical of aspects of gaming culture quickly became a dominant feature of online forums, Twitter, and other media in which discussion about gaming culture occurred. Rape and death threats were sent to prominent game developers and journalists, some of whom were also doxxed (had personally identifiable information such as home addresses disseminated online). This harassment and other extreme expressions of hatred and intolerance quickly eclipsed the alleged ethical issues in gaming culture that supposedly were the impetus for Gamergate.

This chapter focuses on the ethical challenges encountered while working on the difficult topic of Gamergate. We do not present an analysis of the Gamergate phenomenon here, but we do believe that our project's focus on Gamergate-related web content provides a challenging example of dealing with ethics in the (digital) humanities as it relates to gender. The artifacts of the digital age, such as web archives and Twitter streams, are closer to the data sets dealt with by social scientists than to the painstakingly preserved codices of libraries, traditional archives, or the bound volumes of a magazine. As such, digital humanists, including gender scholars, need to engage with the discourse on ethics arising from other disciplines. The humanities have not developed ethical guidelines around contemporary content that is publicly accessible on social media and online forums as they have with archives in libraries and museums. The lessons learned from our experience can, and should, be applied to other digital humanities projects which engage with the digital materials of contemporary culture.

There are multiple reasons why such an examination of the ethical practices in the digital humanities is particularly important for research focused on contemporary subjects. One reason is institution-based. Humanists have traditionally been free from many of the bureaucratic burdens of complying with institutional review board/ethics review board regulations, but there is no guarantee that this condition will last. These considerations include subjectivity or inconsistency from institution to institution in determining if a project should even be classified as research on human subjects. Debates are ongoing, such as whether or not oral history should be subject to institutional oversight. Indeed Chee et al. (2012) have written on how uneven this assessment can be, with consequences for lost time and funding being sustained by investigations that are more reliant on a higher human subject quotient in their research where server-side data collected under blanket consent measures have not had to undergo the same level of scrutiny for potential harm or ethical issues.

As the digital humanities become increasingly interdisciplinary, there are and will be more interactions with social scientists who are already well versed in the institutional protocols related to human subject research. Research ethicists and philosophers have developed sophisticated arguments and guidelines for dealing with research that involves human subjects. Digital humanists can become better collaborators by becoming familiar with the ethical discourses of the fields that commonly undertake human subject research. A familiarity with the key concepts of research ethics is valuable for digital humanists who may be asked to submit an IRB proposal before beginning to work on an internet-based research project.

Another reason is the larger consequences of living in a digital age. The conversation about ethics in digital research relates to larger issues concerning ethics and online activities which are currently being debated in the public sphere (Rockwell and Berendt, 2017). The discussion within digital humanities (Klein, 2015; Rehbein, 2015; Presner, 2015) already intersects with many emergent issues in the digital era, such as surveillance, big data, fake news, and algorithmic bias. We believe that further reflection on these questions may aid humanists in setting an example of thoughtful engagement with these converging topics.

As the evidence of increasing corporate and government surveillance grows, the ethics of work in the digital humanities is not exempt from enhanced scrutiny and questions of power relations. We have considered questions such as: Under what circumstances is it ethical to gather and archive any and all data about people, living or dead? From an academic standpoint, how is our scholarly work not a form of surveillance? In this chapter, we discuss the ethics of datafication and its gendered contours, by which we mean the whole process of gathering, enriching, analyzing and then archiving data (Rockwell and Berendt, 2017) in the context of a particular project, Gamergate Reactions (Rockwell and Suomela, 2015). Here, the nature of our inquiries was shaped by present issues of gendered harassment and systemic bias.

Research ethics as a broad discourse can be seen to draw on many different forms of ethical reasoning, including portions of the deontological, utilitarian, and virtue traditions of ethical argument. For example, in order to assist researchers in making ethical decisions about internet research, the Association of Internet Researchers has published two sets of recommendations that were written in consultation with the collective knowledge and experience of its international membership. One was published in 2002 (Ess and Association of Internet Researchers, 2002), another in 2012 (Markham and Buchanan, 2012). The guidelines were informed by a broad range of challenges researchers were facing in their research projects. This included how context is defined and conceptualized, ethical expectations for researchers and subjects, along with privacy, and care beyond the period of a project's lifespan. Research on informed consent issues by Chee et al. (2012) helped to structure the guidelines as a precursor to our discussion on the Ethics of Care. The most current version, Franzke et al. (2020), is designed to expand how we think about ethics including the broader research life cycle and the ethical protection of researchers themselves. This includes drawing on explicitly feminist renderings of Ethics of Care as detailed in Chee et al. (2021) and the work upon which this chapter is based (Suomela et al., 2019).

In the background section we describe some of the common approaches to online research ethics and the philosophy behind those approaches. We introduce the concept of an Ethics of Care (Held, 2006) as an alternative way of framing discussions of research ethics within digital humanities. The following section presents a case study of how the ideas of an Ethics of Care were applied in the Gamergate Reactions project (Rockwell and Suomela, 2015). The goal was to answer two research questions: What is an appropriate ethical framework for online research conducted within the digital humanities? And how can that framework be put into practice? In the conclusion we discuss how we answered these questions using a feminist praxis-oriented framework and the resultant reflections herein.

BACKGROUND

Ethics of Care and Research Ethics

In this section, we situate Ethics of Care in the wider context of research ethics. In the context and given the space limitations, it is of course impossible to give an exhaustive introduction to these wide fields. However, we consider it necessary to give an overview, for three reasons. First, we want to engage with research ethics debates from other fields; second, digital humanists may not be familiar with the terms of the debate; and third, learning about these debates may make humanities perspectives more relevant to datafication debates.

One approach to ethical judgment is deontological, or the study of ethical duty. An example of such a duty would be Kant's categorical imperative that we should act only on an ethical principle that we would accept becoming universal (Alexander and Moore, 2016). In other words, that we should do only what we would have everyone do. Are there certain duties which we should follow when dealing with data collected from the Internet? One potential way of answering this question is to look at research ethics in general. Many of the guidelines for research ethics proposed in various fields use a form of deontological reasoning. One of the landmark documents in the modern development of research ethics is the Belmont Report (1979). The core principles of the report described three fundamental ethical concerns for human subject research: respect for persons, beneficence, and justice. In a medical context, respect for persons has been interpreted as requiring informed consent, ensuring that patients are not coerced, and that they understand the extent of the study in which they are participating. Beneficence requires the maximizing of benefits and the minimizing of risk for the research subject, as well as guaranteeing that the experimenter will do no harm. Justice ensures that research subjects will not be exploited, and that the study is administered fairly and equally. The institutional support for these principles at most research institutions in the United States are the institutional review boards, which follow the Common Rule established by the 1991 Federal Policy for the Protection of Human Subjects (Department of Health and Human Services, 2009). Canada and many other countries have established similar institutional protections for research subjects through research ethics boards and governmental guidelines (TC3+ 2014).

A second approach to ethical judgment is utilitarian (Sinnott-Armstrong, 2015). The key question for a utilitarian analysis is: who benefits and how much? A utilitarian approach to the ethical questions of internet research might focus on the consequences of such research and attempt to balance the knowledge that would be gained against its potential for harm. Medical and other human subject research boards often ask researchers to assess the potential for harm under the standard of beneficence mentioned above. The difficulty with this approach to humanities or cultural research on the Internet is that there is no easy way to assess potential harm. The biomedical approach to research ethics, which generally assumes data from clinical trials and laboratories, does not adequately fit when addressing the possible nuances of collecting data from social media such as Twitter. The charge to do no harm takes numerous and varied forms when differentiating between drugs and status updates, for example.

A third approach to ethical judgment is virtue ethics. Virtue ethics is based on the assessment of how human actions foster the virtues of human character (Hursthouse and Pettigrove, 2016). It is harder to draw a direct link between virtue ethics and research ethics because most of the discussion within research ethics focuses heavily on the principles of respect for persons, beneficence, and justice, and less on the virtues of researchers or subjects. But we can detect an implicit argument for a type of virtuous behavior on the part of researchers. The injunction of the Hippocratic oath describes a virtuous ideal for behavior in the medical profession. Harder to define, but just as important, is the general idea that a good researcher should not commit fraud of any kind.

Moving Toward the Ethics of Care

The Ethics of Care[1] is a theory that developed from feminist thought about the moral importance of the experiences of caring for children, the elderly, and vulnerable populations. Others have used it in the context of design and in the digital humanities to draw attention to carework (Klein, 2015; Jackson, 2014). Unlike previous ethical theories that start from the position of an independent rational subject thinking about how to treat other equally independent rational subjects, the Ethics of Care starts with the real experience of being embedded in relationships with uneven power relations.

The Ethics of Care framework prioritizes research relationships and focuses attention particularly on the features of ethical decision making which are relevant to project design. What are the relationships between the people involved in a project? Who possesses the power or authority in a given situation? The Ethics of Care recognizes the differences in vulnerability and need among stakeholders between those studying and those being studied. Using our case study, we will outline some of the complicated relationships between researchers, subjects, communities, and discourses, which need to be continually revisited.

The crux of the difference between Ethics of Care and justice-oriented approaches to ethics is the issue of rights. Held (2006) argues that care is a precondition for rights – the social cohesion that makes it possible for political institutions to exist (p. 126) such that there could be protection of rights. Civil society is built on civic virtues that guide how we interact outside of formal legal ways. Care would be one of the most important civic virtues.

If one agrees with Held, then the Ethics of Care emphasis on relationships is not in opposition to rights but is a matter of attention or emphasis. An Ethics of Care guides us in all the everyday situations where rights don't apply or aren't clear. Alternatively, one could argue that respect for rights is a type of relationship. We are in relationships with our research subjects. Considering their rights is part of thinking about the relationship.

Closer to what we are doing, Klein in the Carework and Codework of the Digital Humanities (2015) asks how codework (or work using digital tools and code) can help recover the carework that often gets hidden by the grand gestures of scholarship. She references Blackwood (2014) who reminds us how gendered certain research roles are, like editing and archiving. For that matter, we would argue that codework is also carework that is gendered, though differently. And ethics also has a history of being gendered. Other researchers in digital humanities have engaged in similar efforts to further the discussion of research ethics. Rehbein (2015) presented a paper in 2015 at the Digital Humanities Summer Institute about the dual use problem where resources we develop can be used unethically. Presner (2015) has also written about the ethics of the algorithm and databases.

The research community often treats the work of gathering, editing, and maintaining data as of secondary importance compared to the comparatively more prestigious grant getting, grand theorization or original contributions. The creation of, maintenance of, and enrichment of archives have traditionally been the work of assistants, librarians, and archivists (i.e., feminized work). It is only recently that attention has been turned toward how the stewardship of archives, whether digital or not, is valuable. Jackson (2014), for example, talks about the importance of repair of the infrastructure as being more important than building new infrastructure. The three research councils of Canada and the research infrastructure foundation (TC3+) issued a consultation document in 2013 calling for establishing a culture of stewardship (TC3+ 2013). The Stevens Institute of Technology ran a conference on the Maintainers in 2016 and 2017 (the Maintainers n.d.), a response to Isaacson's (2014) book, *The Innovators*. Klein and D'Ignazio (2020) survey consequences of the neglect of the concrete work on data, and sketch proposals for increasing its recognition. The enhanced valuation of this work is associated with calls for more reflexive practices (Miceli et al., 2021) and their standardization, e.g., in the form of checklists and tools (e.g., Gebru et al., 2018). These are just examples of a clear turn to looking at and valuing the care and repair of our cultural and research infrastructure.

PUTTING THE ETHICS OF CARE INTO PRACTICE: A CASE STUDY

Our choice to use an Ethics of Care to guide our research process arose from several considerations. Our study of Gamergate and our relationships with other researchers pointed to a need to engage with feminist discourses on both the topical and methodological levels. We felt a distinct need to focus on the relationships involved internally and externally to the Gamergate Reactions (Rockwell and Suomela, 2015) project itself. Many of the mainstream approaches to research ethics prioritize the discourse of rights and justice, a deontological approach that is problematized by an Ethics of Care. We also wanted to expand the ethical discussion to include people, such as librarians and archivists, engaged in the carework necessary for development of digital research materials. Finally, an Ethics of Care foregrounded the iterative and dialogic approaches we took to develop an ethically engaged research project (Klein, 2015; Jackson, 2014).

The international research network that is the focus of this case study was composed of a diverse group of students, postdoctoral scholars, and faculty collaborating from fields including computer science, digital humanities, and communication. Relationships between people involved in the research process needed to be consciously and consistently managed throughout the project. Membership changed over time as different people moved in and out of the group because of graduation, promotion, changes in enthusiasm, or acceptance of other jobs. The main research group met regularly to discuss the progress of the project and to manage tasks. During the academic term these meetings occurred weekly, at other times of the year the meetings were less frequent. The typical meeting discussed the immediate tasks that needed to be completed to gather and analyze data, and to write about the project for publication or conference presentations.

We decided that we should be guided by the Ethics of Care after much discussion with members of the research team and our colleagues. The idea for using an Ethics of Care to describe and frame the ethical challenges for the project was suggested by a member of ReFig,[2] another research project with which we are affiliated (some of the authors of this chapter have been involved with ReFig for several years). The Ethics of Care struck us as an appropriate ethical framework for the context, especially given the sustained harassment of researchers studying Gamergate (Chess and Shaw, 2015). Indeed, the more we learn about the Ethics of Care (Held, 2006; Wittkower, 2016) and its relevant applications, the more it seems suited to such complex situations where people have been harassed and ethics itself is at stake. In this chapter we discuss three features of the Ethics of Care and how we applied them: (1) ethics is more about relationships than about rights; (2) a fundamental facilitator of relationships for humanities research is dialogue; and (3) caring is a practice, not a heroic gesture, nor a set of rules for behavior; it is the ongoing activity of being sensitive to others (and oneself).

The next three sections describe the internal and external dialogues we engaged in during our project. We wrestled with many different issues throughout, including how to deal with identification of research team members, toxic data, the potential impacts of using social media content on those who generate it, and archiving research data. Our approach to these issues was informed by our conscious and iterative application of the Ethics of Care.

Identifying Team Members

The focus of most research ethics discussions is usually on the relationship between the researcher and subject(s). The internal dynamics of research groups is an issue that is less often discussed. For our project the internal ethics of the research group were a major concern from the beginning.

From the start of the project, we were aware that other researchers were being threatened and targeted for harassment, particularly if they were women and expressing views about gender and games in the context of Gamergate. Examples included specific targeting of game researchers on YouTube, mobbing tactics on Twitter, and forum posts. We wanted to be proactive about mitigating the potential controversy and harassment that might result for our own research team from our research. We took the following steps to address this risk. First, meetings were arranged with campus staff to discuss the safety and technical security precautions that team members should take to protect themselves online or to respond to harassment. Second, a background literature review was conducted to discover the guidelines used by other researchers (Markham and Buchanan, 2012; Ess and Association of Internet Researchers, 2002; American Anthropological Association, 2012) when dealing with potential harassment. Third, we consulted with key people who had already experienced harassment through these networks for their counsel on best practices for how to proceed with optimal attention to care.

One of the key insights from the Ethics of Care which we applied to our own internal discussions about the project was the recognition that members of the research team were vulnerable for different reasons. Senior researchers and tenured faculty members were typically already protected by their status within the university and the wider research community. But other researchers, such as students, postdocs, and untenured faculty in their various capacities and identities are often in more vulnerable positions. The harassment of student researchers could do significant harm to them. For example, the data sets published through the online repository at the University of Alberta in Canada, listed a subset of people involved in the project and provided a group email account for public comment instead of individual email account names that could be singled out. Given the potential negative publicity and/or doxxing risks of which those involved with the project became increasingly aware, we consulted with the students and gave them the option of limiting the number of times their names were published with the artifacts of the research, such as on papers or data sets. This action, taken by the principal investigators to consider the standpoint of student workers, was guided by the Ethics of Care in how power over the distribution of one's name rested ultimately with the owner. Reflecting upon these remediations several years into the future, it is clear to us that while larger organizational units may provide cover and support for those who may refuse to be named, it also may do a disservice in attributing proper credit to those who have contributed to a project. We realize that the attribution of credit and authorship is also a systemic problem in research more generally, as women's research and teaching tend to be more minimized professionally in general (Frances et al., 2020). As we have learned more in the intervening years on how to deal with weaponized online harassment of women scholars, a possible consideration may lean toward embargos and limited release of research in early phases (to allow for attention level/ harasser attention to die down), rather than outright erasure or redacting.

It is not lost on our research team that there are definitely problematic power differentials associated with the choice to redact one's own name and involvement. Given the disproportionate potential for women to sustain harassment over men, our isolated action could not ameliorate the social tendency for women to also be unfairly penalized professionally as well as a result of being targeted. Rather, we need to state here that this whole generation of scholarship that has been colored by the Gamergate scandal replicates the dynamics where women have worked covertly behind the scenes (sometimes through necessity), and that they should be recognized by institutional bodies that consider authorship credit and promotion. While we do not have the answers when it comes to this systemic and endemic issue, we do consider this situation and discussion as one we wish to improve through increased awareness.

Toxic Data

In addition to protecting the members of the research team from potential external harms, there was a concern about the psychological impact of reviewing the digital artifacts of online harassment and hate speech. Some of the data we collected originated from websites (4chan and 8chan) that have a reputation for producing pornographic and offensive (even illegal) content on a regular basis. Both websites are transient forums in which postings are automatically deleted after a period of time. Some of the campaigns, meme development, and organizing for Gamergate were conducted on these forums. The majority of postings are anonymous, and these postings often contain offensive material. We collected data from these websites for a few months, but then decided to delete that information unused because of the negative impact that analysis of this content would likely have on the researchers, especially students, in the project group. We also concluded that the detriments of archiving toxic data for possible later analysis did not outweigh the potential research benefits.

Indeed, consideration of all the possible influences or effects that a particular decision about data collection or retention may result in for this project was greater than the scope of any one individual's imagination. The instrumental viewpoint that all research merely involves a selection of topics, material, methods, and data was indeed put to the test in unanticipated ways that presented ethical implications, and intersecting issues. Analyzing large-scale events in social media compounds the problem because the collection of data is dependent on factors beyond the control of the researcher, such as the capabilities of the API (Application Programming Interface) for gathering data. The policies of social media companies may also affect the types and completeness of the data collected. The boundary between data collection and analysis is also complicated. An initial analysis may find that more data is needed or that a conclusion is justified based on the data already collected. The decision whether to include data in an analysis is fraught with potential errors and caveats. Works of interpretation, such as in the digital humanities, are even harder to assess. How much evidence do we need to reach the conclusion that many of the participants in the Gamergate controversy were motivated by anti-feminism or misogyny, or that some of them actively engaged in harassment of prominent women gamers (Chituc, 2015).

From Gamergate Reactions (2015), we extracted data from a scrape of #gamergate tweets from October 2014. The data sets released include high-frequency words, links, and hashtags. We initially collected data from 4chan because we knew that 4chan was a major communication channel for Gamergate. But we also knew, based on our own experiences with the website and media reports (Hess, 2017; Lopez, 2018; Nagle, 2017), that 4chan was home to pornography, harassment, doxxing, and other noxious behavior. When it came time to decide whether to analyze the data we had collected from 4chan, the research team faced a complicated decision with no obvious answer and few, if any, clear rules for appropriate behavior. Some researchers may assume that any data collected must be analyzed, but that runs opposite to the Ethics of Care approach we were trying to honor throughout the research process. The Ethics of Care approach underlines the contextual nature of ethical decision making and the relationships in which we are all embedded (Held, 2006). Our research group was composed of graduate students, post-docs, and professors. The bulk of the analysis of data was performed by the graduate students and post-docs involved in the project. Was it fair to ask students to review potentially offensive material in order to make a marginally more forceful case that Gamergate contained misogynists or rape threats? We already had evidence for those behaviors in the social media data and the webpages that had been collected from the rest of the web. The small size of the 4chan data we had collected in comparison to the large numbers of tweets in our data led us to the conclusion that additional analysis of the 4chan material would probably not change or challenge our interpretations significantly. Given the unlikeliness of discovering new information and the likely encounter with offensive material, the research group decided that analysis of 4chan data was not worth the effort.

We also decided to delete the 4chan data because of another concern that was raised by our colleagues in ReFig and at conferences: what was the future impact of preserving the evidence of online harassment? Again, we felt there was no clear standard for making this judgment based on our knowledge of research ethics. The presence of potential or future harms is hard to negotiate in any ethical decision-making process. One might say that trying to provide guidance based on potential outcomes is the core of ethical argument. If something has already happened, then we only need to deal with the consequences. Many discussions of research ethics try to minimize potential harm through such actions as informed consent and transparency. Although there may be no clear standard to which we can appeal to justify our decision to delete data, we believe that transparency can provide an explanation for why we made our decision. There may always be some who find our decision unreasonable or invalid. To them we can only say that this chapter attempts to describe our decision, the reasons why we reached that decision, and how those decisions might help others make better choices in the future.

Finding a balance between the potential for increasing our understanding of online behavior versus doing psychological damage to ourselves is an ongoing dilemma. But we could not know the risk until the data was actually reviewed. In the case of some of the data we collected, we decided that the research benefits did not outweigh the potential damage. Other researchers may have acted differently but we can only report the results of our own analysis and reflection upon the methods which produced that analysis.

Using Social Media Content: Impacts on Content Creators

The relation between researchers and research subjects can be especially challenging for humanists and social scientists to navigate. Humanists have not been typically accustomed to having to contort their thinking to consider ethics in human subjects research because the subjects of their research have more generally concerned the deceased or public figures like authors, politicians or other humanists, whose roles and activities are open to scrutiny and debate. The harms, real or potential, inflicted by research methods on these subjects are often intangible and hard to measure. Social media presents an ethical challenge in research because of the complicated division between public and private activity (Zimmer, 2010). For example, it is not obvious that users of Twitter understand the potential for their tweets to be collected as part of a research project (Fiesler and Proferes, 2018). The potential harms for the misuse of social media content might include threats to marginalized groups, fraudulent manipulation of the data, and reputational harms (Jules et al., 2018).

Given the complicated nature of social media research we felt it was prudent for us to take future potential harms from our data seriously. Our first duty as researchers should be to avoid harm. This is one of the main conclusions of the Belmont report and many other writers on research ethics (Buchanan and Zimmer, 2016; National Commission for the Protection of Human Subjects of Biomedical and Behavioral Research, 1979; Department of Health and Human Services, 2009). At first glance this appeal to duty may appear to conflict with the Ethics of Care approach we have been describing in this chapter. The mention of duty immediately brings to mind the discussion of deontological ethics mentioned earlier. However, we do not think that our invocation of duty contradicts our promotion of an Ethics of Care. The Ethics of Care is based on relationships (Held, 2006), with our fellow research members, the research community, and our subjects. One of our goals is to maintain those relationships over time, and avoiding harm is one way of accomplishing this goal. It is possible to justify avoiding harm to research subjects from multiple philosophical perspectives.

Privacy and reputation were the two biggest harms to research subjects that we discussed and attempted to mitigate in the Gamergate project. The privacy of our subjects was protected in two overlapping ways. First, the results of the research were reported only in aggregate forms, and second, the sources for direct quotes were not identified (A. Budac, 2016; R. Budac, 2016; Gouglas et al., 2016; Kuznetsova and Suomela, 2016; Rehbein, 2015; Wilson et al., 2016). Neither method of privacy protection can completely guarantee that people will not be identified because the activity we collected and analyzed occurred in online forums that anyone can search. A determined person could still recover the original source of a quote by searching Twitter or the Internet, so the results could not be completely anonymized. The question of privacy is highly fraught when it comes to research about topics such as Gamergate, which depend on the Internet as the primary medium for communication. Any quote from a publicly accessible website could potentially be re-identified after a research study has been completed. Furthermore, the ethical boundaries for understanding research on public web material are still openly being discussed and have not reached any consensus (Buchanan and Zimmer, 2016).

Reputation harms to research subjects were the other major risk discussed during the Gamergate project and are even harder to quantify and evaluate. Throughout the project, we discussed the possibility that someone who tweeted in support of Gamergate today might, at a later point in their lives, change their mind and be embarrassed by their prior views. Future harms could extend beyond mere embarrassment to include losing a job, to emotional damage, or to self-harm. There have been media reports about how online behavior might be a detriment to future reputations online (Weber, 2014). We believe that the potential damage from our research revealing the identity of Gamergate supporters is unlikely to be so drastic, but this is only our current perception. The Ethics of Care suggests that this position may be modified as relationships change, and as the research process progresses.

Publication and Preservation of Data

Direct dialogue with a community that includes those who have targeted academics for harassment has several challenges, as written in Chess and Shaw (2015). After the initial public presentation of our early findings at the 2015 Canadian Game Studies Association (CGSA) conference, we posted a subset of our data to the University of Alberta data repository. We did this in line with an ideology of openness and dialogue with a plurality of viewpoints as a general principle. One way of creating an artifact and encouraging others to examine the data and make their own interpretations included publicly posting portions of our data set comprising thousands of raw social media posts. One online comment on our data was made by a Gamergate supporter who argued that the tweets we posted proved the opposite of what we had observed, although the Reddit thread was eventually deleted ([People] The Person behind the Idea for #Deatheaters on Being Asked If Her PhD Is in GamerGate "Literally Yes. That Is What My PhD Is in." • r/KotakuInAction. 2018.). The Ethics of Care does not provide general guidelines for engaging with research subjects, but it does imply that dialogue is an ongoing process that is developed over the course of a research project.

Our relationships with people who have been harassed by Gamergate participants is another example of an external relationship which has shaped our understanding of the project. Regarding the ethical dilemma of preserving the evidence of harassment in various forms, we incorporated the input of those involved with the research project, members of the ReFig project, and attendees at various conferences where we discussed our findings. The input from these stakeholders led to our decision to delete some of the data from our corpus. This contradicts the first impulse for many researchers, which is often to keep as much of the research data or material that is collected during a project because of the immense effort that goes into developing such collections, including: deploying and customizing software in order to continuously scrape data; curating websites or documents based on the subject matter; analyzing the artifacts in order to deepen our understanding; and presenting those results to the research community or the larger public. Given this effort, many researchers would be reluctant to discard such a signifier of time, energy, and emotional labor.

Another cross-cutting imperative regarding the preservation of research data is to the research community as a whole. Do we owe other researchers access to our data so that they may study it themselves? There are considerable efforts being made in the social sciences to preserve research data so that other researchers may attempt to replicate the results of a study. For example, social psychologists have been arguing about a "crisis of replication" in their field for the past decade. Recent examples of researcher fraud pose a serious challenge to the legitimacy of a field (Aschwanden, 2015; Bartlett, 2015; Marcus and Oransky, 2015). Sharing research data is one way to address this problem.

However, in approaching Gamergate from the perspective of the humanities, we may have different grounding assumptions about what types of material should be made available during, or at the conclusion of a research project. Within the humanities, there is a tradition of curation in which particular archives or editions are thoughtfully created as part of the research process. For digital artifacts, such as the websites and tweets which formed the core materials for our project, the value of creating an archive for future researchers may simply be to preserve the historical record, but we must also be conscious that the archive is not innocent. Decisions are continuously being made about the selection of materials for inclusion. Moreover, digital material, especially on the web, may be altered for many different reasons such as a publisher going out of business, an editor revising a previous story, or a hacker maliciously removing information. We cannot be sure of the motivations of these changes, but collecting our own record of a website at a given time is one way of showing how people communicated at a point in time.

In essence, making a decision about whether to keep all of the material we collected during the project involved a series of cross-cutting discussions that implicated the various disciplinary traditions reflected in the project participants, outside groups with which we collaborated, and a larger public who may be interested in the results of our study. Deciding whether to preserve artifacts such as data or documents that demonstrated harassment, was one of the hardest questions we faced because of the multiple relationships we had before, and developed during, the course of the study. We had to ask ourselves challenging questions, such as whether preserving evidence of harassment was just another way of perpetuating or repeating that harassment into the future. The very real and immediately present effects of Gamergate on many people have been harmful. We ultimately decided that the harm of adding to the arbitrary preservation of the harassment we saw occurring, outweighed the loss of any information that might have been gained by analyzing those specific materials. Given that researchers make executive choices from conception of a research question through to the final write-up of a project, our choice was guided by an ethic of care and doing less harm along the way.

The emphasis of the Ethics of Care on relationships instead of rights informed our decisions about preservation of toxic materials. We engaged in an ongoing dialogue both within the research group and with external groups. We decided that our relationships were more important than an abstract duty to preserve everything we collected. There is a real and perhaps unresolvable tension in this decision. We are still sensitive to the different decisions which other researchers, in a similar position, might have taken.

Any move toward understanding a moral position with which we disagree requires some form of empathy. Cognitive and emotional work must be essayed in order to help both ourselves, as researchers and inhabitants in our various communities, as well as others who may be outside of our communities. The goal of our research is less about persuasion than it is about reporting our interpretations of what we have studied.

CONCLUSION

In the introduction to this chapter, we argued that digital humanities should engage in a broader discussion of ethics, especially as it is applied to the research process for digital materials. In this conclusion, we add some of the lessons learned from our case study and suggest how those lessons learned may contribute to the ongoing conversations surrounding research ethics, especially as they pertain to a feminist ethics of care. Let us begin with the lessons for digital humanities research projects.

For digital humanities in particular, the lessons to be drawn from our case study imply that the discussion of research ethics should be incorporated into the research process during the early stages of planning and data collection. Researchers should be aware of the relationships upon which their work depends. Graduate students and postdocs are in an especially delicate position during a research project because they may not feel empowered to affect the direction of a project. Senior researchers need to be aware of this relationship and work to include all parties in the execution of said project.

From the very beginning of a project, we must think of the work in terms of the stakeholders involved and the relationships between them. A basic starting point is examining the project from an internal perspective by describing the relationships between the many people who are actively engaged in data collection, analysis, writing, and archiving. The next step is to check on the relationships between the research project and the subject community that is involved. Are the research subjects being consulted or respected? Are the researchers avoiding potential harm? A final step is to look at external relationships beyond the subjects of the study, often this means looking outward to the wider research community for suggestions and openings for dialogue.

Our discussion of research ethics and philosophy was inevitably too brief to give these subjects their due attention. However, we hope to have piqued the interest of those who may be prompted to delve deeper into these subjects, including digital humanists, archivists, librarians, and data scientists. As automation and datafication of everyday life only increase in prominence, the debates that have emerged because of these ethical questions have provided increased awareness of the issues that warrant further exploration.

The Ethics of Care is a fruitful starting point for discussing ethical issues related to humanistic scholarship because it is rooted in a feminist tradition that asks researchers to pay attention also to the mundane activities of editing and archiving. Relationships with the human subjects who are being studied in a research project are important, but the internal relationships among the research team deserve just as much attention. One of the important ideas touched on in this chapter, which we hope to expand on in future work, is the role of infrastructure and the support personnel who maintain it. The roots of the Ethics of Care in feminist scholarship suggest some interesting directions for further work on the gendered nature of various roles, such as editing and archiving.

Regulation of the research process is another important issue that digital humanists can engage from multiple perspectives. The background information and case study presented in this chapter will help humanists in this process. In a future line of research, we will investigate the commonalities and differences of this approach with current developments in European Union law and practices. European Union data protection and privacy thinking and legal codification have long been concerned with attempts to balance the rights of different stakeholders (such as freedom of contract or scientific interests on the one hand, and privacy on the other hand) through laws that arguably try to mitigate the power differentials that are unavoidable between data controllers and data subjects (Gutwirth and De Hert, 2006). With the risk-based approach and data protection impact assessments built into the General Data Protection Regulation, which became directly applicable law in May 2018, data controllers and data processors are now required to assess, ex ante, possible harms to data subjects and take measures to avoid such harms. These harms will overlap with those we have discussed in this chapter, although they may not cover all of them (the harms inflicted on researchers are an interesting exception), such that Ethics of Care approaches may be able to draw on emerging best practices from impact assessments. On the other hand, even the best-intentioned Ethics of Care approach is not immune to a danger that has been discussed with respect to the risk management approach (e.g., Shapiro, 2014), that through the researcher's, or, in general, the data controller's (re)appropriation of definitional power over what the risks are, pre-existing power differentials actually get strengthened, and the weaker parties' rights weakened.

Learning from other disciplines is another potential benefit from engaging in the issues raised in this paper. Digital humanities is also part of a larger conversation about the structure of digital scholarship in general. Norms for publishing digital data and research artifacts are evolving throughout the research environment.

At the start of this chapter, we mentioned the idea of datafication, i.e., the growing importance of data to all types of human endeavor. How does such a large social transformation overlap with the ethical concerns raised by a single research project? We believe that the issues we are currently wrestling with in our own research work will only become more important to the larger world as datafication continues to grow and impact our daily lives in more and more elaborate ways. Issues of labor compensation and gender should be closely connected to larger societal conversations around rights, especially in the digital realm. Changes to the legal regimes in different countries and contexts also need further attention.

Although concerns about the datafication of digital humanities archives have been raised, the analysis so far in the digital humanities has been retrospective and has provided limited guidance for other projects. Incorporating a feminist Ethics of Care bolsters the understanding of power and praxis in the study of digital culture and contributes to the emerging discussion of ethical issues across projects.

NOTES

1. We choose to capitalize the phrase Ethics of Care throughout this chapter in order to emphasize the specific feminist philosophical tradition upon which we base our analysis.
2. Refiguring Innovation in Games (ReFiG) was a five-year project supported by the Social Sciences and Humanities Research Council of Canada. Composed of an international collective of scholars, community organizers and industry representatives, ReFiG is committed to promoting diversity and equity in the game industry and culture and effecting real change in a space that has been exclusionary to so many (ReFiG, 2018).

REFERENCES

Alexander, L., and Michael M. (2016). Deontological ethics. In E.N. Zalta (Ed.), *The Stanford Encyclopedia of Philosophy*. October 17. Accessed November 21, 2018. https://plato.stanford.edu/archives/win2016/entries/ethics-deontological/.

American Anthropological Association. (2012). Principles of professional responsibility. Accessed November 21, 2018. http://ethics.americananthro.org/category/statement/.

Aschwanden, C. (2015). Science isn't broken. *FiveThirtyEight*. August 19. Accessed November 21, 2018. https://fivethirtyeight.com/features/science-isnt-broken/.

Bartlett, T. (2015). The unraveling of Michael LaCour. *The Chronicle of Higher Education*. June 2. Accessed November 21, 2018. www.chronicle.com/article/The-Unraveling-of-Michael/230587.

Blackwood, S. (2014). Editing as carework: The gendered labor of public intellectuals. *Avidly*. June 6. Accessed November 21, 2018. http://avidly.lareviewofbooks.org/2014/06/06/editing-as-carework-the-gendered-labor-of-public-intellectuals/.

Buchanan, E.A., and Zimmer, M. (2016). Internet research ethics. In E.N. Zalta (Ed.), *The Stanford Encyclopedia of Philosophy*. August 24. Accessed November 21, 2018. https://plato.stanford.edu/archives/win2016/entries/ethics-internet-research/.

Budac, A. (2016). Tweeting obsessives. Paper presented at Canadian Game Studies Association/l'Association canadienne d'études vidéoludiques Annual Conference. Calgary, Canada, June 3.

Budac, R. (2016). Controlling with mini-corpora of comments. Paper presented at Canadian Game Studies Association/l'Association canadienne d'études vidéoludiques Annual Conference. Calgary, Canada, June 3.

Chee, F., Hjorth, L. and Davies, H. (2021). An ethnographic co-design approach to promoting diversity in the games industry. *Feminist Media Studies*. DOI:10.1080/14680777.2021.1905680.

Chee, F.M., Taylor, N.T., and de Castell, S. (2012). Re-mediating research ethics: End-user license agreements in online games. *Bulletin of Science, Technology & Society*, 32, 6, 497–506. December 1. DOI: https://doi.org/10.1177/0270467612469074.

Chess, S., and Shaw, A. (2015). A conspiracy of fishes, or, how we learned to stop worrying about #GamerGate and embrace hegemonic masculinity. *Journal of Broadcasting & Electronic Media*, 59, 1, 208–20. DOI: https://doi.org/10.1080/08838151.2014.999917.

Chituc, V. (2015). GamerGate: A culture war for people who don't play video games. *New Republic*. September 11. Accessed November 21, 2018. https://newrepublic.com/article/122785/gamergate -culture-war-people-who-dont-play-videogames.

Department of Health and Human Services. (2009). Protection of human subjects. Accessed November 21, 2018. www.hhs.gov/ohrp/sites/default/files/ohrp/humansubjects/regbook2013.pdf.pdf.

D'Ignazio, C., and Klein, L.F. (2020). *Data Feminism*. Cambridge, MA: MIT Press.

Ess, C., and Association of Internet Researchers. (2002). Ethical decision-making and internet research: Recommendations from the AoIR ethics working committee. *Association of Internet Researchers*. November 27. Accessed November 21, 2018. https://aoir.org/reports/ethics.pdf.

Fiesler, C., and Proferes, N. (2018). "Participant" perceptions of Twitter research ethics. *Social Media + Society*, 4, 1, 1–14. DOI: https://doi.org/10.1177/2056305118763366.

Frances, D.N., Fitzpatrick, C.R., Koprivnikar, J., and McCauley, S.J. (2020). Effects of inferred gender on patterns of co-authorship in ecology and evolutionary biology publications. *Bull. Ecol. Soc. Am.*, 101, 3, e01705. https://doi.org/10.1002/bes2.1705.

Franzke, A.S., Bechmann, A., Zimmer, M., Ess, C.M. and the Association of Internet Researchers (2020). Internet research: Ethical guidelines 3.0. *Association of Internet Researchers*. Accessed September 27, 2021. https://aoir.org/reports/ethics3.pdf.

Gebru, T., Morgenstern, J., Vecchione, B., Vaughan, J.W., Wallach, H., Daume, H., and Crawford, K. (2021). Datasheets for Datasets. *Communications of the ACM*, 64, 12, 86–92.

Gouglas, S., Rockwell, R., Chee, F., and Suomela, T. (2016). G@merg@te Origins. Paper presented at Canadian Game Studies Association/l'Association canadienne d'études vidéoludiques Annual Conference. Calgary, Canada, June 3.

Gutwirth, S., and De Hert, P. (2006). Privacy, data protection and law enforcement. Opacity of the individual and transparency of power. In E. Claes, S. Gutwirth, and A. Duff (Eds.), *Privacy and the Criminal Law*, 61–102. Cambridge: Intersentia.

Hathaway, J. (2014). What is Gamergate, and why? An explainer for non-geeks. *Gawker*. October 10. http://gawker.com/what-is-gamergate-and-why-an-explainer-for-non-geeks-1642909080.

Held, V. (2006). *The Ethics of Care: Personal, Political, and Global*. Oxford: Oxford University Press.

Hess, A. (2017). How the trolls stole Washington. *The New York Times*. February 28. Accessed November 21, 2018. www.nytimes.com/2017/02/28/magazine/how-the-trolls-stole-washington.html.

Hursthouse, R., and Pettigrove, G. (2016). Virtue ethics. In E.N. Zalta (Ed.), *The Stanford Encyclopedia of Philosophy*. December 8. Accessed November 21, 2018. https://plato.stanford.edu/archives/ win2016/entries/ethics-virtue/.

Isaacson, W. (2014). *The Innovators: How a Group of Hackers, Geniuses and Geeks Created the Digital Revolution*. London and New York: Simon & Schuster.

Jackson, S.J. (2014). Rethinking repair. In K.A. Foot, P.J. Boczkowski, and T. Gillespie (Eds.), *Media Technologies: Essays on Communication, Materiality, and Society*, 221–40. Boston, MA: MIT University Press. DOI: https://doi.org/10.7551/mitpress/9780262525374.003.0011.

Jules, B., Summers, E., and Mitchell, Jr., V. (2018). Ethical considerations for archiving social media content generated by contemporary social movements: Challenges, opportunities, and recommendations. *Documenting DocNow*. July 19. https://news.docnow.io/documenting-the-now-ethics-white -paper-43477929ea3e.

Klein, L.F. (2015). The carework and codework of the digital humanities. Accessed November 21, 2018. http://lklein.com/2015/06/the-carework-and-codework-of-the-digital-humanities/.

Kuznetsova, E., and Suomela, T. (2016). Hashtags make meaning. Paper presented at Canadian Game Studies Association/l'Association canadienne d'études vidéoludiques Annual Conference. Calgary, Canada, June 3.

Lopez, C. (2018). A day before Parkland survivor David Hogg got "swatted," trolls shared his address on 4chan and 8chan. *Media Matters for America* (blog). June 5. Accessed November 21, 2018. www .mediamatters.org/blog/2018/06/05/day-parkland-survivor-david-hogg-got-swatted-trolls-shared-his-address-4chan-and-8chan/220382.

Marcus, A., and Oransky, I. (2015). The lessons of famous science frauds. *The Verge.* June 9. Accessed November 21, 2018. www.theverge.com/2015/6/9/8749841/science-frauds-potti-lacour.

Markham, A., and Buchanan, E. (2012). Ethical decision-making and internet research 2.0: recommendations from the AoIR ethics working committee. *Association of Internet Researchers.* Accessed November 21, 2018. https://aoir.org/reports/ethics2.pdf.

Miceli, M., Yang, T., Naudts, L., Schuessler, M., Serbanescu, D., and Hanna, A. (2021). Documenting computer vision datasets: An invitation to reflexive data practices. FAccT 2021, 161–72. https://dl .acm.org/doi/10.1145/3442188.3445880.

Nagle, A. (2017). *Kill All Normies: The Online Culture Wars from Tumblr and 4chan to the Alt-Right and Trump.* Winchester: Zero Books.

National Commission for the Protection of Human Subjects of Biomedical and Behavioral Research (1979). *The Belmont Report: Ethical Principles and Guidelines for the Protection of Human Subjects of Research.* Accessed November 21, 2018. https://videocast.nih.gov/pdf/ohrp_appendix_belmont _report_vol_2.pdf.

• r/KotakuInAction (2018). [People] The person behind the idea for #Deatheaters on being asked if her PhD is in GamerGate "literally yes. That is what my PhD is in." *Reddit.* Accessed February 20. https://web.archive.org/web/20150612102837/ www.reddit.com/r/KotakuInAction/comments/ 38uday/people_the_person_behind_the_idea_for_deatheaters/.

Presner, T. (2015). The ethics of the algorithm: Close and distant listening to the Shoah Foundation visual history archive. In C. Fugu, W. Kansteiner, and T. Presner (Eds.), *Probing the Ethics of Holocaust Culture*, 175–202. Cambridge: Harvard University Press.

ReFiG. (2018). About ReFiG. Accessed February 20, 2018. www.refig.ca/about-refig/.

Rehbein, M. (2015). On ethical issues in the digital humanities. Paper presented at the Digital Humanities Summer Institute. Vancouver, Canada, June 5.

Rockwell, G., and Berendt, B. (2017). Information wants to be free, or does it?: The ethics of datification. *Electronic Book Review.* December 3.

Rockwell, G., and Suomela, T. (2015). Gamergate reactions. University of Alberta: UAL Dataverse. DOI: https://doi.org/10.7939/DVN/10253.

Romano, A. (2021). What we still haven't learned from Gamergate, Vox, Retrieved at: www.vox.com/ culture/2020/1/20/20808875/gamergate-lessons-cultural-impact-changes-harassment-laws.

Shapiro, S. (2014). The risk of the "risk-based approach." *International Association of Privacy Professionals.* March 31. Accessed November 21, 2018. https://iapp.org/news/a/the-risk-of-the-risk -based-approach-2/.

Sinnott-Armstrong, W. (2015). Consequentialism. In E.N. Zalta (Ed.), *The Stanford Encyclopedia of Philosophy.* October 22. Accessed November 21, 2018. https://plato.stanford.edu/archives/win2015/ entries/consequentialism/.

Suomela, T., Chee, F., Berendt, B., and Rockwell, G. (2019). Applying an ethics of care to internet research: Gamergate and digital humanities. *Digital Studies.* Open Library of Humanities. Vol. 9, 1, 4. DOI: http://doi.org/10.16995/dscn.302.

TC3+. (2013). *Capitalizing on Big Data: Toward a Policy Framework for Advancing Digital Scholarship in Canada.* Ottawa: Social Sciences and Humanities Research Council of Canada. http://publications .gc.ca/pub?id=9.698761&sl=0.

TC3+. (2014). Tri-council policy statement: Ethical conduct for research involving humans. Accessed November 21, 2018. www.pre.ethics.gc.ca/eng/policy-politique/initiatives/tcps2-eptc2/Default/.

The Maintainers. n.d. The Maintainers. Accessed January 8, 2019. http://themaintainers.org/.

Wagner, K. (2014). The future of the culture wars is here, and it's Gamergate. *Deadspin.* October 14. https://deadspin.com/the-future-of-the-culture-wars-is-here-and-its-gamerga-1646145844.

Weber, J. (2014). Should companies monitor their employees' social media? *Wall Street Journal.* October 22. Accessed November 21, 2018. www.wsj.com/articles/should-companies-monitor-their -employees-social-media-1399648685.

Wilson, M., Palmer, Z.S., Budac, A., Pelletire-Gagnon, J., Rockwell, G., and Berendt, B. (2016). G@ merg@te and Military Campaign Discourse. Paper presented at Canadian Game Studies Association/ l'Association canadienne d'études vidéoludiques Annual Conference. Calgary, Canada, June 3.

Wittkower, D.E. (2016). Lurkers, creepers, and virtuous interactivity: From property rights to consent and care as a conceptual basis for privacy concerns and information ethics. *First Monday*, 21, 10. DOI: https://doi.org/10.5210/fm.v21i10.6948.

Zimmer, M. (2010). "But the data is already public": On the ethics of research in Facebook. *Ethics and Information Technology*, 12, 4, 313–25. DOI: https://doi.org/10.1007/s10676-010-9227-5.

21. Longitudinal effects on individual influences in women's pursuit of computer science education
Jeria L. Quesenberry

INTRODUCTION

Our increasingly digital world has brought many exciting career opportunities and an ever-growing demand for skilled workers. The United States Bureau of Labor Statistics (2022) predicts that employment in computer and information technology occupations will grow 13 percent from 2020 to 2030, faster than the average for all occupations. This growth will require an increase in the number of undergraduates graduating with science, technology, engineering, and math (STEM) degrees. At the same time, women's overall representation in STEM has generally risen since the mid-1990s, yet the gains have been limited when considering specific areas of study namely in computer science (CS), information science, and mathematics. Take, for instance, in the United States women earn approximately 57 percent of all bachelor's degrees, but only 22 percent of computer and information science bachelor's degrees. This is a sharp decline from an all-time high of 37 percent in 1985 (DuBow and Gonzalez, 2020; NCWIT, 2022; Zweben and Bizot, 2020).

Addressing the looming worker shortage requires engaging a larger pool of talent. The road ahead is uphill and as the OECD (2018) explains, there are many "hurdles to access, affordability, lack of education as well as inherent biases and sociocultural norms curtail women's and girls' ability to benefit from the opportunities offered by the digital transformation" (p. 5). A significant amount of time and money has been devoted to activities that hope to address the gender gap. The interventions that have demonstrated success, it could be argued, are those that consider influences broadly (including culture and environment) and those that are research-oriented, and data driven.

One such university example that has been successful in recruiting and retaining women in CS is Carnegie Mellon University, a private research university based in Pittsburgh, Pennsylvania, United States. With a strong background in STEM, the university consistently ranks in the top 25 in national rankings and around the top 100 among universities worldwide (US News & World Report, 2022). Carnegie Mellon presents an interesting case study in that it shows how cultural change at the institutional level can improve the representation of women in CS. In 1995 women comprised only 8 percent of the first-year CS class, yet this number increased dramatically to 39.5 percent in 2000. The numbers continued to improve and by 2017 had reached 49 percent which has been sustained since that time. While not immune to national trends, Carnegie Mellon has continued to enroll and graduate women in rates well above national averages for similarly ranked schools across the nation.

Common sense suggests that asking *why* women come to study CS is important in suggesting ways to improve recruitment and retention efforts. Moreover, asking women who are on the pathway to be leaders in CS might provide additional considerations on how their successes can shape interventions for others. In this vein, Carol Frieze and Jeria Quesenberry have written extensively about the Carnegie Mellon story including authoring the book *Kicking Butt in Computer Science: Women in Computing at Carnegie Mellon University* (Frieze and Quesenberry, 2015) and *Cracking the Digital Ceiling: Women in Computing Around the World* (Frieze and Quesenberry, 2019). We bring an extensive and diverse background to our work. Carol Frieze recently retired as the Director of Women@SCS and SCS4ALL in Carnegie Mellon's School of Computer Science where she worked on diversity and inclusion for nearly 20 years. Her publications, teaching, and research interests focus on the culture of computing, stereotypes and myths, unconscious bias, and broadening participation in computing fields. In the mid-2000s, Jeria Quesenberry and Carol began to collaborate on research and assessment activities. Jeria is a teaching professor of information systems at Carnegie Mellon University. Her research interests are directed at the study of cultural influences on information technology students and professionals, including topics of social inclusion, broadening participation, career values, organizational interventions, and work–life balance. She was also a PhD student involved in the early work of Eileen Trauth's theory building of the individual differences theory of gender and IT (IDTGIT).[1]

The goal of this chapter is to present an analysis of individual factors that played a role in shaping women's pathway in their decisions to study CS at Carnegie Mellon. By employing the IDTGIT, I explore which personal influences and characters played a role in shaping women and how these factors have dynamically evolved over the last three decades. This chapter first addresses multiple theoretical underpinnings in the studies of gender and IT. I describe some theoretical limitations and provide a rationale as to why the IDTGIT is a useful lens for understanding individual influences. Next, I provide an overview of the Carnegie Mellon story including a description of six case studies that have been conducted since the 1990s. My analysis considers why women enter the CS pathway, identifies personal influences that played a role in their journey, and analyzes their ultimate decision to major in CS. I conclude with lessons learned from our work and recommendations for future interventions.

BACKGROUND

Marini (1990) argues that there are no universally male or female qualities, but emphasizes that within STEM certain cultural characteristics are gathered on the basis of gender. In this sense, the STEM field is deemed a "male" domain rather than "female." As Wajcman (1991) states "it is evident that men identify with technology and through their identification with technology form bonds with one another" (p. 141). The presumption that women lack technical competence is not merely a sex stereotype but "does indeed become part of feminine gender identity" (p. 155). Building on this argument, Wajcman suggests that programs designed to encourage women to pursue technical careers may face challenges if women are actively resisting these careers because of the implications for feminine identities. Furthermore, Walton (1986) adds that scientific industries are typically dominated by men, most of its practitioners are men and it is said to have a masculine image. Therefore, if a woman enters the industry she is seen as "stepping out of line" in terms of societal expectations.

These concepts attempt to unpack the notions of what it means to be a man versus a woman and largely assumes a gender dichotomy. Although productive at considering social influences, these perspectives are difficult to employ when considering variations in women's behavior. DeCecco and Elia (1993) argue that a purely social construction theory can under emphasize the role of consciousness and intention of women. Trauth (2006, 2002) argues that the shaping of beliefs about gender in a social constructionism perspective operates at a group level, and hence explains the choices of all men or women in the same ways. This focus on monolithic societal messages makes it challenging to investigate the diversity that exists within people. Moreover, these perspectives typically group men and women as having different and opposing socio-cultural characteristics. Hence, it can end up essentializing gender as well. Trauth (2002) has argued that this can be problematic because "the message is that women in the IT profession, as a group, are different from men, as a group, in the profession, albeit for sociological rather than biological or psychological reasons" (p. 102).

In many ways, these theoretical conversations become even more clouded when considering the passing of time and evolving changes. Cockburn and Ormrod (1993) explain that gender segregation, or division of labor by sex, differs by culture and society and changes over time. They maintain that gender stratification is not static, but rather is constantly constructed and reconstructed through individual and collective action. For instance, Marini (1990) highlights the labor force differences between communist and democratic societies and the changes in the gender segregation in the US labor force during World War II. Steiger and Wardell (1995) also explain that changes in skill structure of the service sector from 1950 to 1990 created a more inclusive workforce for women. These historical and cross-cultural variations in gender role differentiation and stratification demonstrate that female absence or presence in male-dominated careers can evolve over time.

Motivated by these concerns, in part, *the individual differences theory of gender and IT* (Trauth, 2006, 2002; Trauth and Connolly 2021; Trauth et al., 2004) explains women's participation (or nonparticipation) in the IT field by considering gender relations at both the social and individual levels of analysis. The theory rejects claims of inherent gender-based difference and it also adds individual influences onto societal explanations for women's relationship to STEM. In this sense, the theory investigates common social messages and the variety of ways that individual women receive and react to such messages. The theory argues that part of the explanation for the under representation of women in STEM can be explained by the variation across women. The theory places focus on differences among women, rather than relying solely on gender stereotypes or societal influences. Hence, the theory focuses on the differences within rather than between the genders through the understanding of personal characteristics, technical talents, inclinations, and social responses that span the gender continuum.

As developed by Trauth (e.g., Trauth, 2006, 2002; Trauth et al., 2004) the individual differences theory of gender and IT is comprised of three high-level constructs including: environmental influences, identity influences, and individual influences. Within each of these constructs are embedded sub-category influences. Again, as described by Trauth, the environmental influence sub-category constructs include cultural attitudes and values (such as attitudes about IT and/or women), geographic data (about the location of work) and economic and policy data (about the region in which a woman works). The sub-category constructs for identity influences include demographics (such as age, racial/ethnic identity, socio-economic class and relationship status), and career items (such as job title and technical level). The individual influence factors sub-category constructs include personal characteristics (such as educational background, personality traits and abilities) and personal influences (such as mentors, role models, experiences with computing, and other significant life experiences).

Trauth's recent work has used the individual differences theory of gender and IT as a lens to consider how the influences can account for changes over time. Trauth and Connolly (2021) analyzed the life histories of women in IT in Ireland conducted over a period between the 1970s and 2010s. Their work extended the theory with the addition of seven additional themes that characterize the nature of change in many factors: environmental (i.e., policy, infrastructural, and cultural), identity (e.g., motherhood) and individual (e.g., family).[2] Their results provide a detailed framework of individual influences that can be considered in many contexts and thus possibly help in developing interventions to improve the representation of women in STEM. Thus, the individual differences theory of gender and IT is a useful lens in probing deeper to understand influences on women and how these influences evolve over time.[3]

The Carnegie Mellon Story

All too often it is assumed that Carnegie Mellon changed the curriculum to be "female friendly" to increase the numbers of women in the undergraduate major. *But this is not the case.* If anything, the academic expectations have become more rigorous and changes to curriculum are made only for pedagogical purposes – *not* as a presumed means to promote gender equity (Blum and Frieze, 2005). We believe Carnegie Mellon's success in reaching gender parity has come from focusing on culture. Our work shows how efforts over nearly three decades improved the culture to allow a more balanced environment to evolve (e.g., Frieze and Quesenberry, 2019, 2015). Our research shows how in a more balanced environment, perceived gender differences begin to fade, and we see students displaying a spectrum of attitudes, including many similarities across men and women. We focus on undergraduates because it is at the undergraduate level that many interventions leading to change took place and because we can compare studies over time.[4]

Six case studies of undergraduate students in CS at Carnegie Mellon were conducted over the past three decades: a pre-1999 longitudinal study, and then studies in 2002, 2004, 2009–10, 2011–12, and 2016–17. The post-1999 case studies represent "snapshot" research points – each motivated by and conducted with specific objectives. A longitudinal analysis of the studies helps to provide an understanding of the ongoing shaping of the culture of computing at Carnegie Mellon.

Pre-1999: Research findings from studies at Carnegie Mellon in the mid-late 1990s, when women comprised between 8 percent and 15 percent of the CS student body, suggested that there was a strong gender divide in men's and women's attitudes towards CS. While poor gender balance was not the only factor at work, in this gender imbalanced environment women reported feeling like they did not fit into the computing culture, their confidence was low, and many smart, capable women left the program without completing their studies (Margolis and Fisher, 2002).

2002 and 2004: By 2002, changes in the culture of computing were already being observed in the School of Computer Science. The situation from 2000 onwards clearly warranted closer examination. So with a grant from the Sloan Foundation we set about conducting an initial case study. Frieze and Blum interviewed 33 seniors (17 women and 16 men) from the class of 2002 and 55 students (32 women and 23 men) from the class of 2004. The findings were so striking that we carried out another case study in 2004. The latter not only confirmed our 2002 findings, that students demonstrated a spectrum of attitudes along with many similarities. We also found evidence that led us to challenge the traditional thinking about women and men in computing – namely we found examples of women who were excited about programming and men who wanted to use technology to help others (Blum and Frieze, 2005).

2009–10 and 2011–12: We continued to monitor the culture and environment carrying out additional interviews and surveys in the academic years 2009–10 and 2011–12. As with our initial study in 2002, our 2011–12 study was prompted by change, in this case a newly revised curriculum providing greater emphasis on theoretical thinking. We conducted a survey with 259 undergraduate students in 2009–10 and another 200 students in 2011–12. Moreover, CS sophomores (first to experience the new 2010 curriculum as first-year students) were invited by email to participate in interviews in 2011–12. We randomly identified and interviewed a cohort of 20 men and 20 women. We wanted to assess the impact of the change on the academic and social fit of CS students, looking for any indicators of problems with the women-CS fit (Frieze and Quesenberry, 2015; 2013; Frieze et al., 2011).

2016–17: Our last case study was intended to monitor the culture and environment by carrying out interviews with a cohort of male and female students in their sophomore year. As in 2011–12, we randomly identified and interviewed a cohort of 20 men and 20 women in their second year of study. We wanted to consider whether academic and social fit of CS students was still positive and if any areas for continued improvement could be identified (Frieze and Quesenberry, 2019).

INDIVIDUAL INFLUENCES ON YOUNG PEOPLE'S INTEREST IN COMPUTING

Prior to 1999, the admissions policies at Carnegie Mellon supported a specific type of student: those who had exhibited great programming proclivity. Research conducted at that time (1995–99) found a gender divide in the CS fit; in sum men were more likely to feel at home in the computer science department among their likeminded peers while the small number of women often felt out of place. The studies also concluded that there were strong gender differences in the way men and women were relating to CS. Men were more likely to be focused on coding and the machine itself, while women wanted to do something useful in terms of applications (e.g., Margolis and Fisher, 2002). Later, significant changes were made in admissions criteria to the CS major – such as minimizing the need for previous CS programming experience. These changes fit well with then Dean Raj Reddy's vision to broaden the scope of the search for potential CS talent which ultimately helped increase the numbers of women entering as first-year CS majors.

When we compared our findings to the 1995–99 study, we found that the perspectives of our students were often more alike than different. Notably the pre-1999 gender divide with men "dreaming in code" and women "computing with a purpose" was blurred. We found that the reasons students wanted to pursue a degree in computer science were largely attributable to individual influences. We also saw students whose views of their field had broadened quite dramatically from seeing CS as *programming* to seeing the field reflecting *an exciting range of possibilities* as influenced by a range of individuals.

Entering the Computer Science Pathway

As we investigated the backgrounds of our cohorts from the multiple case studies, we found that in contrast to the 1990s women were just as likely to grow up with a computer in the house as the men and were also just as likely to be the primary user. Students (men and women) had similar reasons for choosing the CS major and many came from families who encouraged their daughter or son to pursue her/his interests, and even encouraged their daughter's curiosity to just "play around" with the computer. What *had not changed* was the influence of family as a primary force for exposing students to computing and for providing role models. Most of the women in both the pre-1999 and post-1999 studies came to computing through their interest in math science and "were students who enjoyed problem solving, doing puzzles, exercising logical thinking skills" (Margolis and Fisher, 2002, p. 18).

If our students' backgrounds are in any way typical of CS students across the nation it could have serious ramifications for our efforts to broaden participation in CS, especially when CS is largely absent from the K-12 curriculum. The potential for discovering talent and interest in the field will remain elusive if, for the most part, it is nurtured only among students who come from "computing families." This suggests we need to do a better job of helping *many more* parents and guardians understand what CS is, and why computing studies can benefit their girls and boys especially in terms of potential careers. Changing the perceptions of CS among parents is critical for changing the perceptions of their young students.

The Computing Family

During the post-1999 interviews, we asked several questions relating to students' early computing experiences. Most of the students recalled having a computer around as they were growing up. Indeed, most recalled having one in the house before middle school age. During the 2004 interviews only one woman and one man said they had had a computer in the house for as long as they could remember. By the 2011–12 and most certainly in the 2016–17 interviews we saw a shift in that most of the students felt they *always* had a computer in the home or in their lives. What was consistent between the two time periods is that both men and women claimed to have been the most frequent user of the computer in their household. "Probably me" was a common response when students were asked who used the computer most. One woman shared:

> I grew up with a computer in my house so – I mean I always used to play around with it and I was sort of interested in web design so my parents signed me up for a couple of those classes. I was pretty young then so it was basic HTML. (Female Sophomore, Class of 2014)

Past research has found that women entering male-dominated fields tend to come from families where both parents are highly educated and success is considered highly critical (e.g., Jackson et al., 1993; Smith 2000; Trauth et al., 2004). Other studies have noted the strong influence of fathers and other male family members in exposing children to computing (e.g., Margolis and Fisher, 2002; Schulte and Knobelsdorf, 2007; Teague, 2002; Trauth and Connolly, 2021). For instance, a study conducted by Turner et al. (2002, p. 6) found that among the female information technology professionals in their sample "27% of the fathers held jobs that could be considered technical, with technical being interpreted as jobs in engineering, computer science, mathematics, physics or chemistry."

Our findings also reflect existing research regarding the importance of a family that embraces computing. We found during the post-1999 case studies that parents – and in particular father figures – featured prominently in the background of all students, including women, and their early computing experiences. This was a marked change from the pre-1999 studies when fathers "actively engaged" their sons much more so than their daughters (Margolis and Fisher, 2002, p. 25).

By the time of our post-1999 case studies fathers proved to be (and continued to be) major influences on their daughters. One female student explained that she grew up with a computer in her home and while she couldn't recall her first experience, she was certain it involved playing some sort of game with her father. Several females discussed programming on the computer with their fathers. Others mentioned "figuring things out" on the computer with fathers. For example, one female recalled working with her dad to program an electronic clock with the BASIC programming language. Women also spoke often about the positive encouragement and mentoring they had received from their fathers:

> My dad, because he encouraged me. I thought [when I took my first computer programming class], "That's not interesting." He was encouraging. He was like, "Oh, that's a good idea," because he knew that in his line of work, programming was really helpful. And so I think he was pretty supportive of the idea of going into computer science. (Female Sophomore, Class of 2014)

> When I was trying to choose my major, I'm sort of the person who's really interested in everything and it was really hard for me to decide. Then having grown up with my dad as a computer software engineer, I was like, oh, this stuff in computer science can be applied to so many different fields. And it sounds like something that fits me really well. (Female Sophomore, Class of 2019)

> My parents made me do it, to be completely honest. (Female Sophomore, Class of 2019)

> [My dad] encouraged me to join the robotics team at school. And that was sort of my first experience. Then after that I looked more into it and then I realized it was something I wanted to do. (Male Sophomore, Class of 2019)

References to mothers also appear in the comments of female students as they discuss influences and early computing experiences but are typically often absent in the comments of male students. In fact, only three men from the case studies (one from the 2004 interviews, one from the 2011–12 interviews, and one from the 2016–17 interviews) were the exception in specifically mentioning their mom's role in influencing their interest in CS.

"Siblings" and "sharing the computer" were frequently mentioned together. Several women and one man saw the computer as a family computer: "Everyone kind of used it for their own thing." What became clear from these students' early memories was that the computer was *not* a boy's toy, tucked away in his bedroom. The women grew up using the machines as frequently as their male siblings and peers. They did not see themselves as watching from "the sidelines," or becoming interested in computing vicariously by watching a father or brother work. The women in our studies were, for the most part, active computer users whose personal initiative sparked their interest in computers and computing. As one woman remarked:

> I actually started programming in BASIC when I was really little ... I used to get magazines and they'd have little code snippets so I'd put those in, try them out on the computer. (Female Senior, Class of 2004)

> From when I was very little, in like fourth grade, I discovered actually the Alice Program [an educational programming tool] which was made here and then I wanted to go to Carnegie Mellon. (Female Sophomore, Class of 2019)

The students in the case studies often had another extended family member in the technology field. Again, fathers figured largely, and mothers and siblings to a lesser extent. Yet, the students frequently discussed the role of family members, and friends in the field (e.g., uncles, cousins, etc.). During the 2011–12 interviews, only four women, as compared to eight men, mentioned role models or teachers as being an influential person in their decision to pursue a bachelor's degree in CS.

This also ties in with another finding in our case studies, which showed that both men and women were more likely to note a range of people when asked who was most influential in their decision to major in CS, but the order and weight varied between the groups and over time. For example, during the 2004 interviews, women cited fathers, teachers, family, and mothers (in that order) as influencing their decision. Whereas the men interviewed at that time said their interests were their primary motivating factor followed by influences from teachers, dads, and family. This held up for the men in the 2011–12 interviews, *but also emerged for the women*. Out of the 20 women interviewed, 11 of them pointed to themselves as a major influence by stating, "I did" or "I just decided for myself." For instance, this woman stated:

> My parents are supportive, were supportive of any major I chose, and so I mean in that sense, I guess it would be them. But it was really myself who decided this is what I want to do with my life and pursued it. (Female Sophomore, Class of 2014)

These two types of personal characteristics – women's interest and internal motivation continued to be important influences raised during the 2016–17 interviews. Again, out of the 20 women interviewed, 11 felt their individual interest was the primary influencing factor in their decisions even mitigating other influences that might sway them away.

> I chose it mostly because I enjoyed [CS] in high school and I thought that it was interesting. (Female Sophomore, Class of 2019)

> My parents wanted me to be a doctor or to have a medical related job but if someone pushes me to do something I'll say no. I don't want to do that. I was searching what I wanted to do in my high school. I'm from Korea. We had a little field trip to a software company in Korea. There was a seminar. I thought this might be an interesting field to study and work with. (Female Sophomore, Class of 2019)

Initial Interests in Computing

Throughout all the post-1999 case studies, we found students' initial interests in computers and computing ranged from programming to playing games to using applications (graphics, Word, etc.), and combinations of those activities. The word *curiosity* emerged in women's responses, although it is stereotypically thought to be a more common motivator of men. Some females would discuss building basic websites with HTML to show their comic book drawings or interesting projects. Women would say they became interested by just "playing around," "fiddling around," or just being "curious." For instance, this student commented:

> The second my father took the first computer out of the box. I was interested because I didn't know what it was ... I've always been curious about things. When I was younger I used to take things apart to figure out how they worked. (Female Senior, Class of 2004)

> I really liked the process of sitting down and breaking things down and being given a really complicated problem and being able to break it down into little steps or little blocks and build a cool solution. I like the process of programming. And I also think the applications are really fascinating and really interesting. So, I think a combination of those two things. (Female Sophomore, Class of 2019)

Most students recalled having a computer in the house by their middle school years and home was by far the most significant place for instigating their initial interest in computers and computing. For several students early schooling was critical as one woman who graduated in 2004 mentioned that she first learned Logo – a basic programming language – in kindergarten. She used it to draw pictures and really enjoyed the experience. Most students were first introduced to formal CS classes in high school (including some programming). Several students shared their multiple experiences with CS classes in middle and high school:

> (W)hen we started getting into more of the programming side, like actually writing a code and stuff, I just really liked it and so I just kept taking more and more classes and I just kept getting more and more into it. I took a normal CS class and then I took the Advanced Placement computer science class. And afterwards, I actually ended up doing an independent study. I built a chess bot. It was fun. (Female Sophomore, Class of 2014)

These opportunities still played an important role as raised by the students in the 2016–17 interviews.

> I had a really great computer science teacher in high school. I think he definitely did a lot for all of his students to push us more towards the field. I also grew up in Cincinnati, so there was a non-profit that I actually spent a lot of time helping them promote technology among high school students. (Female Sophomore, Class of 2019)

The Decision to Major in Computer Science

The majority of the students we interviewed in the case studies, women and men, came into CS initially because of their interest in the field. For some it was a simple decision. For example, one woman coming into CS felt like it was not a decision, just something she was "going to do." This woman highlights a typical response when she explains "it just seemed like a fit," and this man saw it as a "natural follow up" to high school. For others the decision was more complex, but more often than not the students were already on a math/science track. For some students the career prospects were important. For others the field offered more options than traditional math and science majors. For example, some students shared these thoughts:

> There are so many possibilities with [CS]. It was mostly the fact that it was so versatile that you could include music and art and language and all of it, that I was like, Hmm, I should see what the opportunities are. I've always really liked math, so it seemed like the most right choice I could make. (Female Sophomore, Class of 2014)

> It's a huge field that has a lot of interesting problems that need to be worked on and pretty much really relies on computer science now, whether it's math or any of the sciences or anything. I think it's still a growing field and a really interesting one. (Male Sophomore, Class of 2014)

The field was very broad so there were a lot of things that you could do with computer science. And I think a huge thing was also like it's like a challenging field so just learning that like I feel like it would open up other outlets, whether or not they were in the software engineering industry or not. (Female Sophomore, Class of 2019)

There were few to no gender differences in the reasons students chose to enter the CS major. Cohoon and Aspray (2006) found something very similar "For the most part, the male and female students with whom we spoke felt initially attracted to computing for the same reasons" (p. 218).

CONCLUSION

The many positive changes in the micro-culture of the School of Computer Science at Carnegie Mellon have shown that a more diverse student body, including increased numbers of female students, has enriched the social and academic environment for *everyone*. The work presented in this chapter examined the individual influences that play a role in women's decisions to pursue an undergraduate degree in CS. Drawing from the IDTGIT we identified key influences and considered their dynamic nature – those that remained consistent and those that evolved over time.

Personal influences, especially father figures, and the encouragement they provide play a consistent and positive role in engaging women in CS. This influence was important in the early 2000s and consistently plays an important role nearly 20 years later. Educating *many more* parents and guardians about the benefits of CS degrees could build and reinforce these messages. Early exposure to technology in the home or at school was somewhat limited in the early 2000s but played an important role in building interest at that time. Exposure has improved over the last 20 years with more regular and ubiquitous access to technology and affords many more opportunities for women to engage. Addressing potential gaps in exposures to computing (at home and in school) is an important intervention to keep in focus. Personal characteristics of empowerment and agency were not discussed by women in our studies in the early 2000s but in more recent years have become an important influence. Many women explained their choice to study CS was motivated by their interest despite roadblocks or challenges they may have faced – perhaps indicating that this younger generation of women are empowered to use their personal agency when making career decisions. We should find ways to help encourage and foster their voices.

Interventions to help encourage women's participation in STEM is crucial. Our experiences with *Women@SCS* – an organization that includes all female students in the School of Computer Science at Carnegie Mellon – plays an important role in recruiting and retaining women in our undergraduate CS major. As *Women@SCS* has evolved, the organization has built a strong pipeline of connections at all levels, connecting faculty with students, graduates with undergraduates, and all with a local community of K-12 students, teachers, and families. *Women@SCS* is especially keen to ensure that female and minority computer scientists are actively involved in presenting themselves, their work, and new images of the field to young audiences, their teachers, parents, and others. The Roadshow, TechNights (Creative Technology Nights for Girls), Adventures in Computing at SciTech (Pittsburgh Science and Technology Academy) and OurCS (Opportunities for Undergraduate Research) are the primary outreach vehicles.[5]

While there is still work to be done for building diversity and promoting inclusion, we believe the Carnegie Mellon journey offers an interesting case study of the many issues surrounding gender and CS and specifically how the role of individual influences has evolved over the last three decades. We hope the Carnegie Mellon story of women in computing can help challenge the gender divide in CS and encourage and inspire, parents, teachers, and students to think more broadly about intellectual and academic expectations.

NOTES

1. Much of the work described in this chapter was done in partnership, although Jeria is the sole author. Thus, she uses "we" throughout the chapter to reference the research and interventions conducted by both her and Carol.
2. For a detailed description of the dynamic nature of these themes see Trauth and Connolly (2021) p. 2065.
3. For a detailed summary see Chapter 2, "An overview of the individual differences theory of gender and IT."
4. We acknowledge that Carnegie Mellon is somewhat unique in comparison to other universities. Hence, we recognize that results from our case studies may not be fully generalizable outside of the boundaries of the university. Yet, we believe that this work details examples of the many efforts to support a more balanced situation. Some examples may be replicated or effective at other institutions – others may relate less so. Furthermore, we acknowledge that our case studies use gender as the primary identity component, but do not assume women can be generalized into a single category. Rather, all people are shaped by their different identities – albeit ethnicity, race, class, sexual orientation, family background, etc. For more information see Trauth et al. (2012) for a discussion of intersectionality and a more detailed analysis of the underrepresentation of marginalized people in the IT field.
5. For a detailed description of the Roadshow and the TechNights program see Frieze and Treat (2006) or visit www.cmu.edu/scs/women-scs/.

REFERENCES

Blum, L., and Frieze, C. (2005). The evolving culture of computing: Similarity is the difference. *Frontiers*, 26, 1, 110–25.

Bureau of Labor Statistics (BLS), US Department of Labor. (2022). *Occupational Outlook Handbook*, Computer and Information Technology Occupations. Retrieved from: www.bls.gov/ooh/computer-and-information-technology/home.htm.

Cockburn, C., and Ormrod, S. (1993). *Gender and Technology in the Making*. London, England: Sage.

Cohoon, J.M. and Aspray, W. (Eds.) (2006). *Women and Information Technology: Research on Underrepresentation*. Cambridge, MA: The MIT Press.

DeCecco, J.P., and Elia, J.P. (1993). A critique and synthesis of biological essentialism and social constructionist views of sexuality and gender. *Journal of Homosexuality*, 24, 3/4, 1–26.

DuBow, W., and Gonzalez, J.J. (2020). *NCWIT Scorecard: The Status of Women in Technology*. Boulder, CO: NCWIT.

Frieze, C., and Quesenberry, J.L. (Eds.) (2019). *Cracking the Digital Ceiling: Global Views of Women in Computing*. Cambridge: Cambridge University Press.

Frieze, C., and Quesenberry, J.L. (2019). How computer science at Carnegie Mellon University is attracting and retaining women. *Communications of the ACM*, 62, 2, 23–26.

Frieze, C., and Quesenberry, J.L. (2015). *Kicking Butt in Computer Science: Women and Computing at Carnegie Mellon University*. Indianapolis, IN: Dog Ear Publishing.

Frieze, C., and Treat, E. (2006). Diversifying the images of computer science: Carnegie Mellon Students take on the challenge! *Proceedings of the 2006 WEPAN Conference*. Pittsburgh, PA.

Margolis, J., and Fisher, A. (2002*). Unlocking the Clubhouse: Women in Computing.* Cambridge, MA: The MIT Press.

Marini, M.M. (1990). Sex and gender: What do we know? *Sociological Forum*, 5, 1, 95–120.

National Center for Women & Information Technology (NCWIT). (2022). *By the Numbers.* Retrieved from: https://ncwit.org/resource/bythenumbers/.

Organisation for Economic Co-operation and Development (OECD). (2018). *Bridging the Digital Gender Divide: Include, Upskill, Innovate.* Retrieved from: www.oecd.org/digital/bridging-the -digital-gender-divide.pdf.

Quesenberry, J., and Trauth, E.M. (2012). The (dis)placement of women in the IT workforce: An investigation of individual career values and organizational interventions. *Information Systems Journal*, 22, 6, 457–73.

Steiger, T.L., and Wardell, M. (1995). Gender employment in the service sector. *Social Problems*, 42, 1, 91–123.

Trauth, E.M. (2017). A research agenda for social inclusion in information systems. *The Data Base for Advances in Information Systems*, 48, 2, 9–20.

Trauth, E.M. (2006). Theorizing gender and information technology research. In E.M. Trauth (Ed.), *Encyclopedia of Gender and Information Technology* (pp. 1154–59). Hershey, PA: Idea Group Publishing.

Trauth, E.M. (2002). Odd girl out: An individual differences perspective on women in the IT profession. *Information Technology & People*, 15, 2, 98–118.

Trauth, E.M., Cain, C.C., Joshi, K.D., Kvasny, L., and Booth, K. (2012). Understanding underrepresentation in IT through intersectionality. Proceedings of the 2012 iConference (Toronto, CA).

Trauth, E.M., and Connolly, R. (2021). Investigating the nature of change in factors affecting gender equity in the IT sector: A longitudinal study of women in Ireland. *MIS Quarterly*, 45, 4, 2055–100.

Trauth, E.M., Quesenberry, J.L., and Morgan, A.J. (2004). Understanding the underrepresentation of women in IT: Toward a theory of individual differences. *Proceedings of the ACM SIGMIS Computer Personnel Research Conference* (Tucson, AZ, April).

Turner, S.V., Bernt, P.W., and Pecora, N. (2002). Why women choose information technology careers: Educational, social, and familial influences. Paper presented at the Annual Meeting of the American Educational Research Association (New Orleans, LA, April 1–5).

US News & World Report. (2022). Best Global Universities. Retrieved from: www.usnews.com/ education/best-global-universities.

Wajcman, J. (1991). *Feminism Confronts Technology.* University Park, Pennsylvania: The Pennsylvania University Press.

Walton, A. (1986). Women scientists: Are they really different? An exploration of the significance of attitudes. In J. Harding (Ed.), *Perspectives on Gender and Science.* New York: The Falmer Press.

Zweben, S., and Bizot, B. (2020). 2020 Taulbee Survey: Bachelor's and doctoral degree production growth continues but new student enrollment shows declines. *Computing Research Association*, 33, 5, 2–68.

22. Am I good enough? Sources of IT self-efficacy as key impediments to narrowing the IT gender gap

K.D. Joshi

INTRODUCTION

Among the confluence of factors that collectively conspire to create a gender imbalance in the information technology (IT) profession, the feeling that one is not good enough and thus does not belong in the IT profession is one of the most pervasive and damaging factors. Such feelings of inadequacy breed self-doubt, which creates an internal struggle over whether one is capable of pursuing a career in IT. Although women in IT programs and IT jobs are now performing at the same levels as men, they are more likely than men to experience self-doubt regarding their fit (Liberatore and Wagner, 2020; Sanli, 2019). This chapter discusses the sources that seed, grow, and amplify doubts among women regarding their fit within the IT profession.

Self-doubt erodes women's confidence in their abilities to learn and perform IT skills (Joshi and Schmidt, 2006; Joshi and Kuhn, 2011). Such lack of confidence has been viewed as one of the primary barriers to women's entry into IT careers (Dickhaeuser et al., 2002; Hackett et al., 1981; Hartzel, 2003; Liu et al., 2001). Self-efficacy, which is a psychological construct and refers to an individual's personal belief in their capability to complete a specific task in order to achieve particular attainments, is germane to understanding women's lack of confidence in IT careers. Therefore, the objective of this chapter is to reveal and critically examine the sources of IT self-efficacy and the role it can play in narrowing the technical gender gap.

The remainder of the chapter proceeds as follows. First, in the Background section, the concept of IT self-efficacy is explained. Second, in the same section, my research program, which examines IT self-efficacy within the context of gender and IT career choices, is critically reviewed to highlight the deep and strong stranglehold of hegemonic masculinity that contributes to the lower levels of confidence among women regarding their ability to participate in IT. Next, the four main sources that influence the formation of IT self-efficacy are discussed. Finally, the paper concludes with a call to action to direct our focus on uncovering the role of those who are engaged deliberately and systematically in creating conditions that sow and produce self-doubt about IT careers among women.

BACKGROUND

Self-efficacy is defined as "beliefs in one's capabilities to mobilize the motivation, cognitive resources and courses of action needed to meet given situational demands" (Wood et al., 1989, p. 408). An individual's efficacy expectations are the major determinant of goal setting, activity choice, willingness to expend effort, and persistence (see Bandura, 1997). Self-efficacy as a strong predictor of educational behaviors, such as preferences, performance, and persistence, is confirmed in a comprehensive review and meta-analysis of two decades of research by Lent and his colleagues (see Hackett and Lent, 1992; Lent et al., 1994; Multon et al., 1991; Lent et al., 2000). Specifically, the important role of self-efficacy in influencing IT career choices is also well established (see Joshi and Kuhn, 2011; Akbulut and Looney, 2007; Dickhaeuser et al., 2002; Smith, 2002). However, the nature (positive or negative) and magnitude (the strength) of IT self-efficacy's influence on career choices are determined by beliefs about one's capabilities to learn and perform two distinct sets of IT skills: technical (hereafter referred to as tech) skills and nontechnical (hereafter referred to as nontech) skills.

IT Self-Efficacy, IT and Gender

In order to succeed in IT careers, individuals need a variety of tech and nontech skills (Bailey et al., 2001; Bassellier et al., 2003; Green, 1989; Lee et al., 1995; Litecky et al., 2004; Todd et al., 1995; Vitello, 1997; Huang et al., 2009; Joshi et al., 2010a). The literature classifies these skills into three categories: business skills, humanistic skills, and technical skills (Huang et al., 2009). Business skills comprise skills such as negotiation, leadership, customer relationships, and project management, and particular business area knowledge such as finance and company-specific skills. Humanist skills consist of skills such as communication, the ability to handle ambiguity, adaptability, and dependability. Technical skills consist of programming, system analysis and design, computer security, and networking, and analytics. A comprehensive psychometric analysis of the list of all these skills was conducted to uncover the factor structure underlying the IT self-efficacy construct (for details see Joshi et al., 2010b). The analysis uncovered two distinct constructs (see Box 22.1). The review of the skills in each of the two constructs revealed that one factor consisted of all the skills that were predominately viewed as technical skills (e.g., programming skills, integrating enterprise applications, systems implementation). I refer to this factor as technical self-efficacy (hereafter referred to as tech self-efficacy). The second construct predominately consisted of non-technical skills (such as customer relationship skills, and leadership skills) and is therefore referred to as nontechnical self-efficacy (hereafter referred to as nontech self-efficacy). Tech self-efficacy is defined as one's level of self-confidence about their ability to learn and engage with technical skills. Nontech self-efficacy is defined as one's level of self-confidence about their ability to learn and engage with business, organizational, and human skills.

BOX 22.1 IT SKILLS USED TO CAPTURE IT SELF-EFFICACY

Nontech IT Skills

- Ability to engage in independent learning
- Ability to work in teams
- Ability to work under pressure
- Adaptability
- Communication skills (e.g., verbal, written, presentation skills)
- Creativity

- Critical thinking
- Customer relationship skills
- Dependability
- Initiative
- Leadership skills
- Negotiation skills
- Openness to new experiences
- Problem solving skills
- Sensitivity to organizational culture and politics
- Workplace relationship skills
- Ethics
- Professionalism.

Tech IT Skills

- Business analytics skills (e.g., data mining, online analytics processing systems)
- Database management skills (e.g., managing SQL server, ORACLE DB)
- Design skills (e.g., systems design; ER modeling; dimensional modeling; data modeling)
- Integrating enterprise applications
- IT architecture/infrastructure
- IT security
- Networking skills (e.g., LAN/WAN; setting up networks; wireless networks)
- Process analysis (e.g., gathering systems requirements; systems analysis)
- Programming skills (e.g., C#, XML, VB, Java)
- System auditing and information assurance
- System implementation skills
- Web development skills.

Source: Joshi et al. (2010b) and Joshi et al. (2013).

The extant literature suggests that, despite having reason to be as confident as men about technical skills (Henwood, 2000), women generally have lower self-efficacy expectations than men for traditionally male-dominated occupations (Cassidy et al., 2002; Cross, 2002; Beyer, 1999a, 1999b, 2002; Beyer et al., 2004; Dickhaeuser et al., 2002; Hartzel, 2003; Lips, 2004; Joshi and Schmidt, 2006; Joshi and Kuhn, 2011). In studies of gender differences in IT self-efficacy, one of the major emphases has been to examine the role gender typing of IT skills plays in the formation of IT self-efficacy beliefs. This emphasis is based on the underlying thesis that gender differences in IT self-efficacy are socially constructed. Therefore, studies on gender and IT self-efficacy have often incorporated the influence of gender role socialization (and the resulting stereotypes) to explain the differences in self-efficacy expectations of men and women in male-dominated occupations such as IT. For instance, women receive societal messages that they are better than men in language and communication and men receive societal messages that they are better than women in science and math, resulting in gender stereotypes that shape the self-concept of ability of both men and women students (Kwan et al., 1985; Guimond and Roussel, 2001). Such stereotypical perceptions have also led to gender typing of technical skills as masculine and nontechnical skills as feminine, which are used by individuals as a proxy to evaluate their IT-related self-efficacy. Gender typing of IT skills differentially shapes men's and women's IT self-efficacy (Joshi and Kuhn, 2007; Trauth et al., 2016). Studies show that women lack confidence in their ability to perform IT technical skills, such as programming, which are stereotyped as masculine, whereas there is no difference in men's and women's beliefs about their ability to perform nontechnical skills, such as communication abilities, which are either perceived as feminine or gender neutral (see Cejka et al., 1999; Ritter, 2004; Joshi and Kuhn, 2007; Trauth et al., 2016). Gender confidence gaps, where women underestimate their skills while men overestimate their own abilities in male-dominated occupations (Hamlin et al., 2010), are likely to persist until socio-cultural representations of technology stop equating technical skills with masculinity and men (Henwood, 2000).

IT Self-Efficacy Matters – The Good, the Bad, and the Ugly

As illustrated in the previous section, a vast body of research has established that self-efficacy matters in choosing a career (see Hackett and Betz, 1995; Lent, 2005). Similarly, IT scholars have also established that IT self-efficacy matters and it drives IT career choices for both men and women (see Joshi and Kuhn, 2011). This finding is neither new, nor surprising. It is natural to voluntarily avoid a career pathway where one believes they will not succeed in favor of a career where one expects they will. However, in this section, I will nuance a troubling theme that emerged as I critically examined the IT self-efficacy findings in my research program (see Table 22.1 for a summary of my research program on IT self-efficacy). Here I argue we have not been able to make a noticeable difference on the number of women in IT for the past four decades because of the presence of hegemonic masculinity[1] in the IT profession. Although this chapter is United States focused, as all the studies discussed in this section were conducted there, many themes span multiple cultural contexts.

The good news is that self-efficacy is malleable. IT self-efficacy beliefs determine how people feel and think about IT. By changing IT self-efficacy beliefs so that women have high confidence in their IT capabilities, they are likely to view IT skills as challenges to be mastered rather than as threats to be avoided. High IT self-efficacy is achieved by making women believe that they are capable of mobilizing the necessary motivation, cognitive resources, and courses of action (Bandura, 1994). This can be used as powerful fuel to motivate and drive women into IT careers.

The bad news is that nontech IT self-efficacy, where there are no gender differences, does not matter. The research conducted by my team of colleagues and myself shows that nontech IT self-efficacy does not influence IT career choices. Rather, some results show that it may achieve the opposite. An NSF-funded study (Joshi and Kuhn, 2002) involved 500 business students who had yet to crystallize their career choices. The students were studied to capture shifts in IT self-efficacy beliefs by examining their perceptions both before and after exposure to IT careers via an introductory required course in IT. The study found that there was no significant gender difference for nontech IT self-efficacy. The encouraging finding that men and women were equally confident about their ability to learn and perform nontech IT skills such as problem solving, communication, and team management abilities was overshadowed by the findings that this belief did not significantly influence their attitudes toward IT. Similar findings were revealed in another study where the beliefs of college IT majors and non-majors about their nontech IT self-efficacy were assessed (Joshi and Kuhn, 2011). What was troubling

Table 22.1   *A summary of my research program: IT self-efficacy related studies 2000–16*

Timeline	Research topic	Summary	IT skills and self-efficacy
2000–2004	*IT Self-efficacy and career choices*	*The motivation behind the studies in this research program, which surveyed business students in the United States who had yet to crystallize their career choices, was to understand relevant attitudes and norms of both men and women students to reveal why women students might be less attracted to a career in IT.*	
Joshi, and Kuhn, National Science Foundation Grant No. 0089986, 2000–02		In this study 500 business students in the United States who had yet to crystallize their career choices were surveyed. Using the theory of reasoned action, we captured shifts in related beliefs and values by examining their perceptions both before and after exposure to IT careers via an introductory required course in IT. The results of this study challenge some of the stereotypic beliefs that persist regarding women's negative attitudes toward male-dominated careers. Overall, the results show that when making career decisions, men and women who are interested in business majors are motivated by similar factors, but their beliefs associated with these factors differ.	Findings specifically related to self-efficacy revealed that there were no significant gender differences for nontech self-efficacy before or after the exposure to IT careers. Unfortunately, IT nontech self-efficacy had no significant effect on men's or women's attitudes toward IT before or after the course experience. We expected that this impact would become positive after the course as students become more informed about the value placed on nontech skills for IT careers. In contrast, there existed a significant gender difference with men having higher-tech self-efficacy than women for software/programming-skills at the beginning of the course, which did not attenuate after the exposure to IT careers. This gender difference is problematic because higher tech self-efficacy (software/programming skills) had a significant positive influence on both men and women students' IT career choices at the start and at the end of the course.

Timeline	Research topic	Summary	IT skills and self-efficacy
Joshi and Kuhn (2011)		This study posits an IS-specific career choice model that provides good predictive power and elucidates the nuances of factors underlying attraction to an IS career. Undergraduate students from an introductory IS course at a large public university in the United States, which for most was their first introduction to the field, were surveyed to test the study's model. The data were collected in early 2000 and the actual behavior (i.e., graduating major) was collected during the fifth year. Most of the participating students were freshmen and sophomores, therefore, a lag of five years was necessary to collect the actual behavior information. A total of 508 students (over a period of two consecutive semesters) completed an online questionnaire. From university records, we determined the actual degree earned by each student. From this sample, 65 students completed degrees in IS, and the remaining 443 completed degrees in other majors. Overall, the results show that, when making a career choice decision, individuals who are interested in business majors are collectively motivated by IT self-efficacy, work value congruency, and normative beliefs. More specifically, students who have high confidence in their technical abilities believe that IS careers are compatible with their work values, and receive positive encouragement to pursue IS careers and choose IS as their major.	For all students, higher levels of tech (software/programming) skill self-efficacy led to more positive attitudes toward IS but higher nontech efficacy did not influence students' attitudes toward IS. As expected, the results suggest that tech skill efficacy is a crucial determinant of attraction to an IS career. Although there were no significant group differences in the perceptions of the importance of tech skills to an IS major the individuals in the IS group had significantly higher personal confidence regarding their tech skill competency. Interestingly, despite the non-IS group's lower confidence in their tech skill efficacy, there was not a significant difference in their class performance (grades received) between the IS graduates and non-IS graduates. Conversely, both groups (IS and non-IS graduates) were equally confident about their soft skills, such as problem solving, communication, and team management abilities. However, students' attitudes toward IS were not affected by their beliefs about whether they possessed these nontech skills. Even though nontech efficacy did not significantly impact attitudes and the groups did not differ in their perceived nontech self-efficacy, it is interesting to note that the non-IS group perceived nontech to be less valuable to an IS major. This result illustrates how deeply ingrained are students' misperceptions of IS careers as only technical careers.
2002–10	*Gender typing of IT skills*	*Motivation behind the studies in this research program was to examine the effects of stereotype fit by examining what it takes to succeed in IT careers and what role the gender typing of IT skills plays in defining excellence in IT.*	
Joshi, and Kuhn, National Science Foundation Grant No. 0204469, 2002–07		This grant-funded research stream focused on revealing the skills to succeed in information technology. A multi-level analysis of stakeholders' perceptions of critical attributes was conducted at two IT consulting firms to examine beliefs about the attributes of successful IT professionals, applying social psychological research on gender stereotypes and cognitive processes in performance evaluation. The associated effects at both the individual and collective levels were uncovered.	The key findings of this research stream that relates to IT skills and self-efficacy are published in Joshi and Kuhn (2007) are described below.

Timeline	Research topic	Summary	IT skills and self-efficacy
Joshi and Kuhn (2007)		This study profiled IT consultants' prototype of a "top performer" at three levels – entry, middle, and upper levels. This study represents an exploration of both what attributes are considered the key characteristics of top IT consultants, as well as the gender typing (i.e., masculine and feminine nature) of these attributes. Multiple stakeholder perceptions from a large international IT consulting firm located in the United States were captured to assess prototypes of a "top" consultant. Data were collected using focus group sessions and interviews. The picture that emerged of a "top consultant" was rich and varied, and showed the breadth of skills and personality attributes needed to excel. Overall, the characterization of a top IT consultant is somewhat masculine-typed. Interestingly, an employee's own characterization of a top performer within their own job level was not congruent with other stakeholders' perceptions. Participants, especially females, perceived other stakeholders as strongly emphasizing attributes that are stereotypically perceived as masculine.	A "top" IT consultant is perceived as masculine. Overall, most critical attributes elicited for a top performer in each of the job levels were masculine-typed (aligned with tech skills) rather than feminine-typed (aligned with nontech skills). When all the phrases for all job types are pooled together there is a significantly greater proportion of masculine-typed attributes elicited.
2007–16	*Gender role congruency and career choices*	*Motivation behind this set of studies was to investigate the effect of intersecting group membership – including race, ethnicity, and social class – on the types of societal messages about gender role received as well as the perception of and response to these societal messages. Investigating the intersectionality of gender, race and social class can produce a more nuanced explanation of the underrepresentation of women in IT.*	
Trauth, Kvasny, Joshi, National Science Foundation Grant No. 0733747, 2007–13		In this grant-funded research stream, we investigated the effect of intersecting aspects of group membership – gender, ethnicity, and socio-economic class – that shape one's identity on the perception of and response to societal messages about gender roles. Through a survey of over 6,000 university students, we investigated how these intersecting group memberships shape students' IT career choices to uncover more nuanced explanations of the gender imbalance in the IT field.	The key findings of this research stream that relates to IT skills and self-efficacy which are published in Joshi et al. (2010b) are described below.

Timeline	Research topic	Summary	IT skills and self-efficacy
Joshi, Kvasny, McPherson, Trauth, Kulturel-Konak, and Mahar (2010b)	The objective of this study was to explore the effects of two key factors that shape students' IT career choices (i.e., IT skills self-efficacy and IT skills importance). A survey methodology was used to achieve the aforementioned objective. Undergraduate students enrolled in IT courses at two large US public universities were surveyed. This study developed validated scales for two beliefs that are crucial in capturing students' attraction to IT careers. Second, it provides insights into students' perceptions regarding the importance of various IT skills. Third, it posits a preliminary model, which helps in our understanding of how students' personal perceptions regarding the importance of IT skills affect self-efficacy beliefs and career intentions.	The study reveals that generally (with the exception of individuals who have very high tech self-efficacy) college students who perceive IT tech skills to be very important for the IT profession are less likely to engage in IT careers. Therefore, the college students who have high confidence in their nontech self-efficacy are less likely to choose IT as a career if they perceive IT tech skills to be more important for the IT profession. This finding is particularly troublesome because of the relatively high self-efficacy score for nontech skills in our sample. The students do not perceive nontech business and organizational skill such as leadership, customers, and relationships to be critical to IT career success. The stereotypical perception that business and organization skills are not important to the IT profession discourages students with high levels of confidence in these skills from entering IT careers.
Joshi, Trauth, Kvasny, and McPherson (2013)	This study explains the similarities and differences between the decision making of those who choose to pursue IT-majors and those who do not. Using individual-identity and self-efficacy along with gender role theory we tried to understand students' IT career decision making. By exploring both societal and individual factors, we found a middle ground that avoids the dual problems of social determinism and of seeing college students as completely free agents. By doing so, we were able to compare how these societal and individual factors shape career intentions for both majors/nonmajors of undergraduate students enrolled in two large US public universities.	The negative effects of nontech self-efficacy (confidence in soft skills such as leadership skills, negotiations skills) and positive effect of tech self-efficacy (such as programming) on IT career intentions were also replicated in this study. However, these relationships were tempered by students' perceived gender roles. This study shows that the negative association between nontech self-efficacy and intentions to major in IT does not hold if students' IT technical roles are congruent with their gender roles. However, tech self-efficacy remains a dominant force as its influence does not negate the gender role incongruities created by the societal messages about the importance of technical skills required to perform IT roles.
Trauth, Cain, Joshi, Kvasny, and Booth (2016)	Using the individual differences theory of gender and IT this study takes into account the intersectionality of gender and ethnicity to empirically reveal how gender stereotypes about the skills and knowledge affects the IT profession. A survey of 4,046 university students in the United States was conducted to examine gender stereotypes held by contemporary university students (White, Black and Latino men and women) about the skills and knowledge in the IT profession. These findings revealed the masculine gender stereotyping of skills in the IT profession, but also uncovered differential gender stereotyping of tech vs. nontech skills and variation in gender stereotyping by the intersectionality of gender-ethnic groups.	The findings clearly revealed that masculine gender stereotypes exist with respect to the skills and knowledge that an IT professional should possess. In addition, there is considerable consistency in the masculine stereotyping of the tech skills in the IT profession. This study provides empirical evidence of within-gender variation in gender stereotypes of IT skills resulting from the intersectionality of gender and ethnicity. Tech skills were more consistently stereotyped by both men and women in each of the gender-ethnic groups than were nontech skills. This variation is mostly accounted for by inconsistent gender stereotypes applied to the nontech skills in the IT profession.

in this study's findings was that the group that was not majoring in IT perceived nontech IT skills to be significantly less valuable to IT careers than the group that was majoring in IT (Joshi and Kuhn, 2011).

Subsequent studies that examined a broader group of over 650 US college students (including students of business, engineering, computer science, and information science) revealed that college students who have high confidence in their nontech IT self-efficacy are less likely to choose IT as a career, implying that it acts as a powerful fuel in the opposite direction (Joshi et al., 2010b; Joshi et al., 2013). Joshi et al. (2013) further revealed that this inverse relationship is reversed when individuals' perceptions about technical IT skills are aligned with their perceived gender roles (i.e., men receiving societal messages that men are good at technical skills). Therefore, while it demotivates and drives both men and women who have high levels of nontech self-efficacy away from IT careers (Joshi et al., 2010b), the stereotypical societal messages about tech skills being masculine could help soften the blow for men (Joshi et al., 2013). These findings reveal the presence of hegemonic masculinity in IT careers, which perpetuates the deeply ingrained beliefs (albeit misguided) about IT careers among college students as requiring very strong technical skills, which are stereotypically gender-typed as masculine (Joshi and Kuhn, 2007; Trauth et al., 2016). This brings us to the ugly truth about tech IT self-efficacy's realm of influence over IT career choices.

The ugliest aspect of IT self-efficacy is exposed in what actually matters. Tech IT self-efficacy, in areas such as coding and programming, which are stereotypically gender typed as masculine skills (Margolis et al., 2000), is what matters the most in shaping IT career choices. The same studies that are described above simultaneously uncover significant gender differences in tech IT self-efficacy where men have higher tech IT self-efficacy than women. In addition, these beliefs about their ability to learn and perform tech skills were found to have had a strong influence over how men and women feel and think about IT (Joshi and Kuhn, 2002; Joshi et al., 2010b; Joshi et al., 2013). These findings suggest that efficacy about coding and software/programming skills is a dominant determinant of IT career attitudes for both genders. One study further revealed that these gender differences not only persist but are even stronger after students are made aware of IT careers via a course experience (Joshi and Kuhn, 2002). This positive influence of one's confidence in learning and performing coding and software/programming skills is reasonable because a certain degree of confidence in one's ability to understand and conduct technical tasks is necessary in order to gain and sustain interest in tech careers. However, tech IT self-efficacy's stranglehold on IT career choices is troubling. This is troubling because women on average are perceived to have lower tech IT self-efficacy. The perception that tech skills such as coding and programming are the only gateway to IT careers, rather than one important skill set they need to be exposed to in order to succeed in IT, is at best misleading and at worst faulty. Girls and women are hearing that this is all that matters.

MAKING IT SELF-EFFICACY MATTER IN NARROWING THE IT GENDER GAP – UNCOVERING STRUCTURES THAT ENGAGE IN EXCLUSION

The main thesis of this chapter is that the disproportionate influence of a single type of skill set has hijacked the IT profession and drowned out the more important conversations about the diversity and variety of skills necessary to succeed in IT careers. The number of women in computing is declining across the board (from classrooms to boardrooms) (Conway et al., 2018; Fessenden, 2014). The deliberate and focused efforts, such as DigiGirlz,[2] CSforALL,[3] and government initiatives,[4] aimed at building women's and girls' beliefs about their IT self-efficacy through four main sources of influence, are failing. This narrative about coding, along with its associated biases and stereotypes, I argue, may be contributing to pushing women and girls out of IT spaces instead of keeping them in. A message from an engineer at a top IT company (i.e., Google), suggesting male domination of Silicon Valley is as a result of biological differences between the sexes and that this is one of the reasons why we don't see equal representation of women in tech and leadership, is discouraging (Devlin and Hern, 2017). Although most people were outraged and objected to the implication made in Google's infamous memo[5] that there are biological differences that make women less capable, people were mostly perplexed and unable to answer why, in 2017, such a misguided cultural assumption grounded in gender essentialism[6] still exists and persists in our society's consciousness (Devlin and Hern, 2017). This misconception that one has to be completely obsessed with coding and computers, which boys/men were assumed to be biologically predisposed to, was revealed and debunked in Jane Margolis' research in the 1990s (Margolis et al., 2000). Why, in 2017, were women "… made to think that, if they didn't dream in code and if it wasn't their full obsession, they didn't belong or were not capable of being in the field?" (Jane Margolis quoted in Devlin and Hern, 2017). How do we stop such thinking? One way to do that is to understand the source of IT self-efficacy beliefs that are collectively and perhaps unwittingly pushing this narrative. Understanding the source of such false narratives can help in confronting and combating them. Fortunately, the theoretical context of the self-efficacy construct proposed by Bandura (2004), in his conceptualization of self-efficacy, provides the means for altering beliefs through applications of the four sources of efficacy information (Betz and Luzzo, 1996).

According to Bandura (2004), there are four main sources of self-efficacy beliefs. The first, mastery experiences, refers to people personally experiencing success when they perform a specific task. The second, vicarious experiences, refers to individuals observing others like themselves succeeding in the task. The third, verbal persuasion, refers to receiving regular encouragement. Finally, the fourth is learning to regulate emotional and physiological states to manage negative feelings (e.g., fear of failure).

I illustrate below how these key sources of self-efficacy, which are essential for building high IT self-efficacy, are working against women and thus contributing to pushing women out of IT careers. These four sources need to holistically, jointly, and in unison focus on constructing the beliefs about one's ability to learn and perform both tech and nontech IT skills, not just coding or programming. By focusing primarily on one kind of tech IT skill, which is associated with stereotypes and biases, we weaken the argument that tech career pathways are diverse and are not exclusively for coders.

Mastery Experiences: Women Personally Experiencing Success with Tasks that Require IT Skills

One way to garner mastery experiences is by taking IT coursework. This is not working in women's favor. Women are not taking enough IT courses (in terms of number and content) where they can experience success and gain confidence by repeatedly practicing IT skills.

According to the National Center for Women and Information Technology, girls comprise 57 percent of all advanced placement[7] (AP) test-takers, 47 percent of all AP calculus test-takers, but only 31 percent of AP computer science test-takers (up by 12 percent since 2012) (NCWIT, 2019). Plus, the courses that are offered are predominately focused on technical skills such as math and coding. The other path to mastery experiences in IT is through extracurricular clubs and activities. But only 31 percent of girls participate in such activities.[8] A survey of more than 6,000 girls and young women conducted by Microsoft regarding their interests and perceptions of science, technology, engineering, and math found, "In middle school, for example, 31 percent of girls believe that jobs requiring coding and programming are 'not for them.' In high school, that percentage jumps up to 40. By the time they are in college, 58 percent of girls count themselves out of these jobs" (Nage, 2018). If women/girls are good at coding but are turned off because they don't want to do coding and programming all day long, then why are they not offered alternatives to coding, such as project management, web design, data visualization, and data-driven storytelling, which might be more aligned with their interests? If some women are not good at coding, then a failure in just one kind of skill undermining their whole sense of efficacy in IT is a shame. If women are convinced that coding and programming are just a part of many skills they have to learn to succeed, it will motivate them to persevere in IT even when they face adversity in coding classes. Plus, ample opportunities to learn other skills (e.g., web design, databases, digital marketing) will allow them to rebound from coding-related setbacks (Bandura, 2004).

Vicarious Experiences: Women Observing Others Like Themselves Succeeding in IT Careers

The second way of strengthening self-beliefs of efficacy that are being developed through mastery of experiences is by modeling social influences through vicarious experiences (Bandura, 2004). This influence also works in reverse when people observe others similar to them failing despite high effort. Such experiences lower observers' judgments of their own efficacy (Bandura, 2004). Sixty-four percent of girls and 56 percent of women do not know a woman in STEM.[9] Perhaps the percentages are even lower in IT in the United States. In addition, the majority (72 percent) of women in tech are regularly outnumbered by men in business meetings by at least a 2:1 ratio. Twenty-six percent of women report being outnumbered by 5:1 or more (Sullivan-Hasson, 2021). Furthermore, the percentage of female IT leaders (such as chief information officers, chief technology officers, or vice presidents of technology) has been stuck at 9 percent, according to the 2017 Harvey Nash/KPMG CIO survey (Graham, 2017). Paradoxical recruitment practices, where, on the one hand, companies are looking to hire women in IT leadership positions, but, on the other hand, women are asking but not getting leadership opportunities internally (Fidelman, 2012), compound such matters for women who do pursue technical careers and are met with additional challenges. Seeding positive beliefs about IT capabilities through role modeling becomes extremely difficult given the aforementioned numbers and experiences. What girls are observing is that others like them have to overcome tremendous challenges and the pathway to IT careers is arduous. It is not surprising that girls and women would avoid career pathways where they see few women working, even fewer excelling, and many struggling.

Verbal Persuasion: Women Receiving Regular Encouragement to Pursue IT Careers

The third way of bolstering people's beliefs is via social persuasion. Bandura (2004) makes an interesting point when he suggests that social persuasion alone cannot help, but it can do harm by undermining beliefs that are in their incipient stages. The importance of referent others in choosing an IT career is well documented in the extant literature (see Joshi and Kuhn, 2011). Referent others are parents, peers, career counselors, teachers, and other family members. These referent others can provide regular encouragement. Findings from one of our NSF studies (Joshi and Kuhn, 2000) suggest that the opposite is happening. Our findings show that referent others impact career intentions for both men and women. However, one critical difference that we found in our study was that women received negative encouragement from referent others (indicating that they don't believe that their referent others think that they should major in IT) as opposed to men (men believed that their referent others think that they should major in IT). Our findings suggest that although both men and women are seeking encouragement from referent others, only men are receiving encouragement. What is worse is that women are being discouraged and this would act as a strong verbal persuasion to avoid IT careers. The following quotation from a woman who participated in this study reinforces our finding, "… I was told not to major in MIS [management information systems] from an advisor, saying that it was a major for men" (Joshi and Schmidt, 2006).

In addition, societal messages play a role in social persuasion. However, the aforementioned Google memo, coupled with a reflection by Susan Flower, in a blog about her stint as a software engineer at Uber (Flower, 2017), suggest that the opposite is happening. Her description of one very strange year at Uber, which illustrates the hostile tech-work environment that ended her tech career, is demoralizing (Flower, 2017). In a recent interview, Flower was asked, "Do you miss it? Coding?" to which she replied, "I don't miss it because I associate it with so many of my negative experiences" (Lopatto, 2020).

A survey conducted by TrustRadius in 2021 (a software review site) revealed the following: 78 percent of women in tech feel they have to work harder than their men coworkers to prove their worth. Women in tech are four times more likely than men to see gender bias as an obstacle to promotion. Thirty-nine percent of women see gender bias as a barrier to promotion in 2021. Women of color are less confident than white women about their promotion prospects – and that gap has increased by three times over the past year. Thirty-seven percent of women of color in tech feel that racial bias is a barrier to promotion (Sullivan-Hasson, 2021).

In summary, the aforementioned contribute little to persuading women to enter IT spaces. As a result of these societal messages, if women choose careers where they perceive they don't have to constantly battle biases and stereotypes, then I posit that women avoid IT careers as a rational act of self-preservation and self-care.

Learning to Reinterpret and Regulate Negative Feelings and Messages

The fourth way of recalibrating self-belief of efficacy is to teach people to mitigate their stress reactions by altering their negative emotional proclivities and misinterpretations of their physical states (Bandura, 2004). More than a decade ago, as a professor of information systems at Washington State University, I was sharing with my women students who had been accepted into our department's fellowship program (the top 10 percent of the business students who were majoring in management information systems were invited to participate in this fellowship program) about the gender differences in the self-efficacy findings in my research studies. I indicated to them that even though women in our IT courses performed as well as or better than men, they frequently reported lower levels of IT self-efficacy. I jokingly remarked that perhaps men were overconfident about their abilities while women were under-confident. A few days later, one student from that group of women shared the following with me: "Dr. Joshi, last week when my male peers indicated that they were feeling very confident and pre-pared for the big exam in a (very technical) IT course and that they were not worried about it, I immediately thought about your research about gender differences in self-efficacy. Normally, I would get really very stressed because my level of confidence for this exam did not match my male counterparts' level of confidence. However, I thought about your comments and it alleviated my stress levels." She had learned from the findings I shared with them. She used it to reinterpret the differences in her confidence level as compared to her male counterparts to regulate her stress levels (i.e., due to fear of failure) about the exam.

The term "bro culture" refers to a misogynist culture within an organization or community, such as occupational inequality in Silicon Valley. Seventy-two percent of women in tech have worked at a company where bro culture is pervasive (Sullivan-Hasson, 2021). In order for women to persevere and excel in such a culture, women must learn to reinterpret and regulate negative feelings and experiences. If they succeed in combating their negative feelings, it will empower them, but if they don't, then they will choose to exit. The leaky IT pipeline (Conway et al., 2018), a metaphor used to describe the decrease in the number of women along the IT career pathway, is an indication that perhaps the latter is happening more frequently than the former.

Teaching women to regulate negative feelings to alter their state of mind to overcome chal-lenges they face in IT is perhaps the least researched practice in the IT and gender literature. Given the negative messages and experiences women face in the tech environment, I posit that this skill is the single most powerful mechanism that can help build women's belief in their own ability to succeed in IT.

CONCLUSION

The magnitude of gender imbalance at all levels of IT is a social justice issue (Conway et al., 2018). The accelerating pace with which technology is influencing our work and lives means that the future will be shaped and controlled by people who know how to design and build technology. To ensure that the future of work and life is not decided *for* women in advance, women need to actively participate in designing and building future systems, algorithms, and technologies. We are already seeing gender biases codified into the technology designed, built, and used as a foundation for the digital economy (Devlin and Hern, 2017). Narrowing the tech gender gap is vital to safeguarding against gender biases unwittingly being programmed into the technology of the future (Trauth et al., 2018). The contrarian view of IT self-efficacy presented in this chapter is not intended to be dystopic, but rather intended to illustrate sources of IT self-efficacy threats that contribute to the declining number of women in the IT pipeline despite the efforts directed by governments, businesses, and educational institutions. If we want to reverse this trend, it is not enough to find ways to lower the entry barriers (e.g., cor-recting skill deficiencies, amplifying self-efficacy, and providing better mentoring) that allow women to enter the IT spaces by simply crossing the IT "boundaries," but we also need to examine the process of being "shut out" from IT structures (Walker, 1997) by uncovering the role of those who are engaged in exclusion (Trauth et al., 2018; Trauth and Howcroft, 2006).

NOTES

1. Hegemonic masculinity is defined as a practice that legitimizes men's dominant position in society.
2. DigiGirlz Home. Global Diversity and Inclusion at Microsoft.
3. Home Page. CSforALL.
4. President Obama's Record on Empowering Women and Girls. The White House (archives.gov).
5. Google's_Ideological_Echo_Chamber-FINAL.1.1 (documentcloud.org).
6. Gender essentialism posits that men and women are fundamentally different due to their biology and these differences account for the underrepresentation of women in IT.
7. Advanced placement examinations are tests offered in the United States that allow students to earn college credits.
8. v1DeDjM.png (1100×4091) (imgur.com).
9. Microsoft Educator Center, Girls in STEM – Closing the STEM Gap, Module 3: Provide Role Models https://education.microsoft.com/en-us/course/c3c376f8/2.

REFERENCES

Akbulut, A.Y., and Looney, C.A. (2007). Inspiring students to pursue computing degrees. *Communications of the ACM, 50*(10), 67–71.

Bailey, J.L., and Stefaniak, G. (2002). Industry perceptions of the knowledge, skills, and abilities needed by computer programmers. In M. Serva (Ed.), *Proceedings of the Association for Computing Machinery Special Interest Group on Computer Personnel Research (SIGCPR)*.

Bandura, A. (1997). *Self-Efficacy: The Exercise of Control*. Freeman.

Bandura, A. (1994). Self-efficacy. In V.S. Ramachaudran (Ed.), *Encyclopedia of Human Behavior*, Vol. 4 (pp. 71–81). Academic Press (Reprinted in H. Friedman [Ed.], *Encyclopedia of Mental Health*. Academic Press, 1998).

Bandura, A. (1992). Social cognitive theory of social referencing. In S. Feinman (Ed.), *Social Referencing and the Social Construction of Reality in Infancy* (pp. 175–208). Plenum Press.

Bassellier, G., Benbasat, I., and Reich, B.H. (2003). Development of measures of IT competence in business managers. *Information Systems Research, 14*(4), 317–36.

Betz, N.E., and Hackett, G.T. (1981). The relationship of career-related self-efficacy expectations to perceived career options in college women and men. *Journal of Counseling Psychology* (28), 399–410.

Betz, N.E., and O'Connell, L. (1989). Work orientations of males and females: Exploring the gender socialization approach. *Sociological Inquiry* (59), 318–30.

Betz, N.E., Klein, K.A., and Taylor, K.M. (1996). Evaluation of a short form of the career decision-making self-efficacy scale. *Journal of Career Assessment* (4), 47–57.

Betz, N.E., and Luzzo, D.A. (1996). Career assessment and the career decision-making self-efficacy scale. *Journal of Career Assessment, 4*(4), 413–28.

Beyer, S. (2006). Comparing gender differences in computer science and management information systems majors. In E. Trauth (Ed.), *Encyclopedia of Gender and Information Technology* (pp. 109–15). Idea Group Inc.

Beyer, S. (2002). The effects of gender, dysphoria, and performance feedback on the accuracy of self-evaluations. *Sex Roles*, (47), 453–64.

Beyer, S. (1999a). Gender differences in the accuracy of grade expectancies and evaluations. *Sex Roles*, (41), 279–96.

Beyer, S. (1999b). The accuracy of academic gender stereotypes. *Sex Roles*, (40), 787–813.

Beyer, S., Rynes, K., and Haller, S. (2004). What deters women from taking computer science courses? *IEEE Technology and Society*, (23), 21–8.

Busch, T. (1995). Gender differences in self-efficacy and attitudes toward computers. *Journal of Educational Computing Research*, (12), 147–58.

Cassidy, S., and Eachus, P. (2002). Developing the computer user self-efficacy (CUSE) scale: Investigating the relationship between computer self-efficacy, gender, and experience with computers. *Journal of Educational Computing Research*, (26), 133–53.

Cejka, M.A., and Eagly, A. (1999). Gender-stereotypic images of occupations correspond to the sex segregation of employment. *Personality and Social Psychology Bulletin*, 25(4), 413–23.

Colley, A., and Comber, C. (2003). Age and gender differences in computer use and attitudes among secondary school students: What has changed? *Educational Research*, 45(2), 155–65.

Conway, M., Ellingrud, K., Nowski, T., and Wittemye, R. (2018). *Closing the Tech Gender Gap Through Philanthropy and Corporate Social Responsibility*. McKinsey and Company.

Cross, S.E. (2002). Training the scientists and engineers of tomorrow: A person x situation approach. *Journal of Applied Social Psychology*, (31), 296–323.

Devlin, H. and Hern, A. (2017). Why are there so few women in tech? The truth behind the Google memo, *The Guardian*, www.theguardian.com/lifeandstyle/2017/aug/08/why-are-there-so-few-women-in-tech-the-truth-behind-the-google-memo.

Dickhaeuser, O., and Stiensmeier-Pelster, J. (2002). Gender differences in computer work: Evidence for the model of achievement-related choices. *Contemporary Educational Psychology*, 27(3), 486–96.

Fessenden, M. (2014). What happened to all the women in computer science? Smithsonianmag.com.

Fidelman, M. (2012). Here's the real reason there are not more women in technology. www.forbes.com/sites/markfidelman/2012/06/05/heres-the-real-reason-there-are-not-more-women-in-technology/.

Flower, S. (2017). Reflecting on one very, very strange year at Uber. www.susanjfowler.com/blog/2017/2/19/reflecting-on-one-very-strange-year-at-uber.

Gloria, A.M., and Hird, J.S. (1999). Influences of ethnic and nonethnic variables on the career decision-making self-efficacy of college students. *Career Development Quarterly*, 48(2), 157–74.

Graham, L. (2017). Women take up just 9 percent of senior IT leadership roles, survey finds, CNBC, www.cnbc.com/2017/05/22/women-take-up-just-9-percent-of-senior-it-leadership-roles-survey-finds.html.

Green, G.I. (1989). Perceived importance of systems analysts' job skills, roles, and non-salary incentives, *MIS Quarterly*, 13(2), 115–33.

Guimond, S., and Roussel, L. (2001). Bragging about one's school grades: Gender stereotyping and students' perception of their abilities in science, mathematics, and language. *Social Psychology of Education*, 4(3), 275–93.

Hackett, G., and Betz, N.E. (1995). Self-efficacy and career choice and development. In self-efficacy, adaptation, and adjustment (pp. 249–80). Springer, Boston, MA.

Hackett, G., and Betz, N.E. (1981). A self-efficacy approach to career development of women. *Journal of Vocational Behavior*, 18(3), 326–99.

Hamlin, B., Riehl, J., Hamlin, A.J., and Monte, A. (2010). Work in progress – What are you thinking? Over confidence in first year students. In *2010 IEEE Frontiers in Education Conference (FIE)* (pp. F2H-1). IEEE.

Hartzel, K. (2003). How self-efficacy and gender issues affect software adaptation and use. *Communications of ACM*, 46(9), 167.

Henwood, F. (2000). From the woman question in technology to the technology question in feminism: Rethinking gender equality in IT education. *European Journal of Women's Studies*, 7(2), 209–27.

Huang, H., Kvasny, L., Joshi, K.D., Trauth, E.M., and Mahar, J. (2009). Synthesizing IT job skills identified in academic studies, practitioner publications and job ads. Proceedings of the ACM SIGMIS Computer Personnel Research Conference (Limerick, Ireland, May 28–30).

Hunt, N.P., and Bohlin, R.M. (1993). Teacher education and students' attitudes toward using computers. *Journal of Research on Computing in Education*, 25, 487–97.

Joshi, K.D., and Kuhn, K. (2011). What determines interest in an IS career? An application of the theory of reasoned action. *Communications of the Association for Information Systems*, 29(1), 8.

Joshi, K.D., and Kuhn, K.M. (2007). What it takes to succeed in information technology consulting: Exploring the gender typing of critical attributes. *Information Technology and People*.

Joshi, K.D., and Kuhn, K. (2000). Are women really less interested in information technology? Dispelling gender stereotypes about IT career choices, Unpublished Manuscript, National Science

Foundation Grant No. 0089986. Women and information systems: Modeling the impact of work values, attitudes, and attributes on career choices.

Joshi, K.D., Kuhn, K.M., and Niederman, F. (2010). Excellence in IT consulting: Integrating multiple stakeholders' perceptions of top performers. *IEEE Transactions on Engineering Management*, 57(4), 589–606.

Joshi, K.D., Kvasny, L., McPherson, S., Trauth, E.M., Kulturel-Konak, S., and Mahar, J. (2010). Choosing IT as a career: Exploring the role of self-efficacy and perceived importance of IT skills. *In International Conference on Information Systems* (ICIS) (p. 154).

Joshi, K.D., Kvasny, L., Unnikrishnan, P., and Trauth, E. (2016). How do Black men succeed in IT careers? The effects of capital. In *2016 49th Hawaii International Conference on System Sciences (HICSS)* (pp. 4729–38). IEEE.

Joshi, K.D., and Schmidt, N.L. (2006). Is the information systems profession gendered? Characterization of IS professionals and IS career. *ACM SIGMIS Database: The DATABASE for Advances in Information Systems*, 37(4), 26–41.

Joshi, K.D., Trauth, E., Kvasny, L., and McPherson, S. (2013, December). Exploring the differences among IT majors and non-majors: Modeling the effects of gender role congruity, individual identity, and IT self-efficacy on IT career choices. In *International Conference on Information Systems (ICIS)* (pp. 1613–33).

Kuhn, K, and Joshi, K.D. (2009). The reported and revealed importance of job attributes to aspiring information technology: A policy-capturing study of gender differences. *ACM SIGMIS Database: The DATABASE for Advances in Information Systems*, 40(3), 40–60.

Kwan, S.K., Trauth, E.M., and Driehaus, K.C. (1985). Gender differences and computing: Students' assessment of societal influences. *Education and Computing*, 1(3), 187–94.

Lee, D., Trauth, E., and Farwell, D. (1995). Critical skills and knowledge requirements of IT professionals: A joint academic and industry investigation. *MIS Quarterly*, 19(3), 313–38.

Lent, R.W. (2005). A social cognitive view of career development and counseling. In S.D. Brown and R. W. Lent (Eds.), *Career Development and Counseling: Putting Theory and Research to Work* (pp. 101–27). John Wiley and Sons, Inc.

Liberatore, M.J., and Wagner, W.P. (2020). Gender, performance, and self-efficacy: A quasi-experimental field study. *Journal of Computer Information Systems*, [online] 2380–2057. Available at: https://doi.org/10.1080/08874417.2020.1717397.

Lips, H.M. (2004). The gender gap in possible selves: Divergence of academic self-views among high school and university students. *Sex Roles*, 50, 357–71.

Litecky, C.R., Arnett, K.P., and Prabhakar, B. (2004). The paradox of soft skills versus technical skills in hiring. *Journal of Computer Information Systems*, 4, 69–76.

Liu, J., and Wilson, D. (2001). Developing women in a digital world. *Women in Management Review*, 16(8), 405–16.

Lopatto, E. (2020). To expose sexism at Uber, Susan Fowler blew up her life: The risks and rewards of blowing the whistle at Uber. The Verge, www.theverge.com/2020/2/19/21142081/susan-fowler-uber-whistleblower-interview-silicon-valley-discrimination-harassment.

Margolis, J., Fisher, A., and Miller, F. (2000). The anatomy of interest: Women in undergraduate computer science. *Women's Studies Quarterly*, 28(1/2), 104–27.

Nage, D. (2018). Closing the STEM gap: Mentors, encouragement, hands-on learning boost girls' interest in STEM substantially, Campus Technology, https://campustechnology.com/articles/2018/03/13/closing-the-stem-gap.aspx.

NCWIT. (2019). Women and information technology, by the numbers, ncwit_btn_03252021_full-size.pdf. October 22, www.smithsonianmag.com/smart-news/what-happened-all-women-computer-science-1-180953111/.

Parasuraman, S., and Igbaria, M. (1990). An examination of gender differences in the determinants of computer anxiety and attitudes toward microcomputers among managers. *International Journal of Man-Machine Studies*, 32, 327–40.

Ritter, D. (2004). Gender role orientation and performance on stereotypically feminine and masculine cognitive tasks. *Sex Roles*, 50(7-8), 583–91.

Sanli, L. Impostor Syndrome in Computer Science, *The Guardian*, https://ucsdguardian.org/2019/05/12/impostor-syndrome-computer-science, May 12, 2019.

Serva, M.A., Baroudi, J.J. and Kydd, C.T. (2009). The effects of stereotype threat on MIS students: An initial investigation, *Journal of Computer Information Systems*, *50*(2), 142–50. DOI: 10.1080/08874417.2009.11645393.

Shashaani, L. (1994). Gender-differences in computer experience and its influence on computer attitudes. *Journal of Educational Computing Research*, *11*(4), 347–67.

Sullivan-Hasson, E. (2021). *TrustRadius 2021 Women in Tech Report*. www.trustradius.com/buyer-blog/women-in-tech-report#sources.

Todd, P., Mckeen, J., and Gallupe, B. (1995). The evolution of IT job skills: A content analysis of IT job advertisements from 1970 to 1990. *MIS Quarterly*, *19*(1), 1–26.

Trauth, E.M., Cain, C.C., Joshi, K.D., Kvasny, L., and Booth, K.M. (2016). The influence of gender-ethnic intersectionality on gender stereotypes about IT skills and knowledge. *ACM SIGMIS Database: The DATABASE for Advances in Information Systems*, *47*(3), 9–39.

Trauth, E.M., and Howcroft, D. (2006). Social inclusion and the information systems field: Why now? In E.M. Trauth, D. Howcroft, T. Butler, B. Fitzgerald, and J. DeGross (Eds.), *Social Inclusion: Societal and Organizational Implications for Information Systems* (pp. 3–12). Springer.

Trauth, E., Joshi, K.D., and Yarger, L. (2018). ISJ Special Issue on Social Inclusion in the Information Systems Field, Editorial, 990.

Venkatesh, V., and Morris, M.G. (1997). Why don't men ever stop to ask for directions? Gender, social influence and their role in technology acceptance and usage behavior. *MIS Quarterly*, *24*(1), 115–39.

Vitiello, J. (1997). Hard facts on soft skills. *Computerworld*, *31*(41), 91–2.

Walker, A. (1997). Introduction. In A. Walker, and C. Walker (Eds.), *Britain Divided: The Growth of Social Exclusion in the 1980s and 1990s* (pp. 1–9). Child Poverty Action Group.

Whitley, B. (1997). Gender differences in computer-related attitudes and behavior: A meta-analysis. *Computers in Human Behavior*, *13*(1), 1–22.

Wood, R., and Bandura, A. (1989). Social cognitive theory of organizational management. *Academy of Management Review*, *14*(3), 361–84.

Zeldin, A., and Pajares, F. (2000). Against the odds: Self-efficacy beliefs of women in mathematical, scientific, and technological careers. *American Educational Research Journal*, (37), 215–46.

Index